Antimicrobials in Livestock 2: Choices of Prevention and Treatment

Lucie Pokludová
Editor

Antimicrobials in Livestock 2: Choices of Prevention and Treatment

A European Perspective

Springer

Editor
Lucie Pokludová
Institute for State Control of Veterinary
Biologicals and Medicines
Brno, Czech Republic

ISBN 978-3-031-60809-4 ISBN 978-3-031-60807-0 (eBook)
https://doi.org/10.1007/978-3-031-60807-0

This Springer imprint is published by the registered company Springer Nature Switzerland AG
The registered company address is: Gewerbestrasse 11, 6330 Cham, Switzerland

Cover Image: nobeastsofierce/shutterstock.com

With many thanks to my closest family
Tomáš, Jakub, Barbora
and
Jana and Lubomír

Preface

First Volume, which attempts to merge in complexity view of legal provisions, science together with routine laboratory diagnostic, molecular biology methods and links towards practice, including aspects of use of antimicrobials from the perspectives of consumer and user safety, touching as well risks for the environment is followed by the second Volume targeted on the main infectious diseases' prevention and antimicrobial treatment in the main food producing species farmed in Europe, including horses and, through the first chapter provide insight to essentials of pharmacokinetic and pharmacodynamic.

First Chapter not only covers four essential pharmacokinetic (PK) parameters, but also attempts to elucidate pharmacodynamic (PD) variables and provide basic description of PK/PD modelling important for the attainment of the appropriate dose considering not only efficacy but trying to minimise the risks of selection of antimicrobial resistance. In current days we should do next, more difficult step, as it is not enough to just say that we have to use antimicrobials responsibly and distinguish 1st, 2nd and last choice. As each use of antimicrobial has potential to select resistance, once we use antimicrobial, we should target pathogen in the most efficacious way. It is therefore of importance to know minimum inhibitory (or bactericidal) concentration of antimicrobial used against certain pathogen causing the disease to be treated. Unfortunately, these concentrations are known from in vitro testing only on artificial media. In the future, big challenge is in front of the scientists, to develop tissue mimicking culturing media and infected tissue/organism conditions modelling systems. It is as well critical to know key pharmacokinetic parameters reached at the certain time after dosing at the site of infection. The concentration achieved at the site of infection should be sufficiently high, but still non-toxic, ideally out of mutation selection window, i.e. minimising impact on resistance selection. Even setting such dose is not easy, it is still not enough. Further effort is targeted on proper frequency of dosing (quite important in dairy cattle, especially with current routines of milking) and finally total duration of administration. Sufficient dose and treatment long enough (with proper frequency) are key for reaching clinical and ideally also bacteriological cure of infection and short enough and appropriately frequent to minimise pressure on selection of resistance. Setting the appropriate dosing schedule is even more tricky in veterinary medicine of food producing animals, than I human. Why? Because in these animals we moreover need to establish safe withdrawal period for any commodity intended for human consumption to ensure consumer safety. In food producing animals despite the main goal is healthy animal, it is also animal sufficiently recovered "for being back to full production". Reflecting these facts, dosing schedule and withdrawal period have not only importance from efficacy and safety point of view, but also huge economic impact. Big challenge is "an ideal withdrawal period", whose length should be sufficient for depletion of residues to safe concentration for consumer and at the same time not too long, which would make such a veterinary medicinal product unusable in the real practice of food animals.

Next chapters are dedicated to main food producing animals and horses. While swine, poultry and cattle sectors are rather organised as farming systems and animals are kept for production of food commodities as meat (including offals), eggs and milk, horses are specific, being many times kept individually or in smaller groups and owned by individual persons. Horses, despite being currently kept rather for sport and hobby, are still considered food producing animals, according to EU legal provisions. Second Volume therefore brings information on above described animals´ species, European position in production of meat and food commodities within the world context, consumption of antimicrobials and its trends in Europe (with zero use of antimicrobials for growth promotion in EU), followed by structured info on diseases, preventive tools and antimicrobial treatment possibilities as well as, where available also alternative tools, for three main food producing species—swine, poultry and cattle. In horses, except similar approach related to diseases, prevention and possible alternatives, more attention is paid to antimicrobial stewardship and issues related to use of antimicrobials outside the terms of authorisation (so-called off-label use). Due to the fact that among antimicrobials used in critical care (sepsis, especially in foals; off-label use) also some highly critically important antimicrobials can occasionally occur, a piece of the recent legislation should be highlighted. It is Commission Implementing Regulation (EU) 2022/1255 banning authorisation and use of certain substances in any animals to preserve these antimicrobials for use in serious infections in human medicine. Therefore, treatment with such antimicrobials in horses will not be legally possible any more (please refer to the exact list of substances/pharmacological groups of antimicrobials in chapter dedicated to horses). Very recently, another piece of legislation of importance was released: Implementing Regulation (EU) 2024/1973, establishing list of antimicrobials which shall not be used under the cascade.

During the writing of these two volumes and cooperating with colleagues, I many times fell asleep late in the evening and woke up the next day with a series of new questions. And the questions wait for the answers! The preface of the Volume I therefore started with questions and Volume II preface will end with the questions—to provoke further thinking and might be start of a new era, which will need also innovative business models to ensure availability of antimicrobials in veterinary medicine, as antimicrobials are the only potent tool veterinarians (fortunately) still have for treatment of infectious diseases of animals.

We were able to decode many genomes of bacteria, we "read" and recognised a lot of genetical sequences, genes of resistance, virulence, genes

A set of questions exist that have been many times raised on global fora and leading some expectations on alternatives to antimicrobials. We can have a belief, but we should be as well careful—e.g. herbal-based alternatives, if effective, they contain active substances as well. Do we intend to move from precisely known and described actives of medicinal products to less known and less defined mixtures of actives and undefined other components in alternatives? What are the benefits and what are the safety risks? Are these alternatives as potent as pure antimicrobial substances, especially for acute, clinical infections or should we rather speak about alternative preventive or treatment supportive tools?

Current knowledge shows that minimisation of use of substances, including conventional antimicrobials, disinfectants, heavy metals and generally other (even naturally sourced) substances able to select resistance needs some compensation. Among the most appropriate ways belong better care of the animals. Would it be possible to perform such a care using automatic, "artificially intelligent", equipment and devices, or do we still need human staff (with empathy and understanding animals)? Then in the current world questions arise. How to attract people for everyday hard work able to work with animals and concomitantly having sense for these advanced technologies? Especially as in many EU countries average age of staff taking care of animals as well as veterinarians is increasing, and lack of qualified personnel more and more limits quality of care.

Meeting teams of people in everyday practice who everyday taking care of animals, doing all necessary things that help made biosecurity and welfare and treating of animals as good as possible, shown me, the mosaic of saving antimicrobials for future generations as more plastic and colourful. The responsible use of antimicrobials using evidence and thoroughly described experience and regularly updated guidelines driving the good choice and following antimicrobial stewardship principles also in veterinary medicine can be one very important piece of such mosaic part of the attempt to save the antimicrobials´ power to treat, finally both for human and veterinary medicine.

Lucie Pokludová

List of Abbreviations (Vol II Chapters I–V: All Chapters Together)

3G	Third generation
4G	Fourth generation
AABP	American Association of Bovine Practitioners
AAC	Antimicrobial-associated Colitis
AAEC	Attaching-effacing *E. coli*
ABC	ATP-binding Cassette
ADG	Average Daily Gain
ADI	Acceptable Daily Intake
ADME	Absorption, Distribution, Metabolism, Excretion
AF	Antibiotic Free
AFB1	Aflatoxin B1
AGP	Antimicrobial Growth Promoter
AHAW	Animal Health and Welfare
AIAO	All-in-All-out
AMCRA	Antimicrobial Consumption and Resistance in Animals
AMEG	Antimicrobial Advice *Ad Hoc* Expert Group
AMP	Ampicillin
AMPs	Antimicrobial Peptides
AMR	Antimicrobial Resistance
ANSES	French Agency for Food, Environmental and Occupational Health & Safety
APEC	Avian Pathogenic *Escherichia coli*
App	*Actinobacillus pleuropneumoniae*
APP	Antimicrobial Perioperative Prophylaxis
ASF	African Swine Fever
AST	Antimicrobial Susceptibility Testing
AT	Austria (in table 6)
ATB	Antibiotic
ATCC	American Type Culture Collection
ATM	Antimicrobial
AUC	Area Under the Curve
AVA-AMA	Australian Veterinary Association—Animal Medicines Australia
B&F	Breeder and Fattener
b.w.	Body Weight
BAC	Bronchoalveolar Cell
BAL	Bronchoalveolar Lavage

BBB	Blood-brain barrier
BE	Bioequivalence
BELF	Bronchial Epithelial Lining Fluid
BEVA	British Equine Veterinary Association
BF	Breeder Farm
BG	Bacterial Ghost
BHV-1	Bovine Herpes-virus-1
BOV	Between-Occasion Variability
BQA	Beef Quality Assurance
BRD	Bovine Respiratory Disease
BRSV	Bovine Respiratory Syncytial Virus
BSV	Between-Subject Variability
BTSCC	Bulk Tank Somatic Cell Count
BVDV	Bovine Viral-diarrhoea Virus
BW	Body Weight
C	Concentration
CA	Competent Authority
CAMP	Christie–Atkins–Munch–Peterson test
CBP	Clinical Breakpoint
CCC	Colon Content Concentration
CDA	Concentration-dependent Antibiotic
CdAD	*Clostridioides difficile*-associated disease
CDDEP	Centre for Disease Dynamics, Economics & Policy
CDI	*Clostridioides difficile*
CECT	Spanish Type Culture Collection
CEF	Ceftiofur
CFU	Colony-Forming Unit
CG	Control Group
CGMCC	China General Microbiological Culture Collection Centre
CHMP	Committee for Medicinal Products for Human Use
CIA	Critically Important Antimicrobial
CIAs	Critically Important Antimicrobials
CIP	Ciprofloxacin
CIWF	Compassion in World Farming
Clf	Clumping factor
CLSI	Clinical and Laboratory Standards Institute
Cl_{β}	Total Body Clearance
CM	Clinical Mastitis (cattle)
CM	Coliform Mastitis (swine)
C_{max}	Maximum Plasma Concentration
CMT	California Mastitis Test
Cna	Collagen-binding protein
CNS	Central Nervous System
COX	Cyclooxygenase
C_p	Plasma concentration
CpA	*Clostridium perfringens* type A
CpC	*Clostridium perfringens* type C
CRD	Chronic respiratory disease

CRISPR	Clustered Regularly Interspaced Short Palindromic Repeats
CRP	C-reactive Protein
Css	Concentration steady state
CT	Co-dependent
CVMP	Committee for Veterinary Medicinal Products
CXCL8	C-X-C motif Chemokine Ligand 8
CZ NMTP	Czech National Monitoring of Target Pathogens
DANMAP	Danish Integrated Antimicrobial Resistance Monitoring and Research Programme
DC	Dry Cows
DCD	Defined Course Dose
DCT	Dry Cow Therapy
DDD	Defined Daily Dose
DK	Denmark (in table 6)
DMEM	Dulbecco's Modified Eagle Medium
DMSO	Dimethyl Sulfoxide
DNA	Deoxyribonucleic Acid
DSM	German Collection of Microorganisms and Cell Cultures
DT	Definitive (phage) Type
EC	European Commission
ECDC	European Centre for Disease Prevention and Control
ECOFF	Epidemiological Cut-Off Value
ED_{50}	Median Effective dose
EE	Exudative Epidermitis
EEA	European Economic Area
EFSA	European Food Safety Authority
ELISA	Enzyme-linked Immunosorbent Assay
EMA	European Medicines Agency (previously EMEA)
E_{max}	Maximum Killing Effects
ENR	Enrofloxacin
EO	Essential Oils
EP	European Parliament
ERIC-PCR	Enterobacterial Repetitive Intergenic Consensus-PCR
ES	Spain
ESBL	Extended-Spectrum β-Lactamase
ESC	Extended-Spectrum Cephalosporin
ESI-MS	Electrospray Ionisation Mass Spectrometry
ESVAC	European Surveillance of Veterinary Antimicrobial Consumption
ETEC	Enterotoxigenic *Escherichia coli*
EU	European Union
EUCAST	European Committee on Antimicrobial Susceptibility Testing
EXPEC	Extraintestinal Pathogenic *Escherichia coli*
FAO	Food and Agriculture Organisation
Fba	Fructose-1,6-bisphosphate aldolase
FBOp	Food Business Operator
FCR	Feed Conversion Ratio

FDA	Food and Drug Administration
FF	Fattening only Farms
FINRES	Finnish Veterinary Antimicrobial Resistance Monitoring and Consumption of Antimicrobial Agents
FLF	Florfenicol
FMD	Foot and Mouth Disease
FnBP	Fibronectin-Binding Protein
FP	Food Producing
FR	Front-right
FVE	Federation of Veterinarians of Europe
GFR	Glomerular Filtration Rate
GI	Gastrointestinal
GIT	Gastrointestinal Tract
GLP	Good Laboratory Practice
Gps	*Glaesserella parasuis*
GVA	German Veterinary Association
HAI	Hospital-Acquired Infections
HBS	Haemorrhagic Bowel Syndrome
HPAI	High Pathogenic Avian Influenza
HRM	High-resolution Melt
HS	Histophilus somni
IB	Infection Bronchitis
IBK	Infectious Bovine Keratoconjunctivitis
IBV	Infectious Bronchitis Virus
IC	Infectious Coryza
ICC	Ileum Contents Concentration
IF	In Feed
IgA	Immunoglobulin A
IgG	Immunoglobulin G
igPBPK	Interactive Generic Physiologically Based Pharmacokinetic
IgYs	Egg Yolk Immunoglobulins
IL	Interleukin
IM	Intramuscular
iMBC	Intracellular Minimum Bactericidal Concentration
IMI	Intramammary Infection
iMIC	Intracellular Minimum Inhibitory Concentration
IMM	Intramammary
INJ	Injectable
IROMP	Iron-regulated Outer Membrane Proteins
IS	Infectious Sinusitis
ISF	Interstitial Fluid
IU	International Unit
IV	Intravenous
IW	In Water
JIACRA	Joint Inter-agency Antimicrobial Consumption and Resistance Analysis
JVARM	Japanese Veterinary Antimicrobial Resistance Monitoring System

LA	Long-Acting
LAMP	Loop-Mediated Isothermal Amplification
LC	Lactating Cows
LI	*Lawsonia intracellularis*
LPS	Lipopolysaccharide
LT	Heat-labile toxin
LTB4	LeukoTriene B4
MAG	alpha 2-Macroglobulin, serum Albumin and immunoglobulin G binding protein
MAH	Monitoring Animal Health
MALDI-TOF	Matrix-Assisted Laser Desorption/Ionisation—Time-of-Flight
MBC	Minimum Bactericidal Concentration
MDA	Maternally Derived Antibodies
MDH	Meat Producers from Dairy Herds
MDR	Multidrug Resistance
MDR	Multidrug-Resistant
MG	*Mycoplasma gallisepticum*
MH	*Mannheimia haemolytica*
Mhp	*Mycoplasma hyopneumoniae*
MI	*Mycoplasma iowae*
MIC	Minimal Inhibitory Concentration
MIRD	Mycoplasma-Induced Respiratory Disease
MLM	Modified Lacks Medium
MLS_B	Macrolide-lincosamide-streptogramin B
MLST	Multilocus Sequence Typing
MM	*Mycoplasma meleagridis*
MMA	Mastitis-Metritis-Agalactia
MMP	Matrix Metalloproteinase
MPC	Mutant Prevention Concentration
mPCR	Multiplex PCR
MPEC	Mammary Pathogenic *E. coli*
MRL	Maximum Residue Limit
MRP	Multidrug resistance-associated protein
MRSA	Methicillin-Resistant *Staphylococcus aureus*
MRT	Mean Residence Time
MS	*Mycoplasma synoviae*
MSW	Mutation Selection Window
MY	Mycoplasmas
NA	Not Available
NAD	Nicotinamide Adenine Dinucleotide
NAE	No Antibiotic Ever
NAGase	N-acetyl-beta-D-glucosaminidase
NAPQI	N-acetyl-p-benzoquinone-imine
NAS	Non-aureus Staphylococci
NCA	Non-Compartmental Analysis
NCD	Neonatal Calf Diarrhoea
NCP	National Control Programme

ND	Neonatal Diarrhoea
NEO	Neomycin
NetB+	Necrotic enteritis B-like toxin
NF	Needle-free
NGS	Next-generation Sequencing
NL	The Netherlands
NLME	Nonlinear "Mixed-effect" Models
NMEC	Neonatal-meningitis *E. coli*
NMI	Non-medically Important
NORMVET	Norwegian monitoring programme for antimicrobial resistance in the veterinary and food production
NSAID	Non-steroidal Anti-inflammatory Drug
NSAIDs	Non-steroidal Anti-Inflammatory Drugs
NSSRF	National Union of Swine Breeders of Russia
NWD	Newcastle Disease
NWT	Non-wild Type
OD	Oedema Disease
OMI	Otitis Media Interna
ORT	*Ornithobacterium rhinotracheale*
OVH-2	Ovine Herpes Virus-2
PAE	Post-Antibiotic Effect
PAEP	Post-Antibiotic-Effect-associated Permeabilisation
PALE	Post-Antimicrobial Leucocyte Enhancement Effect
PBPK	Physiologically Based Pharmacokinetic
PCR	Polymerase Chain Reaction
PCU	Population Correction Unit
PCV-2	Porcine Circovirus-2
PD	Pharmacodynamic
PDT	Pharmacodynamic Target
PE	Porcine Proliferative Enteropathy
PEC	Predictable Environmental Concentration
PED	Porcine Epidemic Diarrhoea
PEG bG-CSF	Pegylated Bovine Granulocyte Colony Stimulating Factor
PELF	Pulmonary Epithelial Lining Fluid
PI-3	Parainfluenza type-3 Virus
PK	Pharmacokinetic
PL	Package Leaflet
PLF	Precision Livestock Farming
PM	*Pasteurella multocida*
PMMA	Polymethylmethacrylate
PMQR	Plasmid-Mediated Quinolone Resistance
PMWS	Post-weaning Multisystemic Wasting Syndrome
PNEC	Predictable No Effect Concentration
PO	Per Oral
PPB	Plasma Protein Binding
PPDS	Postpartum Dysgalactia Syndrome
PPLO	Pleuropneumonia-like Organisms
PRDC	Porcine Respiratory Disease Complex

PRRSV	Porcine Reproductive and Respiratory Syndrome Virus
PTA	Probability of Target Attainment
PWD	Piglet Wasting Disease
PWEC	Post-weaning *E. coli*
qPCR	Quantitative PCR
QRDR	Quinolone-Resistance Determining Regions
QS	Quorum Sensing
QTLs	Quantitative Trait Loci
R	Resistant
RA	*Riemerella anatipestifer*
rAg	Recombinant antigens
RAPD-PCR	Random Amplified Polymorphic DNA-PCR
RCT	Randomised Clinical Trial
REP-PCR	Repetitive Element Palindromic PCR
RESAPATH	French network for surveillance of antimicrobial resistance in bacteria from diseased animals
RFID	Radiofrequency Identification
RL	Rear-left
RONAFA	Working Group on the Reduction of the Need to Use Antimicrobials in Food-producing Animals
RPMI	Roswell Park Memorial Institute
RQ	Risk Quotient
RSPCA	Royal Society for the Prevention of Cruelty to Animals
RT PCR	Reverse Transcription Polymerase Chain Reaction
rtPCR	Real-Time PCR
RT-qPCR	Reverse Transcription-Quantitative polymerase chain reaction
RVA	Rotavirus type A
RWA	Raised Without Antimicrobials
S	Susceptible
SA	Short-acting
SA	*Staphylococcus aureus*
SAA	Serum Amyloid A
SAAC	Slime Associated Antigenic Complex
SC	Subcutaneous
SCC	Somatic Cell Count
SCFP	Fermentation Products of *Saccharomyces cerevisiae*
SD	Swine dysentery
SDCT	Selective Dry Cow Therapy
SDQ	bacteriophage vB_SauM_SDQ
SIRS	Systemic Inflammatory Response Syndrome
SNPs	Single-nucleotide Polymorphisms
SOP	Standard Operating Procedure
SPC	Summary of Product Characteristics
SPF	Specific Pathogen Free
SSC	Somatic Cell Count
ST	Heat-stable toxin
STEC	Shigatoxigenic *E. coli* strains

enabling utilisation of different substrates, genes encoding properties, which in complexity enabled bacteria to effectively survive. But certain genes of unknown purpose still remain. What are they encoding? Do we know which are signal molecules and how communication works among bacteria? The time of movement from reading of genomes towards explaining the roles of gene products in their complexity could help, together with other knowledge to better understand roles of bacteria and their potential not only in infection processes, but their symbiosis as well as competing. The most challenging part of such investigation will be transferring and use of such knowledge into practice.

We have certain portfolio of antimicrobials in different pharmaceutical forms for different routes of administration, but still indications and target sites of infection remain, which we are not able to reach with sufficiently high concentration to inhibit or kill the pathogens. Is there any space to recognise new delivery systems applicable (considering also economy) in veterinary medicine?

We intensify pressure on minimising the use of antimicrobials, which is beneficial from the perspective of minimising the pressure on resistance selection, for load of antimicrobials and their metabolites to the environment, for keeping antimicrobials functioning. But what amount will be sustainable for producers to ensure availability of medicines with antimicrobials? Is a European market big enough to keep "stocks of essential antimicrobials for veterinary use" readily available across Member States? We urgently need to set innovative business models for antimicrobial medicinal products in veterinary medicine. But question is—how—once we have already recognised how difficult it is in human medicine.

With pressure of responsible and rational use of antimicrobials, including accepting principles of antimicrobial stewardship, more and more questions arrived on correct dosing schedules updated according to the situation with decreased pathogen susceptibility. Critical questions are: Who will perform and pay studies for such updates? Who will consequently perform and pay the residue studies for new dosing schedules? And will not be the withdrawal period so long to disqualify food producing animals´ treatment?

With question of adjusting proper dosing schedules, also question of modelling is of importance, especially nowadays when mathematical or even artificial intelligence modelling becomes more used. Pressure on stop of use of animals for testing of new medicines and utilisation of modelling only is linked with burning questions. On which data we will build these models? Will the quality of data coming from old studies fulfil current standards and demands? Do we have input data for modelling in cases of completely new molecules of antimicrobials? Is the modelling sufficient to cover clinical conditions, biological variability and real practice scenarios? Can we release broadly new medicine for thousands or millions of animals, without previous ethically performed clinical trial on limited (statistically sufficient) numbers of animals? Very thorough evidence-based considerations and need of keeping of justified balance among modelling and ethical trials with animals are here in front of us.

STS	Standard Two-Stage
Stx2e	Shigatoxin
SUAM	*Streptococcus uberis* Adhesion Molecule
SUL	Sulfamethoxazole
SWEDRES	Swedish Antibiotic Sales and Resistance in Human Medicine
T max	Time taken to reach the maximum concentration
T	Time
$T_{1/2}$	Half-life Time
TDM	Therapeutic Drug Monitoring
TIL	Tilmicosin
TLM	Tilmicosin
T_{max}	Time taken to reach the maximum concentration
TMP	Trimethoprim
TNF	Tumor Necrosis Factor
TP	*Trueperella pyogenes*
TTW	Transtracheal Wash
TUL	Tulathromycin
UE	Ulcerative Enteritis
UK	United Kingdom
UNEP	United Nations Environment Programme
UPEC	Uropathogenic *E. coli*
US	United States (of America)
USA	United States of America
USDA	United States Department of Agriculture
USKVBL	Institute for State Control of Veterinary Biologicals and Medicines (Czech abbreviation)
UV-C LED	Ultraviolet-C light-emitting diode
V_d	Volume of Distribution
V_d	Volume of Distribution
$V_{dß}$	Apparent Volume of distribution
VETCAST	Veterinary Committee on Antimicrobial Susceptibility Testing
VFD	Veterinary Feed Directive
VMD-DEFRA	Veterinary Medicines Directorate—Department for Environment, Food and Rural Affairs
VMP	Veterinary Medicinal Product
VRE	Vancomycin-Resistant Enterococcus
WHO	World Health Organisation
WOAH	World Organisation for Animal Health
WOAH (OIE)	World Organisation for Animal Health (previously OIE)
WOAT	Whisper On Arrival Technology
WP	Withdrawal period
WSD	Weighted standard deviation
ZnO	Zinc Oxide

List of Abbreviations (Vol II Chapters I–V: By Chapters)

Chapter 1: Choice of the Antimicrobial Drug and Dosing Schedules Optimisation Using PK/PD Characteristics

ADME	Absorption, Distribution, Metabolism, Excretion
AMEG	AntiMicrobial Advice *Ad Hoc* Expert Group
AMR	Antimicrobial Resistance
ANSES	French Agency for Food, Environmental and Occupational Health & Safety
APEC	Avian Pathogenic *Escherichia coli*
App	*Actinobacillus pleuropneumoniae*
AST	Antimicrobial Susceptibility Testing
AUC	Area Under the Curve
b.w.	Body Weight
BAL	Bronchoalveolar Lavage
BE	Bioequivalence
BEVA	British Equine Veterinary Association
BOV	Between-Occasion Variability
BRD	Bovine Respiratory Disease
BSV	Between-Subject Variability
CBP	Clinical Breakpoint
CEF	Ceftiofur
CFU	Colony-Forming Unit
CIAs	Critically Important Antimicrobials
CIP	Ciprofloxacin
CLSI	Clinical and Laboratory Standards Institute
Cl_{β}	Total Body Clearance
C_{max}	Maximum Plasma Concentration
CNS	Central Nervous System
C_p	Plasma concentration
CRP	C-reactive Protein
CVMP	Committee for Veterinary Medicinal Products
CXCL8	C-X-C motif Chemokine Ligand 8
DC	Dry Cows
DMEM	Dulbecco's Modified Eagle Medium
DMSO	Dimethyl Sulfoxide
EC	European Commission
ECV, ECOFF	Epidemiological Cut-Off Value
ED_{50}	Median Effective dose

EMA	European Medicines Agency (previously EMEA)
E_{max}	Maximum Killing Effects
ENR	Enrofloxacin
EP	European Parliament
ESVAC	European Surveillance of Veterinary Antimicrobial Consumption
EU	European Union
EUCAST	European Committee on Antimicrobial Susceptibility Testing
FDA	Food and Drug Administration
FLF	Florfenicol
FP	Food Producing
GFR	Glomerular Filtration Rate
GI	Gastrointestinal
GIT	Gastrointestinal Tract
GLP	Good Laboratory Practice
igPBPK	Interactive Generic Physiologically Based Pharmacokinetic
IL	Interleukin
IM	Intramuscular
ISF	Interstitial Fluid
IU	International Unit
IV	Intravenous
LA	Long-Acting
LC	Lactating Cows
LPS	Lipopolysaccharides
LTB4	LeukoTriene B4
MBC	Minimum Bactericidal Concentration
MDR	Multidrug-Resistant
MIC	Minimal Inhibitory Concentration
MLM	Modified Lacks Medium
MPC	Mutant Prevention Concentration
MRL	Maximum Residue Limit
MRSA	Methicillin-Resistant *Staphylococcus aureus*
MRT	Mean Residence Time
MSW	Mutation Selection Window
NAPQI	N-acetyl-p-benzoquinone-imine
NCA	Non-Compartmental Analysis
NLME	Nonlinear “Mixed-effect” Models
NSAIDs	Non-steroidal Anti-Inflammatory Drugs
PAE	Post-Antibiotic Effect
PAEP	Post-Antibiotic-Effect-associated Permeabilisation
PALE	Post-Antibiotic Leucocyte Enhancement Effect
PBPK	Physiologically Based Pharmacokinetic
PCR	Polymerase Chain Reaction
PD	Pharmacodynamic
PDT	Pharmacodynamic Target
PEC	Predictable Environmental Concentration
PELF	Pulmonary Epithelial Lining Fluid

PELF	Pulmonary Epithelial Lining Fluid
PK	Pharmacokinetic
PL	Package Leaflet
PNEC	Predictable No Effect Concentration
PO	Per Oral
PTA	Probability of Target Attainment
RFID	Radiofrequency Identification
RPMI	Roswell Park Memorial Institute
RQ	Risk Quotient
RT-qPCR	Reverse Transcription-Quantitative polymerase chain reaction
SC	Subcutaneous
SCC	Somatic Cell Count
SPC	Summary of Product Characteristics
STS	Standard Two-Stage
SUL	Sulfamethoxazole
$T_{1/2}$	Half-life Time
TDM	Therapeutic Drug Monitoring
TLM	Tilmicosin
T_{max}	Time taken to reach the maximum concentration
TMP	Trimethoprim
TNF	Tumor Necrosis Factor
TUL	Tulathromycin
V_d	Volume of Distribution
$V_{dß}$	Apparent Volume of distribution
VMP	Veterinary Medicinal Product
WP	Withdrawal period

Chapter 2: Swine

3G	Third generation
4G	Fourth generation
ADG	Average Daily Gain
AGP	Antimicrobial Growth Promoter
AHAW	Animal Health and Welfare
AIAO	All-in-All-out
AMCRA	Antimicrobial Consumption and Resistance in Animals
AMEG	AntiMicrobial Advice *Ad Hoc* Expert Group
App	*Actinobacillus pleuropneumoniae*
ASF	African Swine Fever
AT	Austria (in table 6)
ATB	Antibiotic
AUC	Area Under the Curve
BAC	Bronchoalveolar Cell
BELF	Bronchial Epithelial Lining Fluid
CCC	Colon Content Concentration
CDA	Concentration-dependent Antibiotic
CdAD	*Clostridioides difficile*-associated disease
CDDEP	Centre for Disease Dynamics, Economics & Policy

CDI	*Clostridioides difficile*
CHMP	Committee for Medicinal Products for Human Use
CIA	Critically Important Antimicrobial
CLSI	Clinical and Laboratory Standards Institute
CM	Coliform Mastitis
C_{max}	Maximum Plasma Concentration
CpA	*Clostridium perfringens* type A
CpC	*Clostridium perfringens* type C
Css	Concentration steady state
CVMP	Committee for Veterinary Medicinal Products
DANMAP	Danish Integrated Antimicrobial Resistance Monitoring and Research Programme
DK	Denmark (in table 6)
DT	Definitive (phage) Type
EC	European Commission
ECOFF	Epidemiological Cut-Off Value
EE	Exudative Epidermitis
EFSA	European Food Safety Authority
EMA	European Medicines Agency
ES	Spain (Table 6)
ESBL	Extended-Spectrum β-Lactamase
ESVAC	European Surveillance of Veterinary Antimicrobial Consumption
ETEC	Enterotoxigenic *Escherichia coli*
EU	European Union
FAO	Food and Agriculture Organisation
FCR	Feed Conversion Ratio
FMD	Foot and Mouth Disease
GIT	Gastrointestinal Tract
Gps	*Glaesserella parasuis*
HBS	Haemorrhagic Bowel Syndrome
ICC	Ileum Contents Concentration
IF	In Feed
IM	Intramuscular
iMBC	Intracellular Minimum Bactericidal Concentration
iMIC	Intracellular Minimum Inhibitory Concentration
Inj	Injectable
IV	Intravenous
IW	In Water
JVARM	Japanese Veterinary Antimicrobial Resistance Monitoring System
LA	Long-acting
LI	*Lawsonia intracellularis*
LPS	Lipopolysaccharide
LT	Heat-labile toxin
MBC	Minimum Bactericidal Concentration
MDA	Maternally Derived Antibodies
MDR	Multidrug-Resistant

Mhp	*Mycoplasma hyopneumoniae*
MIC	Minimal Inhibitory Concentration
MIRD	Mycoplasma-Induced Respiratory Disease
MMA	Mastitis-Metritis-Agalactia
MRSA	Methicillin-Resistant *Staphylococcus aureus*
MRT	Mean Residence Time
NA	Not Available
ND	Neonatal Diarrhoea
NF	Needle-free
NSAID	Non-steroidal Anti-inflammatory Drug
NSSRF	National Union of Swine Breeders of Russia
OD	Oedema Disease
PAE	Post-Antibiotic Effect
PALE	Post-antimicrobial Leucocyte Enhancement Effect
PCR	Polymerase Chain Reaction
PCU	Population Correction Unit
PCV-2	Porcine Circovirus - 2
PD	Pharmacodynamic
PE	Porcine Proliferative Enteropathy
PED	Porcine Epidemic Diarrhoea
PELF	Pulmonary Epithelial Lining Fluid
PK	Pharmacokinetic
PLF	Precision Livestock Farming
PMWS	Post-weaning Multisystemic Wasting Syndrome
PPB	Plasma Protein Binding
PPDS	Postpartum Dysgalactia Syndrome
PRDC	Porcine Respiratory Disease Complex
PRRSV	Porcine Reproductive and Respiratory Syndrome Virus
PWD	Piglet Wasting Disease
PWEC	Post-weaning *E. coli*
RONAFA	Working Group on the Reduction of the Need to Use Antimicrobials in Food-producing Animals
RVA	Rotavirus type A
SA	Short-acting
SC	Subcutaneous
SD	Swine dysentery
SPF	Specific Pathogen Free
ST	Heat-stable toxin
STEC	Shigatoxigenic *E. coli* strains
Stx2e	Shigatoxin
T max	Time taken to reach the maximum concentration
UNEP	United Nations Environment Programme
US, USA	United States of America
VETCAST	Veterinary Committee on Antimicrobial Susceptibility Testing
VMP	Veterinary Medicinal Product
WHO	World Health Organisation
WOAH (OIE)	World Organisation for Animal Health (previously OIE)

WSD	Weighted standard deviation
ZnO	Zinc Oxide

Chapter 3: Poultry

ABC	ATP-binding Cassette
ADI	Acceptable Daily Intake
AF	Antibiotic Free
AFB1	Aflatoxin B1
AGP	Antimicrobial Growth Promoter
AMCRA	Antimicrobial Consumption and Resistance in Animals
AMEG	AntiMicrobial Advice *Ad Hoc* Expert Group
AMP	Ampicillin
AMR	AntiMicrobial Resistance
APEC	Avian Pathogenic *Escherichia coli*
ATCC	American Type Culture Collection
AUC	Area Under the Curve
BE	Belgium
BG	Bacterial Ghost
CA	Competent Authority
CBP	Clinical Breakpoint
CECT	Spanish Type Culture Collection
CFU	Colony-Forming Unit
CGMCC	China General Microbiological Culture Collection Centre
CIA	Critically Important Antimicrobial
CIWF	Compassion in World Farming
Clf	Clumping factor
C_{max}	Maximum Plasma Concentration
Cna	Collagen-binding protein
CRD	Chronic respiratory disease
CRISPR	Clustered Regularly Interspaced Short Palindromic Repeats
DCD	Defined Course Dose
DDD	Defined Daily Dose
DNA	Deoxyribonucleic Acid
DSM	German Collection of Microorganisms and Cell Cultures
EC	European Commission
ECDC	European Centre for Disease Prevention and Control
EEA	European Economic Area
EFSA	European Food Safety Authority
ELISA	Enzyme-linked Immunosorbent Assay
EMA	European Medicines Agency
ERIC-PCR	Enterobacterial Repetitive Intergenic Consensus-PCR
ESBL	Extended-Spectrum β-Lactamase
ESC	Extended-Spectrum Cephalosporin
ESVAC	European Surveillance of Veterinary Antimicrobial Consumption
ExPEC	Extraintestinal Pathogenic *Escherichia coli*
FAO	Food and Agriculture Organisation
Fba	Fructose-1,6-bisphosphate aldolase

FBOp	Food Business Operator
FDA	Food and Drug Administration
FnBP	Fibronectin-Binding Protein
FVE	Federation of Veterinarians of Europe
GIT	Gastrointestinal Tract
HAI	Hospital-Acquired Infections
HPAI	High Pathogenic Avian Influenza
IB	Infection Bronchitis
IBV	Infectious Bronchitis Virus
IC	Infectious Coryza
IgA	Immunoglobulin A
IM	Intramuscular
IS	Infectious Sinusitis
IU	International Unit
JIACRA	Joint Inter-agency Antimicrobial Consumption and Resistance Analysis
LAMP	Loop-Mediated Isothermal Amplification
LPS	Lipopolysaccharides
MAH	Monitoring Animal Health
MG	*Mycoplasma gallisepticum*
MI	*Mycoplasma iowae*
MIC	Minimal Inhibitory Concentration
MLST	Multilocus Sequence Typing
MM	*Mycoplasma meleagridis*
MPC	Mutant Prevention Concentration
mPCR	Multiplex PCR
MRL	Maximum Residue Limit
MRSA	Methicillin-Resistant *Staphylococcus aureus*
MRT	Mean Residence Time
MS	*Mycoplasma synoviae*
NAD	Nicotinamide Adenine Dinucleotide
NAE	No Antibiotic Ever
NCP	National Control Programme
NEO	Neomycin
NetB+	Necrotic enteritis B-like toxin
NL	The Netherlands
NMEC	Neonatal-meningitis *E. coli*
NMI	Non-medically Important
NORMVET	Norwegian monitoring programme for antimicrobial resistance in the veterinary and food production
NWD	Newcastle Disease
NWT	Non-wild Type
ORT	*Ornithobacterium rhinotracheale*
PAE	Post-Antibiotic Effect
PCR	Polymerase Chain Reaction
PCU	Population Correction Unit
PD	Pharmacodynamic
PK	Pharmacokinetic

PMQR	Plasmid-Mediated Quinolone Resistance
qPCR	Quantitative PCR
QRDR	Quinolone Resistance-Determining Regions
QS	Quorum Sensing
RA	*Riemerella anatipestifer*
rAg	Recombinant antigens
RAPD-PCR	Random Amplified Polymorphic DNA-PCR
REP-PCR	Repetitive Element Palindromic PCR
RESAPATH	French network for surveillance of antimicrobial resistance in bacteria from diseased animals
RSPCA	Royal Society for the Prevention of Cruelty to Animals
RT PCR	Reverse Transcription Polymerase Chain Reaction
RWA	Raised Without Antimicrobials
SC	Subcutaneous
SPC	Summary of Product Characteristics
$T_{1/2}$	Half-life Time
T_{max}	Time taken to reach the maximum concentration
UE	Ulcerative Enteritis
UK	United Kingdom
UPEC	Uropathogenic *E. coli*
US, USA	United States of America
USDA	United States Department of Agriculture
USKVBL	Institute for State Control of Veterinary Biologicals and Medicines (Czech abbreviation)
UV-C LED	Ultraviolet-C light-emitting diode
VFD	Veterinary Feed Directive
VMD-DEFRA	Veterinary Medicines Directorate—Department for Environment, Food and Rural Affairs
VMP	Veterinary Medicinal Product
VRE	Vancomycin-Resistant Enterococcus
WHO	World Health Organisation
WOAH (OIE)	World Organisation for Animal Health (previously OIE)
WP	Withdrawal Period

Chapter 4: Cattle

AABP	American Association of Bovine Practitioners
AAEC	Attaching-effacing *E. coli*
AGP	Antimicrobial Growth Promoter
AMCRA	Antimicrobial Consumption and Resistance in Animals
AMPs	Antimicrobial Peptides
AST	Antimicrobial Susceptibility Testing
ATCC	American Type Culture Collection
ATM	Antimicrobial
AUC	Area Under the Curve
AVA-AMA	Australian Veterinary Association—Animal Medicines Australia
B&F	Breeder and Fattener
BAL	Bronchoalveolar Lavage

BF	Breeder Farm
BHV-1	Bovine Herpes-virus-1
BQA	Beef Quality Assurance
BRD	Bovine Respiratory Disease
BRSV	Bovine Respiratory Syncytial Virus
BTSCC	Bulk Tank Somatic Cell Count
BVDV	Bovine Viral-diarrhoea Virus
BW	Body Weight
C	Concentration
CAMP	Christie– Atkins–Munch–Peterson test
CIA	Critically Important Antimicrobial
CLSI	Clinical and Laboratory Standards Institute
CM	Clinical Mastitis
C_{max}	Maximum Plasma Concentration
CMT	California Mastitis Test
COX	Cyclooxygenase
CT	Co-dependent
CZ NMTP	Czech National Monitoring of Target Pathogens
DC	Dry Cows
DCT	Dry Cow Therapy
DDD	Defined Daily Dose
DNA	Deoxyribonucleic Acid
EC	European Commission
EFSA	European Food Safety Authority
ELISA	Enzyme-linked Immunosorbent Assay
EMA	European Medicines Agency
EO	Essential Oils
ESBL	Extended-Spectrum β-Lactamase
ESI-MS	Electrospray Ionisation Mass Spectrometry
ESVAC	European Surveillance of Veterinary Antimicrobial Consumption
ETEC	Enterotoxigenic *Escherichia coli*
EU	European Union
EXPEC	Extraintestinal Pathogenic *Escherichia coli*
FDA	Food and Drug Administration
FF	Fattening only Farms
FR	Front-right
FVE	Federation of Veterinarians of Europe
GIT	Gastrointestinal Tract
GVA	German Veterinary Association
HS	*Histophilus somni*
IBK	Infectious Bovine Keratoconjunctivitis
IgG	Immunoglobulin G
IgYs	Egg Yolk Immunoglobulins
IL	Interleukin
IM	Intramuscular
IMI	Intramammary Infection
IMM	Intramammary

INJ	Injectable
IROMP	Iron-regulated Outer Membrane Proteins
IU	International Unit
IV	Intravenous
LA	Long-Acting
LAMP	Loop-Mediated Isothermal Amplification
LC	Lactating Cows
LPS	Lipopolysaccharide
MAG	alpha 2-Macroglobulin, serum Albumin and immunoglobulin G binding protein
MALDI-TOF	Matrix-Assisted Laser Desorption/ Ionisation—Time-of-Flight
MDH	Meat Producers from Dairy Herds
MDR	Multidrug-Resistant
MH	*Mannheimia haemolytica*
MIC	Minimal Inhibitory Concentration
MPEC	Mammary Pathogenic *E. coli*
MRL	Maximum Residue Limit
MRSA	Methicillin-Resistant *Staphylococcus aureus*
MRT	Mean Residence Time
MY	Mycoplasmas
NAGase	N-acetyl-beta-D-glucosaminidase
NAS	Non-aureus Staphylococci
NCD	Neonatal Calf Diarrhoea
NGS	Next-generation Sequencing
NSAID	Non-steroidal Anti-Inflammatory Drug
OMI	Otitis Media Interna
OVH-2	Ovine Herpes Virus-2
PCR	Polymerase Chain Reaction
PD	Pharmacodynamic
PEG bG-CSF	Pegylated Bovine Granulocyte Colony Stimulating Factor
PELF	Pulmonary Epithelial Lining Fluid
PI-3	Parainfluenza type-3 Virus
PK	Pharmacokinetic
PM	*Pasteurella multocida*
PPLO	Pleuropneumonia-like Organisms
qPCR	Quantitative PCR
QTLs	Quantitative Trait Loci
R	Resistant
RAPD-PCR	Random Amplified Polymorphic DNA
RL	Rear-left
rtPCR	Real-Time PCR
S	Susceptible
SA	*Staphylococcus aureus*
SAA	Serum Amyloid A
SAAC	Slime Associated Antigenic Complex
SC	Subcutaneous
SCFP	Fermentation Products of *Saccharomyces cerevisiae*

SDCT	Selective Dry Cow Therapy
SDQ	bacteriophage vB_SauM_SDQ
SNPs	Single-nucleotide Polymorphisms
SOP	Standard Operating Procedure
SPC	Summary of Product Characteristics
SSC	Somatic Cell Count
SUAM	*Streptococcus uberis* Adhesion Molecule
T	Time
$T_{1/2}$	Half-life Time
TIL	Tilmicosin
T_{max}	Time taken to reach the maximum concentration
TNF	Tumor Necrosis Factor
TP	*Trueperella pyogenes*
TTW	Transtracheal Wash
UK	United Kingdom
US, USA	United States of America
V_d	Volume of Distribution
VMP	Veterinary Medicinal Product
WOAH	World Organisation for Animal Health
WOAT	Whisper On Arrival Technology

Chapter 5: Horses

AAC	Antimicrobial-associated Colitis
AHAW	Animal Health and Welfare
AMR	Antimicrobial Resistance
APP	Antimicrobial Perioperative Prophylaxis
AST	Antimicrobial Susceptibility Testing
BBB	Blood-brain barrier
CFU	Colony-Forming Unit
CG	Control Group
CNS	Central Nervous System
CVMP	Committee for Veterinary Medicinal Products
DNA	Deoxyribonucleic Acid
EC	European Commission
ESBL	Extended-Spectrum β-Lactamase
EU	European Union
FINRES	Finnish Veterinary Antimicrobial Resistance Monitoring and Consumption of Antimicrobial Agents
HRM	High-resolution Melt
IM	Intramuscular
IU	International Unit
IV	Intravenous
LAMP	Loop-Mediated Isothermal Amplification
MDR	Multidrug Resistance
MIC	Minimal Inhibitory Concentration
MLS_B	Macrolide-lincosamide-streptogramin B
MMP	Matrix Metalloproteinase
MRP	Multidrug resistance-associated protein

MRSA	Methicillin-Resistant *Staphylococcus aureus*
NSAIDs	Non-steroidal Anti-Inflammatory Drugs
PCR	Polymerase Chain Reaction
PD	Pharmacodynamic
PK	Pharmacokinetic
PMMA	Polymethylmethacrylate
qPCR	Quantitative PCR
RCT	Randomised Clinical Trial
RESAPATH	French network for surveillance of antimicrobial resistance in bacteria from diseased animals
SAA	Serum Amyloid A
SC	Subcutaneous
SIRS	Systemic Inflammatory Response Syndrome
SWEDRES	Swedish Antibiotic Sales and Resistance in Human Medicine
T_{max}	Time taken to reach the maximum concentration
UK	United Kingdom
US	United States (of America)
WHO	World Health Organisation

Acknowledgments

Many thanks to all the members of my family for their patience and great support.

Big thanks also belong to the contributing authors who shared their experience, expertise, effort and time, David G Burch, Keith Edward Baptiste and Daniel Sperling.

My big regrets and sadness to express that Prof David G Burch passed away and we cannot share with him the book release.

Thanks also to my colleague Pavla Novotná, who helped me with technical issues.

My gratitude also goes to the colleagues I have met in the institute, at the university and in everyday practice, from whom I have learned a lot.

Despite it is very hard to name and not to forget anybody from the long list of nice, helpful and supportive colleagues, still I would like to mention some of them, Alfred Hera and Věra Billová, whom I am grateful for kind leadership from my carrier beginning in the Institute I am working for; Jiří Bureš, for sharing of views, discussions and support. I cannot forget to sincerely thank my colleague and friend Eva Vernerová as well as other colleagues and friends who helped me especially within the last two years, overcome health difficulties, which cause also delay in the second Volume issue.

My sincere and specific thanks go as well to my colleagues, who shared with me their knowledge, views and expertise, from everyday veterinary practices on farms, especially Petr Stolář, Alena Návarová, Tomáš Beránek, Marek Žižlavský, Václav Osička and Petr Slavík.

From the international community, except the co-authors of the book first Volume, I have to mention also close and kind cooperation and sharing of the views, experiences and pool of knowledge with Christina Greko, Kari Grave and Cedric Műntener.

Disclaimer

Although the information in this publication has been properly searched in scientific bibliography and cited, or is based on the personal professional experience of the authors, it was valid at the time of completion of this publication. There is no universally applicable treatment that works in every farm or in every breed of species described within this publication. The correct approach tailored for the exact conditions must be used for each treatment, based on the specific circumstances of the given case, and thus responsibility cannot be assumed for possible misinterpretation or acceptance of general data for the treatment of a specific animal or group of animals. Therefore, the data presented here must always be considered and compared with the up-to-date level of expert or scientific knowledge. In the case of the doubt regarding the exact use, administration and dosing schedule (dose-frequency-duration) and route of administration of a specific veterinary medicinal product containing a certain antimicrobial (or fixed combination of them), it is necessary to rely primarily on the information submitted to the specific VMP via an authorised summary of product characteristics or product leaflet (i.e. in line with the terms of marketing authorisation), as for European Union is also stipulated by the Regulation of EP and Council (EU) 2019/6 article 106/1. In the case of suspicion of adverse effects, it is necessary to report them relevantly, according to the rules stipulated by above-mentioned Regulation and in line with national legal provisions related to pharmacovigilance system. Neither the publisher nor the authors of the individual chapters assume any responsibility for possible misinterpretation or use of the information provided in the publication in cases of specific treatment of a specific animal or group of animals.

Contents

About the Editor

Lucie Pokludová had graduated as microbiologist at the Faculty of Science at Masaryk University in Brno, Czech Republic, and finished PhD studies in analytical microbiology and computer image analysis. She worked as microbiologist and then residue assessor. During the last 15 years, she works as an assessor of efficacy of antimicrobials and AMR at the Institute for State Control of Veterinary Biologicals and Medicines and is also responsible for the data reporting and analysis of consumption of antimicrobials in animals in the Czech Republic. Work in the team responsible for setting National Antibiotic Programme and Action Plan against AMR helps her to implement One Health perspective to her expert activities. She also represents the Czech Republic on different global and European fora, as e.g. EMA—ESVAC/ESUAvet, EU JAMRAI project EARS-vet, AACTING and VETCAST. She teaches part-time at Masaryk University in Brno, but also performs individual lectures on antimicrobials in Veterinary University and regular lectures in European AMR BTSF programmes as well as invited lectures on different occasions, enjoying teaching as well as communication and sharing the views with students, veterinarians and further audience bringing her different experiences from practice.

She loves her jobs but also her family, being married and having two children.

Within leisure time she is keen on photography, especially old-timer cars and during her trips, especially to the mountains, nature as well.

Selecting Antimicrobial Drugs and Dosing Schedule Optimisation Using PK/PD Characteristics

Lucie Pokludová

Abstract

Although successful therapy is the primary purpose of the antimicrobial treatment of diseased animals, the food safety of products made from animals once they are healthy and animals' production statuses are also vital. Veterinarians, in their practices, therefore must make decisions while considering and weighing the merits of several medical views: assessments of clinical signs, bacteriological investigation, the expected efficacy of the selected drug, the animals' welfare, risks of resistance development and a consideration of residues and their depletion (altogether influencing also the cost-effectiveness of treatment, i.e. a nonmedical view). For antimicrobial therapy to be considered effective, it must deliver clinical cure (but not necessarily in all cases achieving also bacteriological cure), and for this, selecting the optimal antimicrobial, taking the appropriate route of administration and using the best possible dosing schedule are vital. The three principal factors of this schedule are dose, frequency and duration. To determine the best therapy according to the above requirements, this chapter provides an overview of the interactions among antimicrobials and disease-causing microorganisms, including the potency and efficacy of antimicrobials, and the mechanisms of inhibiting/killing the pathogen, in the case of the susceptibility to concentrations achievable in the target tissue. Attention is also paid to pharmacological and toxicological effects and emerging resistances (pharmacodynamic perspective). To take the most comprehensive picture possible, characteristics important for the antimicrobial administration, together with the antimicrobial's action and behaviour inside the bodies of diseased animals (pharmacokinetic perspective) and the key variables related to the drug, its medicinal properties, its route of administration, the dose and interlinked absorption, distribution, metabolism and excretion (ADME) parameters, ensuring the achievement of an effective concentration at the site of infection for a sufficient period of time are described. The essential facts of the pharmacodynamic (PD) and pharmacokinetic (PK) indices and their uses are briefly summarised, including comments on the use of PK/PD models working with robust results reached via well-designed clinical trials and can help to set up dosing schedules for the effective therapy, once modelled data is verified by clinical studies.

Keywords

Pharmacokinetic · Pharmacodynamic · Dosing regimen · Route of administration · PK/PD indices · PK/PD modelling

L. Pokludová (✉)
Institute for State Control of Veterinary Biologicals and Medicines, Brno, Czech Republic
e-mail: pokludova@uskvbl.cz

L. Pokludová (ed.), *Antimicrobials in Livestock 2: Choices of Prevention and Treatment*,
https://doi.org/10.1007/978-3-031-60807-0_1

1 Selecting a Drug for the Purpose of Treating the Diseased Animal

1.1 General View

The effectiveness and success of the treatment via antimicrobials depends on the selection of the appropriate antimicrobial, one that can reach the site of infection and maintain the needed concentration at that site. When the mechanism of action of an antimicrobial can kill or significantly inhibit the infection-causing microorganism(s), two variables—concentration and time—are especially important. These general presumptions are crucial across the animal species and categories to be treated with antimicrobials.

The health aspects and welfare of the diseased animal are of high importance in the veterinarian's decision selecting the appropriate therapy for the disease. But when using antimicrobials, antimicrobial resistance must be factored into such a decision as well (Constable 2017). This is linked directly to the pathogen causing the disease, but at the same time, it also affects microbiomes (especially the intestinal one), which can serve as a reservoir/hot spot of antibiotic resistance and a site of its spread. Therefore, each use of antimicrobials brings selection pressure and should therefore be properly justified. Resistance development and spread through clinical pathogens are linked with clinical cure, such that the total amount of the causative pathogenic microorganism drops significantly (e.g. from 10^9 CFU/mL to 10^2 CFU/mL) and the immune deffence mechanisms can finally help to recover fully, but the pathogen is not eradicated, which means that the bacteriological cure using the medicinal product was not successful (Lees et al. 2008). The final battle with microbes can be won by immune defence; otherwise, the infection can become recurrent or start to be chronic. When the bacteriological cure for the diseased animal is not achieved, the animal can be a source of the pathogen bearing and potentially also transferring resistance. In relation to selecting an antimicrobial and determining the efficacy of this treatment while bearing resistance in mind, in vitro susceptibility testing in the laboratory plays a role. In a European survey (De Briyne et al. 2013), veterinarians indicated that laboratory services and susceptibility tests need to be improved (as for the correlation between research results and clinical outcomes) and that susceptibility testing and resistance testing need to be more accessible, faster and cheaper.

Despite the abovementioned essential prerequisites of treatment success, in practice veterinarians should also pay attention to other considerations. The cost of the treatment, including the immediate cost of the veterinary medicinal product (VMP); any related costs of additional care (the ease, frequency and overall duration of VMP's administration); the withdrawal period (e.g. economic losses from discarding the milk from treated dairy cows or from discarding eggs from treated laying hens); and the treatment's economic effectiveness (when and whether treated animals can be returned to the production status), is also important to consider (Constable 2017; De Briyne et al. 2013). Also, the availability of the proper antimicrobial in the proper pharmaceutical form that is most suitable for administration in a specific animal species or production category can be a concern.

When selecting the appropriate veterinary medicinal product containing an antimicrobial for the treatment of a disease in a specific animal, the following steps should be followed:

(A) Consider the site of infection and the clinical signs to help to determine the disease-causing pathogen (ideally confirmed via sampling and culture results).

(B) Treat life-threatening infections, such as in the cerebrospinal area (where immune defence is minimal) by using bactericidal antimicrobials or combinations of them, ideally based on knowledge of the minimum inhibitory concentrations (MICs) of the respective causative pathogens. Treat mild infections with bacteriostatic antimicrobials; the immune-defence mechanisms of the animal are expected to contribute these antimicrobials for an effective treatment. In cases where the clinical status should start to be solved immediately, a more empirical approach needs to be taken (ideally one based on susceptibility data: local/national data or data that as much as possible represent current status of susceptibility and resistance). In life-threatening infections and polymicrobial infections, more broad-spectrum and bactericidal antimicrobials need to be chosen, but some diagnostic tools (e.g. sonography and biochemical or other tests, such as those based on Polymerase Chain Reaction (PCR) tests, can help to indicate the site and extent of the infection or to confirm the pathogen). Attention should be paid to cases when the primary cause of the pathology is the toxins' release by the bacteria. In such cases, the antimicrobial affecting proteosynthesis can have a better effect than bactericidal antimicrobials can (Böttcher et al. 2004). Controversy persists in severe infections; e.g. septicaemia with endotoxin production could be a factor significantly contributing to the mortality rate, and the prognostic value of endotoxemia remains unresolved (Hurley et al. 2012). Some studies in piglets have also described that the immune response intensified with increased endotoxin exposure and/or an increased rate of delivery (Lipcsey et al. 2008). Metanalyses and individual studies in human medicine have indicated that primarily bacteriostatic antimicrobials may have better clinical cure rates than those of their bactericidal counterparts. The sufficient antibacterial action of antibiotics (i.e. a susceptibility of the pathogen) in such a patient group is therefore crucial (Nemeth et al. 2015). In human medicine, the method by which the endotoxins neutralise peptides, has been described (Wong and Luk 2009) and some new molecules are being researched. Despite will hardly become a standard treatment in veterinary medicine, at least chimeric peptidomimetic antibiotics in combination with LPS neutralisation for the prevention of unwanted inflammation are showing promise (Javed et al. 2023).

(C) From the preselected portfolio of antimicrobials (active substances), select the optimal available VMP(s). Considerations should include the route of VMP administration and the appropriate strength so that the proper dose can be safely administered. Taking these considerations into account helps to ensure the good and timely penetration of the antimicrobial to the site of infection.

(D) Attention should be paid to achieving a sufficiently high concentration for treatment that at the same time poses a low probability of adverse reactions (e.g. toxicity) and minimal pressure on selection and spread of antimicrobial resistance.

(E) Consider the cost and other nonmedical factors of the treatment when choosing VMPs/routes of administration.

(F) Establish the detailed parameters of the treatment:

a. The VMP, route of administration, dose, frequency and expected duration of the treatment. The treatment duration might change, and in some cases, prolongation might be necessary but should be thoroughly deliberated on.

b. The on-label vs off-label use of a VMP, especially in light of the indication and species of the animals. Further parameters as specified in the summary of the VMP product's characteristics should be taken into account because they affect not only the efficacy of the treatment but also the appropriate duration of the withdrawal period. Using a VMP in a way that differs from its SPC/PL (i.e. a way that it has not been authorised) can have a significant influence on the treatment. For example, exceeding the recommended volume of an antimicrobial to the injection site can cause swelling, abscesses or even necrosis, and it can affect other tissues and affect both the efficacy of the treatment and the depletion of residues from the injection site. Any deviations from the approved dose, frequency and total duration of the antimicrobial administration are key variables to be thoroughly considered when establishing the final withdrawal period. Also, please be aware of the laws in the European Union (in Regulation (EU) 2019/6, especially articles 106/1; 112–114; 115).

(G) The ease of administration and probability that the proper dosing schedule can be appropriately followed in practice by the vet and/or the farmer.

(H) The mechanisms for announcing a treatment failure and the need to change the therapy: Ideally, the sample should be taken from diseased animal/s before the first dose of the treatment, to have in vitro laboratory results ready for the cases of treatment failure. This can help to change therapy effectively considering the pathogen's resistance and/or decreased susceptibility to the antimicrobial originally administered and to better determine which antimicrobial to switch to.

Not all variables/parameters that should be considered are easily accessible. The first obstacle that clinicians face is the uncertainty of the exact causative agent linked with the presumptive disease (using differential diagnostic is important here). In some diseases and circumstances (e.g. a closed turnover of the herd), clinicians are aware of the prevalence of certain diseases, the disease-causing pathogens and—if susceptibility testing is at least occasionally performed at the farm or holding level—the pathogens' susceptibility profiles and trends. Even more complicated is the import of animals (e.g. one-day-old broiler chicks), where the situation is greatly influenced by the prevalence of pathogens, their susceptibility within the whole production pyramid (parent and even grandparent flocks) and the level of stress throughout the animals' transportation to, and their settling into, the new farm facility conditions. In the above cases, taking samples immediately prior/at starting the empirical treatment, which can be switched to narrower-spectrum or more-appropriate antimicrobial selected once the pathogens, their susceptibility to treatment options is detected and the development of the disease have been clinically determined. In the worst-case scenario (e.g. in veal calves' sector), animals from different sources are mixed together such that the same disease from the one species of pathogen, but different strains with, variable susceptibility profiles are present and start to circulate among the animals newly housed in the stable. Empirical treatment without susceptibility testing causes the wrong antimicrobial to be selected in up to 50% of cases (Constable 2017), and the risk of increased selection of antimicrobial resistance (AMR) dramatically rises. The first and most important step should be to avoid following the wrong practices. New rapid diagnostic tools (e.g. multiplex RT-qPCR) can also help and be used to gather respiratory disease data and thus improve diagnosis, and once necessary to use antimicrobial, then the disease can be better targeted (Goto et al. 2023; Pardon and Buczinski 2020).

To thoroughly consider especially the clinical parameters linked with the appropriate concentration of antimicrobial at the site of infection, the PK parameters at the site of infection and sometimes the basic PK parameters ($C_{\max}$, $T_{\max}$, $T_{1/2}$, V_{d}, F and the area under the curve (AUC); see section 1.6 and Table 5 in certain animal species for certain antimicrobials, the route of administration) are vital. Unfortunately, in many VMPs, sufficiently complete and high quality information cannot be found in product texts reflecting the pharmacokinetic properties of the individual VMP. In scientific bibliography, such data are limited, are not linked to the pharmacokinetic behaviour of exact product (product group with identical features) or are controversial. Thus, following all the above steps is difficult to do in everyday clinical practice.

1.2 Dosing Schedule as a Key Factor in Rational Use of Antimicrobials

In the European Union, the evaluation of doses for each new veterinary medicinal product (VMP) should be in accordance with the requirements of Directive 2001/82/EC, as amended in 2022 with Regulation 2019/6/EC, replacing this directive (EP/EC 2019, 2001). The revised Guideline for the demonstration of efficacy for veterinary medicinal products containing antimicrobial substances, namely EMA/CVMP/627/2001-Rev.1 (EMA 2016a), specifies the data required to demonstrate the therapeutic efficacy of veterinary medicinal products (VMPs) containing an antibacterial agent for a given indication (or indications) by using an appropriate therapeutic regimen. This means that in the European Union, for the approved veterinary medicinal products, two of the key data that are given in the SPC[1] are the dosing schedule and the route of administration. Here, the dose is set for the approved indication and target species (categories) and is set according to the results of studies that confirmed it as appropriate for certain conditions (including susceptibility patterns) and in line with the above-cited guideline. Such a dose is considered as safe for the target animal species (and, if defined, also for the age and production-specific category), and the withdrawal period that ensures consumer safety is calculated to cover the worst-case scenario, which is then included in the dosing schedule proposed for the VMP (the maximum dose is taken from the range listed in the SPC and the defined frequency and maximum duration are also taken from the range defined in the SPC). However, in practice, such a dose may be the "baseline dose," i.e. the minimum dose. In cases that require a higher dose and/or prolonged administration, the toxicity of the antimicrobial, adverse reactions from overdoses or repeated doses and prolonged administration need to be considered. Some antimicrobials (e.g. benzylpenicillins, not considering possible hypersensitivity) have broad margin of safety and low toxicity, so their doses can be relatively safely increased in the target animals. Other antimicrobials (e.g. specifically injectable gentamicin) have narrow margin of safety, and especially in specific categories of animals (e.g. newborn foals), increasing the dose can have fatal consequences. Gentamicin also has other toxic effects (especially nephrotoxicity) that can be potentiated by the concomitant use of, for example, non-steroidal anti-inflammatory drugs (NSAIDs)—where it poses a high-level risk to neonate foals (EMA 2015a).

Additional indications, severe infections, pathogens classified as "susceptible, increased exposure," later phase of infection at the point of the start of a treatment, necrotic tissues and debris-affecting penetration belongs among the factors influencing the various degrees of a treatment's success. Along with other possible reasons, the dose may need to be increased in real clinical situations. In practice, once a veterinarian considers increasing the dose, other points listed in the SPC should be taken into account, including the aforementioned key parts of the SPC: indications, dosing schedule (dose, frequency and duration), the route of administration (including volume at the injection site for intramuscular (IM) and subcutaneous (SC) administration) and the withdrawal period. Of importance is the information in the SPC of each VMP, such as contraindications (e.g. those that can warn clinicians about the toxicity to certain tissues/organs—such as the kidney, liver or central nervous system (CNS), special precautions for use in animals (where specific target species/categories and rules/warnings for use in them can be highlighted), adverse reactions and overdose levels (with descriptions of symptoms, emergency procedures and antidotes). Use during pregnancy, lactation or lay and the safety for breeding animals—due to possible effects on reproduction—cannot be omitted. Sometimes possible interactions (including those with nonantimicrobial VMPs or substances/additives administered concomitantly in feed or water) are forgotten, and this can significantly hamper the final treatment effect (e.g. a combination of antagonistically acting antimicrobials). In the SPC, if well elaborated, the parts related to pharmacodynamics/pharmacokinetics (PD/PK) contain useful information on the specific active substance and PD and PK parameters for the appropriate routes of administration for a certain VMP. The SPCs of VMPs that were authorised a long time ago often have limited information, as for PK/PD characteristics. In practice, veterinarians are faced with changing situations for some antimicrobials and

[1]SPC, summary of product characteristics: the essential and comprehensive document listing the characteristics of each authorised VMP.

pathogens, where the prevalence of resistance and MICs has increased. Therefore, especially over the past several years, the intention to revisit old VMPs and their dosing schedules is more and more often discussed as part of attempts to set methodological principles of PK/PD modelling that can help to update dosing schedules (ANSES 2017; EMA 2018a). Despite the benefits of and need for updates, modelling, even by using advanced technologies, should be verified with clinical data, especially if harmonisation among the VMP authorisations across Europe is intended. Conducting clinical trials to confirm such modelling outcomes seems to be the hardest part, because of a lack of investment into revisiting old molecules/VMPs. Of importance is also following the current scientific consensus on information about the target species/categories, target disease/pathogen, local resistance patterns, safety and efficacy. Especially in food-producing animals, the portfolio of VMPs used by single veterinarian in everyday practice is based on several substances/VMPs for which, however, the level of knowledge should be continually improved, despite SPCs remain unchanged. Because of the increased pressure to minimise the use of antimicrobials, new veterinary products have started to arrive on the market as "alternatives" to antimicrobials. Although more and more attention is being paid to these products to minimise the use of "conventional" antimicrobials, not only their efficacy but also their safety and potential interactions with conventionally used antimicrobials (e.g. some phytogenic feed additives can cause significant interactions) are of concern. Given the strong need to reduce the volume of antimicrobials used in animals, whenever they must be used, antimicrobials should be used according to proper treatment regimens: at the sufficient dose, frequency and duration of the treatment and by appropriately selecting the active substance and exact VMP. The decrease in the overall use of antimicrobials should not be achieved via the unjustified shortening of the treatment duration, by decreasing the dose or by using critically important antimicrobials (CIAs), for which lower doses per kg of body weight are set because they are "more potent" than some older antimicrobials. The replacement of legacy antimicrobials (with higher doses, and therefore higher total quantity) with critically important antimicrobials or with newer, long-acting molecules is worrisome (Bokma et al. 2019).

According to studies on setting the dose of an antibacterial agent/VMP, the susceptibility data of the target bacteria should receive the utmost consideration. Therefore, for all active substances administered as VMPs with systemic effect, the in vitro susceptibility data—minimum inhibitory concentration (MIC) and a collection of the basic pharmacodynamic (PD) parameters—should be compared with the concentration of the actve substance at the relevant target site of infection and with the relevant parameters of the pharmacokinetics (PK) gained by the route of administration at the therapeutic dose of individual VMP. According to the available MIC data (which should be collected from a sufficiently representative and recent sampling of the causative pathogens) and target animal PK data (after the exact administration route, dose, frequency duration have been determined), an analysis of the PK–PD relationship may be used to support dose regimen selection and interpretation criteria for resistance. The overall assessment of the PK–PD relationship should be sufficiently comprehensive to assess with reasonable confidence whether the antibacterial agent under investigation would show the expected clinical efficacy against the target pathogens that appear to be susceptible in vitro when that agent is used at the selected dose regimen and while following the selected route of administration. According to guideline EMA/CVMP/627/2001-Rev.1 (EMA 2016a), the PK–PD relationship can be used to justify the dosages in dose-determination studies (EMA 2018a).

Despite above briefly described scheme, input data that are used in modelling also have their weaknesses. Ideally, PK datasets are gathered from animals of the proper age and in the production category of the species that has the disease and from those for which the antimicrobial is intended. Unfortunately, a lot of PK studies are performed mostly on healthy animals. Once animals are infected, their body temperature, blood flow, capillary permeability, metabolic ability and changes in plasma protein binding affect PK. This can be seen, for example, in a study on tylosin, where healthy pigs' pharmacokinetic parameters were compared to the sick animals infected by *Streptococcus suis*. The following results for C_{max}, T_{max} and $T_{1/2}$ were achieved: In diseased pigs (2.372 μg/mL), C_{max} was higher than that in healthy pigs (2.056 μg/mL), and the results for T_{max} and $T_{1/2}$ indicated that tylosin was absorbed faster, was eliminated faster, and reached its peak concentration earlier in pigs with *S. suis* infection than in healthy pigs. This affects the dosing schedule of tylosin against *Streptococcus suis* in pigs (Huang et al. 2018). Of high importance is the target site of the infection because the concentrations achieved in plasma, respiratory compartments (lungs) and gastrointestinal tracts (colon and ileum) often differ significantly.

1.3 Antimicrobials' Essential Pharmacokinetic Parameters

Essential pharmacokinetic parameters include the absorption, distribution, metabolism and excretion (elimination) of each antimicrobial in its route of administration and pharmaceutical form, which are vital sources of information when selecting a treatments. They are influenced not only by the character of the antimicrobial's active substance (or prodrug) but also by the medicinal product's authorised formulation, including the excipients in the specific VMP. The following subsections detail these essential pharmacokinetic parameters and their characteristics.

1.3.1 Absorption

Absorption, one of the key pharmacokinetic parameters, factors in all routes of administration except the intravenous one (which delivers the formulation of veterinary antimicrobial directly into the bloodstream and ensures that the drug is distributed and has the desired systemic effect). Simplistically expressed, the extent of absorption is called bioavailability. This parameter (F) is calculated with "area under the curve" parameter and is expressed as a ratio of AUC extravascular to AUC intravenous.

1.3.1.1 Bioavailability (F)

The rate and extent to which the drug enters the systemic circulation unchanged. It can be assumed to be complete only for IV-administered drugs. Bioavailability is influenced by the active substance's absorption characteristics, the pharmaceutical form/formulation of the exact medicinal product, and the route of administration.

The rate of absorption is estimated according to evaluations of samples, in which concentrations are measured at certain time points from which two key PK variables are identified: maximum concentration C_{max} at time T_{max} (i.e. T_{max} is the time point when C_{max} has been reached). Both variables are influenced by elimination, and C_{max} is also influenced by the rate of absorption. A sufficient number of samples should be taken around the anticipated T_{max} to make a reliable estimate of peak exposure (C_{max}). C_{max}/AUC, which can be calculated and can be expressed in units of reciprocal time (h^{-1}), is an additional parameter that can be used to help determine the rate of absorption.

Absolute bioavailability compares the bioavailability of the active substance in the systemic circulation following extravascular (non-intravenous) administration (mostly after oral, subcutaneous and intramuscular administration) with the bioavailability of the same active substance following intravenous administration. It can be described as the fraction of the active substance absorbed through extravascular administration compared with the corresponding intravenous administration of the same active substance. A dose-normalised approach should be used. A well-designed, performed and validated proper study on the pharmacokinetics should yield plasma drug concentrations vs time to determine the volume of the active substance after intravenous (iv) and extravascular (e.g. oral) administrations. Absolute bioavailability is the dose-corrected area under the curve extravascular (AUC_{PO}) divided by the area under the curve intravenous (AUC_{IV}).

$$F_{absolute} = AUC_{PO} / AUC_{IV} \times Dose_{IV} / Dose_{PO}$$

where AUC refers to the area under the curve. The area under active substance concentration vs. the time curve serves as a measure of active substance exposure.

Relative bioavailability compares the bioavailability of the AUC_A of a test formulation (A), e.g. oral solid forms that contain (according to precise measurement) a certain amount of the active substance to that of the AUC_B of another established standard formulation (B), e.g. oral solution that contains the same amount of the same active substance or that is administered via a different route.

$$F_{relative} = AUC_A / AUC_B \times Dose_B / Dose_A$$

Relative bioavailability is one of the measures used to assess the bioequivalence (BE) between two medicinal products. For both aprovals of the Food and Drug Administration (FDA) and European Medicine Agency (EMA), a generic VMP should demonstrate (at a 90% confidence interval) that the ratio of the mean responses (usually of the AUC and the maximum concentration, C_{max}) of the generic product to the "reference product" is within the limits of 80–125%; in extraordinary cases of greater intraindividual variability, a maximal widening of the limits to 70–143% could be acceptable if it has been prospectively defined in the protocol and is justified considering efficacy and safety perspective (EMA 2018c). Whereas $AUC_{0\text{-}\infty}$ refers to the concentration of the drug in the blood over time $t = 0$ to $t = \infty$, C_{max} refers to the maximum concentration of the drug in the blood at the moment quantified as T_{max} (EMA 2018c).

Factors affecting absorption can be classified as drug-specific factors, animal-specific factors and interaction-specific factors.

Drug-specific factors, also known as pharmaceutical-specific (physicochemical and formulation) factors, include physicochemical and pharmaceutical variables. Among the physicochemical ones, the following are the most important: drug solubility; the effect of pH and pKa, where most drugs act as weak acids or bases in solutions in both ionised and nonionised forms; lipophilicity (nonionised antimicrobials, which penetrate well via simple diffusion); and hydrophilicity (ionised antimicrobials). On one hand, weak electrolytes cross membranes in a pH-dependent manner and are also influenced by the drug's pKa. Weakly acidic drugs are easily absorbed in stomach conditions with low pH values. On another, weakly basic drugs will not be absorbed until they have reached the high-pH conditions of the small intestine. Variables such as particle size (inversely related to the dissolution rate), amorphism and polymorphism characteristics, surface area, dissolution rate and molecular weight can play roles in absorption, to different extents. For example, chloramphenicol palmitate has three polymorphic forms: A, B and C. Among these, form B has been found to have the highest absorption rate and bioavailability (Alagga and Gupta 2020). The exact formulation of a VMP (including the characteristics of its excipients) influences the absorption and therefore also the bioavailability of the VMP; even two VMPs with the same active substance and in the same form (e.g. injection suspension) can have very different pharmacokinetics, starting with their absorption levels from the site of injection. VMP formulation and its influence on pharmacokinetics (e.g. the long release of the antimicrobials after an intramuscular or subcutaneous administration of suspension for an injection containing fractioned oils, aluminium stearate and fractions of particles of different sizes) are described later on in this chapter.

Animal-specific factors affecting drug absorption (physiological variables) include breed, age, pregnancy, gastrointestinal tract characterisation (monogastric vs ruminant, especially in orally administered VMPs), gastric emptying time, intestinal transit time, disease status, blood flow at the absorption site, the presystemic metabolism and gastrointestinal (GI) content. Critically ill animals may exhibit reduced blood flow to the GI tract, which will result in reduced drug absorption. Generally, intestinal absorption is more critical for most drugs than any other site in the GI tract because of the increased surface area of the intestinal mucosa. For orally administered drugs, before they reach circulation, they can be metabolised within the gut wall or the liver. This is known as the first-pass metabolism, which will result in a decreased amount of the active drug to be absorbed. Food content appears to affect the absorption rate of many orally administered drugs. However, the absorption of albendazole, an antiprotozoal agent, is enhanced in the presence of lipid-containing food (Alagga and Gupta 2020).

Interaction-specific factors, in orally dosed VMPs and especially those that depend on GI tract content, can also influence absorption. An example includes the effect of the pH of the stomach (e.g. acidifiers in drinking water), the GI-emptying process and the formation of complexes (e.g. kaolin and tetracyclines).

Several mechanisms of drug absorption have been identified, including passive diffusion, carrier-mediated membrane transport (e.g. active and facilitated diffusion) and other nonspecific drug transporters, such as P-glycoprotein (Alagga and Gupta 2020).

Oral administration can be, among the other factors, influenced by interactions with ions or the components of ingested feed and drunk liquids within the gastrointestinal tract. Other important factors are pH, physiological processes (e.g. gastrointestinal motility) and pathological processes (e.g. inflammatory processes changing the permeability of the biological membranes/barriers or the processes of bacterial invasion causing lesions on intestinal mucosa). In orally administered antimicrobials, the enteric epithelium represents one of the major limiting factors for drug absorption because the efflux transporters on the intestinal membrane or mucous layer, enzymatic degradation and the tight junctions along the intestinal linings can limit the absorption of hydrophilic drugs (especially), high-molecular-weight drugs and generally drugs that are substrates of the efflux transporters (Azman et al. 2022). Antimicrobials' high solubility in lipids is a prerequisite for passive diffusion across the mucosal epithelial barrier, followed by their moving in hepatic portal venous blood to the liver, where most drugs undergo biotransformation before entering systemic circulation. The first-pass effect decreases systemic availability and in specific cases activates prodrugs; most importantly, it can occur in the liver, gut lumen or mucosal epithelium.

Especially in the case of *oral administration*, differentiating between monogastric and ruminant animal species and between herbivorous, omnivorous and carnivorous animal species is crucial for defining comparative pharmacokinetics. Jerzsele (2012) provided an example of amoxicillin (frequently used beta-lactam) which presents with various differences in several species: in descending order, the best bioavailability of amoxicillin was documented in dogs and cats (60–80%), poultry (50–68%), swine (approx. 30%) and horses (5%, where better rates were achieved in foals). Moreover, as also mentioned above, interactions with, for example, acidifiers (administered via drinking water) or sanitisers (present even in residual amounts, after cleaning of water delivery equipment) can have significant effects (Ledesma et al. 2018). After administration of amoxicillin (orally dosed at 20mg/kg b.w. to broiler chickens), its relative bioavailability, C_{max} and mean residence time (MRT) significantly increased ($p < 0.05$) in the presence of concentrations of the citrate-based drinking water sanitiser (8 and 16 mg/mL) (Ledesma et al. 2018). Most aminoglycosides and colistin (sulphate) are not absorbed after oral administration in the gastrointestinal tract and should therefore be used for treatment of noninvasive pathogens only, which are mainly located in the lumen of the intestine (e.g. *E. coli diarrhoea* in piglets and calves). Some aminoglycosides differ (e.g. apramycin or paromomycin with approximately 30% bioavailability in pigs, or neomycin, where normally a maximum of 11% is absorbed), but in very young calves, a slightly higher absorption rate has been documented, especially in cases where the gut motility was slowed and/or the bowel wall was damaged (Aschbacher and Feil 1994). In colistin methanosulphonate, the parenteral form should be administered once the systemic effect has been anticipated. Acid-susceptible penicillins (e.g. benzylpenicillin), unlike amoxicillin, are destroyed by acidic pH values, and significant blood levels are not attained following oral administration (without specific substance protecting drug delivery pharmaceutical forms). Certain sulphonamides (e.g. phthalylsulphathiazole) are not absorbed in the gastrointestinal tract. In pigs, preruminant calves and lambs, most antimicrobials are generally absorbed after oral administration. In ruminating animals, the oral administration of antimicrobials can cause serious issues with an animal's microbiome. In horses, the bioavailability of orally dosed antimicrobials poses great interindividual variability (e.g. metronidazole, 60–90%). Moreover, age dependence can be seen (e.g. orally dosed cefadroxil in neonatal foals is 100% bioavailable, but until 5 months of age, this decreases to 15%). Some difficulty can be caused by the low oral availability of aminopenicillins, where the persistence of antimicrobials in the gut content can cause serious microbiota disturbances in horses. Similarly, the use of the colistin, especially in foals, is not recommended, because they disrupt the gut microbiome and cause colitis X, with possibly fatal outcomes (EMA 2015b). However, generally, the dose administered via the oral route should be higher because the levels of the drugs reached in blood and tissues are in many cases significantly lower than those gained once the same drug dose is administered parenterally. Recently improved pharmaceutical forms—such as drug-delivery systems (e.g. microencapsulation)—could improve the bioavailability of antimicrobials that are more fragile when passing through the gastrointestinal tract. Such formulations are used, for example, in amoxicillin, where the oral availability in pigs is relatively low, probably due to attacks from intestinal enzymes, acid-catalysed hydrolysis and the carrier-uptake-mediated mechanism (Reyns et al. 2007). Microencapsulated granules, sometimes with a taste-masking coating, improve both palatability and stability, especially in acidic conditions (Keshawani et al. 2017). Bioavailability after oral administration in poultry has been considered to be similar to that in humans (Jerzsele 2012). For amoxicillin, one of the most frequently used approaches is medication via drinking water, where one study reported its bioavailability to be at 61% and clavulanic acid (currently not authorised in food-producing birds in the European Union because no maximum residue limits (MRLs) have been established) at 66%, and the bioavailability of another frequently used antimicrobial, enrofloxacin, has been reported to be 72.5% in broiler chicks experimentally infected with *E. coli* and 69.78% in healthy broilers (Soliman 2000).

Absorption after *IM and SC administration* is rapid and complete for the majority of antimicrobial agents once nonirritating solutions have been used. Peak plasma concentrations are reached within 1 h. The barrier to absorption is the capillary endothelium (the passive diffusion of most drugs, where small molecules pass through intercellular pores) (Baggot 2001). Although drug absorption from IM injection sites is generally assumed to be a first-order process, this is not always valid for oil-based suspensions, mostly long-acting (LA) and unbuffered aqueous solutions, because they may cause irritation and tissue damage at the injection site (from mild swelling to serious irritation from a necrotic process). Slow and erratic absorption therefore occurs in such cases and the systemic availability of the antimicrobial can be incomplete (Baggot 2001). Factors determining active substance absorption from the injection site include primarily the VMP formulation (e.g. in amoxicillin, LA injectable suspensions with excipients such as aluminium stearate and fractioned oils), the vascularity of the injection site, the volume of VMP injected into one injection site, the concentration of the active substance in the VMP and physicochemical properties (especially lipid solubility and the pKa of the active substance). The formulation of the VMP and the amount administered to one injection site influence not only absorption but also the persistence and the kinetics of residues at the injection site core and the injection site surrounding. Finally, sampled at slaughter, the concentration of the antimicrobial could exceed maximum residue limits in the muscle at or near the injection site.

Except in small animals, subcutaneous administration is used most often in cattle, where absorption is rapid (e.g. macrolides are used relatively frequently in young cattle). After a single

SC injection of 4 mg of tildipirosin/kg b.w., bioavailability is 78.9%, and the average C_{max} of 0.7 μg/mL is reached within 23 min; a high concentration is achieved in the lung and the pulmonary epithelial lining fluid (PELF). Similarly gamithromycin has even higher bioavailability (>98%) after 6 mg of gamithromycin/kg b.w., and C_{max} is reached within 30–60 min, where high concentrations are reached in BAL and PELF (EMA – ZACTRAN 2020; EMA – ZUPREVO 2016; Giguère et al. 2013). In another VMP containing florfenicol as a solution for injections indicated for cattle, after a single subcutaneous dose of 40 mg of florfenicol/kg b.w., the average C_{max} (4.3 μg/mL) is reached in 5.5 h. Ceftiofur in the form of crystalline-free acid is also administered subcutaneously to the ear base in cattle, with rapid absorption, but its metabolisation to microbiologically active desfuroylceftiofur starts immediately, and its plasma protein binding is approximately 70–90%. One hour after a single administration, a concentration exceeding 1 μg/mL was reached in the studies performed for marketing authorisation. A C_{max} of approximately 5 μg/mL can be detected after 12 h after administration, where a minimal 0.2 μg/mL of ceftiofur + desfuroylceftiofur is present for 7 days (EMA – NAXCEL 2020).

For absorption after *intramammary infusion*, the administration of specific VMPs via specific intramammary injectors directly into the teat canal allows for the delivery of the active substance as close as possible to the site of infection. The high absorption of lipophilic substances is due to the large inner surface of the mammary gland and is even enhanced by the inflammatory process because of the higher permeability of the epithelial cells' layers. Lipophilic substances are absorbed to a lesser extent. Of importance is the exact composition of the intramammary VMP because factors such as the release of the active substance(s) from the formulation, and the physicochemical properties of the active substance(s) and the excipients should be considered because they might influence the availability of the product in milk or dry udder secretions and in udder tissue. In this respect, parameters such as composition, particle-size distribution, viscosity and dissolution in milk affect the efficacy (and residue depletion) of each individual intramammary VMP (EMA 2017). The milk is a suspension of fat droplets in an aqueous phase that contain proteins (i.e. some protein-bound fraction can be present in certain antimicrobials—in penicillin G, it is low), carbohydrates and electrolytes, in which proteins can extensively bind to some antibiotics, resulting in the uneven distribution of the drug in different components of the milk (Ziv and Rasmussen 1975). The location of the pathogen is also important. Pathogens in milk cistern are better cured than those in intracellular areas (e.g. *S. aureus* and *S. uberis* with virulent host-specific factors) because of the limited intracellular penetration of most antimicrobials. Also biofilm formation remains an issue. Although it is not a "true" pharmacokinetic, biofilm formation affects the kinetics of antimicrobial/s and thus reduces effectiveness of intervention against bioflm forming target pathogens (e.g. non-aureus Staphylococci). In biofilm conditions, many antimicrobials have difficulty reaching bacteria present in the deeper layers (gradual or heterogeneous exposure to antimicrobials risks having subinhibitory concentrations). Transcriptional responses at low doses, potentially promoting antimicrobial resistance mechanisms, can be activated via gradual exposure (Ruiz-Romero and Vargas-Bello-Pérez 2023; Wei and Yun 2022), what pose risks, especially in antimicrobials with a single step/high resistance potential.

The compositions of intramammary VMPs for lactating cows (LCs) and those for dry cows (DCs) significantly differ because of the different purposes of using them. In DC VMPs, the benzathine salts of antimicrobials (e.g. cloxacillin benzathine) and specific excipients that cause long persistence in udders are used. Rapid release is typical for aqueous formulations, and a slow release takes place, for example, in formulations containing certain percentages of aluminium monostearate in peanut oil (about 3%) or aluminium distearate and tristearate (sole or in combinaton) ranging between 2.5 and 4% and combined with different excipients (e.g. liquid paraffin), some DC VMPs also contain specific components that act as teat sealant, mechanically blocking the entrance of microorganisms after drying off.

Percutaneous absorption consists of three distinct phases: dissolution, penetration into the stratum corneum and the epithelial layer and finally entrance into the bloodstream. Lipophilicity is a major determining factor for percutaneous absorption. Absorption can be enhanced via formulation, namely by using surfactants or organic solvents like dimethyl sulphoxide (DMSO). In food-producing animals spray or foam formulations are most often used for cutaneous administration, where high rates of absorption are not expected in cases where the skin surface is not heavily damaged. They are used to a smaller extent for skin injuries, after small surgical invasions or after castration and act mainly on the surface in a preventive manner rather than after absorption into deeper parts of the skin.

1.3.2 Distribution

Another key PK parameter is the volume of distribution (V_d), which refers to the amount of the active substance in the body, assessed through its concentration in the plasma. The volume of distribution of the active substance that stays in the plasma is smaller than that of the active substance that is widely distributed in tissues. It provides an estimate of the extent of distribution of the active substance by quantifying the apparent space, both in the systemic circulation and in the distribution tissues, available to contain the active substance. When the pattern of distribution is important (when the distribution to the target tissue of infection is known), it can only be determined by measuring the level (amount) of the antimicrobial in the various organs and tissues of the body (Giguère et al. 2013).

Antimicrobials with high lipid solubility (nonpolar drugs), low rates of ionisation or low plasma-protein-binding capabilities can be roughly assumed to have higher volumes of distribution than more polar, highly ionised and high plasma-protein-binding antimicrobials. The volume of distribution may decrease because of dehydration or increase because of kidney failure (due to fluid retention) or liver failure (due to altered body fluid and plasma protein binding) (Baggot 2001).

The complexity of the factors influencing the distribution of antimicrobials in body fluids towards the target site of infection is due to multicompartmental system involvement. The main influencers to be considered are the concentration of the active substance in the blood, the intensity of the capillarisation of different tissues and inflammatory processes that can change both penetration and the amount of blood circulated at the site of inflammation. Unfortunately, infection processes often take place in the extravascular tissue compartments, in many cases crossing biological barriers via simple diffusion, and in the end, reaching the target compartment rendered impossible. Among those barriers is the blood–brain barrier, where the capillary membrane between the plasma and brain cells is much less permeable to water-soluble antimicrobials than is the membrane between the plasma and other tissues. Only antimicrobials with a high lipid–water partition coefficient are able to penetrate the tightly apposed endothelial cells of the capillary. Partially ionised antimicrobial molecules with only moderate lipid solubility are characterised by considerably slower rates of penetration. Lipid-insoluble or highly ionised antimicrobials mostly fail to enter the brain in amounts that can be effective for antimicrobial treatment. Weak organic bases can concentrate in the cerebrospinal fluid, whereas weak organic acids are excluded. Characterisations of various antimicrobials according to their lipophilicity and pH are described in Table 1. When an unbound drug is available for diffusion, extensive plasma protein binding influences in high extent antimicrobial transfer into the brain. Inflammation (bacterial meningitis or encephalitis) increases the permeability of the blood–brain barrier such that ionised lipid-insoluble compounds (e.g. antimicrobials such as penicillins and aminopenicillins) are able to penetrate it (a similar situation is the penetration of these molecules from blood to milk and blood to prostate). Most macrolides penetrate and allow for sufficient therapeutic levels in lung tissues, but their penetration to the brain is limited (with cerebrospinal fluid concentrations 2% - 13% of plasma concentration unless the meninges are inflamed (Mercer (2022). Another barrier to be considered when using antimicrobials, both for therapeutic reasons

Table 1 *Characterisation of antimicrobials and effects on tissue distribution* (modified and amended according to Lees et al. (2006), Giguère et al. (2013) and data available from the EMA webpage (at VI 2021) for individual antimicrobials)

Characteristics of an antimicrobial	Antimicrobial	Effects on tissue distribution
Very low lipophilicity		
Strong bases or polar bases	**Polymyxins** (e.g. Colistin, Polymyxin B)	* do not readily penetrate membranes
Low lipophilicity		
Strong bases or polar bases	**Aminoglycosides** (e.g. Gentamicin, Streptomycin, Dihydrostreptomycin, Neomycin, Apramycin, Paromomycin)	* do not readily penetrate biological barriers (usually therefore not achieved effective concentrations in intracellular fluids)
	Aminocyclitols (e.g. Spectinomycin)	* certain concentrations in CSF and milk obtained in inflammatory processes
Strong acids	**Beta-lactams** (e.g. Penicillin G, Cloxacillin, Ampicillin, Amoxicillin, selected Cephalosporins)	* effective concentrations may be attained in synovial, pleural, peritoneal fluids
	Beta-lactamase inhibitors (e.g. Clavulanic acid)	
Moderate to high lipophilicity		* penetrate membranes readily
Weak acids	**Sulphonamides** (e.g. Sulfadiazine, Sulfamethazine)	* available in intracellular and transcellular fluids
Weak bases	**Macrolides** (e.g. Tylosin, Tilmicosin)	* penetration to CSF differ (protein binding influence, level of lipophilicity)
	Triamilides (Tulathromycin)	* macrolides, lincosamides, tetracyclines negligible/no penetration to CSF
	Lincosamides (e.g. Lincomycin, Pirlimycin, Clindamycin)	* weak bases (e.g. Macrolides) - effect of "ion trapping", good penetration to fluids more acidic than plasma (i.e. milk, urine of carnivores, prostatic fluid)
	Diaminopyrimidines (Trimethoprim)	* weak bases (e.g. Macrolides) - effect of "ion trapping", good penetration to fluids more basic than plasma (i.e. urine of herbivores)
Amphoteric	**Selected tetracyclines** (Tetracycline, Oxytetracycline, Chlortetracycline)	* excellent penetration to lung and PELF (Macrolides)
High to very high lipophilicity		* penetrate membranes very easily
Weak bases	**Macrolides** (e.g. Erythromycin, Gamithromycin, Tildipirosin)	* excellent penetration to intracellular fluids
	Phenicols (Chloramphenicol, Florfenicol)	* excellent penetration to lung and PELF (Macrolides)
Amphoteric	**Ansamycins** (e.g. Rifampin)	* penetrate CSF readily (especially Phenicols)
	Selected tetracyclines (e.g. **Doxycycline**)	* penetrate transcellular fluids (e.g. prostatic fluid)

and to minimise toxicity, is the placental barrier, consisting of a number of tissue layers separating the blood vessels of the foetus from those of the pregnant female animal. The placental barrier is selective, and the factors that regulate the passage of drugs through any membrane (e.g. pKa, lipid solubility or protein binding) are applicable here. The penetration of the blood–milk barrier is important in acute mastitis treatment in cattle. Although inflammation increases the level of antibiotics in the milk, a limited number of antimicrobials can still reach therapeutically effective concentrations that exceed the MICs of the target pathogens for a sufficient period of time in all compartments of the udder, where target pathogens can be located. Excellent blood–udder tissue penetration ensures the effective treatment of mastitis caused by Gram-positive bacteria by a narrow-spectrum penicillin, namely penethamate hydroiodide (which is metabolised to penicillin G and rapidly ionised in udders (pKa = 2.8), where it remains "trapped" and act effectively) (Mestorino and Errecalde 2012). In contrast, the intramuscular administration of ceftiofur sodium injection has resulted in tissue and milk concentrations that are below detectable limits (Owens et al. 1990), and the authors of another study concluded that intramuscular administration of ceftiofur sodium did not affect the outcome of severe clinical mastitis when all etiologic agents were included in the analysis (Erskine et al. 2002). Another beta-lactam of importance is amoxicillin (either alone or in combination with clavulanic acid), where the average time of penetration into milk, 1.24 ± 0.09 h, is considered fast enough for this agent, but on the other hand, the AUC_{milk}:AUC_{serum} ratio was 0.18, which means that the amoxicillin had a low level of availability in milk. Because of the acidic character of this beta-lactam structure, its components dissociate more in plasma than in milk, so milk penetration is severely restricted. After 48 h after administration, C_{max} in milk was only 0.44 ± 0.13 μg/mL (intramuscular administration of 15 mg amoxicillin/kg b.w. in healthy cows), which is considered too low to treat *S. aureus*, for example. In mastitic animals, a greater penetration of amoxicillin in the gland can be expected because raising the pH permits a greater dissociation of amoxicillin in the cisterna and a process of ion sequestration with higher drug concentrations in milk (Mestorino and Errecalde 2012). Certain lipophilic and alkaline substances, however, can reach high concentrations in the milk/mammary gland thanks to their excellent lipophilicity and the ion-trapping mechanism. Milk:plasma concentrations ratios have been shown to be low for beta-lactams—0.1 in penicillin G and 0.3 in first-generation cephalosporin, whereas they were shown to be 4.6 and 8.7 in the cases of spiramycin and erythromycin, respectively (Giguère et al. 2013). Specific attention will be paid to other antimicrobials for mastitis treatment in Chapter IV dedicated to cattle.

Other places that are difficult to penetrate are joints and synovial fluid, eyes, teeth and serous cavities. For example, aminoglycosides generally achieve effective levels in synovial fluid and the pleural and peritoneal fluid, but not in the brain or the eyes. Fluoroquinolones well penetrate into joints, synovial fluid and eyes (but retinopathy has been documented in cats treated with enrofloxacin). These above commented barriers can usually be more easily crossed once the inflammatory process is ongoing; the only tissue to which this general rule does not apply is eye tissue. To attain effective levels in infections in eyes, high circulating levels of the antimicrobial are required, and IV injection can therefore be a solution. Parenterally administered erythromycin, oxytetracycline or gentamicin may achieve bacteriostatic concentrations in tears. Local—intraarticular—administration can help in cases of joints and intraperitoneal administration (Giguère et al. 2013; Reeves et al. 2015).

Interesting information as for penetration to bones and joints can be obtained from a review article using meta-analysis data (in human medicine), where >30 antimicrobials were evaluated. Most of them (listed groups with veterinary use)—namely amoxicillin, cloxacillin, cephalosporins, aminoglycosides, fluoroquinolones, doxycycline, clindamycin, trimethoprim/sulphamethoxazole and rifampin—showed good penetration into bone and joint tissues (the authors also compared the concentrations that they achieved with the MICs of selected pathogens). A few exceptions include penicillin and metronidazole, which showed a lower than optimal penetration into bones (Thabit et al. 2019).

Of significant importance is also the protein binding of the antimicrobial. Antimicrobials, and drugs in general, can bind to the proteins in plasma, milk and the proteins in debris in necrotic processes. Competition for plasma albumin is one of the most frequent causes of pharmacokinetic interactions in veterinary medicine. Drugs are reversibly bound primarily to plasma protein, maintaining approximately equal values of free and bound molecules. Natural processes (metabolism, redistribution and elimination) cause decrease of the concentration of available free fraction. Fraction bound to the protein prolongs the elimination phase, which means an increase of the antimicrobial elimination half-life ($T_{1/2}$). This has been recognised as the characteristic of the cephalosporin of the third generation, which is used to treat infections in dogs and cats, where protein binding is 96–98.7% in dogs and 99.8% in cats, and the elimination half-life and the action is therefore prolonged (EMEA 2006). Similarly, in one study, ceftiofur metabolised to desfuroylceftiofur (which is also highly protein bound), which caused prolonged elimination in cattle and horses (Constable 2017). Once the rate of the protein binding is high, it may be necessary to increase the dose. Changes in the levels of free drug influence the volume of distribution because this free active substance may distribute into the tissues, leading to a decrease in the plasma concentration profile. Additional crucial factors for drug distribution are pH differences in the compartments, the molecular weight of the substance, the degree of ionisation, the lipid solubility of the drug and the blood flow rates in tissues. Generally, a small molecule that has a lipophilic character and is less ionised has more complete distribution in the whole body and its compartments. Among the pharmacokinetic factors, of importance is the so-called volume of distribution (V_d), which can be explained as the volume (amount A(t)) necessary for the drug to distribute throughout the body according to its plasma concentration (C_p). $V_d = A(t) / C_p$ (where t is time). A large V_d means extensive drug distribution and the achievement of high tissue concentrations. After distribution, redistribution can occur. Another phenomenon is accumulation in different compartments of the body. In the case of antimicrobials, aminoglycosides, for example, have an affinity for the cortex of the kidney (due to its phospholipid content). Certain antifungals accumulate either in the stratum corneum or in keratin. Another accumulation case can be explained by the effect of pH or pKa. Alkaline drugs accumulate in milk (pH 6.5–6.8), inside cells (pH ~7) or in the reticulomen of ruminants (pH 5.6–6.5). The reason for that is the pH in these compartments is similar to that in plasma (pH ~7). Lipophilic, alkaline antimicrobials with the ability to penetrate into cells thanks to an ion-trapping phenomenon are those that are effective against intracellular pathogens that infect various animal species—e.g. *Rhodococcus equi* in horses, *Mycoplasma* in various species and *Actinobacillus pleuropneumoniae* in pigs. These properties are characteristic of macrolides (erythromycin and, from human medicine, both azithromycin and clarithromycin); also new macrolides, e.g. tulathromycin, have been proven to penetrate into, accumulate in and persist in the pulmonary epithelial lining fluid, and macrophages and neutrophils have been shown to have an immunomodulatory effect (Plumb 2018; EMA–DRAXXIN 2020). Macrolides, as weak bases, become ion trapped within acidic intracellular compartments of lysosomes and are accumulated in phagosomes. The unique delivery system of the phagocytic cells is important for the slow release of the drug at the site of infection. Tulathromycin (triamilide) and two macrolides—gamithromycin and tildipirosin—belong to the new generation of antimicrobials that have plasma concentrations considerably lower than MICs established for in vitro susceptibility testing but concentrations in PELF (pulmonary epithelial lining fluid) or in bronchoalveolar cells are higher. Moreover, they "cooperate" with the immune-defence mechanisms of animals, where the most frequently and consistently reported immunomodulatory effect of macrolides and related substances in both human medicine and veterinary medicine is a reduction in neutrophilic inflammation. In several studies, tulathromycin has been shown to inhibit CXCL-8 and LTB4 production in stimulated neutrophils and macrophages (Fischer et al.

2013; Duquette et al. 2015) and induces the production of anti-inflammatory and pro-resolving lipid lipoxin A4. In addition, tulathromycin promotes the apoptotic death of neutrophils and their phagocytic clearance by macrophages—a phenomenon known as efferocytosis—both of which are crucial processes in the resolution of inflammation. Lincosamides (e.g. clindamycin, pirlimycin and lincomycin) penetrate cell membranes well; for example, pirlimycin has exhibited an intracellular concentration in udder tissue that is two- to threefold higher than that in extracellular fluids (Plumb 2018). Florfenicol is also a suitable drug for the treatment of intracellularly located bacteria, and it is used, for example, in the treatment of *Mycoplasma bovis*–caused infections.

Lastly, can be noted that the majority of antimicrobials diffuse relatively easily in extracellular fluids, but sulphonamides, the phenicol group, tetracyclines, fluoroquinolones and macrolides have properties that allow them to distribute into cells, usually enhanced by inflammatory processes.

1.3.3 Metabolism

Metabolism refers to processes that lead to the biological transformation of the parent active substance distributed after its administration in the body, which usually involves a series of reactions. These enzymatic transformations usually lead to a decrease in lipophilicity and an increase in polarity. Solubility in water therefore increases, which influences both distribution and elimination, so all these pharmacokinetic processes should be considered as complex. Biotransformation is also conventionally divided into two phases:

Phase I is called the nonsynthetising phase, and it involves chemical reactions as oxidative processes (oxidation or reduction) and dealkylation or hydroxylation, converting functional groups and preparing molecules for conjugation in phase II. The antimicrobial parent substance can be inactivated or, on the contrary, become active. Those drugs that become active after biotransformation are called prodrugs. Typical examples of antimicrobials that undergo metabolisation are enrofloxacin and ceftiofur. Enrofloxacin is metabolised to ciprofloxacin (to different extents in different species), ceftiofur to desfuroylceftiofur, penethamate to benzylpenicillin, cephapirin to desacetyl cephapirin, tylvalosin to 3-O-acetyltylosin, clarithromycin to 14-hydroxyclarithromycin, difloxacin to sarafloxacin and metronidazole to hydroxymethyl metabolite, all of which are microbiologically active.

Phase II is called the synthetising phase, and it results in the conjugation of the molecule with a polar substance, most often glucuronic acid (glucuronidation) or acetic acid (acetylation) and less often sulphate, glutathione or glycine. Finally, water-soluble and almost exclusively inactive metabolite is the product of the phase-II conjugative processes, which can be excreted. Liver tissue and enzymatic equipment are considered the primary sites of metabolism, but several other organs also exhibit metabolic activity. For example, the intestinal tract exhibits metabolic activity, where several drugs are transformed into active or inactive forms in the intestinal wall. Data on antimicrobials have shown that pivampicillin is hydrolysed to active ampicillin and that cefuroxime-axetil is converted to active cefuroxime. The kidney also has large metabolic capacities. Cyclosporin, an immunosuppressive subsubstance with antimicrobial properties, is metabolised by renal cytochrome P450 enzymes and also causes enzyme induction (Nakamura et al. 1994). Enzyme induction and enzyme inhibition are crucial for drug metabolism. Additional factors include a decrease in plasma protein binding or a decrease in hepatic blood flow (Jerzsele 2012). Enzyme induction in humans has been experienced, for instance, in rifampin. But the effects of rifampin differ depending on the animal species that it is administered in. In rats, for instance, rifampin has a negligible effect on CYP3A enzymes (Lu and Li 2001). Enzyme inhibition is peculiar to several antimicrobials, such as macrolides (erythromycin, clarithromycin), antimycotic ketoconazole, certain fluoroquinolones and chloramphenicol. Erythromycin and clarithromycin increase the risk of toxicity in cases of terfenadine or theophylline. Some fluoroquinolones have also been shown to increase theophylline plasma levels by inhibiting CYP1A2 in dogs (Fink-Gremmels 2008). One of the most significant metabolic interactions, which comes with a significant increase in mortality, was described for macrolides (erythromycin) or pleuromutilins (tiamulin, valnemulin) with the ionophore anticoccidials (monensin, salinomycin and narasin). The main reason is the decreased elimination of the anticoccidials. Well known is interaction between monensin and tiamulin: Tiamulin inhibits the biotransformation of monensin on the CYP3A subfamily, and the very low margin of safety associated with monensin can increase mortality (Nebbia et al. 1999; Szucs et al. 2004). Some other processes are clinically important (e.g. glucuronidation); for example, paracetamol is metabolised primarily via glucuronidation or sulphate conjugation and in smaller amounts by microsomal enzymes (catalysed by cytochrome P450) into reactive intermediates (mainly N-acetyl-p-benzoquinone-imine (NAPQI))—which when used under normal conditions in pigs is rapidly detoxified by reduced glutathione and removed via urine after conjugating with cysteine and mercapturic acid. In cats, however, in the absence of glucuronic acid conjugation, NAPQI accumulates and causes methemoglobinemia and liver necrosis, leading to death (Jerzsele 2012).

1.3.4 Excretion

More polar, water-soluble compounds are readily eliminated. Certain antimicrobials are polar and hydrophilic enough to be excreted unchanged (e.g. penicillins and aminoglycosides are excreted with urine in active forms). Elimination, in its complexity, consists of metabolism and excretion. Polar drugs are eliminated mainly by excretion only. Excretion with first-order kinetics depends on the concentration of the antimicrobial, and a constant fraction of the antimicrobial is eliminated per unit of time. The rate of elimination is proportional to the amount available in the body. The majority of drugs are excreted/eliminated by the first-order kinetic. Elimination with zero-order kinetics means only a limited constant amount of antimicrobial is eliminated within a certain unit of time. For example, 10 mg of the antimicrobial per hour might be eliminated. This rate of 10 mg elimination is constant and is independent of the total antimicrobial concentration in the plasma. Zero-order kinetics are rare. In the second case, excretion mechanisms become saturated, such as in severe renal insufficiency. Wang and Kestenbaum (2018) indicated that renal excretion is the most important route of elimination, which involves passive *glomerular filtration* (significantly hindered by protein binding if >80%) and active tubular secretion (mainly via a proximal tubule, requiring energy and carrier molecules-, so called "organic anion transporters"). The process can be saturated. Polar, hydrophilic antimicrobials can be eliminated via the urine also in unchanged (not metabolised) forms. This is important especially when treating urinary-tract infections. Some antibacterial agents (penicillins and aminoglycosides) are practically not metabolised, and short-acting tetracyclines and fluoroquinolones are metabolised to a low extent and eliminated mainly via urine. All these antimicrobials can be effectively used in the treatment of urinary-tract infections provided that the causative pathogen is susceptible. Also, active secretion plays an important role in the excretion of several substances, such as most of the beta-lactams. Once the active secretion has been inhibited, elimination is significantly reduced and thus their half-life increases. Reabsorption in the distal tubule prolongs the half-life of drugs. Nonionised, lipophilic substances can passively diffuse from the tubular fluid back to plasma. Urine pH is a limiting factor when predicting the tubular reabsorption of weak acid or weak bases. Urine acidification increases the ionisation of weakly alkaline substances, while alkalisation increases the ionisation of weakly acidic ones, and these polar molecules are ion trapped in the tubular fluid. Thus, both alkalisation and acidification govern the elimination of some potentially toxic substances via urine (Jerzsele 2012).

The *biliary excretion* of antimicrobials depends mostly on their molecular weight. Molecules heavier than 500 Da are usually excreted with bile in all animal species, including humans (except in dogs and chickens, where the threshold starts at 300 Da). For example, in rats, the biliary excretion of cephalosporins is highly dependent on molecular weight: less than 15% of the dose is excreted into bile for cephalosporins with a molecular weight of less than 450 Da, but those with a molecular weight of more than 450 Da exhibit 15–100% recovery in bile (Wright and Line 1980). In addition, their elimination pathway is mainly excretion into bile and/or urine, with minimal metabolism in the body. Certain antimicrobials, such as erythromycin, lincomycin, clindamycin, chloramphenicol, ketoconazole and griseofulvin, are excreted with bile. Conjugated forms of these substances can undergo deconjugation processes in the small intestine thanks to the activity of bacterial β-glucuronidase enzymes and can be reabsorbed. This enterohepatic circulation can significantly increase the half-life of certain drugs (Jerzsele 2012). Nafcillin and cefoperazone, because of their extensive binding to plasma proteins and excretion via bile, are exceptions among the beta-lactams, as other antimicrobials from this group are eliminated mostly by the kidneys. Macrolide antibiotics and their metabolites are excreted mainly in bile (>60%) and often undergo enterohepatic cycling. Urinary clearance may be slow and variable (often <10%) but may represent a more significant route of elimination after parenteral administration. Biliary excretion is the main elimination route of both macrolides and lincosamides, where some are excreted unchanged (erythromycin, lincomycin) and some in metabolised form (clindamycin, clarithromycin). Antimicrobials from lincosamide and also some molecules from macrolide groups can cause serious colitis and therefore should not be administered by any routes to adult horses (Baggot 2001).

Considering the residues of antimicrobials in milk and eggs, safety for human consumer is a priority. From the animal health perspective also the potential for antimicrobial resistance selection and possible influences on the microbiomes of suckling young mammals is important in determining the proportion of antimicrobial excreted via milk. For example, 3.8% and 6.8% of the dosage of erythromycin and of spiramycin, respectively, two lipophilic macrolide antibiotics, were shown to have been excreted via the milk (Giguère et al. 2013). These above commented facts are reflected in the warning introduced in the texts of veterinary medicinal products′ authorised for lactating food producing animals: The feeding of waste milk containing residues of "antimicrobials" to calves should be avoided up to the end of the milk withdrawal period (except during the colostral phase), because it could select antimicrobial-resistant bacteria within the intestinal microbiota of the calf and increase the faecal shedding of these bacteria. Elimination via eggs is of high practical importance in hens laying eggs for human consumption (especially in selected anticoccidials, like robenidine, which is not approved for egg-producing animals, because robenidine reaches high concentrations and gives an unpleasant taste in the egg).

Elimination half-life ($T_{1/2}$) is the time when the plasma levels of the drug decline to half. Because the half-life of a drug differs depending on the animal species and because it also influences setting of the dosage and dosing interval and allows at a certain scale for predicting toxic effects in animals, knowing this pharmacokinetic characteristic is beneficial. Oxytetracycline is one of the examples of interspecies differences in elimination half-life: horses (9.6 h), donkeys (6.5 h), pigs (6.0 h), sheep (5.2 h), cattle (4.0 h), goats (3.4 h) and rabbits (1.3 h) (Jerzsele 2012). Also, the elimination half-life of oxytetracycline in rainbow trout is very specific: 80.3 h at 12 °C. Sulphonamides and trimethoprim are good examples of combinations that demonstrate differences in drug elimination half-lives among species. Trimethoprim, for instance, has elimination half-life of 4.6 h in dogs, > 3.2 h in horses and 1.25 h in cattle. Sulphamethoxazole has elimination half-life of 4.8 h in horses and 2.3 h in cattle. Differences in these parameters show why the dose and dosing schedule need to be adapted to each different species. Interestingly, in humans, both substances—sulphamethoxazole (10.6 h) and trimethoprim (10.1 h)—have practically identical elimination half-lives, which makes combining them excellent for pharmacodynamics and pharmacokinetics. In animals, however, the elimination half-lives of the sulphonamides and trimethoprim are infrequently similar, so the efficacy of combinations of them in veterinary products depends on their having

Table 2 Groups of routes of administration during treatment, according to the risks of AMR (modified according EMA 2019)

Type of use	Route of administration and examples of pharmaceutical forms used
Treatment	
Local individual	Administration of the drops, ointments, etc. to eyes, ears and limited area of skin or wounds
Specific local individual	Intramammary administration of suspensions, solutions and ointments in intramammary applicators Intrauterine administration of foam, intrauterine tablets or boluses For these two specific administrations, please find the comments in the text below this table.
Parenteral individual	Intravenous, intramuscular or subcutaneous administration via injection to individual animal.
Oral individual	Oral administration of tablets or bolus to individual animal.
Metaphylaxis[a]	
Parenteral group	Parenteral administration using injection (e.g. gun devices) to the group of animals (e.g. piglets and calves).
Oral group	Oral administration via drinking water/milk replacer (powders, granules or concentrates for use in drinking water/milk/milk replacer). Oral administration via individually mixed feed or "top dressing" (powders for use in feed—either solid or liquid, individually prepared on the farm and administered to the small groups of animals).
Oral group/herd	Oral administration via medicated feed (medicated premix incorporated into the licensed feed mills according to a veterinary prescription).

[a] Since January 2022, Regulation 2019/6/EC (on veterinary medicinal products) and Regulation 2019/4/EC (medicated feed) have come into effect, so stricter rules are in force in the European Union. The routine use of antimicrobials is not allowed to mask any wrong practices (Art. 107/1), and prophylaxis (Art. 107/3) and metaphylaxis (Art 107/4) in antibiotics/antimicrobials are significantly restricted. Moreover, prophylaxis using antimicrobials is banned by Regulation 2019/4/EC (medicated feed)

well-balanced ratios of the substances; veterinary-fixed combinations in VMPs usually have ratios of 1(TMP):5(SUL).

1.4 The Effect of Routes of Administration on PK

In veterinary medicine, both scientists and practitioners are faced with different animal species (including categories) as well as many types of husbandry systems. Some antimicrobials show across the species/categories of animals little variety in their pharmacokinetic characteristics, but others show great differences. These characteristics are influenced by the properties of the active antimicrobial substance, and some of the PK parameters are significantly influenced by the routes of administration and types of formulations.

Table 2 lists the pharmaceutical forms in the same order as that appearing in (EMA 2019), and it reflects the estimated impact on the selection of antimicrobial resistance (AMR), which can be also one of the factors that influences prescription decisions. The first group has the lowest risk, and the last group has the highest risk.

These types of uses and routes of administration, as listed in Table 2, should be considered not only in the context of AMR selection but also in the context of pharmacokinetics. For example, local treatment and the pharmaceutical forms and devices used for this type of administration, once properly used, should ensure that a sufficient dose is delivered directly to the site of infection (e.g. ear), a dose that well exceeds the MICs of the pathogen. *Intramammary administration* is more complex. Here, the medicinal product is delivered directly to the teat canal, but the properties of the antimicrobial molecule and the excipients and the character of the inflammatory process influence its penetration. Especially in pathogens that are located intracellularly (e.g. *Staphylococcus aureus*), an effective concentration that is above the MICs of the pathogens and that persists for a sufficiently long period may not be reached. Therefore, despite the in vitro susceptibility of the strain of *Staphylococcus aureus* isolated from this exact case, if antimicrobial does not in the end reach a sufficient concentration in the target location of the pathogen, the infection will be neither clinically nor bacteriologically fully treated. Also, depending on the extent of the inflammatory process, the tissues' barrier properties might be affected (in some cases, the penetration of the drug to the surrounding tissues is easier), and broader distribution via the blood stream can be expected. For the sake of efficacy and to eliminate antimicrobial residues—for consumer safety and for economic reasons—those above commented facts and both the type and exact location of the pathogen should be therefore considered for each individual antimicrobial, each individual VMP and each individual case of intramammary infection.

The local administration of antimicrobials to the uterus may not always be the preferred route, regardless of the historical precedence. In cases of genitourinary tract infections, the parenteral administration of antimicrobials can achieve effective tissue concentrations of the active substance in all areas of the genitourinary tract, whereas intrauterine infusion results in comparable concentrations only in the endometrium, lochia and uterine secretions. Only local, but sometimes also comcomitant local and parenteral or only parenteral, administration may be indicated in specific cases of genitourinary tract infection.

As for the *parenteral administration,* intravenous (IV), subcutaneous (SC) and intramuscular (IM) are the three most frequently used injection routes in medication administration, in both human medicine and veterinary medicine. The preferred route of administration differs not only because of PK characteristics but also across species because of local tolerances and sometimes because of the ease of administration (e.g. in pigs, mostly IM administration is used; in horses/foals, intravenous administration is used more frequently than it is in other "large" animals; in racehorses, it is used to avoid muscular soreness; and in cattle, the initial administration of the first dose as IV is preferred in severe mastitis caused by Gram-negative microbes because of speed attainment of the necessary concentration in the system, which can be maintained through intramuscularly administered follow-up doses). Adjusting the amount of injectable VMP to be administered according to the body weight of the animal is important, and it enables the administration of precise doses. The main aim of such adjustment is to avoid both underdosing because of its effect on efficacy and overdosing because of potential toxicity and also of its effect on exceeding the residue limits for food commodities. From the perspective of local tolerance (with a possible influence on efficacy and residues), the volume administered to one injection site is of concern and following the instructions in the approved product text is recommended whenever possible.

Intravenous administration (IV) is characterised by the achievement of high blood and tissue concentrations. The initial high concentration in blood drops rapidly owing to the distribution to the body tissues, including organs of elimination (kidney and liver). Characteristic disposition curve can be calculated with the so-called biexponal equation, consisting of (α) phase (distribution, mostly via passive diffusion) and (β) phase (elimination) after a single bolus dose of the drug. Because passive diffusion is the process of entering the antimicrobials to the cells and crossing cellular barriers, the physicochemical properties of the active molecule is important. Lipid solubility, the degree of ionisation of weak organic acids or bases and the concentration gradient are determining factors of the concentrations attained in cells, transcellular fluids (e.g. cerebrospinal, synovial and ocular) and glandular secretions (e.g. milk, saliva and prostatic fluid) (Giguère et al. 2013). The advantage of IV administration is the systemic availability and the high concentration reached in a relatively short time compared to other routes of administration. On the other hand, administration should be more frequent, and the concentrations fluctuate to a greater degree. This is the case for individual injections. But there is also possibility (limited by practical difficulties) of using an IV infusion of a solution of antimicrobial at a fixed concentration, where a precise dose that systemically attains a steady state can be ensured. The time taken to reach a steady state is limited by the drug's elimination half-life (i.e. the rate of elimination). A steady-state concentration is a function of the rate of infusion. An increase in the infusion rate will not shorten the time required to reach a steady-state drug concentration (Giguère et al. 2013; Ahmed 2015). It is the preferred route to treat acute, life-threatening, systemic infections such as in septicaemia. It can be also used in some cases of chronic infections, such as pneumonia caused by *Corynebacterium* spp. in foals (Constable 2017). This route of administration is also beneficial in infections where the causative microorganism is "intermediary susceptible" and where an increased dose can help (especially in bactericidal antimicrobials with concentration-dependent bacteria-killing effects). Diffusion concentrations at the site of infection are significantly higher than those in the IM route, which are successfully used in cases where necrotic processes slow down the achievement of the concentration necessary for the treatment of infection. Intravenous administration can also be preferred because of the volume that is needed to be injected, especially for some antimicrobials (low concentration, high volumes—e.g. sulphonamide and trimethoprim in injectable solutions) administered to large animals (cattle) that are expected to be slaughtered for human consumption (to avoid damaging the muscles because they are marketed commodities). Despite the above-described benefits, intravenous administration also has weaknesses. Among them are (especially in large animals) the need for animal fixation. Possible adverse reactions can also be associated with an intravenous administration. These include acute toxic reactions (due to the drug or the excipient), so carefully following the instructions for each individual veterinary medicinal product is highly recommended. The instructions usually recommend slow administration and warming intravenously dosed fluids near to core body temperature; small volumes are administered first, and then, if no shock reaction occurs, they can be continued via the slow administration of the rest of volume (for products that have expected adverse effects and that should be used in life-threatening infections only, where the benefits exceed the possible risks)—as indicated, for example, in several VMPs containing combinations of sulphonamide and trimethoprim for horses. The administration of intravenously dosed injectables generally requires veterinarians to be skilled at administration. Administration to the jugular vein can be complicated by inadvertent intracarotid injection. Administration to the jugular vein is used in large animals, with the exception of pigs (whose administration uses ear veins instead). Perivascular reactions and intravascular thrombosis are among the other complications, especially from irritant drugs (Constable 2017).

Intramuscular administration (IM) is the most common route of administration for antimicrobials. The bioavailability of the exact VMP after IM administration is influenced by the formulation (including the excipients and particle sizes) and the rate of irritability (once the product is irritating, it can cause not only microscopic lesions and changes at the injection site but also abscesses, cysts and other signs of irritation influencing the pharmacokinetics of the active antimicrobial substance). Particular attention should be paid to the appropriateness of the drug concentration in the VMP administered to each different animal species. This is because the drug concentrations in the respective VMPs and the total dose required determine the volume to be administered (Giguère et al. 2013). The location of the injection can heavily influence the rate of absorption and the bioavailability (Constable 2017; Baggot 2001). In cattle and horses, injection into the neck should therefore be preferred over injection into the gluteal or shoulder muscles, probably owing to the different distribution among/within muscles and the penetration into the system via the blood supply. One example can be given for horses. In one study, following the IM injection of procaine penicillin G (20,000 IU/kg bodyweight) in various parts of the neck, the highest systemic availability was obtained after administering it to *m. serratus ventralis cervicitis*, followed in descending order by administering it to *m. biceps*, *m. pectoralis* and *m. gluteus*, and the lowest availability was administered via subcutaneous injection (Firth et al. 1986). Horses are very sensitive to intramuscular injection, especially when an irritant VMP is administered; some clinical guidelines therefore recommend administration to the bunchy muscles in the lower half of the chest between the tops of the forelegs (milder reactions, less spread and easier handling when treating by using drainage). In food producing animals, the influence on the residues at the injection site and its surroundings, which can persist for a very long time (even longer than a hundred days after administration), is an issue in some VMPs. To avoid local irritation that influences also residue depletion, following the recommended volumes to be administered to one injection site for each specific VMP is vital. Special attention should be also paid to the specific pharmacokinetic in so-called long-acting (LA) medicinal products, where not only the active substance and its salt but especially the excipients (e.g. oily fractions, aluminium stearate, triglycerides and ratios of them) in these nonaqueous pharmaceutical forms, mostly suspensions for injection (which are very different from water solutions for injection), influence the pharmacokinetics. These long-acting VMPs enable the slow absorption of the active substance over an extended period thanks to their either gradual or staged availability for absorption (Baggot 2001). An example is given for injectable VMPs containing ceftiofur in different salt forms and crystalline-free acid (see also the example in Table 3 for PK in pigs). Sodium salt of ceftiofur (reconstituted from a lyophilised powder into an aqueous solution) is rapidly and completely absorbed (systemic availability of 100%) after IM injection in different species, whereas ceftiofur crystalline-free acid in an oil suspension is more slowly absorbed.

One of the most frequently used types of antimicrobials administered intramuscularly in many

Table 3 Examples of differences in the pharmacokinetics of various VMPs (differences in the form of the active ingredient ceftiofur and in the qualitative and quantitative content of the excipients)

Target species	Route of administration	Active substance (salt/ester)	Excipients	Recommended dose according to the SPCs (VI/2023[a])	Selected pharmacokinetic data from the SPC (to VI/2023)
Pig	Intramuscular	Ceftiofurum hydrochloridum	Polysorbate 80 Triglycerides Medium-chain water for injections	3 mg/kg b.w.	C_{max} of active metabolite desfuroylceftiofur 11.8 ± 1.67 μg/mL reached within 1 h; t½ of active metabolite desfuroylceftiofur reached within 16.7 ± 2.3 h
	Intramuscular	Ceftiofurum sodium	Potassium acid Phosphate Sodium hydroxide Solution 10% (for pH adjustment)	3 mg/kg b.w.	C_{max} of active metabolite desfuroylceftiofur reached within 1 h. t½ of active metabolite desfuroylceftiofur reached within 13 h.
	Intramuscular	Ceftiofurum as crystalline-free acid	Oily vehicle (vegetable) Triglycerides Medium-chain cottonseed oil	5 mg/kg b.w.	One hour after a single administration, plasma concentrations > 1 μg/mL. C_{max} of active metabolite desfuroylceftiofur 4.17 ± 0.92 μg/mL reached at approx. 22 h; t½ was reached within 49.6 ± 11.8 h

C_{max} = maximum plasma concentration; $t_{1/2}$ = terminal elimination half-life
[a] The SPCs of the respective VMPs in English can be viewed at https://www.vmd.defra.gov.uk/productinformationdatabase/Default.aspx (last accessed 29.06.2023)

EU countries are aminopenicillins, usually specifically amoxicillin. The intramuscular bioavailability in pigs is 83% (amoxicillin sodium, aqueous, IV; amoxicillin trihydrate at 15% in oily LA suspension with IM administration to neck behind the ear). Slow absorption is characteristic of LA suspension, with two peaks (C_{max}1.7 μg/mL (1.3 h) and C_{max} 0.8 μg/mL (6.6 h)), and the elimination of the active substance is significantly prolonged. A short-acting (SA) VMP reached C_{max} 5.1 (2 h). The MRT for the SA one was 8.8 h and for the LA one, 66.8 h; in the SA one, the $AUC_{0-\infty}$ was 33.1 μg.h/mL, and in the LA one, 47.6 μg.h/m (Agersø and Friis 1998). The long-acting formulation had a lower maximum concentration (C_{max}), a greater area under the curve to infinity ($AUC_{0-\infty}$) and a longer mean residence time (MRT) than the SA formulation did, which makes it more suitable for a time-dependent antibiotic (Burch and Sperling 2018). For the LA formulation, a repetition of administrations after 48 h can be used effectively.

Some studies have revisited some first-line antimicrobials, paying attention to the dosing schedule (dose, frequency and total duration) and to the salts, pharmaceutical formulations and assessments of all these variables as they influence the treatment of certain bacteria in certain target species/category. Table 4 summarises the data from a study by Sjölund et al. (2020), which covered the dosing of different penicillin-containing medicinal products (potassium penicillin and procaine benzylpenicillins) in different forms (aqueous solution, aqueous suspension or oily suspension) and at different dosing schedules for the treatment of *A. pleuropneumoniae* (App) infection in pigs. The data from healthy pigs and pigs after being infected with App were investigated. According to the data, for some dosing schedules, the concentrations of penicillin over the duration were not sufficient to be used to treat App once the microorganism had reached MIC 500 μg/L (Sjölund et al. 2020).

Subcutaneous administration (SC) is another route that is frequently used in some species or categories of target animals (e.g. cats, dogs and small piglets). This route has also started to be used in livestock animals (namely cattle), especially because of concerns related to the residues that remain at the core and surrounding of injection site after intramuscular administration. Depending on the administration route—subcutaneous (SC) or intramuscular (IM) injection of the same VMP—the rate and speed of the absorption of the active drug differs. This is mainly because far more blood vessels are in the muscle than under the skin, so an intramuscularly administered antimicrobial is taken up faster into the blood stream. This can be seen as an advantage or a disadvantage, depending on the desired effect. If the animal has an especially poor systemic health status, the shortest T_{max} is desired, so, if possible, VMP suitable for IV administration should be used (via intravenous route), the next best choice is the IM administration route and the last choice is the SC administration route. If it is a *long-acting* antimicrobial, then SC should be selected even though C_{max} and therapeutic levels will be reached later than they would if IM were used, and the nature of the VMP (usually a specific formulation of a suspension and/or the active ingredient) allows therapeutic levels to be maintained in the animal tissues for a longer period. At the same time, residues after SC administration usually do not persist for as long a time as they do after IM injection (when using the same VMP at the same dose and administering it with the same frequency and for the same total duration of treatment).

Oral administration is most often used in animal species kept in groups or herds (among major livestock animals, it is used most often in pigs, in young nonruminating calves and in poultry). Usually, especially in large animals, it is used when the infection is not so serious, though it is also used in the treatment/metaphylaxis/prevention of infections in the gastrointestinal and respiratory tracts. However, in some species (poultry, rabbits and fish), especially those in extensive farming systems, this is practically the only route of administration for antimicrobials. Oral administration can be dosed via medicated water (in preruminating calves via a milk replacer); this method is preferred for treatment and metaphylaxis because diseased animals are more likely to drink than to accept feed.

Table 4 Comparison of various veterinary medicinal products (salts, pharmaceutical forms and excipients) and their dosing regimens in the context of whether the achieved PK parameters treated *Actinobacillus pleuropneumoniae* (the summary of the data was based on Sjölund et al. (2020))

Veterinary medicinal product	Active substance	Pharmaceutical form	Route of administration	Maintenance dose (mg/kg b.w.)	Dosing interval/duration	C_{max} plasma	T_{max}	Note
Benzylpenicillin Meda	Potassium benzylpenicillin	Solution	IV	10	Single dose	1860 to 9318 μg/L	0.25 h	At 4 h, plasma concentrations dropped to 15–76 μg/L, which represents 2–3 orders of magnitude
Benzylpenicillin Meda	Potassium benzylpenicillin	Solution	IV	20	Single dose			
Ethacilin	300 mg/mL of procaine benzylpenicillin	Aqueous suspension	IM	20	Single dose	1000 to 4270 μg/L	1 h	C_{max} delayed until 4 h in one pig
Ultrapen	300 mg/mL of procaine benzylpenicillin	Oily suspension	IM	30	Single dose	910 to 3220 μg/L	1 h	In three of the pigs, this was the C_{max}, but in three other pigs, there was a second peak at 4 or 24 h, which was the C_{max}
Ethacilin	300 mg/mL of procaine benzylpenicillin	Aqueous suspension	IM	20	Every 12 h/3 days			Adequate exposure (above the MIC for wild-type *A. Pleuropneumoniae* of 500 μg/L, total concentration ≈ 1000 μg/L) for more than 50% of the dosing interval
Ethacilin	300 mg/mL of procaine benzylpenicillin	Aqueous suspension	IM	20	Every 24 h/3 days			Did not give adequate exposure
Ethacilin	300 mg/mL of procaine benzylpenicillin	Aqueous suspension	IM	30	Every 24 h/3 days			Did not give adequate exposure
Ultrapen	300 mg/mL of procaine benzylpenicillin	Oily suspension	IM	30	Every 24 h/3 days			Adequate exposure (above the MIC for wild-type *A. Pleuropneumoniae* of 500 μg/L, total concentration ≈ 1000 μg/L) for more than 50% of the dosing interval

Ethacilin	300 mg/mL of procaine benzylpenicillin	Aqueous suspension	IM	30	Every 12 h/5 days			All these dosing schedules were administered to pigs experimentally infected by *A. pleuropneumoniae*. Among these dosing schedules, the concentrations achieved were approx. two times above the MIC 500 μg/L, which indicates successful treatment. The level 1000 μg/L was counted as the total concentration, which should be twice the MIC for 50% of the dose interval, given that the plasma protein binding for benzylpenicillin in pigs is up to 45%
Ultrapen	300 mg/mL of procaine benzylpenicillin	Oily suspension	IM	30	Every 24 h/5 days			
Ultrapen	300 mg/mL of procaine benzylpenicillin	Oily suspension	IM	30	Every 24 h/3 days			
Ultrapen	300 mg/mL of procaine benzylpenicillin	Oily suspension	IM	Loading single dose Ultrapen 60 mg/kg, after 24 h for 2 days 30 mg/kg/day	Every 24 h/3 days			
Ultrapen + Ethacilin	300 mg/mL of procaine benzylpenicillin		IM	Loading single ethacillin 30 mg/kg, after 12 h repeated 3 times Ultrapen 30 mg/kg	Every 24 h/3.5 days			

Benzylpenicillin Meda: No excipients, solution for injection
Ethacillin: Lecithin 3.3 mg, povidone K30 1.5 mg, dinatriumedetate4 0.1 mg, sodium citrate dihydrate 17.60 mg, potassium diphosphate 0.54 mg, methylparahydroxybenzoate 1.1 mg, sodium hydroxide and phosphoric acid (both pH adjusted), water for injection at 1 mL
Ultrapen: aluminiumdistearate, butylhydroxyanisol, butylhydroxytoluen, propylenglykoldikaprylokaprat

Medicated feed (medicated premixes incorporated into feed in feed mills or, e.g., oral powders administered as "top dressings" at the farm level) has been used for years, especially for growth promotion (banned in the European Union since 01.01.2006) or for prevention (prophylaxis). Medicated feed use for prophylaxis is banned by Regulation EP and Council 2019/4/EC and significantly limited by Regulation EP and Council 2019/6/EC and by a commission-delegated regulation. These core regulations have been in force since January 2022. Metaphylaxis practices are also restricted (Article 107/4), but the exact conditions (criteria of metaphylaxis initiation, appropriate alternatives to metaphylaxis and guidelines promoting the risk factors associated with metaphylaxis) may be set at the national level.

The main difference between the parenteral types of administration (IV, IM and SC) is that the blood and tissue levels attained following the oral administration of the same dose (same antimicrobials) are considerably lower than those achieved after parenteral administration. Therefore, the oral dose of the same antimicrobial should be (in general) higher than the parenteral dose (in some substances, the ratio can be twofold to fivefold higher). The rate of absorption varies with the formulation/pharmaceutical form; oral solutions in medicated water provide rapid absorption.

Dissolution is the first step, and a crucial one, for all solid pharmaceutical forms, and it limits the rate of antimicrobial absorption. As for absorption rate, oral suspensions and pastes generally stand somewhere between solutions and solid pharmaceutical forms. Of importance are also the technical parameters, such as maximum solubility. Therefore, for any new marketing authorisation for a VMP intended for administration via medicated water, the data on its solubility should be submitted when a metering pump is used (with a stock solution that is up to 20 times more concentrated) in two types of water with different pH and hardness values and at two temperatures (4 and 20 °C): soft water/low pH, with a pH between 5.0 and 7.0 and 60 mg/L or less of calcium carbonate (0–6 °D), or hard water/high pH, with a pH between 8.0 and 9.0 and 180 to 350 mg/L of calcium carbonate (18–35 °D). In practice, farmers also use solubility enhancers (sodium carbonate or a complex of maltodextrin, sodium carbonate, lactose and vitamin C), as reported by Vandael et al. (2019). Stability after the reconstitution of a VMP in medicated water (at the lowest nominal concentration for batches of different ages, in the two water qualities and at the two temperatures) is important. The impact of exposure to light should be also studied. In any case, regardless of the results, acceptable expiration in drinking water must not exceed 24 h (which means that each day, a new solution in water should be prepared). Despite the aforementioned qualitative proof, some data are still lacking: the impact of the physicochemical and bacteriological quality of the water, interactions with the various biocidal products used in animal husbandry, interactions with acidifying substances added to the water (e.g. ascorbic acid) and efficacy in the presence of biofilms. Also, many questions remain about the equipment used: several studies (Hémonic et al. 2016; Zangaro 2018) have indicated differences between the pump's types in how they affect the accuracy of the dosing. Of importance are also regular maintenance (cleaning and removing sediments and biofilm) and the supply of quality drinking water, to ensure to the greatest possible extent that the correct dosage is administered. Although the aforementioned prerequisites are considered more technical than medicinal, they greatly influence the dose achieved in animals: They can cause underdosing, which negatively affects both the treatment's efficacy and the development of resistance (Zangaro 2018).

Orally dosed antimicrobials' pharmacokinetic also in more or less depends on the volume and nature of the ingesta (e.g. some divalent cations can create complexes with tetracycline antimicrobials, so chlortetracyclines have much lower bioavailability in calves when administered in milk replacer rather than in water, because of the high degree of binding to calcium cations in milk). Some evidence has indicated that the oral administration of antibiotics to calves in glucose (glycine) electrolyte solutions is associated with more-favourable absorption characteristics

(Constable 2017). Generally, the big molecules are not absorbed or are absorbed to a limited extent in the gastrointestinal tract (most aminoglycosides and colistin). The systemic availability of orally administered antimicrobials is limited by the highly acidic content of gastrointestinal tract compartments (pH 3–4). For example, some penicillins (e.g. benzylpenicillin) are destroyed by the acidic pH levels in the stomach. Like other routes of administration, oral administration also has weaknesses. First, oral administration experiences difficulty in delivering the proper dose. Although the dose can be calculated according to the body weight of the animal, real consumption of water/feed per day can be limited and therefore influence how much of the dose the animal receives, such as in the case of a decreased appetite/water intake in diseased animals or social hierarchy in the group of animals (influenced also by the weakness of the diseased animals limiting their approach to medicated water/feed). Moreover, some antimicrobials can influence the immunostatus of the intestine. A prolonged administration of antimicrobials can lead to a significant gut microbiome imbalance, leading to superinfections caused by ubiquitary pathogens like *Clostridium difficile* or to a disruption in the barrier and udder defence mechanisms, which itself leads to fatal mastitis caused by *Pseudomonas aeruginosa*. Well known is that ruminants cannot be orally dosed with antimicrobials, because this would disrupt the rumen microbiome, causing significant growth depression or even death.

Some specific examples to be considered in the oral dosing of antimicrobials:

Preruminating calves: Attention should be paid to administration via milk replacer (difficult to check the full solubility, complexation of certain antimicrobials with divalent cations and palatability). Further impacts can be on very young animals (e.g. described in neonate calves); orally administered antimicrobials (e.g. neomycin, tetracyclines and aminopenicillins) can cause villous atrophy within the intestine and malabsorption diarrhoea (Constable 2017).

Pigs (piglets, fattening pigs, pregnant sows and lactating sows): In piglets, the initiation of the treatment can be carried out via the parenteral route (using either individual injection or automatic guns to administer treatment to groups of animals) and then continued via oral medication. Properly adjusting the amount to be administered is necessary (attention should be paid, e.g., to changes in climatic conditions in housing facilities—especially in cases where they have not yet installed air conditioning/humidity-controlling systems because the cold or hot weather can significantly influence the intake of medicated water). Palatability also has an effect (some substances, e.g. florfenicol in older VMPs in premix form, have lower palatability values than newer VMPs, which are used in appropriate concentrations in medicated water). In fattening piglets/pigs, infections greatly influence the intake of feed. Also, the role of the feed and interactions with the VMP, if orally administered, are crucial. Morthorst (2002) described the oral bioavailability of amoxicillin in weaned piglets after oral administration. The dose of 20 mg/kg b.w./day of amoxicillin that was incorporated into feed or drinking water resulted in very low amoxicillin levels in the serum (bioavailability of 6–29%). After the intragastrical gavage of 200 mL in drinking water with the respective dose directly into the stomach in fastened pigs, much higher but broad values of pharmacokinetic variables were achieved (serum concentration of C_{max} (21.55 μg/mL ± 34.55), a bioavailability of 91% and a short T_{max} of approx. 30 min). Data from this study shows that the significant reduction in bioavailability after the oral administration of amoxicillin was caused by interactions with the feed.

Water medication should be controlled (sometimes animals can be kept without water for a certain amount of time to ensure a faster intake of medicated water). Generally, water intake in pregnant sows is among 5–8 L per day (depending on the duration of pregnancy at the point of medication). Hierarchy behaviour should not be forgotten in the groups of pregnant sows, and whenever possible, the animals should be monitored for medicated water intake. The intake of water dramatically increases, to 15–20 L per day, in lactating sows (depending on the number of suckling piglets, the stage of lactation and breed-specific characteristics).

As mentioned above, in the group/herd where medication is administered via the oral route in pigs, the input dose can be greatly influenced by social hierarchy. This effect was described in a study by Soraci et al. (2014) on pigs after the oral administration of fosfomycin, where differences in feeding behaviour resulted in varying amounts of ingested drug between animals, and this resulted in variability in the measured PK parameters. Future directions in pig production should allow for individualisation among doses, such as by using radiofrequency identification (RFID) technology (Neethirajan 2020). To honour the antimicrobial stewardship principles applied in using antimicrobials in animal husbandry and to minimise the volume of antimicrobials used, water consumption monitoring (e.g. via RFID) and early detection of disease in pigs can be used to selectively treat diseased animals (Matthews et al. 2017).

Poultry (broilers, laying hens, turkeys, geese, ducks and quails): Mainly the oral administration of antimicrobials is used in poultry, predominantly via medicated water and less via feed—mostly in geese and ducks). The main advantages are the ease and the speed of preparing the medication (compared to mixing it into feed). The main disadvantages are the variability in drug uptake by individual animals and some technological issues. Among these technological issues are the solubility in different pH conditions (optimum 5–7), the hardness degrees of the water (different max concentrations per certain amount of water are used to avoid sediments and/or precipitation), incompatibility issues (the material of water delivery systems), the proper system of delivery (obstruction in the nipples) and lightening periods in the hall (as birds are drinking only when lightening system is on and different lightening regimens/colours support water intake) (Vermeulen et al. 2002; Wu et al. 2022). Physiologically, the most important factor influencing the rate and amount of drug absorption is the gastric emptying rate. In the case of the administration of a stable drug solution, the drug quickly passes through the crop and stomach and is readily available for absorption in the intestine within a few minutes following administration (Vermeulen et al. 2002). A poultry flock's actual water intake should therefore be thoroughly monitored and the concentration of antimicrobial in drinking water should be properly adjusted (by factoring in both water intake the body weight of the group of treated birds). In the poultry industry, only 1% of the total veterinary treatments administered to birds is estimated to use parenteral administration; other possible routes include IM (pectoral or leg muscle) and SC (slow release and low absorption).

1.5 The Antimicrobial Pharmacodynamics and View on Selection of Antimicrobial Resistance

Two principal categories should be considered when describing the effects of an antimicrobial: bactericidal (i.e. the effect of killing bacteria) and bacteriostatic (i.e. inhibiting bacterial cell division). Most of the antimicrobials can have a bactericidal effect via one of the following mechanisms of action, provided that a sufficiently high concentration and a sufficient amount of time for bactericidal action is achievable in vivo at the site of infection or systemically, and this depends on how broad the margin of safety is for the drug. New methods (using dynamic bacterial morphology imaging) for the antimicrobial mechanism of action classification are available (Ouyang et al. 2022). The well-known mechanisms of action at the level of the molecular interactions of the antimicrobial moiety with the target in bacterial cells are briefly summarised as follows:

- Cell wall synthesis: inhibition or attack by enzymes on the integrity of the cell wall, mostly with bactericidal effects.
- Plasmatic membrane located inhibition with influence on the permeability and/or integrity of the membrane, leading to an osmotic imbalance and the loss of the intracellular molecules, mostly with bactericidal effects.
- Protein synthesis inhibition by binding and interacting with different locations of 30 S or 50 S ribosomal subunits, either with bacterio-

static effects (macrolides, lincosamides, tetracyclines, amphenicols, etc.) or bactericidal effects (aminoglycosides or aminocyclitols).
- Nucleic acid synthesis and replication inhibition; mostly with bactericidal effects.
- Metabolic pathway inhibition (e.g. folic acid); the bacteriostatic effects in individual substance and the bactericidal effect of combination (e.g. sulphonamide with trimethoprim).

The mechanism of action and the characteristics of the antimicrobial influence the spectrum of activity and the resistance mechanism, including constitutive/innate or acquired type of resistance. The above noted information should be considered together with the immune and clinical status of the animal, which are key for selection of antimicrobial/s of choice for use in clinical practice.

Table 5 lists key PD parameters that are explained in more detail here. As listed in the guidelines on the use of pharmacokinetics and pharmacodynamics for the development of antimicrobial medicinal products (EMA 2016b), for most antimicrobial agents, the specific nonclini-

Table 5 Parameters used in pharmacodynamics, pharmacokinetics and PK–PD integration (modified from Ahmad et al. 2016; Mouton et al. 2005; Drlica 2003; Shimizu et al. 2013)

Abbreviation	Definition	Unit
PD	*Pharmacodynamic parameters*	
MIC	Minimum inhibitory concentration = the lowest concentration of antimicrobial that completely inhibits the growth of the tested microorganism	mg/L µg/mL
MBC	Minimum bactericidal concentration = the lowest concentration of antimicrobial that kills the tested microorganism	mg/L µg/mL
MPC	Mutant-prevention concentration = the lowest concentration that prevents the emergence of mutants after 120 h of incubation	mg/L µg/mL
MSW	Mutant-selection window = drug concentration zone in which resistant mutants are selectively amplified, reducing drug susceptibility	Interval: mg/L µg/mL
PAE	Postantibiotic effect is the time during which bacterial growth is suppressed after the bacteria have been exposed to antibiotic for a short time	h (time in hours)
PK	*Pharmacokinetic parameters used in predicting antimicrobial dosing schedules (PK/PD modelling)*	
C_{max}	The highest concentration of drug reached or estimated in the compartment of the reference (usually in plasma, but can be in any target tissue)	mg/L µg/mL
T_{max}	The time when peak/maximum concentration is achieved in the compartment of the reference	h (time in hours)
AUC	The area under the concentration–time curve over 24 hours in steady-state unless otherwise stated; is equivalent to a single dose $AUC_{0\text{-}\infty}$	µg.h/mL
$T_{1/2}$	Half-life of the drug to be considered especially in antimicrobials with time dependence bacterial killing effect (e.g. florfenicol, beta-lactams)	h
	PK–PD integration used for predicting clinical efficacy	
C_{max}/MIC[a]	Maximum concentration of the antimicrobial (ideally at the target tissue) divided by MIC of the causative pathogen; more recently, fC_{max} should also be calculated (free, unbound max concentration of microbiologically active substance)	Ratio, numerical value only
T > MIC[a]	The cumulative percentage over a 24 h period in which the antimicrobial concentration (ideally at the target tissue) exceeds the MIC of the causative pathogen in a steady-state pharmacokinetic condition; more recently, fT should also be calculated (time during which the free, unbound concentration of microbiologically active substance exceeds the MIC)	%
AUC/MIC[a]	The area under the concentration -time curve divided by MIC of causative pathogen; more recently, $fAUC$ should also be calculated (the area under the *free* concentration–time curve)	Ratio, numerical value only

[a] MIC is being replaced by MPC, where known combinations of antimicrobial/bacterial species not only ensure efficacy but also minimise the risk of resistance development

cal pharmacodynamic targets (PDTs) for each pathogen or group of pathogens that result in the following effects on organism counts over a specific time period should be able to be identified:

- A net static effect, i.e. no log10 drop in colony forming units (CFUs)
- Killing/inhibiting effect 1-log10 drop in CFUs
- Killing/inhibiting effect 2-log10 drop in CFUs

Generally, as a meaningful quantitative variable that indicates the observed in vitro concentration, in which complete bacterial growth inhibition is detected, is the minimum inhibitory concentration (MIC). One MIC represents inhibition for one isolate tested (in fact, only a subpopulation of cells isolated from a sample). But within the bacterial population of, for example, *Staphylococcus aureus*, the causative pathogen isolated from udder tissue in clinical mastitis in dairy cows at one farm/district/country allows for many isolates/strains with different MICs to be obtained. Plotting the results of "many" isolates onto a histogram, we can see the distribution of MICs throughout the population of *Staphylococcus aureus* mastitis isolates to certain antimicrobials (normal, lognormal, bimodal or multimodal). With the set of MICs from a certain number of isolates from several farms that are interconnected by the movement of the animals, MIC_{50} and MIC_{90} can be calculated. Here, MIC_{50} represents the concentration, when $\geq$ 50% from the set of the tested isolates of *S. aureus* (from samples from farms) is inhibited by this concentration, and similarly for MIC_{90}, it is the concentration inhibiting $\geq$ 90% of isolates tested. According to the EMA/CVMP guideline (EMA 2016b) on the data required to support dosage claims for marketing authorisation applications, MIC_{90} should relate solely to susceptible strains. For the purpose of identifying potentially efficacious dose regimens to treat pathogens with MICs of the test agent at the upper end of the wild-type distribution (e.g. including the MIC_{90} and/or the epidemiological cut-off values (ECV), the proposed dose regimen (dose–dose interval, where validly establishing duration is also possible) is generally expected to provide a probability of target attainment (PTA) > 90% based on the selected PDT (a magnitude for a PK-PD index at which a desired level of predicted response is achieved). The value of MIC_{90} can be used to calculate the dosing schedule in such cases. In some cases, having not only the MICs but also the minimum bactericidal concentration (MBC), which is considered a concentration that kills bacteria, leads to a 99.9% decrease in the total bacterial count of the pathogen. In the population of bacteria at the stage of growing, in most bacterial species mutations, resistance can occur spontaneously at rates of 10^6 to 10^{10} per gene per generation (Lees et al. 2008). For resistance to emerge, usually several mutations are needed. The smaller the bacterial population (e.g. thanks to antibacterial or organism immune-defence effects), the smaller the likelihood of mutation and the smaller the number of the mutants in the population will be. The variable linked to this effect, which has been described in antimicrobials with concentration-dependent effects (e.g. fluoroquinolones), is the mutant-prevention concentration (MPC). MPC-based assumptions should not be the only considered, but broader view of containment strategies that maintain a subpopulation of sensitive cells, which can slow the expansion of resistant bacteria should be investigated (Hansen et al. 2020).

Originally, the MPC was considered a concentration that does not allow any mutant to be recovered from a population of more than 10^{10} microorganisms (Blondeau et al. 2001), but later on Blondeau and Shebelski (2016) have amended the definition: The MPC is the antimicrobial concentration that would require a microorganism to simultaneously possess two resistance mutations for growth in the presence of a given drug. MPC prevents selection of the first step resistant mutants. The MIC of the most resistant cell in a population is tested and applies to microorganisms deemed susceptible (according to internationally recognised guidelines—e.g. CLSI-vet). For MPC to be a therapeutically useful parameter, its value must be below the plasma and target tissue concentrations of antimicrobials reached following the administration of therapeutic doses (Blondeau et al. 2001). In 2019, Blondeau and Fitch published a study determining and reporting MPC values for *A. pleuropneumoniae* and *P.*

multocida (swine clinical isolates) for beta-lactam (ceftiofur, or CEF), fluoroquinolone (enrofloxacin, or ENR), phenicol (florfenicol, or FLF), macrolide (tilmicosin, or TLM) and triamilide (tulathromycin, or TUL). The MPC values were lowest for ceftiofur and enrofloxacin. The authors of the study reported the MIC_{90}/MPC_{90} values (mg/L) for 67 isolates of *A. pleuropneumoniae*—CEF: 0.031/0.5, ENR: ≤0.016/0.5, FLF: 0.5/2, TLM: 4/32, TUL: 2/32; for 73 isolates of *P. multocida*—CEF: 0.004/0.25, ENR: 0.016/0.125, FLF: 0.5/0.5, TLM: 8/16, TUL: 0.5/1; and for 33 isolates of *S. suis* with MIC_{90} values (mg/L)—CEF:1, ENR:0.25, FLF:4, TLM and TUL: ≥8. In total, 16 isolates of *S. suis* with MIC values of CEF 0.063–0.5 mg/L and ENR 0.25–0.5 mg/L were tested for MPC, which yielded 0.5 mg/L of CEF and 1 mg/L of ENR. MPC concentrations can be assumed to be dosing targets which may serve to reduce the amplification of bacterial subpopulations with reduced antimicrobial susceptibility. Utilising the aforementioned in vitro study's dataset, expected drugs' potencies based on MIC_{90} values were CE F > ENR > FLO = TUL > TLM, based on MPC_{90} values were ENR > CEF > TUL > FLO ≥ TLM (Blondeau and Fitch 2019).

The mutation selection window can be identified within the interval between the minimum inhibitory concentration and the mutation prevention concentration. The larger the window between MIC and MPC is, the greater the probability of selecting resistant mutants is. Given the pragmatic uses and the PK of the antimicrobials, the moments when concentrations adequate to the MSW will appear during the treatments should be minimised. Long-term administrations of subinhibitory concentrations within the MSW should therefore be avoided.

In fact, preventing resistant mutants requires measuring the MPC for each causative bacterial pathogen in an individual animal, which in most cases is unrealistic. However, some activities already exist in human medicine (see e.g. project AutoKinetics, taking a computerised approach to drug dosing, as described by Roggeveen et al. (2020)), that use data uploaded to a database, which are further improved and can be used to personalise dosing schemes for critically ill patients (Roggeveen et al. 2022). In the future, therefore, MPCs should be divided into three basic PK/PD indices, both in human medicine and in veterinary medicine, not only to optimise the efficacy of antimicrobials but also to minimise the risks of resistance selection. But first the MPC distribution values for each antimicrobial/bacterial species combination must be defined. Such definitions would allow for obtaining three new PK/PD indices, by replacing the MIC with the MPC. These new indices would be *f*AUC/MPC; *f*T > MPC; and fC_{max}/MPC.

The mutation selection window concept originated from considerations about the mechanisms of resistance that are generated by spontaneous mutations of the bacterial genome and that are transmitted vertically to successive generations. Cantón and Morosini (2011) applied the MSW concept to mechanisms of resistance involving the horizontal transfer of resistant genes. The MSW concept can be used for heterogenous bacterial populations that coexist with a susceptible majority subpopulation and one or more minority subpopulations with lower levels of susceptibility. The traditional concept of "subinhibitory" concentrations favouring the emergence of resistance should be therefore conditioned by the MIC of the least susceptible pathogenic subpopulation, which in fact corresponds to the MPC (ANSES 2017).

The MIC-based approach is limited in several ways, given its role in PK/PD. Although MIC testing has been highly internationally standardised (using mostly CLSI and EUCAST assays), it still uses growth media that are balanced towards both good conditions for growth of microorganisms and good conditions to keep antimicrobials stabile and effective. The main factors affecting the physiological conditions at the site of infection are pH, osmolarity, redox potential, cations and protein concentrations. Therefore, the values of MICs may differ if they are tested in conditions simulating in vivo conditions (e.g. by using broth and components mimicking, for example, the binding to plasma proteins, tested with human albumin (Beer et al. 2009)), or if they are influenced by the presence of erythrocytes (Nussbaumer-Pröll et al. 2019). The host factors (phagocytic cells, antibodies and complements) and other blood components can influence the MIC determined for some antimicrobials (with high protein-bound fraction, such as some cephalosporins or ciprofloxacin, whose interactions with human erythrocytes have been proven). More recent studies have explored the influence of media types: more host-mimicking media, such as cell culture media (RPMI), and media-buffered ions, such as bicarbonates (which are physiologically closer to in vivo status) (Ersoy et al. 2017, 2019). Some recent studies on respiratory pathogens have also run MIC tests using the tissue-cage fluid as a matrix (Zeng et al. 2017). Except for the analysis of antimicrobial activity of the drug, host-defence peptide interactions can be studied as well. Interestingly, some of the recent studies have shown that in the selected antimicrobials, once they were diluted in conventionally used bacterial growth media (Mueller-Hinton), resistance (azithromycin against *Acinetobacter baumannii* and *Pseudomonas aeruginosa*; nafcillin and cefazolin against selected MRSA isolates) was detected, while RPMI and other host-mimicking media illustrated their susceptibility (Ersoy et al. 2017; Lin et al. 2015). Five host-mimicking media are currently used in different experiments in human medicine: (1) Dulbecco's modified eagle medium (DMEM), a tissue culture medium supporting mammalian cell growth; (2) Lacks medium, supporting pneumococcus growth; (3) modified Lacks medium (MLM), simulating the nasopharynx for invasive pneumococcal carriage; (4) a low-phosphate, low-magnesium medium (LPM, pH 5.5), simulating the macrophage phagosome in which many intracellular pathogens reside/replicate; and (5) a tissue culture medium from the Roswell Park Memorial Institute 1640 (RPMI) (ThermoFisher Scientific), that allows growth of non-fermenting rods as, for example, *Stenotrophomonas maltophilia* (Kumaraswamy et al. 2016).

Conventional in vitro susceptibility test methods also cannot describe the impact of an antimicrobial on pathogen virulence factors (Barczak and Hung 2009), responsible for invasion into or attachment to tissues, quorum sensing (the bacterial production of autoinducers that regulate gene expression in biofilm formations), the production of toxins or the factors that affect host immune functions (Kumaraswamy et al. 2016; Martinez et al., 2013; Giguère et al. 2013).

Also, MIC testing requires a twofold dilution approach, which provides only an approximate inhibitory value and in fact can in some cases overestimate the true MIC for individual isolates. The conditions of in vitro MIC testing represent individual concentrations kept for fixed periods of time, which requires exposure to a constant concentration for a fixed duration (usually 18–24 h) and is therefore a more "static" than "dynamic" concentration curve, which is in reality achieved in vivo (in different tissues of the animal after drug administration).

Owing to the aforementioned weaknesses, discussions have started, and some studies attempting to overcome them have appeared. One option that seems promising for the future is among the specific "in ex vivo" PK/PD models that are suitable for the study of interactions between a drug and bacteria as they combine in vitro and in vivo PK/PD model characteristics (Luo et al. 2019). An in ex vivo PK/PD combination model uses data on the PK in target animals and data on the PD in plasma, tissues and fluids (taken from samples at different times) (Aliabadi et al. 2003; Sidhu et al. 2014). These models reflect the real PK processes in vivo; at the same time, they can reduce the number of animals needed for experiments and thus reduce the cost of experiments (Sang et al. 2016) and reflects also recently promoted 3Rs principles (Replacement, Reduction and Refinement, where replacement alternatives refer to methods which avoid or replace the use of animals or at least number of animals can be reduced). In the study of enrofloxacin in swine colibacillosis, Wang et al. (2016) used a combination of PK data with in ex vivo PD data in pig ileum; similarly, Sang et al. (2016) performed modelling for broiler colibacillosis and pointed out the need to obtain samples from the appropriate target tissue (infected animals). Plasma, bronchial lavage and intestinal fluid were identified as more appropriate than other samples for studying PK and PD in the infected systemic models of respiratory or gastrointestinal tract infections, respectively. But challenges remain in trying to gather sufficient data of appropriate quality to use them in the above modelling system (Luo et al. 2019).

One more thing that cannot be overlooked is the inoculum size effect. Several studies have examined the influence of inoculum size and the effectiveness of killing bacteria (bactericidal effect 3-log kill) on the MIC value (Martinez et al. in Giguère et al. 2013). Antimicrobial treatment regimens against bacterial pathogens mostly use in vitro MICs for combinations of pathogens and antimicrobials, measured at a bacterial density of 5.7 log 10 CFU/mL. To reflect the influence of inoculum size, making realistic mathematical models of the relationships is important. Salas et al. (2020) compared two Gram-negative enterobacteria—*Escherichia coli* and nontyphoidal *Salmonella enterica* subsp. *enterica* and two Gram-positive cocci—*Staphylococcus aureus* and *Streptococcus pneumoniae* (where $n = 4$ drug-susceptible strains per (sub) species and with densities of 1–8 log10 CFU/mL). The modelling was performed for bactericidal antimicrobials: β-lactams (ceftriaxone and oxacillin), fluoroquinolones (ciprofloxacin), aminoglycosides (gentamicin), glycopeptides (vancomycin) and oxazolidinones (linezolid). Six candidate mathematical models were used, none fully describing the relationships among the pathogen–antimicrobial combinations. Gompertz and logistic models (rather than a previously proposed Michaelis-Menten model) fitted best most often. The inoculum effect of a selected bacterial pathogen has been detected to vary depending on which antimicrobial was tested (i.e. variation among classes was likely related to the mechanism of action). Inoculum density caused a sharp increase in MICs, which can complicate interpretations. Capturing the dependencies for disease–pathogen–antimicrobial combinations could help determine the MICs, for which bacterial densities are most informative for designing treatment regimens (Salas et al. 2020).

Once within the curve, the concentration drops below the MIC, but some effect (especially in some bactericidal antimicrobials), called the postantibiotic effect (PAE) can still remain. This effect is defined as a period of the inhibition of bacterial growth occurring after exposure to concentrations exceeding the MICs, when the concentration of the drug decreases mostly because of metabolisation (to inactive metabolites) or excretion. Although this variable is frequently mentioned, the importance of its role in clinical/bacteriological treatment outcomes within in vivo conditions has not been fully elucidated. The PAE was shown in most antimicrobials used against susceptible Gram-positive bacteria. A review by Onufrak et al. (2016) summarises the sources of the PAE in different antimicrobials against Gram-positive bacteria and Gram-negative bacteria. The values for PAE/G+ range from < 2 h for beta-lactams to nearly 5 h for glycopeptides against representatives of Gram-positive *Staphylococcus aureus* (but individual point estimates vary considerably). A prolonged PAE has been described in susceptible bacteria in antimicrobials targeting protein or nucleic acid synthesis, such as aminoglycosides and fluoroquinolones. The PAE values derived from animal models were within the range of 1.2–12.8 h for aminoglycosides and within the range 1.9–7.5 h for fluoroquinolones. Administration for longer intervals between doses might therefore maintain treatment efficacy. On the other hand, most antimicrobials influencing cell wall synthesis (beta-lactams) have been described as having a short PAE (< 1 h), which is one of the factors contributing to the need of a higher frequency of administration per day. The exception among antimicrobials belonging to the beta-lactam group is the subgroup of carbapenems, for which the PAE range is approx. 2–4 h (enterobacteria, e.g. *P. aeruginosa*). Recent studies on pigs and the PAE of amoxicillin in respiratory pathogenic *Actinobacillus pleuropneumoniae* has shown a significant PAE, which might explain both in vitro results (Nedbalcova et al. 2019) and the results from experimentally infected pigs, where the effect of the subinhibitory concentration protects against mortality in infected pigs (Tanigawa and Sawada 2002). As for pathogens in pigs, the prolonged postantibiotic effect of fluoroquinolones (marbofloxacin and enrofloxacin) and of amoxicillin has been confirmed for important Gram-negative swine pathogens (*Actinobacillus pleuropneumoniae*, *Glaesserella* (previously *Haemophilus*) *parasuis* and *Pasteurella multocida*), with a statistically significant difference in the duration of the PAE of amoxicillin compared to both fluoroquinolones (Nedbalcova et al. 2019).

Another described phenomenon is postantibiotic leucocyte enhancement (PALE), and studies have indicated that this effect functions through the increased susceptibility of antibiotic-damaged bacteria to leukocytes' intracellular killing mechanisms. Studies were performed on Gram-negative bacteria (*E. coli*) by McDonald et al. (1981), where after a shorter exposure (10–20 min) to high doses of antibiotics (i.e. quinolones and chloramphenicol), the tested strains became more susceptible to the killing activity of phagocytes. PALE has also been investigated for use in pulse therapy regimens. Horgen et al. (1998) described the PALE effect in *Mycobacterium avium*, as well. Interestingly, new, recent studies on PALE are rarely found, for unknown reasons. One of the exceptions is recent study of Wu et al. (2024) on PALE-mediated reduction of intracellular bacteria by macrophages.

With the emergence of resistance, especially in some Gram-negative pathogens in human medicine (e.g. *Pseudomonas aeruginosa*), the need to bring (even temporarily) back the susceptibility of bacteria has become important. Certain antimicrobial peptides and membrane-disturbing agents are known to be able to sensitise bacteria to antibiotics. A so-called postantibiotic-effect-associated permeabilisation (PAEP) has been described for *P. aeruginosa* and several antimicrobials from different classes, in experimental conditions (Rázquin-Olazarán et al. 2020). PAEP is defined as the sensitisation of a culture to sub-inhibitory concentrations of an antimicrobial agent caused by a previous treatment with a PAE-inducing agent that is no longer present (Rázquin-Olazarán et al. 2020). The authors of this study determined in in vitro conditions that a multidrug-resistant strain of *P. aeruginosa* can be rendered susceptible for several hours to subinhibitory concentrations of selected antimicrobials upon treatment with an antimicrobial peptide (P4–9) that causes lesions in the bacterial cell wall.

Assessing the optimisation of the antimicrobial therapy can be further complicated by the so-called Pollyanna phenomenon (Lees et al. 2008; Jones 2014), when the efficacy of an antimicrobial is measured via symptomatic clinical responses. In principle, drugs/dosing schedules that are expected to have high antimicrobial activity appear not as clinically efficacious as anticipated, in contrast with antimicrobials with poor antibacterial activity (minimum bacteriological cure rate); sometimes this can also be the case in antimicrobials with higher potentials to select resistance. It can be summarised also by using other wording: Weak bacteriological cure may be associated with good clinical responses, while complete bacteriological cure may not yield 100% clinical responses/successes. Studies in human medicine described it first. Well described is also a paediatric study on otitis treatment in children by using cephalosporin that came to the position of a narrower spectrum than amoxicillin or even amoxicillin/clavulanic acid (Cocco et al. 2010). But studies are also available on the veterinary side. A large-scale study performed by Yancey et al. (1990) on several hundred neonatal piglets and ceftiofur (hydrochloride) has also shown this phenomenon. The dose effect occurred according to identified bacteriological criteria (shedding of bacteria), but clinical end points (mortality or abnormal stool) seem to have failed. Assessing the efficacy study data for quinolone/chickens/*Escherichia coli* infection from Charleston et al. in 1998 reveals that very different ED_{50} values for mortality (8 mg/kg) and bacteriological treatment (13 mg/kg) were obtained (Toutain et al. 2002). Once similar cases started to be collected, the phenomenon seemed to appear in antimicrobals both with bacteriostatic or bactericidal effects, where both groups also interacted with/stimulated the inflammatory or immune-defence mechanisms.

As for the importance of interactions with immune-defence mechanisms, although new studies have been conducted, some gaps and controversies remain in the literature. Zimmermann et al. (2018) published a systematic review in which an analysis of metadata was performed on the antimicrobial immunomodulatory properties of selected macrolides (primarily for human medicine). The review included four macrolides: one 15-memebered ring (azithromycin) and three with 14-membered rings (clarithromycin, erythromycin and roxithromycin). The authors concluded that the investigated macrolides exhibited immunomodulatory effects through the inhibition of neutrophilic inflammation and macrophage activation, where the most frequent actions were as follows: a decrease in the number of neutrophils and the concentrations of neutrophil elastase, interleukin (IL)-8, IL-6, IL-1beta, tumour necrosis factor (TNF)-alpha, eosinophilic cationic protein and matrix metalloproteinase 9. According to the analysed dataset, the macrolides' ability to accumulate in neutrophils and macrophages was identified as important. Van Bambeke in Vinks et al. (2014) summarised the data on seven macrolides' cellular accumulation in macrophages, polymorphonuclear neutrophils and epithelial cells/fibroblasts, whose conclusion was that, in general, accumulation at the highest level was described in ketolides (highly lipophilic) and azithromycin (a hypothesis of dicationic nature). The authors (Zimmermann et al. 2018) hypothesised that such accumulation in immune cells may result in immunomodulatory effects at lower doses and be longer lasting than antimicrobial effects. Other studies, such as those featuring intracellularly located *S. aureus*, have indicated a minor reduction in the cell counts of this pathogen. It therefore seems that parameters other than accumulation and distribution therefore need to be taken into account in the intracellular activity of macrolides (but, generally speaking, antibiotics). One of the recent complex review by Blondeau (2022) brings together published evidence on immunomodulatory effects of macrolides considering data coming from human and veterinary medicine. In this review up to seventeen pathways where macrolides have immunomodulatory effects were noted, most of them related defences against respiratory infectons. The expression of activity in the intracellular environment, the bacterial responsiveness and the cooperation with cell defence mechanisms seem to be among the key properties influencing the final effect (Carryn et al. 2003; Van Bambeke et al. 2006). The poor efficacy of azithromycin was confirmed within in vivo PD model studies of *S. aureus* peritonitis in mice, where only 0.2 log drop in intracellular counts was detected for the target pathogen (Sandberg et al. 2009). This fact contradicts the hypothesis that intracellular breakpoints could be higher because of the high accumulation of these drugs (Amsden 2001), the wide tissue distribution of macrolides and related compounds (azalides, ketolides) and support from the formulation of the newest medicinal products do not necessarily mean high efficacy (especially for intracellular pathogens). Two limiting factors are the bacteriostaticity that most of them have and the deleterious effect that acidic pH has on their activity (Van Bambeke in Vinks et al. 2014).

Therefore, clinical data outcomes and their correlations with laboratory results should be collected and analysed to establish "predictor values" for certain combinatons of all these factors: MIC/drug/microbe/disease/animal species/ animal status (selecting from the appropriate variables is critical for certain diseases). Predictive MICs can be used for PK/PD models when setting/supporting an appropriate dose within a specific dosing schedule. Computerisation with metadata analysis working with information from existing experimental studies as well as artificial-intelligence technologies and software applications for analysing of AMR can be used for search for and assess, for example, combinations of antimicrobials but also the proper dosing schedules to effectively treat infections and to overcome current issues with AMR (Lv et al. 2021; Chai et al. 2020). These techniques could be applied not only in human medicine, but also in veterinary medicine, at least in treatment of hospitalized animals like horses or dogs, where highly individualised veterinary medicine has been already performed.

1.6 Completing PK/PD Indices to Model Dosing Schedules

The PK and PD principles and modelling systems can be used when determining doses. Some crude estimates of the dose required for an antimicrobial can be obtained (especially for antimicrobials such as fluoroquinolones, aminoglycosides, whose bactericidal effect is dependent on concentration) by comparing the MICs detected in the target pathogen and concentrations achieved at the site of infection. Some authors have mentioned that concentrations that are three to five times the MIC are considered effective (Constable 2017), but for a more precise analysis, three ratios for the assessment of effective antibiotic concentrations are commonly used, according to Ahmad et al. (2016): AUC/MIC; C_{max}/MIC; and T > MIC. The pharmacodynamic (PD) part of the PK:PD ratio includes the MIC, which refers to the number that indicates the concentration that inhibits bacterial growth under in vitro well-defined testing conditions (the exact fixed time, the fixed concentration of the antimicrobial, the standardised "artificial" medium and the growth conditions).

By using Table 5 and the three main PK/PD indices of presumptive efficacy, ratios can be characterised: C_{max}/MIC determines the intensity (bactericidal/bacteriostatic concentration pressure), T > MIC determines the duration (time necessary for killing/inhibition) and AUC/MIC determines the combination of intensity and duration. The same PK parameter dataset is usually used for calculations, with the exception of the models that account for PK in the target tissue (e.g. the respiratory tract, gastrointestinal tract, genitourinary tract or intramammary gland), which compares this mostly to the PK in plasma. The modelling should be based on precise combination of antimicrobial and species of bacteria. The best PK/PD index for the aforementioned combinations is determined by plotting the value of a specific end point (typically $\log_{10}$ CFU/mL after 24 h of treatment) against the magnitude of each of the three PK/PD indices (EMA/CHMP/594085/2015).

EMA (2016a) recommends determining PK/PD indices and PD targets (citaton):

- "When a *concentration-dependent* pattern of bactericidal activity is observed in time-kill studies:
  - The area under the plasma concentration time curve representing the period from zero to 24 h $AUC_{0\text{–}24}{:}MIC$ ratio and/or,
  - The maximum plasma concentration, $C_{max}{:}MIC$ ratio, is/are usually found to be predictive of effcacy in PK-PD model systems (e.g. enrofloxacin).
- When a *time-dependent* pattern of bactericidal activity is observed in time-kill studies:
  - The percentage of the dosing interval during which the plasma concentration exceeds the MIC $\%Time > MIC$, through MIC and/or,
  - The $AUC_{0\text{–}24}{:}MIC$ ratio is/are usually found to be predictive of efficacy in PK-PD model systems" (e.g. phenicols (Pelligand et al. 2019) and beta-lactams (Nielsen et al. 2011; Kristoffersson et al. 2016).

Moreover, for long-acting formulations is of importance a recent knowledge on the half-life of the antibiotic. Mathematical physiological models show that the AUC_{24h}/MIC index is at least as effective as the T > MIC index at predicting antibacterial activity for the antibiotic with the long half-life (e.g. 1.5-3.5 hours) (Toutain et al. 2017a).

Toutain et al. (2002) described the concentration–effect relationship as the sigmoidal E_{max} model (also known as the Hill equation):

$$E_{(t)} = E_0 + \frac{E_{max} \times C^h(t)}{EC^h{}_{50} + C^h(t)}$$

The shallower the curve, the greater the relationship between the rates of bacterial killing and the antimicrobial drug concentration—which is typical of antimicrobials with "concentration-dependent killing" relationship because the degree of bactericidal activity increases as concentration increase, up to the point where the maximum killing effects (E_{max}) are achieved. Such behaviour is typical of fluoroquinolones (Martinez et al. in Giguère et al. 2013).

The steeper the curve, the less difference among antimicrobial concentrations needed to achieve the maximum killing effect—so such antimicrobials are considered to act via "time-dependent killing" because the only way to further increase the bactericidal effect is to increase the duration (the dosages remain above the pathogens' MICs, but longer total durations or more-frequent doses are needed over a 24 h period). Such behaviour is typical of beta-lactams (Martinez et al. in Giguère et al. 2013).

Overview of the pharmacological classes of antimicrobials used in veterinary medicine classified according to predominant type of bacterial killing action and listing the PK/PD indices used in correlation to efficacy is presented in Table 6.

The following data supporting PK-PD analyses are particularly important for any new antimicrobial and its intended clinical uses (EMA 2016; EMA 2018a, amended):

(A) *Minimum inhibitory concentration* (MIC) and related data: (MICs: ideally determined by using internationally agreed and standardised methods; Note: recently some discussions indicate that MICs determined in media mimicking as close as possible host and target compartment (plasma, gut, respiratory tract) can contribute to catch more precisely conditions closer to those in the tissues).

- MIC distributions composed of clinical isolates data obtained from infections in target animal species within the indications relevant for the antimicrobial of concern (e.g. set of MICs from *S. aureus* clinical isolates from the acute mastitis of dairy cows, under condition that sufficient number of isolates from epidemiologically unrelated cases/units was collected).
- For more-precise dosing, when designing models and when running PK-PD analyses, more information should also be taken into account: Even one bacterial species (e.g. *E. coli*, where strains differ in their susceptibility patters (MICs ranges), virulence factors and adaptations to specific host tissues) can be linked to different types of infections (neonatal sepsis, diarrhoea, metritis and mastitis) in one animal species (cattle) but in various categories (neonates, nonruminating calves and heifers/dairy cows). Despite above noted details, normally, one dose of injectable antimicrobials (sometimes using different routes of administration or a combination of routes first injection IV and then continuing with SC or IM routes) is authorised for many antimicrobials containing VMPs.
- The generated MIC data should be presented by species, genus or microorganism groups (e.g. enterobacteria or the beta-haemolytic streptococci of groups A, B, C and G) and also separated into subsets with and without acquired resistance. Especially in new molecules; in substances that have some specificity in their respective mechanisms of action; and in cases in which the substances or related substances are not used in any other sectors—such as "new" veterinary antimicrobials from a group of antimicrobials already used in human medicine—acquired resistance may not be detected.
- The MICs data of antimicrobials of concern should be evaluated for the presence of a range of known resistance mechanisms.
- The MICs used for PTA[2] analyses should always include the range of MICs of each causative pathogen/group of pathogens: values at the upper end of the distribution (the distribution should represent the whole range of MICs). MIC_{90} should be used preferably and/or the epidemiological cutoff values (ECOFFs[3]) for each species of interest.

[2]PTA = probability of target attainment: For reporting of outputs from simulations, including Monte Carlo simulations, the PTA is the probability that at least a specific value of a PDT is achieved at a certain MIC, where PDT is a magnitude for a PK-PD index at which a desired level of predicted response is achieved (EMA 2016).

[3]ECOFFs = Epidemiological Cut OFF values: if used an adequate collection of isolates (ideally consisting of

- The indentification and comparison of (sub)types (genotypes, serotypes and clonal complexes) of causative pathogens that have higher or lower rates of resistance, which may mechanism-specific, should be performed.
- For indications where several target pathogens are involved, defining the least susceptible pathogen, i.e. "the dose limiting" bacterial species, is important if it is possible.
- For the future, MPC values' importance will grow due to the resistance issues increases.

(B) *Time-kill studies* including studies with different inocula sizes and studies performed in the defined time period for which change in organism numbers is documented. The selected time period should be justified.

(C) *Specific data* (examples, amended to those listed by the EMA).
- Once the target causative pathogen is mainly within the host cells, the intracellular antimicrobial activity should be evaluated.
- Specificities linked to the target tissue of infection should be considered.
- Interactions with host-defence mechanisms should be taken into consderation in the MICs and dose assessment.
- PAEs effects should be considered.
- The potential for the antimicrobial to accumulate in certain tissues after repeated doses needs to be determined.
- Any potential effects on microbiome or disturbances in its balance should be accounted for.
- Consideration related to healthy animals vs diseased animals (rate of affecting of the PK) must be included, as in organisms suffered from the disease PK can differ signifcantly, e.g. inflammatory process affection of penetration across haematoencephalic barrier.
- Effect of bactericidity on endotoxins'release should be considered.

Although the PK/PD index is calculated by using real in vivo experimental data for C_{max}, T_{max} and the calculated AUC after a certain dose, the route of administration and for certain group of the target animals, still these make up a limited set of variables, so that is highly recommended to use clinical data that will confirm how the calculations for determining optimal doses and dosing regimens reflect clinical conditions. Moreover the same PK/PD index can be calculated for different dosing schedules.

PK/PD assessments are based on the MIC for the target pathogen and the *unbound* antibiotic concentration in the host plasma, because only the free fraction exhibits antibacterial activity. Therefore *f* (free = unbound) is added to the parameters when they are based on unbound product concentrations. For certain tissues with a strong barrier (e.g. the haematoencephalic barrier), free plasma concentration is not applicable and different pharmacokinetic approaches may be necessary to model infections in such tissues. Also, doses for intracellularly located bacteria or bacteria-forming biofilms can barely be predicted by using PK/PD, especially once the PK parameters are derived from plasma concentrations. The notation of the three PK/PD indices has been standardised (Mouton et al. 2005) into *f*AUC/MIC, *f*Cmax/MIC and *f*T > MIC. If no subscripts indicate a time interval, the calculations for AUC and *T* > MIC are assumed to be based on a 24-h interval under pharmacokinetic steady-state conditions.

The results of the AUC/MIC ratio for fluoroquinolones have been documented for representatives of Gram-positive pathogens and Gram-negative pathogens. The AUC/MIC ratio should be within or exceed the range of 25–30 for immunocompetent animals to better treat infection caused by Gram-positive pathogen. In the case of immunocompromised animals, a ratio of AUC/MIC exceeding 100–125 has been considered as sufficient to treat infections caused by Gram-negative pathogens (Lees 2010; Ahmad

"wild-type" organisms that are known rather than assumed to lack mechanisms of resistance to the antimicrobials tested). Simple visualisation of histograms or a mathematical approach, how the ECOFF was calculated should be provided.

et al. 2016). For concentration-dependent antimicrobials, reaching or exceeding the MICs (ideally those at the sites of infection) is vital, and usually, the C_{max} in plasma is known for antimicrobials at certain doses/routes of administration in different target species/categories. For some antimicrobials (e.g. injectable aminoglycosides like gentamicin), one dose per day ensures to achieve concentration high enough to well exceed the MIC and is considered as appropriate because of the ototoxicity and nephrotoxicity potential of aminoglycosides (Turnidge 2003). Data from one study (Wen et al. 2016) has demonstrated that susceptibility cannot simply be extrapolated from the antimicrobial's MIC only. The rate of the isolate's response (an inhibitory effect) correlates with steepness (which depends on a combination of PD parameters). Complete PD profiles for differently susceptible isolates of the target pathogen should be considered along with the antimicrobial drug's pharmacokinetic and toxicological parameters in the host to design effective treatment regimens. Isolates with different susceptibility values seem to have different responses and need different dosing intervals as well as different doses (Wen et al. 2016). Intraspecies differences are also of interest, including the maximal population growth rate during the exponential phase in the absence of exposure and the percentage of pathogenic persisters in the population of the bacteria causing the infection (Wen et al. 2016).

Differences among antimicrobials with primarily bacteriostatic action have been proven in cases with concentrations in tissues < the MIC, where the bacterial count decrease mainly depends on host-defence mechanisms (e.g. macrolide, like tylosin). For bactericidal drugs, an obvious effect of the sterilisation of the tissues from the pathogen has been described for concentrations > the MBC (even after only single administration), and they have been shown to have a bacteriostatic effect when the concentration was equal to or higher than the MIC but lower than the MBC (Huang et al. 2018).

Although the preceding predictable factors introduce uncertainty into simple mathematical modelling, some rules have been generally accepted for use when determining the probability of an effective cure rate (>80–90%). For antimicrobials with rather time-dependent bacterial killing, maintaining this concentration over certain part of the interval between the doses and over the entire treatment period is vital. Generally, maintaining this dose above MICs for at least 80% of the time is recommended for Gram-negative bacteria and at least 40% for Gram-positive bacteria (ANSES 2017). This approach does not include the postantibiotic effect or the individual characteristics of different antimicrobials.

However, all these foregoing values (factors and percentages expressing generally possible thresholds) provide only a general guide to setting clinically effective dosages, for the following reasons:

- Target numerical values are specific to drug and microorganism combinations, which can vary depending on the specific strain (e.g. *E. coli* strains from the gastrointestinal tract, genitourinary tract or the udder are not the same, including their MIC values, which may vary with the isolates from different diseases, target tissues or species).
- The type of killing action depends on the species of microorganism (time or concentration dependency may vary).
- An effective dose depends on the bacterial load and the phase of infection (and therefore the log phase of the development of bacterial population at the site of infection and can be significantly influenced by, for example, biofilm formation).
- Final clinicall effect of the dose will depend on the immunostatus of the host animal.
- The required dose varies depending on its intended aim or priority; e.g. a clinical cure vs a bacteriological cure (eradication) or avoiding the emergence of resistance (Table 6).

Example of beta-lactams (penicillins, cephalosporins), which exhibit a time-dependent bactericidal effect, where maximising *f*T > MIC, will increase the probability of bacterial killing (if the

Table 6 Overview of the pharmacological classes of antimicrobials used in veterinary medicine, which are classified according to the predominant type of bacterial killing action and which lists the PK/PD indices used in correlations with efficacy

Predominant bacterial killing action	Pharmacological class	Examples of antimicrobials	PK/PD indices correlated with efficacy	References
Dependent on concentration, usually accompanied by PAE	Aminoglycosides	Streptomycin, dihydrostreptomycin, gentamicin, neomycin, amikacin	fAUC/MIC, C_{max}/MIC	Lees et al. (2006), Martinez et al. (2013), Toutain et al. (2017a)
		Apramycin, paromomycin, neomycin (oral)	$C_{max\ intestine}$/ MIC	
	Fluoroquinolones	Enrofloxacin, marbofloxacin, danofloxacin, difloxacin	AUC/MIC, C_{max}/MIC	Martinez et al. (2013)
	Nitroimidazoles[a]	Metronidazole	AUC/MIC, C_{max}/MIC	Lees et al. (2006)
	Polymyxins	Colistin (inj)	AUC/MIC, C_{max}/MIC	Lees et al. (2006), Guyonnet et al. (2010)
		Colistin (oral)	$C_{max\ intestine}$/MIC	
	Pleuromutilins	Valnemulin, tiamulin	AUC/MIC	Luo et al. (2019)
	Macrolides	Gamithromycin, tulathromycin[1]	AUC/MIC	Zhou et al. (2017), Villarino et al. (2013)
		Tildipirosin[2]	fAUC$_{0–24\ h}$/MIC	Zeng et al. (2018)
		Tilmicosin[3,4]	C_{max}/MIC, AUC $_{0–96}$/MIC	Huang et al. (2020), Luo et al. (2020)
Dependent on time, usually accompanied by weak or no PAE	Amphenicols	Florfenicol, chloramphenicol[a]	T > MIC, fAUC/MIC	Pelligand et al. (2019)
	Beta-lactams: penicillins	Benzylpenicllin, amoxicillin, cloxacillin	fT > MIC, AUC/MIC	Onufrak et al. (2016), Kuti (2016), Craig (1998)
	Beta-lactams: cephalosporins	Cefalexin, cefapirin, ceftiofur, cefquinome	T > MIC	Craig (2003, p. 7)
	Lincosamides	Clindamycin, Lincomycin	T > MIC, AUC/MIC (lung, plasma)	Lewis et al. (1999), Craig (2003)
	Macrolides	Tylvalosin, erythromycin, tylosin,	T > MIC, AUC/MIC (lung, plasma)	Burch (2012)
	Sulphonamides	Sulphamethoxazole, sulphadiazine, sulphadoxine	T > MIC	ANSES (2017)
	Diaminopyrimidines	Trimethoprim	T > MIC	ANSES (2017)
	Sulpho/Diamino	Sulphonamide/ Trimethoprim	AUC/CBP (hours)	Burch (2012)

(continued)

Table 6 (continued)

Predominant bacterial killing action	Pharmacological class	Examples of antimicrobials	PK/PD indices correlated with efficacy	References
Dependent on time and concentration	Glycopeptides	Vancomycin[b]	AUC/MIC	ANSES (2017)
	Tetracyclines	Oxytetracycline, chlortetracycline, doxycycline	AUC/MIC	Burch (2012)

[a] Not allowed to be used in food-producing animals (no MRL has been established)
[b] The AMEG category AVOID; based on Regulation (EU) 2022/1255 should not be used/authorised in any animals
[1]tulathromycin concentration dependence + PAE determined for *Pasteurella multocida*, *Histophilus somni* and *Glaesserella parasuis*
[2]tildipirosin concentration dependence + PAE for *Pasteurella multocida*
[3]tilmicosin concentration dependence for *Mycoplasma hyopneumoniae* (Huang et al. 2020)
[4]tilmicosin concentration dependence *Lawsonia intracellularis* (Luo et al. 2020)

prerequisite of reaching the MIC in the target tissue is achieved). In general, for beta-lactams to have bactericidal activity, they require 40–80% of the dosage interval when the *f*T is higher than the MIC, depending on the individual subclass and the target bacterial species (Ambrose et al. 2007). Kuti (2016) summarised the *f*T > MIC for subclasses of beta-lactams to be effective against Gram-negative pathogens as follows: 40% for carbapenems, 50–70% for cephalosporins (namely ceftriaxone, ceftazidime, cefepime) and 50% for combinaton of aminopenicillin/beta-lactamase inhibitor. Very recent work in human medicine (De Velde 2020) has shown that the amoxicillin absorption rate appears to be saturable, which results in a nonlinear increase in C_{max} and a later T_{max} for higher doses. Increasing the dose results in a larger %*f*T > MIC, as a consequence of delayed absorption (despite the non-proportional increase in C_{max}). However, a higher dose increases the risk of adverse events. Increased frequency, i.e. shorter intervals, leads to a larger *f*T > MIC as well. The balance between dose and frequency should be optimal to maximise the antimicrobial efficacy (*f*T > MIC) and to minimise the risk of adverse reactions (De Velde 2020). More recently, mathematical physiological models (taking the dynamics of bacterial growth into account) explored in silico the correlations between the PK/PD indices and antibacterial activity of beta-lactams. The work on benzylpenicillin (Nielsen et al. 2011) and more recently on meropenem (Kristoffersson et al. 2016) has shown that when the half-life of the antibiotic is longer, the AUC_{24h}/MIC index is at least as good a predictor of antibacterial activity as the T > MIC index. Also, the T > MIC index seems to yield better predictions of efficacy when the half-lives are approximately half an hour (30 min, common in rodents). The index consisting of AUC_{24h}/MIC ratios gives the best results for the simulations carried out with longer half-lives, such as those observed in humans (1.5–3.5 h) (ANSES 2017). These new insights in PK–PD relationships could be important for VMPs formulated with long-acting effects (Toutain et al. 2017b). A comparison of various formulations containing procaine benzylpenicillin is given in Table 4.

Example of tylosin: A mix of time-dependent and concentration- and PAE-dependent actions (Huang et al. 2018) have been described in macrolide antibiotic tylosin, which showed obvious time-dependent (and weak concentration-dependent) antimicrobial activity. For such a case, the best PK/PD surrogate marker is AUC/MIC. For gastrointestinal tract infections (e.g. *Lawsonia intracellularis*-caused ileitis), the concentrations in ileum content were estimated to be 4 and 10 μg/mL for 40 and 100 ppm tylosin, respectively. The AUC/MIC figures were 96 and 240 h. These figures are quite high but confirm the likely efficacy of tylosin at these levels against susceptible isolates of *L. intracellularis* (Burch and Hammer 2013). Macrolides in general, and tylosin in specific, interact with the immune system and this fact should also be taken into consideraton, when setting the dosing schedule.

Example of tulathromycin: The results of intramuscular administration to piglets at 2.5 mg/kg b.w., C_{max} (0.74 µg/mL) and T_{max} (0.25 h) suggest that tulathromycin is rapidly absorbed and slowly eliminated. Tulathromycin accumulates to a high extent in neutrophils and macrophages in blood (Siegel et al. 2004), and the intracellular concentrations are higher than plasma concentrations (Cox et al. 2010). In a study performed by Zhou et al. (2017), the MIC and MBC values for tulathromycin were determined not only in Mueller-Hinton broth but also in serum and transudate by using the Clinical & Laboratory Standards Institute (CLSI) methodology principles. The average MIC and MBC of tulathromycin against *P. multocida* in serum were 0.04 and 0.10 µg/mL, respectively. To calculate the PK/PD value of MIC_{90} for *P. multocida*, a sufficient number of epidemiologically unrelated clinical isolates of *P. multocida* is needed. The use of AUC/MIC as a PK/PD predictor of the efficacy of tulathromycin is feasible for both *P. multocida* and *M. haemolytica* (Toutain et al. 2017b). Zhou et al. (2017) calculated the theoretical doses for bactericidal eradication and for the bacteriostatic effect. A dosage of 13.25 mg/kg was computed as necessary to achieve a total eradication of *P. multocida* by using MIC_{90} in the serum, but it is several times higher than the recommended dose (2.5 mg/kg). Because MIC_{90} can be considered as the worst-case scenario, the dose would be lower if it were based on *P. multocida*'s MIC distribution when using a Monte Carlo simulation (MCS). When the concentration for the bacteriostatic effect on *P. multocida* was calculated, it resulted in lower doses, in the range of 3.32–6.39 mg/kg, for transudate and serum PK/PD breakpoints. As mentioned above, because of interaction with immune-defence mechanisms, computations and modelling alone cannot give the correct answer for the exact dose needed for a treatment, but in clinical practice, practitioners need to confirm such computations, ideally by running clinical trials using clinically diseased animals. Moreover, end points that detect clinical cure and bacteriological cure could also improve evaluation and analyses of a therapy success.

Example of fluoroquinolones: For fluoroquinolones, which are concentration dependent, $fAUC_{24h}$/MIC predicts efficacy against Gram-negative bacilli once a target value of 70 has been reached (Ambrose et al. 2007). However, it is important, which antimicrobial of the fluoroquinolone group is used and whether the parent substance has been metabolised to another microbiologically active substance. This is the case for the most frequently used fluoroquinolone in veterinary medicine- enrofloxacin. It is metabolised—to varying degrees, depending on the animal species—to ciprofloxacin. Also, the amount of free fraction available is important. Last but not least, Wen et al. (2016) urged researchers to give consideration to low/high MIC vs concentration/time dependency for enrofloxacin (PD against the isolate with low MIC exhibited the expected concentration-dependent characteristics, but PD against the less susceptible isolates demonstrated time-dependent characteristics). Below are two examples of recent studies performed on pathogens of importance.

From the perspective of clinical practice useful study was performed with clinically diseased 7-day-old chicks infected with *E. coli* 078 (causing perihepatitis and pericarditis, with a death rate of 10%). Chicks were treated with enrofloxacin. This antimicrobial is typically administered within the first week of life to counteract *E. coli* infections (Xiao et al. 2018). The bacterial load in the blood, liver and lungs were 7.2 ± 0.92 log10 CFU/mL, 6.4 ± 0.14 log10 CFU/g, and 6.1 ± 0.17 log10 CFU/g, respectively (Xiao et al. 2018). Enrofloxacin in this study was administered by gavage in a range of doses from 1 to 20 mg enrofloxacin/kg b.w. The data for key PK parameters are listed in Table 7. Because enrofloxacin is metabolised to ciprofloxacin to different extents depending on the species, the concentrations of both substances were measured. Finally, the AUC_{0-24}/MIC ratio of the combination of enrofloxacin and ciprofloxacin that produced bactericidal and elimination effects were 21.29 and 32.13 in blood; 41.68 and 58.52 in the liver; and 27.65 and 46.22 in the lungs, respec-

Table 7 Pharmacokinetic parameters and PK/PD ratios calculated for enrofloxacin and its metabolite ciprofloxacin in broiler chickens (35.2–45.6 g b.w./at 1 day, for a mix of sexes, sampled at day 7, fed a standard diet) infected with *E. coli* O78 and dosed orally with 1 mg, 10 mg or 20 mg of enrofloxacin/kg b.w. (modified from Xiao et al. 2018)

Parameters	Oral dose of 1 mg/kg b.w.	Oral dose of 10 mg/kg b.w.	Oral dose of 20 mg/kg b.w.
$T_{1/2\beta}$ [h] (ENR)	**10.75 ± 1.8**	**11.44 ± 1.41**	**9.78 ± 1.26**
$T_{1/2\beta}$ [h] (CIP)	*27.25 ± 7.76*	*19.51 ± 2.60*	*18.77 ± 4.11*
T_{max} [h] (ENR)	**3.35 ± 1.8**	**3.4 ± 2.12**	**3.3 ± 1.77**
T_{max} [h] (CIP)	*5 ± 1.05*	*4.3 ± 1.98*	*4.4 ± 1.58*
C_{max} [µg/mL] (ENR)	**0.16 ± 0.041**	**1.76 ± 0.32**	**2.86 ± 0.52**
C_{max} [µg/mL] (CIP)	*0.03 ± 0.001*	*0.10 ± 0.03*	*0.37 ± 0.05*
AUC_{0-24} [h. µg/mL] (ENR)	**1.64 ± 0.28**	**17.95 ± 2.45**	**30.07 ± 4.64**
AUC_{0-24} [h. µg/mL] (CIP)	*0.61 ± 0.03*	*1.42 ± 0.15*	*4.93 ± 0.76*
CL/F [mL/h/kg] (ENR)	**488 ± 114**	**420 ± 85**	**548 ± 114**
CL/F [mL/h/kg] (CIP)	*0.67 ± 0.11*	*3.96 ± 0.78*	*2.26 ± 0.46*
$AUC_{0-24\,h}$ /MIC (ENR)	**5.27**	**32.11**	**61.93**
$AUC_{0-24\,h}$ /MIC (CIP)	*0.36*	*4.50*	*13.24*

Note: For each value, 10 samples were analysed to establish mean ± standard deviation

tively. For a systemic infection by bacteria (such as in this case), it is esential to acurately describe the relationship between PK and PD. The reduction of bacterial load in various organs should be the PD endpoints, that are included in PK/PD analysis. It is always neccesary to confirm the dose proposed by PK/PD modelling (using the target tissue endpoints) in a well-designed clinical trial.

Another study on healthy broiler chicks, administered 50 mg of enrofloxacin/kg b.w. to chickens through PO and IV routes with a dosing interval of 24 h, provided the data (see Table 8), with which the authors calculated a proposal for an effective treatment of infections caused by *Salmonella enteritidis*. According to the results, the authors concluded that a fivefold increase in the recommended dosage of enrofloxacin in chicken is needed to eradicate this *Salmonella* species. The MICs of enrofloxacin were in range 0.0625–1 µg/mL against *S. enteritidis* strains, and the MIC_{50} was 0.50 µg/mL (Kang et al. 2019).

The differences found in the studies (Kang et al. 2019; Xiao et al. 2018; Mekala et al. 2015; Da Silva et al. 2006) illustrate that the disease status affects the total amount of the antimicrobial in the blood stream. After oral administration of 10 mg/kg of enrofloxacin, the AUC_{0-24} value (17.95 h . µg/mL) in the Xiao et al. (2018) study was much lower than the values reported in previous studies on healthy broiler chicks—e.g. 35 h . µg/mL in Da Silva et al.'s (2006) study and 25.35 h . µg/mL in Mekala et al.'s (2015) study. Biotransformation to ciprofloxacin cannot be forgotten when considering the PK of enrofloxacin. Xiao et al. (2018) also showed the ratio of ciprofloxacin as a metabolisation product: When the doses were 10 and 20 mg of enrofloxacin per kg b.w., the concentrations of ciprofloxacin were 7.9 and 15.3%, respectively. Trouchon and Lefebvre (2016) summarised the results of various studies on the metabolisation of enrofloxacin to ciprofloxacin and the plasma protein binding of both of these fluoroquinolones (Table 9).

The PK characteristics of the most frequently used fluoroquinolone, namely enrofloxacin, differ depending on the species (specificities such as the rate of metabolisation: ENR to CIP) and are influenced by the animals' health status, the age of the animals and the target organ's parameters. These bring to light the difficulties that need to be considered in PK/PD modelling. Some extrapolations have been possible, and models/simulations have improved, but the doses proposed by modelling still need to be confirmed by clinical trials in the respective animal species.

Table 8 Pharmacokinetic parameters and PK/PD ratios calculated for enrofloxacin and *Salmonella enteritidis* infection, based on the PK data from healthy broiler chickens (1 kg b.w., 3–4 weeks old, fed a standard diet) dosed orally and intravenously with 50 mg of enrofloxacin/kg b.w. (Kang et al. 2019)

Parameters	Unit	Intravenous administration	Oral administration
$T_{1/2}$	h	12.84 ± 1.40	25.84 ± 1.40
T_{max}	h	0.22 ± 0.10	0.65 ± 0.12
C_{max}	μg/mL	6.74 ± 0.03	3.82 ± 0.59
AUC_{0-24}	μg . h/mL	21.13 ± 0.90	20.84 ± 5.00
F	%	–	98.60 ± 8.90
C_{max}/MIC_{50}		13.48 ± 0.7	7.64 ± 0.2
$AUC_{0-24\,h}/MIC_{50}$		42.26 ± 0.3	41.68 ± 0.1

Note: For each value, eight samples were analysed to establish mean ± standard deviation (WinNonlin software used)

Table 9 The rate of metabolisation of enrofloxacin to ciprofloxacin and their plasma protein binding in pigs, chickens, rabbits and cattle (dairy cows and steers) (Trouchon and Lefebvre 2016, amended from sources)

Species	Percentage of CIP in ENR plasma concentration	ENR: Percentage of protein binding	CIP: Percentage of plasma protein binding	Notes
Pig fatteners[1]	51.5		35	2.5 mg/kg b.w. IM
Weaned piglets[2]	<10%	31.1–37.1	ND	7.5 mg/kg b.w. SC
Chickens[3,4]	7.9–15.3[3] 10[4]	23[4]	ND[4]	[3]10 and 20 mg/kg b.w. oral [4] 5 mg/kg b.w. IV
Dairy cows[5]	59	59.4	33.7	5 mg/kg b.w. IV bolus jugular catheter; ENR in milk detected beyond 17 h; CIP in milk detected until 131 h
Steers[4]	64	60.8	49.6	5 mg/kg b.w. IV bolus jugular catheter
Rabbits	26.5			10 mg/kg b.w./day orally 5 consecutive days

CIP: ciprofloxacin; ENR: enrofloxacin [1]Anadón et al. 1999; [2]Messenger et al. 2011; [3] Xiao et al. 2018; [4] Ovando et al. 1999; [5] Idowu et al. 2010

Example of aminoglycosides: For aminoglycosides, fC_{max}/MIC and AUC_{0-24}/MIC can be used as predictors of therapeutic efficacy (Craig 2011; Rea and Capitano 2007). Clinical trials for human medicine have produced most of the data on aminoglycosides and only a limited number of the data have come from preclinical trials using laboratory animals as in vivo models for human PK/PD considerations (Andes and Craig 2002). To obtain a clinical response of > 90% in patients and reduce the risk of developing resistance to aminoglycosides, the C_{max}/MIC needs to be 8–10 (Krause et al. 2016; Zelenitsky et al. 2003; Toutain et al. 2002; Kashuba et al. 1999, Moore et al. 1987). A *f*AUC/MIC > 50 was reported for tobramycin (Mouton et al. 2005); an AUC_{0-24}/MIC ≥ 110 was demonstrated as effective in a clinical study on tobramycin conducted by Smith (2001 in Krause et al. 2016); an AUC/MIC target of 100 was reported to be associated with a 1- to 2-log10 kill (*K. pneumoniae*, amikacin) in a neutropenic thigh infection model for mice (Craig 2011). For certain enterobacteria and *P. aeruginosa*, however, the values for an AUC_{0-24}/MIC ratio in preclinical trials were 82 and 91, respectively (Muller 2017). According to the results of studies using PK and PD variables via mathematical modelling but considering also clinical experiences, it is recommended to administer aminoglycosides intravenously. Data on the recently authorised aminoglycoside plazomicin administered as IV injections to humans include also information on toxicity issues. Therefore, for aminoglycosides, primarily in human medicine, a high dose exceeding the MIC of the target pathogen in the target tissue (system) and administered once a day (24 hours interval) is currently recommended for injectable medicinal products administered intravenously (Van Bambeke and Tulkens 2011; FDA 2018). Accordingly, in neutropenic and non-neutropenic (rodent) models of infection, significantly more animals survived a potentially lethal challenge of bacteria when undergoing the following regimen: high dose, once per 24 h, than the same (high) dose/8 h schedule. Reduced nephrotoxicity was proven for a once per day high-dose regimen. Situations in veterinary practice differ because many administrations are delivered via the oral route (e.g. VMPs containing neomycin, apramycin or paromomycin) for the treatment of gastrointestinal tract (GIT) infections in different animal species. Although absorption rates differ depending on the animal species and on age, nearly the entire dose of the administered antimicrobial remains in the GIT and a minimal fraction is absorbed. Their antibacterial activity is therefore exerted in situ. For veterinary purposes, human-derived PK/PD parameters cannot be applied to orally administered VMPs and may be used in parenteral administrations, but only to a limited extent (EMA 2018b).

Another route of administration used in veterinary practices for dairy cows is the intramammary route (neomycin, kanamycin, dihydrostreptomycin or framycetin), in which case the extent of (systemic) toxic effects is also limited. In the administration of injectable aminoglycosides, special attention should be paid to neonates (for toxicity issues), because especially gentamicin is known to be one of the most nephrotoxic aminoglycosides, with a very narrow margin of safety, which is of particular concern in young animals, especially foals. Some years ago, after the so-called referral procedure was released, the assessment by and recommendations of the EMA Committee for Veterinary Medicinal Products (CVMP) have led to modifications of the product information for VMPs indicated for the treatment of serious infections in horses. A single daily dose of 6.6 mg of gentamicin/kg b.w., administered intravenously for 3–5 days (linked to the approved indications), provides an effective dose; it takes into account the desired PK/PD characteristics for a concentration-dependent antibiotic—i.e. a ratio of the optimal maximum serum concentration (C_{max}) to MIC, specifically 8–10 providing that a bacterial MIC ≤ 2 µg/mL. To reduce the risk of nephrotoxicity in adult horses, a single administration is recommended because it is less risky than more-frequent daily administrations are (and the latter comes with a risk of the theoretical concept of adaptive resistance). However, the situation is different in neonate foals in that the PK of gentamicin changes considerably in the first 2 weeks of a foal's life. In the context of efficacy, plasma concentrations of gentamicin are more difficult to achieve because gentamicin is retained for a long time, especially in the kidneys—which increases the risk of nephrotoxicity. So even if a dosing schedule can be proposed in adult horses, administration in foals is not recommended, because of the high risk of gentamicin-induced nephrotoxicity (EMA 2015a).

Example of florfenicol. A single dose is generally administered for LA injectables of florfenicol, where a duration of action of several days is expected. Florfenicol administered in calves as a LA injection has been claimed to maintain effective levels for 96 h (Toutain et al. 2019). Using an in silico dose fractionation approach, *f*AUC(0-96 h)/MIC cut-off values for *P. multocida* of 115 h and *M. haemolytica* of 127 h were proposed by Pelligand et al. (2019). The preferred approach is to determine the scaling factor by dividing the *f*AUC (0–96 h)/MIC by 96 h, which yields scaling factors of 1.19 for *P. multocida* and 1.32 for *M. haemolytica*. Therefore, to elicit an efficacious clinical response, at least 1.19- and 1.32-

fold MICs are necessary for *P. multocida* and *M. haemolytica*, respectively, and the average free plasma concentration of florfenicol over 96 h should be equal to or greater than these values (Toutain et al. 2021).

2 PK Principles and Dosing Schedules

The PK/PD approach allows for calculating a dose that is assumed to provide a sufficient therapeutic effect considering altogether the pharmacokinetic properties of a medicinal product (MP), the pharmacodynamic properties of the active substance of this product and all the variables influencing both the PK and PD for this medcinal product (Lees 2010).

Before conducting a study on dose determination and dose confirmation for new marketing authorisation (or even when revisiting the doses of a legacy antimicrobial used for a long time in clinical practice, where a strong need for an update to dosages has been identified), incorporating robust preclinical data is vital. PK–PD integration is a useful tool for initial predicting an effective dose of antimicrobials, mainly for those that achieve systemic distribution. Several individual authors but teams of experts as well (Lees 2004; Toutain et al. 2002, ANSES 2017; EMA 2018a) have recommended PK/PD modelling as the possible next stage in dosage regimen prediction or verification. Dosing regimens depend on three key variables:

- The dose of the antimicrobial, defined mostly via mg of active substance/kg b.w. and the time period in the case of antimicrobial VMPs for systemic use (exceptions include intramammary VMPs with one applicator, containing a certain amount of the antimicrobial, per teat).
- The frequency of the treatment, defined as a dosing interval usually 24 h (or much less frequently 12 h); in LA injectables given as a single dose, with a possible second dose after 48 or 72 h.
- The duration of the treatment, defined mostly in days as the total period for the administration of the antimicrobial containing VMPs for systemic use (in intramammary VMPs, duration can also be defined as the total number of administrations—e.g. three times, 12 h apart, one applicator per teat).

Finally, the dosing regimen should therefore be precisely defined (examples):

- X mg of active substance/kg b.w./day for 5 consecutive days (= 5 doses) (e.g. generally oral, parenteral treatment).
- Y mg of active substance/kg b.w./12 h for 7 consecutive days (=14 doses) (e.g. generally oral treatment (mostly tablets)).
- Z mg of active substance/kg b.w. as a single dose (or the dose can be repeated after 48 h) (e.g. treatment schedule for long-acting, injectable veterinary medicinal products). Generally in injectable VMPs: it is of importance to precise the route of administration (intravenous, intramuscular, subcutaneous).

Exceptions to these schedules include those for intramammary, intrauterine and topical VMPs (e.g. otologicals, ophthalmologicals and dermatologicals).

For pragmatic reasons, the amount of active substance in mg/kg b.w. can be also expressed as weght or volume units of VMP, i. e. as follows:

- in injectable VMPs in equal amounts to the volume of the VMP = mL VMP/kg b.w.
- in orally, via water, administered VMPs in equal amounts = g (powders) or mL (concentrates) of VMP/L of drinking water
- in orally, via feed, administered VMPs in equal amounts = g of VMP/kg of feed (usually small group/individual treatment)
- in medicated premixes in equal amounts = kg of VMP/tonne of feed (usually treatments for herds/flocks)

For some antimicrobial substances, expressing the active part as a salt or a base is important because it could create non-negligible differences in the dose (and consequently also in the amount to be administered in one injection site, which could affect both local tolerance and residue depletion, especially at the injection site). Examples of the differences can be seen when using a gentamicin base vs gentamicin sulphate or an oxytetracycline base vs oxytetracycline hydrochloride. It matters most in large animals with weights heavier than 500 kg (namely horses, which are quite sensitive to the volume injected into one site).

PK-PD modelling, utilising well-designed software and appropriate data, can generate datasets for the whole sweep of plasma parameters (or other biological fluids) to determine concentration–effect relationships. This helps to determine the drug concentrations and dosages required to achieve specific levels of bacterial growth inhibition (e.g. for bacteriostasis, bacterial killing and the eradication of bacteria) when using in vitro, in ex vivo and in vivo techniques (Lees et al. 2008). Also, new mutation-prevention concentration (MPC) data can be taken into account in calculations. Approaches for determining doses via PK/PD integration mostly use one or more of the indices listed in this chapter: C_{max}/MIC_{90} (primarily for concentration-dependent drug classes); $fAUC/MIC_{90}$ (primarily for concentration- and time-dependent drug classes) and $f\ T > MIC_{90}$ (primarily for time-dependent drug classes). For these indices, both PD variables (mostly MIC, but MBC and MPC can be also used) and PK data on routes of administration (preferably taken from clinically diseased animals) are important. Despite the advances in PK/PD approaches (PK/PD integration and modelling) and population, PK should generally be viewed as an alternative to or as supplemental to dose titration studies in disease models and PK preclinical studies on healthy animals, respectively. Both PK/PD modelling and titration studies targeting doses have advantages and disadvantages. Importantly, PK/PD integration and modelling cannot work with "virtual" data. Robust clinical data and data from in vivo infection models on routes of administration for delivering precise doses to animals, as well as input from appropriate datasets from microbiological laboratories on PD, are needed. Any extrapolations used (e.g. among animal species, animals age categories or microbes) should be made by taking great caution and met with critical assessment. Whenever possible, outcomes of the modelling should be validated and confirmed by the results of clinical studies.

The following are some of the weaknesses of using input data for modelling and what should be avoided when designing dose titration and dose-determination studies (Lees 2010—modified/updated).

– *A lack of sufficiently robust PK data*: plasma/tissue concentrations achieved after certain doses administered via certain routes should be measured (especially from diseased target animal species, for which a dosing schedule is required). These data could serve as inputs for PK/PD modelling and can help to determine doses. With help of the PK/PD integration and the mode of action and character of killing of an antimicrobial (especially time dependency), they can also help to determine dosing intervals.
– *Plasma concentration was most often used in PK-PD modelling and has been considered as superior to administered doses* from the perspective of prediction of the optimal dosing schedule. Liu et al. (2002) concluded that concentrations of antimicrobials in the interstitial fluid (ISF) are responsible for the antimicrobial effect at the target location and are more relevant in predicting therapeutic efficacy than plasma concentrations are. Despite that, even very recent studies still use plasma parameters, such as those determining the PK-PD cutoff values of florfenicol in long-acting formulations aiming to support clinical breakpoint settings for cattle bovine respiratory disease (BRD) (Toutain et al. 2019). This is because of broad availability of plasma data and because such data are available for both the already authorised and the future VMPs and therefore allows comparison if the same conditions of dosing were followed. But possible impacts should be thoroughly considered, case by case, on the basis of the active substance, its behaviour in the body/target tissue of the certain animal species/category, indications and the pathology of the infectious process. Moreover, the importance of evaluating unbound drug concentrations, not total concentrations, was highlighted as key for evaluation of the efficacy of antimicrobials.

- When focusing on the specific site of infection, e.g. the respiratory tract or especially the lungs, most traditional measures of antibiotic concentrations especially in plasma have been gradually replaced by measurements of the PK indices, as exemplified in macrolide investigating studies: on murine models (Zeng et al. 2018), guinea pig models (Zhao et al. 2018), tissue cages (Zhou et al. 2017), homogenised lung tissues and pulmonary epithelial lining fluid (PELF) (Cox et al. 2010). For macrolides and azalides, pulmonary infections caused by extracellular pathogens, measuring the concentration in PELF is a better predictor of efficacy than measurements of concentrations in plasma or the lungs (Giguère and Tessman 2011). Two methods were tested and introduced for measuring the PK parameters in the PELF, the initial site of bacterial colonisation and the target site of respiratory tract infections (Foster et al. 2016). The concentration of drug that penetrates the blood–alveolar barrier can be assessed by collecting the PELF and can be relevant in calculations for PK-PD models at the beginning phases of bovine respiratory disease, before the inflammation process has disrupted the blood–alveolar (blood–bronchus) barrier and before the infection has spread to the interstitium. The PK parameters gained from PELF seem to be better predictors of drug concentrations in the airway secretions and may predict the phase when pathogenic bacteria colonise the airways, because they are adequate to meet preventive PK-PD targets. For the phase of inflammation and the later phase of the disease, interstitial fluid (ISF) drug concentrations best predict the respiratory concentrations needed to treat infections (Foster et al. 2016; Baldwin et al. 1992). In the veterinary field, the tissue-cage model has also been used to study the PK/PD integration of antimicrobial drugs in several species, such as goats (Sidhu et al. 2010a), sheep (Sidhu et al. 2010b), rabbits (Cui et al. 2006; Xiong et al. 2016), calves (Potter et al. 2013; Sidhu et al. 2011) and pigs (Gu et al. 2015).
- When infection is localised in gastrointestinal tract, intestinal content from the relevant part of intestine, which is affected by the target microorganism and its pathology, should be sampled to gather PK data. This is important for orally administered antimicrobials and those that have low absorption rates from the gastrointestinal tract (e.g. colistin sulphate and apramycin sulphate). Such models have been well established in pigs (see the Chapter 2 - Swine). In such cases, the PK data from the target tissue should be used for PK/PD integration and modelling (Guyonnet et al. 2010).

- *Indices of clinical cure* only are considered for various reasons—e.g. either no data are available on bacteriological cure or these data have not been sufficiently taken into account. Evaluations of efficacy may be also influenced by the Pollyanna phenomenon (covered earlier in this chapter). Evaluations of both clinical cure endpoints and bacteriological cure endpoints and of their relationships should be investigated and considered essential for determining the dose of antimicrobials being effective not only from clinical but also from bacteriological cure perspective.
- *Studies on diseased animals*, preferably well-designed and -defined experimental infection models, should be performed for clinical study intended for dose determination. If those are not possible, then well-controlled, naturally infected and clinically diseased animals can be used for such studies. Diseased animals should be sampled to obtain PK data.

In dose-determination studies, the number of tested dosing schemes (the usual recommendation is three levels of dose), the frequency of administration (based on the drug characteristics, such as PK) and the total number of administrations are key parameters to be assessed. Combinations of one dose value, frequency, and total number of administrations should be tested in dose confirmation study or via one or more clinical trials. For injectable products, the exact route of administration (IV, IM or SC) is also important. The PK of the active moiety can be more or less influenced also by the volume administered to one injection site because larger volumes (especially in VMPs with irritation potential) can cause local reactions influencing the distribution of the active moiety with antimicrobial properties to the target tissue. After the final assessment of all the steps, including clinical trials, assessments of resistance and also trends in resistance development, in the case of positive outcomes and positive benefit-risk ratio, the final dosing schedule is to be authorised and is finally ready to be used in practice. Information on the PK characteristics of a VMP is also useful for designing residue studies (in cases where, for example, meat (pigs), meat and milk (dairy cows) or meat and eggs (laying hens) withdrawal periods need be established).

2.1 PK/PD and Establishing Doses

Providing that essential variables will be known in sufficient quality and will also represent scenario, for which dose should be calculated (e.g. antimicrobial pharmaceutical form, route of administration, claimed indication and target tissue), the relationship between the dose and the PK/PD parameters can be, according the ANSES (2017), set by using the following calculation:

$$\mathbf{Dose}_{\text{perunitoftime}} = \frac{\mathbf{Plasma\,clearance}}{\mathbf{Bioavailability}} \times \mathbf{Target\,concentration}$$

Dose $_{\text{per unit of time}}$ means the dose of the antimicrobial expressed per unit of time.

Bioavailability means the rate and extent to which the drug enters the systemic circulation unchanged. It can be assumed as complete only for IV-administered drugs.

Plasma clearance expresses the overall ability of the body to eliminate a drug by scaling the drug elimination rate (amount per time) by the corresponding plasma concentration level (and is expressed as a volume per time).

Target concentration means the mean plasma concentration associated with the desired therapeutic effects.

In the case of antimicrobials, the target concentration must be sufficient to reach the threshold value (or critical value) of the PK/PD index correlated with their efficacy. As mentioned in Table 6, the AUC_{24h}/MIC index can be used for certain antimicrobials. Again, good-quality pharmacokinetic data measured in animals and representing the target species/category in preclinical trials are key to the calculation.

When the efficacy of these antimicrobials is correlated with the AUC_{24h}/MIC index, the following equation can be used to calculate the relationship between the target concentration and the threshold value of the PK/PD index (ANSES 2017):

$$\text{Target concentration} = \frac{\left(\text{AUC}/\text{MIC}\right)_{\text{criticalvalue}}}{24\ \text{h}} \times \frac{\text{MIC}}{\text{fu}}$$

$(\text{AUC/MIC})_{\text{critical value}}$ means the critical value of the PK/PD index expressed in hours.

fu means the free fraction of the antimicrobial in plasma that can be utilised effectively against a pathogen.

MIC means the minimum inhibitory concentration of the antimicrobial as determined for the target pathogen (to be treated by the dose calculated by using the target concentration).

AUC is an abbreviation for the area under the curve.

Most of the antimicrobials used in veterinary practice in food-producing animals are administered either orally or parenterally at 24 h intervals (i.e. a once-daily dosing scheme). In light of this, the following formula was proposed as useful for the calculation of the daily dose (ANSES 2017):

$$\text{Dose}_{\text{daily}} = \frac{\text{Plasmaclearance}}{\text{Bioavailability}} \times \frac{\text{MIC}}{\text{fu}} \times \left(\text{AUC}/\text{MIC}\right)_{\text{criticalvalue}}$$

For the abovementioned formula's variables, four PK parameters should be available from a preclinical PK study on the respective antimicrobial administered by known route of administration in defined dose (or set of doses): bioavailability, plasma clearance, the free fraction of antimicrobial available in plasma and the area under the curve (AUC_{0-24}). Those variables should be available for the active substance/moiety that can be identical (salt, ester) with the substance—e.g. amoxicillin (as amoxicillin trihydrate), doxycycline (as doxycycline hyclate or doxycycline hydrochloride), ceftiofur (as ceftiofur sodium or ceftiofur crystalline-free acid or ceftiofur hydrochloride)—or can be the sum of the administered parent substance (enrofloxacin) and its microbiologically active metabolite (ciprofloxacin). Calculations using a single value of the PK parameter, represented by the mean value of this parameter, will lead to the single dose setting, whereas calculations using a distribution for each PK parameter will lead to the dose distribution setting. The data used for such calculations should preferably come from a researcher's own studies, but it could also come from bibliographical references. Data gained from references should be thoroughly selected from well-designed, well-performed and precisely described trials containing a sufficient level of details. Checking source data for validity and suitability for incorporation is crucial, and if some important parameter is missing, the data cannot be used as input data for the modelling. Simply merging datasets, without pre-evaluation and validation, should be avoided. For the processing of such data, relevant software tools can be used, e.g. Oracle Crystall Ball software; the calculations for the PK and PD indices, a noncompartment model can be used, e.g. Phoenix™WinNonlin software (Certara Corporation), Simulx® (Lixoft) or other appropriate software tools; and for population analysis, population modelling and simulation software can be used, e.g. Phoenix™ NLME software. As the softwares have been continuously advancing, new softwares and approaches appears.

The PK/PD approach can be also used to determine the threshold MICs (PK/PD cutoffs) for the $\text{Dose}_{\text{daily}}$ value, by using the above equation. The selection of the critical concentrations delineating the I/S/R categories of antibiotic susceptibility testing requires determining three threshold MICs: the epidemiological cutoff (ECOFF); the clinical breakpoint (CBP); and the PK/PD cutoff (ANSES 2017). When taking the PK/PD approach (PK/PD cutoff), the threshold MIC can be obtained from the following equation, in which the recently described parameters are used:

$$\text{MIC}_{\text{PK/PD}} = \text{Dose}_{\text{daily}} \times \frac{\text{Bioavailability}}{\text{Plasmaclearance}} \times \frac{\text{fu}}{\left(\text{AUC}/\text{MIC}\right)_{\text{criticalvalue}}}$$

The calculation can be performed with the critical MIC values breakpoints of the "susceptible" category defined for the tested antibiotic-pathogen combination, according to the clinical breakpoints (CBPs) from the CLSI (veterinary- or human-derived CLSI/EUCAST) and the epidemiological cut-offs (ECOFFs) from EUCAST.

For the antibacterial to have the desired effect, the following principles and the PK/PD index calculation (utilising MIC values) can be used in the predictive inhibitory sigmoid $E_{\max}$ model. Such a model can assess the PK/PD index, with the most predictive value, for each index of each dosing regimen.

$$E = E_{max} - (E_{max} - E_0) \times C^N_e / (EC^N_{50} + C^N_e)$$

Here, E is the antibacterial effect; E_0 represents maximum drug effect 24 h after administration. E_{max} is the change in bacterial load ($\log_{10}$ CFU/mL) in the infected tissue cage with no drug present. C_e is the PK/PD index magnitude. EC_{50} is the magnitude required for achieve 50% of E_{max} and N is the Hill coefficient that describes the sigmoid shape (Choe and Lee 2017).

2.2 PK/PD Principles in Setting the Frequency and Duration of Administrations

Even though the above PK/PD modelling can help to determine the dose and establish the values of MICs at the critical threshold value of a pathogen's susceptibility, the frequency and duration parameters cannot be solved by this approach. The lack of scientifically based data confirming the correct frequency and duration of a treatment means that both human medicine and veterinary medicine lack a solid rationale for setting the frequency and duration of antimicrobial treatments when also aiming to both maximise treatment efficacy and minimise the risk of antimicrobial resistance selection. Therefore, especially in orally dosed antimicrobials, the duration is set within a range (e.g. mostly 3–5 days or 5–7 days), and because of the threat of antimicrobial resistance, a maximum duration is also set—e.g. orally dosed colistin-containing VMPs have maximum durations of 7 days (EMA 2015b). A long duration period is still valid for several VMPs used throughout Europe (e.g. for orally administered tylosin-containing premixes has been approved duration of treatment within a range of 4–5 weeks (EMA 2011); for the recently valid treatment duration in VMP containing tylosin, dosing schedule in the Summary of Product Characteristic valid in the respective EU member state can be found and checked via Union Product Database (UPD). Long duration of treatment might occur because some of the tylosin-contaning VMPs were long time ago originally used for growth promotion or as feed additives. To the best of the author's knowledge, high quality publicly available, randomised, double-blind clinical studies comparing two treatment durations for veterinary antimicrobials (where the same active substance is evaluated) have not yet been released. Specific simulations comparing various dosing regimens could be performed, for which appropriately robust and valid data are available to help determine more-suitable dosing schedules. Some data on approaches aiming to optimise the treatment duration come from human medicine and are based mostly on meta-data analyses (Wilson et al. 2019; Onakpoya et al. 2018), some of which also compare practices and treatment guidelines (Pouwels et al. 2019). Trends on shortening the duration of treatment have also caused controversy and have led to recommendations for researchers to use artificial-intelligence assessments of the data, such as using prescribing software to link diagnoses with prescriptions for the appropriately reduced antibiotic quantities (by default), to justify the shortening of the duration and to lay out the possible consequents of doing so (Hudson 2019). Because of the absence of reliable data, because uninvestigated possibilities for modelling remain and because of fears of therapeutic failure, the treatment durations approved for many VMPs are probably longer than necessary. One way of filling this knowledge gap would be to run clinical trials assessing the efficacy of shorter treatment durations in veterinary medicine. However, such trials carried out in the field would need to include relatively high numbers of animals to control for the many sources of bias and will be costly. In the absence of properly designed comparative trials, clinicians could be asked to shorten treatment durations when possible and in accordance with data from thorough clinical follow-ups on diseased animals under their care, to try to find reasonable compromises. In the future, some "scoring" systems defining the parameters of serious infections and those of mild infections, considering also factors as timely start of the treatment and the course of the disease (e.g. the speed of disappearance of clinical symptoms) can be used. Finding biomarkers such as a serum marker like procalcitonin or

CRP assessment-based algorithm can belong among the tools to reduce antibiotic treatment durations (Coelho 2009; Petel et al. 2018). Carrying out scientific studies to assess the therapeutic success by using of biomarkers is vital for any science-based recommendation, which can be used in veterinary medicine to shorten the treatment course reasonably. Such shortening will reduce adverse effects and the length of time that the commensal (mainly intestinal) microbiome is exposed to antimicrobials, which is essential for minimising the emergence and spread of antimicrobial resistance (Toutain et al. 2016).

2.2.1 Frequency of the Treatment Versus PK in Renal and Hepatic Function Failures

The frequency of administration is of importance for an effective treatment, but is a big issue for the practicablity reasons. As has already been discussed, the pharmaceutical form and the route of administration also have significant influences on frequency. Single-shot injectable antimicrobials are favoured because of their ease of administration; in orally dosed antimicrobials, once-daily administration is preferred. Even so, the characteristics of certain molecules of antimicrobials require shorter intervals than 24 h (sometimes 12 h or even 8 h) to maintain the concentration of the antimicrobial at the site of infection at a level at least equal but ideally higher than the MIC of the target pathogen. To optimise the intervals and to correctly divide the daily dose so that it is administered in proper intervals, PK/PD principles that model dosing for pathogen-specific treatments in the target animal species/categories can be used. Such an approach may increase efficacy, help to reduce the overuse of antibacterial agents and minimise exposure—and may therefore help to minimise the emergence of resistant pathogens (Ahmad et al. 2016). Although PK/PD modelling can be used to set the optimal frequency, some factors can complicate the assumptions of such models. One of the complicating factors could be the postantibiotic effect (PAE). Dosing strategies based on PK/PD modelling are pathogen specific; they can make antibacterial therapies safer and more successful by adjusting according to factors such as kidney function, the primary pathogen and the limited patterns of resistance (Owens and Shorr 2009). Thankfully, this approach has already been developed for life-threatening infections in humans (e.g. intensive care units, aminoglycoside dosing and the monitoring of the blood PK parameters, especially clearance). Such modelling needs the timely availability of laboratories; consultations with infectiologists, clinicians and microbiologists; and contributions from their expertise. Such intensive care is currently very limited, reserved mostly for human medicine, and in the veterinary field, it is available for the treatment of highly valuable animals only (e.g. mostly horses) in specialised veterinary clinics and hospitals. In food producing animals, kept in big herds or flocks, with antimicrobials mostly administered via drinking water is hardly achievable to use more than twice daily dosing schedules. Even there will be automatic dosing of medicated water, animal behaviour frames should also be considered, especially drinking habits, to ensure that water intake with the adequate dose of antimicrobial.

PK parameters are important not only in promoting efficacy in that they can be used to set dosing intervals (the time during which the dose is above the target pathogen's MIC) but also in ensuring the safety of the target animals and impact on the consumer safety (e.g. damage at the injection site influences PK). These parameters could also help to determine the rates at which residues deplete in certain tissues important for the drug elimination (e.g. kidney, liver), where residues can persist for a long time, especially in diseased animals with impaired renal or liver functions. Polar antimicrobials and metabolites have restricted extravascular distributions and their respective passive diffusions across membranes are limited, and they are eliminated via renal excretion. Lipid-soluble antimicrobials are excreted mainly via the hepatic route, except for systemically available fractions, which can be eliminated through the renal route (to a smaller extent in herbivores than in carnivores). Although some references still categorise drugs as having either renal clearance or hepatic clearance, the reality is that nearly all drugs undergo a combination of major, minor and co-dominant elimination pathways; therefore antimicrobials' clearance routes can be via renal elimination, hepatic clearance (metabolic inactivation and/or biliary secretion) or via passing directly into the GIT lumen with final intestinal clearance (Fissell 2013; Toutain et al. 2016).

Renal elimination is also influenced by the extent of protein binding, where antimicrobials with extensive binding to plasma proteins limit the glomerular filtration rate (GFR). This parameter differs depending on the species but is measurable because endogenous creatinine clearance is mostly used as an indicator of renal functionality (at least in companion animals). Another possible route of renal elimination is through proximal renal tubule secretion (Baggot in Giguère et al. 2013).

The following antimicrobials undergo renal elimination:

- Aminoglycosides (parenterally administered or where the fraction remains systemically available after limited absorption after oral or intramammary administration) via glomerular filtration
- Most beta-lactams (penicillins, aminopenicillins and cephalosporins) via glomerular filtration and proximal tubular secretion (with the exception of intramammary administered e.g., nafcillin and cefoperazone that are excreted via the liver in bile)
- Clavulanic acid similarly to beta-lactams
- Oxytetracycline (parenterally administered) via renal excretion, in urine
- Ciprofloxacin (main metabolite of enrofloxacin) via glomerular filtration and proximal tubular secretion

The foregoing antimicrobials' rates of elimination decrease when renal function is impaired, and therefore the usual dosing interval (the frequency of administration) may need to be modified. For example, the aminoglycoside group often needs to have its dosing interval tailored to clinical status, especially in the seriously diseased animal. In the case of uraemia associated with chronic renal failure, the binding of acidic drugs to plasma albumin is reduced and the rate of certain biotransformation pathways (with reactions such as reduction or hydrolysis) is decreased. Exact extent of this process and influence on the dosing schedule has not been yet fully determined, but the activation of prodrugs (such as pivampicillin) would likely be decreased; similarly the hydrolytic conversion of ceftiofur to desfuroylceftiofur might be decreased (Baggot in Giguère 2013). When the renal functional reserve (RFR) is large, 75% of the glomerular filtration rate (GFR) must generally be lost before clinical evidence of renal impairment appears. Adjustments to dosage regimens generally account only for decreased GFR (mostly because therapeutic drug monitoring (TDM) is not used to observe drug clearance or the volume of distribution). Nephrotoxic antimicrobials (e.g. aminoglycosides) are the exceptions. Modifying the dosing schedule (including frequency) is usually not necessary for drugs with wide margins of safety (e.g. penicillin). The construction of modified dosage regimens (with either dose or interval modified) in cases of renal failure assumes that renal drug clearance directly correlates with clinical estimates of GFR (e.g. creatinine clearance or if not available can be substituted by 1/SCR (serum creatinine) or 1/BUN (blood urea nitrogen) in the following formulas for interval extension and dose reduction), according to Baggot (in Giguère 2013). **Interval extension** can be calculated by using the following equation: **interval = normal interval × (normal creatinine clearance/patient creatinine clearance)**. **Dose reduction** can be calculated by using the following equation: d**ose = normal dose x (patient creatinine clearance/normal creatinine clearance**). As a result, half or one-third of the dose can be administered at half or one third the interval for some antimicrobials. Despite the benefit of lower toxicity for the treated animal, prolonging the phase runs the risk of delivering subinhibitory concentrations of the antimicrobial into the target tissue. Once the PAE is not present for the respective antimicrobial, the threat of treatment failure emerges. Some controversy exists in such above-mentioned approach. In aminoglycosides interval extension seems beneficial, so currently, the most recommended regimen is to administer a higher dose once a day (see also comment above in this Chapter).

Hepatic metabolism (hepatic microsomal oxidation and glucuronide formation) is typical for most lipid-soluble antimicrobials, from which some are excreted unchanged (as parent drugs) in bile. A list of antimicrobial agents whose main route of elimination is via the liver includes some fluoroquinolones (enrofloxacin, which is partly metabolised to ciprofloxacin and is to various extents, depending on the species, excreted via urine); difloxacin and marbofloxacin (partly eliminated via bile and urine); trimethoprim, sulphonamides, amphenicols, lincosamides, metronidazole, rifaximin and doxycycline (which are orally dosed, are lipophilic, undergo mainly hepatic elimination and are finally excreted through faeces as inactive conjugates or chelates); and tetracyclines (which are dosed orally, are excreted in bile and undergo the enterohepatic cycle, leading to systemic recirculation) (Baggot in Giguère 2013; Karachalios and Charalabopoulos 2003). Liver blood flow and the capacity of the liver to eliminate lipid-soluble antimicrobials by metabolism or via bile excretion are the key parameters that determine hepatic clearance. Hepatocytes' metabolising enzymes might have limited access to the antimicrobials bound in a large extent to plasma proteins. Differences in the rates of hepatic microsomal oxidation or glucuronide conjugation are responsible for species variabilities. Although establishing the level of hepatic dysfunction is hard, the dosage interval for a drug that is eliminated mainly via hepatic metabolism should be increased, and the use of bactericidal antimicrobials should be favoured over others (Baggot in Giguère 2013).

The frequency needed for proper dosing can also be influenced by the age of the target animals. Neonates (within first 24 hours of life and at age about 8 weeks in calves, kids, lambs, and piglet and 1–2 weeks in foals) in particular can have significant differences in their pharmacokinetics from those of their adult counterparts. Some characteristics of the neonatal period include better absorption from the gastrointestinal tract, lower binding to plasma proteins, lower ratios of body fat to fluids, increases in extracellular volumes of distribution for antimicrobials (fluids or total body water), increases in the permeability of the hematoencefalic barrier, slower elimination (longer half-life) and prolonged hepatic metabolisms within the first 24 h for most mammals (Baggot in Giguère 2013; Brown 1992). Some antimicrobial agents (e.g. trimethoprim) that undergo extensive first-pass metabolism (microsomal oxidations in liver) are expected to have higher systemic availability in neonatal animals, e.g. foals (Baggot and Short 1984). Juvenile piglets can serve as models for nonclinical paediatric PK testing to provide preliminary proof-of-concept results that support the use of certain antimicrobials in infants, so some data can also be found from such studies (on rifampin or cefpodoxime), according to some researchers (Roth et al. 2013; Ayuso et al. 2020). Within the first postnatal week, calves' differences in plasma protein binding and body fat–fluid ratios influence the PK of enrofloxacin. Newborn calves metabolise enrofloxacin to ciprofloxacin at a rate slower (C_{max} lower and T_{max} longer) than that of week-old calves (Kaartinen et al. 1997), so the metabolisation rate of cattle depends on their species, breed, sex (in part) and age. The pharmacokinetic parameters of neonatal calves and older cattle were studied for another quinolone—flumequine, showing that both $V_{dß}$ and Cl_B increased with age, resulting in a $T_{1/2}$ that changed very little as calves matured. Mevius et al. (1990) reported that after the oral administration of 10 mg/kg of a 2% flumequine formulation mixed with milk replacer, the $V_{dß}$ and Cl_B of week-old calves (1.07 L/kg, 1.78 mL/min/kg) were significantly lower in five-week-old (1.89 L/kg and 3.23 mL/min/kg) and in 18-week-old calves (1.57 L/kg and 3.10 mL/min/kg). The C_{max} was highest in week-old calves (9.27 µg/mL) and lowest in 18-week-old calves (4.47 µg/mL). The absorption was rapid (T_{max} of approximately 3 h) and complete. The pharmacological parameters of neonatal calves were also discussed in review by Brown (1992), which determined that parenteral absorption is variably altered in neonates owing to an alternation in blood flow to intramuscular and subcutaneous injection sites. Neonates have lower muscle mass, as their body weight is lower compared to that of adults, and the altered activity level was lower in veal calves and sometimes higher in beef calves than that in adults. Brown (1992) added that because the blood–brain barrier is not fully developed at birth, many antimicrobials that would not be expected to have activity there in adults may in fact be active there at effective concentrations in neonatal calves. The renal excretion mechanisms (glomerular filtration and active, carrier-mediated tubular secretion) are incompletely developed at birth in all mammals. During the neonatal period, renal excretion mechanisms develop independently at rates that are specific to the species (Baggot in Giguère 2013). Foal and calf neonates have relatively matured renal function compared to the neonates of most other species. The renal functional reserve (RFR) of neonates is also affected by hydration. The elimination rates of lipid-soluble antimicrobials and their polar metabolites are lower in neonates than those of older/adult animals, due to slow hepatic microsomal oxidation, slow glucuronide conjugation and relatively inefficient renal excretion (Table 10).

2.3 PK Modelling, Analyses and Simulations: Individuals Versus Populations

The essential parameters of pharmacokinetics characterise the fate of antimicrobials and their metabolites in the body in terms of absorption, distribution, metabolism and elimination. The dose received and the individual's characteristics influence concentrations–time courses. Next is a brief description of the principles and methods of individual PK analysis and of population PK analysis, which differ depending on the approach taken and which can be further categorised into parametric, nonparametric, maximum likelihood and Bayesian methods. Simulations of population PK models are used to evaluate these models (internal or external validation) or to evaluate dosing regimens. The probability of target attainment (PTA) can be calculated and used to evaluate dosing regimens (De Velde 2020). The calculation for AUC/MIC is simple to perform, and allows back calculation to set a dose. The parameters for PK modelling, such as $T_{1/2}$, AUC, C_{max} and T_{max}, can be derived from noncompartmental analyses of PK data (time and concentrations). High-quality pharmacokinetic datasets, preferably from studies that are based on validated and internationally accepted methods and that follow good laboratory practices (GLPs)—ideally on diseased animals or animal models—are needed to estimate AUCs. In reality, usually such data are rarely available (especially those on old antimicrobials, because those studies were performed decades ago mostly on healthy animals and because whether they followed GLPs or were at least conducted in GLP-like conditions is not known). Moreover, the above parameters should ideally be taken from sites of infection (e.g. systemic infections in plasma; respiratory infections in the PELF; and ileitis in the ileum). The calculation for the necessary duration above the MIC requires robust PK data (time and concentrations), expertise in pharmacometrics and the use of nonlinear mixed-effects models. The time to maintain concentrations ≥ the MIC is not a simple parameter but instead a variable function of different conditions that depends not only on the dose but also on the shape of the time concentration curve, so it needs to be computed for Monte Carlo studies. The time to maintain concentrations ≥ the MIC, moreover, depends highly not only on the dose administered but also on the frequency, the formulation of the indvidual VMP, the route of administration and the interindividual PK variability (e.g. depending on body weight, sex, age and hierarchical position in the herd—at least for orally dosed group medications) (EMA 2018a).

Population PK methods are performed as concentration–time course analyses of the whole defined population. This technique allows for exploring the relationships between clinical factors (e.g. weight, age, gender and species) and drug disposition so that the efficacy and safety of drugs can be determined (De Velde et al. 2018; Martín-Jiménez and Riviere 1998). The modelling process integrates evaluation of several models,

Table 10 Antimicrobials and adjusting their dose and frequency according to their routes of elimination (modified according to data in Baggot in Giguère 2013)

Class/subclass	Substance	Renal route of elimination, urine	Hepatic route of elimination, faeces, bile	Dose/frequency adjustment[7]
Penicillins	Penicillin G	Mainly renal		Half dose or 2x interval in severe renal failure
Aminopenicillins	Amoxicillin	Mainly renal, urine excretion	Partially bile excretion	
	Amoxicillin/clavulanic acid	Mainly renal, urine excretion		
Antistaphylococcal penicillins	Nafcillin, Oxacillin, Cloxacillin, Dicloxacillin[1]		Small fraction hepatic (as IMM mostly via milk)	No change
Cephalosporins	Cefalexin	Mainly renal, urine excretion		Interval extension in renal failure
	Ceftiofur	Mainly renal, 55% - 60% urine excretion as desfuroylceftiofur	Partly faeces	Interval extension in renal failure
	Cefoperazone[2]	Partly urine	Mainly hepatic, bile and faeces	No change
	Cefquinome	Mainly renal, urine	Up to 5% faeces	Interval extension in renal failure
Aminoglycosides	Dihydrostreptomycin	Mainly renal		Interval extension, modified dosing schedule in renal failure
	Gentamicin	Mainly renal		
	Neomycin[3]	Mainly renal		
	Neomycin[4]		Mainly faeces, unchanged	No change
	Apramycin[4]		Mainly faeces, unchanged	No change
	Paromomycin[3]	Mainly renal		Interval extension
	Paromomycin[4]		Mainly faeces, unchanged	No change
Lincosamides	Lincomycin[3]	49% renal, urine	38 % hepatic, faeces	3x interval in severe renal failure

(continued)

Table 10 (continued)

Class/subclass	Substance	Renal route of elimination, urine	Hepatic route of elimination, faeces, bile	Dose/frequency adjustment[7]
	Lincomycin[4]	Partly renal	Enterohepatic circulation, mainly in bile and faeces	No change
	Clindamycin	Renal	Hepatic	No change
Macrolides and related compounds	Erythromycine[4]	Small proportion urine	Mainly hepatic, bile	No change
	Tulathromycin[5], Tildipirosin, Gamithromycin	Proportion in urine less than in faeces	Mainly hepatic, majority faeces	No change
	Tylosin[6]	Proportion in urine less than in faeces	Mainly bile, faeces metabolites	No change
	Tilmicosin	Proportion in urine less than in faeces	Mainly hepatic, majority faeces	No change
Pleuromutilins	Tiamulin	Small proportion urine: 15-30% pigs as well as chickens	Mainly hepatic, bile and faeces: 70 – 85 % pigs; 55-65 % chickens	No change
	Valnemulin	Small Proportion urine	Mainly hepatic, via bile	No change
Fluoroquinolones	Enrofloxacine[3,4]	Proportion in urine - species specific (depending also on the rate of metabolization to Ciprofloxacin)	Proportion in faeces - species specific (depending also on the rate of metabolization to Ciprofloxacin)	Dosage reduction
	Marbofloxacine[3]	2/3 renal in urine	1/3 faeces	
Polymyxins	Colistin[3]	Renal, urine		Modified dosing schedule in severe renal failure

(continued)

Table 10 (continued)

Class/subclass	Substance	Renal route of elimination, urine	Hepatic route of elimination, faeces, bile	Dose/frequency adjustment[7]
	Colistin[4]		Mainly faeces, unchanged	No change
Tetracyclines	Oxytetracycline[3]	Mainly renal		Do not use in severe renal failure
	Doxycycline[4]		Hepatic, mainly faeces, conjugate or chelate	No change
	Tetracycline		Enterohepatic circulation, mainly faeces	
Sulphonamide		Renal, glomerular filtration, urine		Sulphonamide/trimethoprim Do not use in severe renal failure
Trimethoprim		Metabolites and unchanged trimethoprim are excreted in urine	Partly oxidised and conjugated in liver	

[1] Intramammary administration in veterinary medicine; [2] species-specific ratio among urine, bile and faeces excretion, depending also on the route of administration; [3] parenteral administration; [4] oral administration; [5] species-specific ratio of faeces and urine excretion; [6] oral and parenteral forms, different proportions of metabolites in urine, bile and faeces[7]

which have various numbers of compartments, types of elimination and levels of variability. The final model provides the mean population PK parameters, among which the volume of distribution and clearance are usually key (Li et al. 2014). It also describes the variability between subjects, known as interindividual variability (IIV) or between-subject variability (BSV), and the variability between the doses of an individual, called intraindividual variability or between-occasion variability (BOV) (Li et al. 2014; De Velde 2020). The observed variability is explained by differences in covariates (animal characteristics such as body weight or age). Another variable, which is called residual variability, should not be omitted (it covers, e.g., sampling uncertainties and assay variance). In the past, the standard two-stage "individual" approach was used in population PK analysis, which cannot cover all the types of variability. Later, nonlinear mixed-effect (NLME) models were developed, and they use nonlinear PK equations to characterise the sources of variability in drug dispositions and responses (Bon et al. 2018; De Velde 2020). Systems based on NLME are well suited to analyses of unbalanced data because they can handle multiple levels of variability and can identify covariate relationships for systematic differences between individuals (Fidler et al. 2019). The NLME approach can be used to explore the impact of population-associated variables on the relationship between drug administration, systemic exposure and the levels of drug residues in tissues, and it can help to optimise dosing schedules. NLME is also useful for analysing the sparse data collected in situations where intensive blood collection is technically challenging, as in small animal species that have limited blood volumes, such as poultry and fish (Bon et al. 2018). NLME-PK models typically have three components: structural models, statistical models and (if data are available) covariate models. Because NLME modelling is not a physiologically based mechanistic approach, it is limited in its usefulness in extrapolating beyond the inference range of the experimental data, although that range is considerably wider because data from many sources and studies can be combined (Lin et al. 2016a).

A mixed-effect model uses both fixed (identical and nonrandom) effects and random (individual-specific) effects. For example, the nonlinear mixed-effect model was a meta-analysis tool that used the PK data for penicillin G in cattle and swine in a study by Li et al. (2014); it was used to gather data on residue depletion, which can be used to confirm the withdrawal period. Data uploaded to this model were collected from previously published PK studies (several formulations of penicillin G, diverse populations of cattle and swine data on weight and age). Plasma (368 concentrations for cattle and 443 concentrations for pigs), liver, kidney and muscle residue data were also used in this study. Compartmental models on first-order absorption and elimination took a nonlinear mixed-effect modelling approach to determine plasma and tissue concentrations. A three-compartment model with extra tissue compartments has been selected to describe the pharmacokinetics of penicillin G (Li et al. 2014). A few years later, Li et al. followed this with further studies on penicillin G in cows and pigs and to establish milk withdrawal periods mainly for extra-label (off-label) use of VMPs. Physiologically based pharmacokinetics (PBPK) have been developed and used with compartments in the mammary gland, where those on penicillin G in dairy cows were used to predict withdrawal periods and tissues (including milk) residue depletion rates. Residue behaviour modelling for milk and tissues predicted withdrawal periods for the respective commodities on the basis of comparing data from previous pharmacokinetic studies (Li et al. 2018). After updating these modelling techniques in the more complex interactive generic physiologically based pharmacokinetics (igPBPK) modelling platform for antimicrobials, they were used in studies on penicillin G, florfenicol and flunixin in cattle and swine (Chou et al. 2022) and on other antimicrobials, such as cefquinome in pigs (Zhao et al. 2013), valnemulin in pigs (Zhao et al. 2014) and tulathromycin in lactating goats (Lin et al. 2016b).

For population PK methods, parametric models or nonparametric models (used in the statistical approach) are both used, and they can be further divided into the maximum likelihood approach and the Bayesian approach. Parametric models assume a normal distribution of the data; examples of software programs for maximum likelihood models using one or more mathematic algorithms in the software packages include Monolix, NONMEM and Phoenix NLME. For parametric models, Bayesian MW\PHARM, KinPoP and Pmetrics/ITB software programs are available. Nonparametric methods make no assumptions about the underlying parameters or distribution shapes and are therefore able to detect subpopulations. Different algorithms are used as background methods, such as Pmetrics/NPAG software. For nonparametric Bayesian models, a complete software package was missing, but quite recently artificial intelligence modelling and machine learning software based on Bayesian methodology was announced to be available shortly. Scientists still hesitate on the appropriateness of using specific models to address specific research questions. Thorough model evaluation and validation and thorough comparisons of the models' results with the results gained from clinical studies are important in developing robust and reliable models.

Individual PK methods are performed as concentration–time courses per individual subject. These methods require that many samples be collected and measured. They do not provide detailed information (e.g. covariates) on the variation in PK parameters within a population. One of the most simplistic methods among the individual PK methods is noncompartmental analysis (NCA), which without any modelling connects the observed individual concentrations via linear interpolation. On the other hand, compartmental analysis is used for analysis of individual's PK data via compartmental models. Another model uses a standard two-stage (STS) method, where in the first stage, the data of each individual are included in the process of compartmental model evaluation. In the second stage, it combines individual parameter estimates to generate the mean (population) parameters and standard deviations. Parameters such as AUC and C_{max} are frequently explored by these individual PK methods. The software programs used for PK methods include Phoenix WinNonlin and PKSolver (though some software packages are suitable for both individual and population methods and modelling) (De Velde 2020).

2.4 Benefits and Limitations of Mathematical Modelling and Simulations: The Method and Its Potential Consequences on Clinical Practice

Despite the remarkable progress made in software development, certain challenges and limitations remain. When revisiting old antimicrobials to adjust the doses considering among the other factors higher MICs (and possibly also scenario of new categorisation of in vitro tests interpretation "susceptible, increased exposure"), we should be extremely careful as for new dose setting, especially when updating marketing authorisations (e.g. via referral procedures—for the whole EU). Therefore, very thorough case-by-case decision-making is vital, and the modelling cannot be used in a "unified" way but instead can be used only after carrying out a critical assessment and with big caution. This is especially true for the following cases (modified from ANSES 2017; EMA 2018a, De Velde 2020, Chou et al. 2022):

1. The antibiotic's efficacy cannot be linked to its level of plasma exposure, and proper PK parameters are not available for compartments whose PK/PD characteristic better fit with the infection site and the fate of the antimicrobial (e.g. PELF, ISF and milk).
2. The total free concentration available for an antimicrobial effect is not known and cannot even be reasonably extrapolated (protein binding levels are important here).
3. The PK parameters were gathered from a very small sample, so its representativeness is limited.
4. The PK parameters differ depending on the species, age (and therefore weight as well), production categories and breed because the different genetics and the data gathered will not be fully representative of the claimed population or indication. Despite those limitations, if the data for different species/categories are available, computational models capable of predicting input parameter values for analysis via already-validated PBPK models can allow for comparisons and a better understanding of differences and can inform more-advanced modelling techniques in the future.
5. The PK parameters can be influenced by several variables (Bon et al. 2018, modified).
   (a) The actual individual dose when medication is dispensed for a group in feed/water
   (b) Interspecies or interbreed variability
   (c) Age variability (can be covered in a specific (NLME) model as a covariate)
   (d) Bodyweight (can be covered in a specific (NLME) model as a covariate)
   (e) Hepatic and renal clearance (diseased vs healthy animals)
   (f) The disease severity–influencing ADME parameters
   (g) Particular disease differences and their influences on the ADME parameters in different compartments
   (h) Lactation
   (i) Others, e.g. chronobiology
6. A PTA acceptance level of 90% is sufficient from the perspective of claimed indications and dose selection (though such a level may not achieve the therapeutic concentration sufficient to treat a life-threatening infection for 1 in 10 animals),
7. Insufficient scientific evidence supports setting PDI and PDT in animals for target bacteria.
8. The MICs are not predictive of antibacterial activity, for example, for intracellular pathogens or in a biofilm environment.
9. The MIC does not include the other modes of action of certain antibiotics, e.g. anti-inflammatory and immunomodulatory activities, even when the substance has been identified (Berti et al. 2020).

10. The bioaccumulative properties of the substance in certain compartments is known.

In January 2021, the final version of a reflection paper (EMA 2018a) was published on dose review and adjustments to established veterinary antibiotics in "legacy" veterinary medicinal products that have long been on the market. It was released in the context of SPC harmonisation, as proposed by the new on regulation on VMPs: Regulation 2019/6/EC. In reality, pharmacovigilance announcements reporting failures of efficacy are extremely rare. Nevertheless, in practice, the dose is often increased, especially in cases of severe disease courses, due to concerns about antimicrobial resistance in the target pathogens and on the basis of the experiences of veterinarians/farmers who've encountered previous cases featuring the same clinical symptoms, where treatment failure was due insufficient dose or short duration of administration. With the submission of the draft of the reflection paper, thoughts on potential impacts on animal health and public health as a consequence of a revision of dosing schedules by using modelling have emerged. For practising veterinarians who are considering increasing doses of VMPs or increasing the frequency/duration of dosing, the following list can help inform them on the possible consequences not only on efficacy but especially on safety (target animal tolerance and consumer safety).

1. In clinical practice (based on EMA 2018a, amended and commented), a few factors need to be considered when increasing doses (or changing the dosing schedule) taking into account different perspectives:
   (a) *Animal tolerance* (the safety of the target animal): If the revision of the dosage implies an increase in the dose (and/or the frequency and/or total duration) based on the nature of the individual antimicrobial and on certain VMP qualitative and quantitative active and excipient compositions and physicochemical properties, a reassessment of the animal's tolerance is necessary.
      (i) New studies might be required, especially on injectable VMPs where the injection volume will increase significantly (to obtain the data that confirm/refuse presence of lesions, swelling and local tissue damage but also to quantify PK and residue depletion rates at the injection site after IM and SC administration). Even if the volume delivered to one injection site is limited and the injections should be administered to several injection sites, the animal's welfare must still be considered as well as practicability of administration. For example, if the volume administered was covered by a dose of 10 mL/ per animal (and one injection site with a maximum volume of 10 mL) and now the dose is being increased threefold, the new volume of 30 mL should be divided to three injection sites, each with 10 mL, for three consecutive days (it means nine injections in total). Issues of animal welfare, the ease of administration and possible residues at (multiple) injection sites and their surrounding areas will arise. In such cases, the VMP may need to be reformulated (e.g. a more-concentrated solution/ suspension), which means that the new VMP will need to undergo a completely new set of studies and that those studies will need to be assessed.
      (ii) In the case of orally dosed antimicrobials, once the data on an overdose are available for the VMP or at least for VMPs with the same qualitative and quantitative composition, assessment is easier then in previous case of injectable VMP (in those cases, solubility or palatability could be an issue). Strengthening the monitoring of animals treated with the new dosages should nonetheless be recommended.
   (b) *Antibiotic resistance*: Revising the dosages (dose, frequency and/or dosing

schedule) of older antibiotics by taking a PK/PD approach that integrates both pharmacokinetic variables (clearance and bioavailability) and pharmacodynamic variables (MICs and now also MPCs) must establish dosing schedules that guarantee, in the majority of animals treated, the exposure of the target bacterial population to an effective antibiotic concentration (with minimum pressure on the selection and spread of resistance). On the other hand, veterinarians are faced with a lack of information on how this increased dose might disrupt the balance of the gut microbiota. Therefore, new genetic technologies (e.g. whole genome sequencing) can reveal information on the extent of microbiota modification (especially the gut microbiota) of the target animal, taking into account the microbes present there before and after the treatment—and their proportions—and the prevalence of genes of resistance.

2. *The safety of the consumer* (residues in edible tissues taken from treated animals):
   (a) For the purpose of updating marketing authorisation, the withdrawal period for a new dosing schedule (orally administered antimicrobials) can be determined from tissue or plasma data, provided that they were supplied in the original dossier (using e.g. NLME modelling or (ig) PBPK). However, the most critical part of any modelling is the quality of the dataset inputted, which must be sufficient to carry out modelling and simulations. The need to increase the safety factor can arise to cover uncertainties/weaknesses of the modelling. If the conditions (as linear kinetics and the complete distribution and log-linearity of the elimination phase) are met, WP modelling can be used. Once the inputted dataset is not sufficiently robust, the value of the withdrawal period should be confirmed through an in vivo study. Increasing the dose in injectable antimicrobials (IM- and SC-administered antimicrobials) is even more complicated for the cases, in which the increased volume is injected into one injection site, which may have negative consequences on the target animal's local tolerance and on PK and residue depletion from the injection site. Finally, in these cases, even though the injection site was originally the target tissue, which determines the duration of the withdrawal period, given the increase in the dose, other tissue could now become target tissue too (worse consequences are expected for substances that have cumulative effects in certain tissues, for which MRLs are set—as e.g. kidney or liver), especially in diseased animals, whose renal and liver functions are often impaired. This should be thoroughly considered as it could lead to the finding residues exceeding MRLs (and in practice, it could pose risks to the consumer and may come with a fine imposed on the farmer or attending veterinarian). In addition to possible effects on meat and offal, the possible effects on the milk and eggs in the animal species concerned should be also reassessed. Examples of how to adjust the withdrawal period according to the new dosing schedule are given, for example, in EMA 2018a, ANSES 2017 and a series of articles on modelling and the citations contained in them (Riviere et al. 2015; Lin et al. 2016a; Bon et al. 2018; Li et al. 2018, Chou et al. 2022).
   (b) *In clinical practice*, each time a veterinarian decides to increase a dose or change the frequency or duration of a treatment, the total duration of the withdrawal period should be reconsidered. According to the legislation, namely Regulation 2019/6/EC, Article 115 (in effect since January 2022), after medicinal products have been used in a way outside the terms of authorisation, the withdrawal period should be prolonged, as indicated in Table 11.

Taking into account the health status of the treated animal is also critical. The dose is

Table 11 Summary of withdrawal periods for medicinal products used outside the terms of marketing authorisation for food-producing (FP) animal species, according to Regulation (EU) 2019/6, Article 115

Article 115/1	For the purpose of Art. 113 (terrestric FP) and Art. 114 (aquatic FP), unless a medicinal product used has a withdrawal period provided in its summary of the product characteristics for the animal species in question, a withdrawal period shall be set by the veterinarian following the below criteria:
Food commodity	The withdrawal period (WP) should *not be less than*
Meat and offal	The longest WP provided in the SPC of the product used for meat and offal, multiplied by 1.5
	28 days if the medicinal product is not authorised for food-producing animals
	One day if the medicinal product has a zero WP and is used in a different taxonomic family than the target species authorised
Milk	The longest WP for milk provided in the SPC of the product used for any animal species, multiplied by 1.5
	Seven days if the medicinal product is not authorised for animals producing milk for human consumption
	One day if the medicinal product has a zero WP
Eggs	The longest WP for eggs provided in the SPC of the product used for any animal species, multiplied by 1.5
	Ten days if the product is not authorised for animals producing eggs for human consumption
Aquatic species	The longest WP for any of the aquatic species indicated in the SPC of product used, multiplied by 1.5 and expressed as degree-days
	If the medicinal product is authorised for terrestrial food-producing animal species, the longest WP for any of the food-producing animal species indicated in the SPC, multiplied by a factor of 50 and expressed as degree-days, but not exceeding 500 degree-days
	500 degree-days if the medicinal product is not authorised for food-producing animal species
	25 degree-days if the highest withdrawal period for any animal species is zero
Bees	For bees, the veterinarian shall determine the appropriate withdrawal period by assessing the specific situation of the particular beehive or beehives on a case-by-case basis and in particular the risk of residue in honey or in any other foodstuffs harvested from beehives intended for human consumption.
Article 115/2	If the calculation of the withdrawal period according to the above stipulated rules leads to a fraction of days, the withdrawal period shall be rounded up to the nearest number of days.

usually increased in severe cases, where renal functions (and finally clearance), liver functions (and finally rate of metabolisation) and the animal's whole health status can complicate and prolong the depletion of the residues from the tissues of the treated animals, especially for antimicrobials with cumulative properties. On the other hand, the kinetics are also influenced by inflammatory processes, influencing tissue penetration rates and the crossing of biological barriers, which even further complicates considerations and increases the complexity of any model used for predicting effective doses or withdrawal periods. In injectable VMPs, the residues at the injection site and its surrounding tissues can be significantly higher if a larger volume is administered. In light of these facts, the final withdrawal period may need to be prolonged or multiplied and thoroughly assessed case by case, but it must always follow the legislation, whose key words are "not be less than." Especially for injectable VMPs that are in suspension and that have different compositions of excipients, even if the active substance of several VMPs is identical, with identical concentrations of the active substance per mL of the product and where the dose (volume injected) is therefore also identical for same weight category of certain animal species, the behaviour of excipients and the suspension's pharmacochemical characteristics may vary in each individual VMP, and this can affect residue depletion (especially from the injection site) and local tolerance. Moreover, attention should also be paid to toxicity, mainly in antimicrobials with narrow margins of safety—e.g. aminoglycoside gentamicin—especially in neonates lacking fully developed renal functions. Once the dose in cows, for example, has been increased during parturition or immediately thereafter, the feeding of newborn animals via colostrum/milk containing high levels of residues can be an issue, especially in that it may cause an imbalance in the gut microbiota of neonates, which could affect, among the other things, their immunostatus.

3. *Environmental safety*:

Environmental risk assessment outcomes can differ when the dose is increased, especially for the orally dosed antimicrobials used in herd or flock medications and especially for antimicrobials that persist in the environment. For many older medicinal products, it is therefore appropriate to refine phase I (which calculates exposure) or even going to phase II by providing experimental data to define the predictable no-effect concentrations (PNECs) and the data for calculating a refined predictable environmental concentration (PEC) is therefore appropriate as well. The PECs originally proposed by standard scenarios can be highly dependent on the character of the antimicrobial and on the dosage selected, leading to calculations of the risk quotient, or RQ: risk quotient = PEC/PNEC. In cases with an RQ > 1, a potentially high risk to the environment is indicated. Additional studies will be needed in such scenarios, first to refine the PEC by using the data on the degradation of antibiotics in livestock manure and second to refine the PNEC by used data from chronic toxicity studies (ANSES 2017).

2.5 AMR Considerations and the One Health Concept

The essential principles for minimising or preventing the selection and spread of resistance to antimicrobials in connection with the administration of antimicrobials are linked directly to pharmacokinetic and pharmacodynamic aspects. There can be mentioned several principal areas more or less touching one or both of these parameters:

(a) *Avoid the routine preventive use of antimicrobials.* Administering antimicrobials to groups/herds of animals on a routine basis to substitute for or mask poor-quality breeding conditions (having impact and finally leading to weakness or illness of neonates due to poor hygiene), insufficient care of animals, improper transportation conditions, stressful conditions, the mixing of animals from different sources (e.g. veal calves) or disturbances in animal welfare should be avoided (as per the legal rules stipulated by the Regulation (EU) 2019/6, Article 107/1) or at least minimised. Many times, routine preventive doses have been very low doses (subinhibitory concentrations), moreover administered over long periods. Such practices are considered unacceptable and not in line with the prudent and rational use of antimicrobials.

For example, the preventive (i.e. prophylactic) use of polymyxins in medicated feed (e.g. colistin) for broilers and pigs should be avoided. Importantly, according to the EMA-AMEG, colistin is classified as B-restrict because of its importance in human medicine. According to Regulation (EU) 2019/4, such prophylactic use of antibiotics via medicated feed has already been banned, as of January 2022. Other nonantimicrobial tools should be implemented to avoid as much as possible the use of colistin (e.g. via medicated drinking water).

Similarly, the routine use of dry cow intramammary VMPs for each cow in a herd during the dry period is no longer considered an acceptable practice in dairy farming, and farmers and veterinarians should invest in efforts to improve herd and individual animals' health status, care and welfare and to promote biosecurity and hygienic rules on the farm to minimise the use of antimicrobials at drying off.

(b) *Exceptional preventive use of antimicrobial*, if needed, should be limited to well-defined conditions, which should be carefully evaluated to determine how to minimise the risk of repeating similar or the same situations. Exceptional cases are those linked with justifiably expected serious consequences on animal health and welfare and on the spread of pathogens and AMR. These situations can be linked not only to poor animal health with suffering of animals but also to the spread of zoonotic, resistant or multiresistant causal microbial agents.

Whenever possible, the situation should be analysed and measures carefully set and implemented into routine practice on the farm to avoid the repetition of critical situations requiring the preventive use of antimicrobials in groups/herds of animals.

Many measures are more costly than the routine administration of antimicrobials, and a tailored approach featuring good cooperation among veterinarians and farmers is essential. For these reasons, improving training and education on such situations and exchanging examples of good practices would be beneficial.

For example, after an accident in a ventilation system, caused by a power outage, the level of ammonia increased in a housing facility for prefinishing pigs. Respiratory infections caused by, for example, *A. pleuropneumoniae* (App) would be expected. Antimicrobials could be used in this scenario for disease prevention, but another preventive measures, such as installing an alarm system to detect when the concentration of ammonia has exceeded acceptable levels and by using an alternative power source that connects to all the devices/equipment necessary to run the various farm technologies to improve air quality and to overcome such critical situations, would be more beneficial. Biosecurity measures and the risk of App should be monitored within herds. Vaccination strategies are also powerful tools. These all above noted measures need innovations in management and investments into the new or improved technologies, so they are more costly than individual interventions using antimicrobials, what cause one of the most critical obstacle to succeed.

(c) *Minimise metaphylactic use* and limit it to well-defined cases of the first clinical signs of a disease occurence within defined proportion of animals in the herd or flock, where the proportion is based on the most recent data on the risks associated with the disease's course and its spread.

For example, regularly check the health status, at intervals necessary for the age/production category and the type of housing facility and available farm technology. Smart technologies can be used to indicate disturbances in environmental conditions and in the comfort of animals and to detect early signs of the disease in animals in the group (especially in extensive farming operations). Experienced and well-trained staff should be responsible for such checks. Whenever possible, immediately remove animals showing signs (e.g. move them to quarantine boxes/sections); in poultry flocks, culling may be necessary to minimise the speed at which the infection spreads. After euthanasia, a postmortem clinical examination should be done, and if appropriate, samples should be taken for further laboratory examination. If the signs are typical for a disease, administer antimicrobials at an early stage, before the signs or the disease becomes more serious. Nevertheless, take into account the fact that samples should be taken at just the moment before the first administration so that the laboratory can confirm both the etiological agent and its susceptibility, which can be helpful in the case of treatment failure and from a longer-term perspective helps in identifying trends in resistance development in the pathogen of concern, on the original farm or on epidemiologically related farms.

(d) *Antimicrobials should be used only* after a thorough diagnosis has been performed by the attending veterinarian and only *if infection is reasonably expected but ideally* when it has been *firmly confirmed*. Any use of antimicrobials poses a risk of resistance selection.

For example, although some diseases can exhibit the same signs or very similar signs, veterinarians should be consulted when establishing the correct diagnose. Veterinarians, based on their expertise, can indicate whether the start of antimicrobial treatment is urgent. Once feasible rapid diagnostic techniques able to indicate the presence of infection or even in more detail, on farm culturing systems to detect pathogen involved are favoured (e.g. in mastitis, except for somatic cell counting, farm culturing systems or multiplex PCR methods can help to indicate the causative pathogen). Once culturing is positive, antimicrobial susceptiblity testing (AST) is recommended to be performed by an accredited laboratory before starting treatment (or at least this can serve as advice in the event of treatment failure, or it can provide evidence of susceptibility and resistance in a dairy herd over time). Correct sampling techniques and sampling before the first treatment administration are crucial. Without samples, no overview can be conducted either for a given region or for a given farm, and antimicrobial use therefore cannot be based on evidence or properly targeted.

Perioperative prophylaxis use in horses during highly invasive surgery is standard, but it should not be used without using every other tool for ensuring the highest hygienic standards and following standard operational procedures for proper disinfection and guaranteeing aseptic conditions in the operating facilities and the postsurgical care facilities. Conditions in horse clinics should be thoroughly monitored, by conducting bacteriological sampling not only of the animals also of the local environment and the staff if necessary. Trends in resistance should be thoroughly monitored and evaluated and appropriate measures taken, especially if nosocomial multi-drug resistant (MDR) strains are found in an equine clinic. Rules on the responsible use of antimicrobials in equine clinics (BEVA 2018) and antimicrobial stewardship rules should be followed (Isgren 2021).

(e) Pay specific attention to cases where antimicrobials are frequently used on a farm and analyse them. *Try to find and implement preventive measures tailored precisely to the farm and husbandry conditions.* Prioritise measures that can be carried out immediately; motivate the people taking care of animals and show them why. Analyse and follow the appropriate steps.

For example, start scoring mastitis before starting any measures. Start to improve udder and milking hygiene; identify critical points of care and management for an individual herd; write simple manuals; buy devices, materials and disinfectants and train staff on how to use them; incentivise staff to perform better; evaluate staff hygiene; identify points where people frequently fail to adhere to these hygienic measures; and re-educate staff on these critical points. At various intervals, score staff performance of this new scheme. Check the herd status, such as by conducting bulk tank tests for microbiological quality and conducting somatic cell counts (SCCs) on farm testing and the culturing of mastitis pathogens, and cooperate with laboratories to detect antimicrobial susceptibility. Collect data on mastitis cases, such as those on the use of antimicrobials and herd health status. The preceding examples are not exhaustive, and dairy husbandry management, hygiene, feeding, housing and care conditions are complex. For more information, see also Chapter Cattle (on cattle and mastitis).

(f) *Early start of the treatment.* Thanks to improved diagnostics (and with help of smart farming sensors and well-experienced and -trained staff and with cooperation with the attending veterinarian), treatments can be started even earlier than before. Identifying the logarithmic growth phase of microbial infections is crucial. Some classes of antimicrobials (e.g. beta-lactams) have a bactericidal effect on growing bacteria. In some cases, preventing biofilm formation is particularly important because some biofilms are significantly difficult or even impossible to effectively treat by trying to deliver the necessary concentrations of antimicrobials at the site of infection.

For example, mastitis and lameness are two frequently occurring diseases in dairy cattle that require the administration of antimicrobials. Systems that can detect indicators of the onsets of such diseases in their early stages have been already introduced (Post et al. 2020, 2021). In combination with the experience and expertise of the staff who take care of the animals, this can be a powerful tool for decision-making in general and for other specific decisions, such as those on early treatment starts (important for certain pathogens with high host specificity and an ability to form biofilms), especially if cross-linked with, validated by and evaluated with sampling, culturing and tests on antimicrobial susceptibility. Even today, although "robots" and artificial intelligence are very helpful, they cannot fully replace the empathetic, experienced people who take care of animals.

(g) *Set the proper dosing schedule (dose, frequency and duration), and use the most appropriate route of administration*. For that, having a recent overview of the MICs for pathogens causing certain (serious, acute) infections, which cannot wait for laboratory results on actual cases, is vital. Guidelines usually advise being prepared to implement this knowledge once the outbreak of an infection occurs in the flock/herd. It means that a *sufficiently high dose* (ideally one factoring in the MICs and being above the mutation selection window and at the mutation prevention concentration) must be set. *Underdosing (low doses) usually select for resistance*. Such an approach to determining the proper dosing schedule and best route of administration is very difficult in practice, because in veterinary medicine for food-producing animals, each increase in the dose requires the veterinarian to reconsider the withdrawal period. Moreover, in many cases, especially in "legacy" VMPs that contain "old" first-line molecules and that were authorised a long time ago, it leads to "off-label use," which requires significantly prolonging the withdrawal period. When medicating groups of animals with, for example, medicated water, ensuring that each individual animal receives a sufficient dose is important.

The *proper frequency* of individual doses should be followed whenever possible. This can also be challenging when following a drinking regimen (especially because of the environmental conditions) and because of how sick animals behave towards drinking (or, even worse, towards feeding). Sufficient intake of medicated water or medicated feed in appropriate interval should be ensured to avoid underdosing. Also, stability data and the instructions for a specific VMP (especially for medicated water) should be strictly followed to avoid the decrease of the amount of the active substance, such as after stock solution has been stored.

Avoid prolonging administration (especially that of low doses).

Avoid overly shortening administration (especially that of bacteriostatic antimicrobials, because that would allow for the regrowth of the bacterial population, from which bacteria that were not sufficiently inhibited or that became persisters are selected, increasing the risk of AMR). Also, keep the administration sufficiently long in immunocompromised animals, chronic infections and severe infections that tend to recur.

Take, *for example*, the case of enrofloxacin administered in poultry flocks for APEC infections. Take an overview of the $MICs_{50}$ and $MICs_{90}$ in recent isolates from day-old chicks from recent stocks from same parent flock that the chicks (regularly) come from. Take samples from dead/diseased birds for bacteriological examination and antimicrobial susceptibility testing (which not only will serve as information on the treatment failure but also will generate results in the literature and encourage investigations into trends in resistance, including developments in MIC values). Determine which dose of enrofloxacin will exceed the MIC and be outside the MSW. For enrofloxacin, for which the kill of pathogens is concentration dependent, calculating the dose on the basis of the average body weight of the animals and their water intake is important. Some basic rules for this are as follows:

1. Use the proper water quality, check for dissolution and avoid incomplete dissolution or precipitation.
2. Be sure that the drinking devices are clean—free of algae and biofilms.
3. Set up drinking lines in appropriate positions so that drinking is comfortable for all animals.
4. Regularly check the health status of animals for any improvement.
5. If no improvements are seen or if the health status worsens, look at the results of antimicrobial susceptibility testing and use this information when considering which antimicrobial to switch to.
6. Analyse the possible reasons for the disease, the course of the disease, the doses administered and the antimicrobial susceptibility tests to be better prepared for similar situations.
7. Try to find tools to avoid such disease, if possible.

(h) *Evaluate the clinical efficacy* of the treatment against the disease. Depending on the seriousness of the disease, follow the whole clinical status and factors such as fever, appetite and the consistency of stools. In food-producing animals, the health and welfare status of the animals together with the clinical endpoints should be evaluated to help to return the animals to production (milk and eggs) or growth performance as quickly as possible after they have recovered from the disease. Depending on the type of infection and the antimicrobial used, improvement should be visible after 3–4 days after initiating the treatment.

For example, in the case of clinical mastitis in dairy cows, the infectious and inflammatory process is associated with pain, discomfort, an increased risk of culling and death and also economic losses for farmers. Despite clinical successes, in one study, local clinical symptoms were detected to persist after antibiotic treatment (mild status: not full recovery) for at least a week, which may have affected the welfare of the cows (Fogsgaard 2015). Milk yield and inflammatory markers from 174 mastitic lactations were compared against 858 control lactations. It was shown that milk yield, interquarter milk yield ratio and lactate dehydrogenase activity stabilised within one to three weeks after the antibiotic treatment. On the other hand, premastitis production levels were not achieved during the observation period of eight weeks after mastitis (Fogsgaard 2015). Using pain-relief VMPs (e.g. nonsteroids) can be concomitant with or followed after a treatment using antimicrobials and together with improving the care of animals and using specific feed can improve both the well-being and the immune-defence and healing processes to help return animals to (near) original production status.

(i) Prefer the use of as narrow as possible-spectrum antimicrobials (over broad-pectrum antimicrobials and over unnecessarily broad-spectrum combinations of antimicrobials).

For example, use narrow-spectrum penicillins preferentially over amoxicillin (which can select for extended-spectrum beta-lactamases), or prefer amoxicillin alone (being narrower-spectrum) to, combination of amoxicillin and clavulanic acid (as a choice for pathogens producing penicillinases).

3 Conclusion

Antimicrobials have been used in veterinary medicine mostly for prophylaxis and metaphylaxis (with oral forms of VMPs for group and herd/flock medication), as documented by the data on total volumes of antimicrobials in the ESVAC report containing the data from 31 European countries on the veterinary antimicrobials consumption (available on: https://www.ema.europa.eu/en/documents/report/sales--veterinary-antimicrobial-agents-31-european-countries-2022-trends-2010-2022-thirteenth-esvac-report_en.pdf), and less part of the total volume was used more individually, for the targeted treatment of bacterial diseases (for more context and deeper information and explanation please refer to the report). In addition to two key factors, firstly the choice of the correct antimicrobial VMP and secondly the exact dose, frequency and overall duration of treatment, it is essential for practicing clinicians to consider when to start the treatment with antimicrobial. According to the trends in susceptibility and antimicrobial's MICs—at least those available for certain pathogens isolated from certain diseases affecting certain animal species—and according to some clinical studies, current doses as approved in authorised VMPs are sometimes far lower than those needed to effectively treat the pathogen. Despite that, reports of therapeutic failures from everyday clinical practice remain rare. This apparent contradiction could be related to the proportion of prophylactic and metaphylactic uses comparing to treatment use. Another factor contributing to the low reporting of failure might be the fact that the outbreak of a disease does not occur or progress of the diseases is minimal in animals under prophylaxis or metaphylaxis. In metaphylaxis, the disease is caught in the starting, milder phase, when the success of using antimicrobials corresponds with a low load of bacteria and a low degree of damage to an animal's normal physiological status, where mechanisms of immune defence in animals can contribute to the clinical treatment. In some diseases, relatively high proportions of spontaneous recoveries can also be seen. A question therefore remains, if the current lower dose can still be maintained e.g. for metaphylaxis, but for the treatment of more-serious diseases, in the later stage of progress, an increase in the dose should be considered. For such cases, PK/PD modelling based on the data gathered from diseased animals already dosed with increased doses will be more appropriate to confirm the optimised dose. The PK/PD approach provides no insights into the duration of the treatment, so the increased dose does not automatically mean that the treatment course can be shortened. This should be confirmed by results from clinical studies confirming bacteriological and clinical cure effectiveness. New software and new approaches in PK/PD modelling could improve dosing, model or simulate scenarios to increase dose efficacy and at the same time minimise mutation rates. Although more-advanced modelling systems have been developed and although machine-learning and artificial-intelligence technologies are planned to be used for the (re)assessment or establishment of doses (Chou and Lin 2023), several factors influencing clinical efficacy still cannot be fully incorporated into this modelling. New models also need to be validated with already well-established dosing regimens, with available appropriate PK (in the target tissues of infection) and PD parameters, with the other important characteristics of animals and diseases and with all the factors influencing the efficacy of antimicrobial treatments. Results from such models still need to be supported by data from clinical studies, which remains the main barrier because veterinary pharmacokinetic data are relatively limited or scattered, and some of such data are old or difficult to find; collecting sufficient data to develop reliable models for different species thus remains difficult (Lin and Wong 2017). Therefore, publicly available, comprehensive veterinary pharmacokinetic databanks would be very helpful and need to be developed. In addition to these datasets, the availability of guidelines on modelling, user-friendly software tools and graduate education programmes would enhance the use of modelling to optimise doses in the future. Finally, even if all these advancements are achieved, experienced and continuously educated veterinarians, who have a "sense of diagnosis and art of medicine," are key to the appropriate and responsible use of antimicrobials in practice, and PK and PD characteristics should serve as helpful tools for them.

References

Agersø H, Friis C (1998) Penetration of amoxycillin into the respiratory tract tissues and secretions in pigs. Res Vet Sci 64(3):245–250

Ahmad I, Huang L, Hao H, Sanders P, Yuan Z (2016) Review article application of PK/PD modelling in veterinary field: dose optimization and drug resistance prediction. BioMed Res Int 5465678:1–12

Ahmed TA (2015) Pharmacokinetics of drugs following IV bolus, IV infusion, and oral administration. Basic pharmacokinetic concepts and some clinical applications. https://www.intechopen.com/books/basic-pharmacokinetic-concepts-and-some-clinical-applications/pharmacokinetics-of-drugs-following-iv-bolus-iv-infusion-and-oral-administration. Accessed 20 Mar 2023

Alagga AA, Gupta V (2020) Drug absorption. [Updated 2020 May 14]. In: StatPearls [Internet]. Treasure Island (FL): StatPearls Publishing. https://www.ncbi.nlm.nih.gov/books/NBK557405. Accessed 20 Mar 2023

Aliabadi FS, Ali BH, Landoni MF, Lees P (2003) Pharmacokinetics and PK-PD modelling of danofloxacin in camel serum and tissue cage fluids. Vet J 165:104–118

Ambrose PG, Bhavnani SM, Rubino CM, Louie A, Gumbo T, Forrest A, Drusano GL (2007) Pharmacokinetics-pharmacodynamics of antimicrobial therapy: it's not just for mice anymore. Clin Infect Dis 44(1):79–86

Amsden GW (2001) Advanced-generation macrolides: tissue-directed antibiotics. Int J Antimicrob Agents 18(Suppl 1):11–15

Anadon A, Martınez- Larranaga MR, Dıaz MJ, Fernandez-Cruz ML, Martınez MA, Frejo MT, Martınez M, Iturbe J, Tafur M (1999) Pharmacokinetic variables and tissue residues of enrofloxacin and ciprofloxacin in healthy pigs. Am J Vet Res 60:1377–1382

Andes D, Craig WA (2002) Animal model pharmacokinetics and pharmacodynamics: a critical review. Int J Antimicrob Agents 19(4):261–268

ANSES (2017) Methodology for revising the dosages of older antibiotics, ANSES Opinion Collective Expert Report. https://www.anses.fr/en/system/files/ANMV2014SA0080RaEN.pdf. Accessed 28 Mar 2023

Aschbacher PW, Feil VJ (1994) Neomycin metabolism in calves. J Anim Sci 72(3):683–689

Ayuso M, Buyssens L, Stroe M, Valenzuela A, Allegaert K, Smits A, Annaert P, Mulder A, Carpentier S, Van Ginneken C, Van Cruchten S (2020) The neonatal and juvenile pig in pediatric drug discovery and development. Pharmaceutics 13(1):44

Azman M, Sabri AH, Anjani QK, Mustaffa MF, Hamid KA (2022) Intestinal absorption study: challenges and absorption enhancement strategies in improving oral drug delivery. Pharmaceuticals (Basel) 15(8):975

Baggot JD (2001) Physiological basis of veterinary clinical pharmacology. Blackwell Science

Baggot JD (2013) In Giguère S, Prescott JF, Dowling PM (eds) Antimicrobial therapy in veterinary medicine, 5th edn. Wiley

Baggot JD, Short CR (1984) Drug disposition in the neonatal animal, with particular reference to the foal. Equine Vet J 16(4):364–367

Baldwin DR, Honeybourne D, Wise R (1992) Pulmonary disposition of antimicrobial agents: in vivo observations and clinical relevance. Antimicrob Agents Chemother 36(6):1176–1180

Barczak AK, Hung DT (2009) Productive steps toward an antimicrobial targeting virulence. Curr Opin Microbiol 12(5):490–496

Beer J, Wagner CC, Zeitlinger M (2009) Protein binding of antimicrobials: methods for quantification and for investigation of its impact on bacterial killing. AAPS J 11(1):1–12

Berti A, Rose W, Nizet V, Sakoulas G (2020) Antibiotics and innate immunity: a cooperative effort toward the successful treatment of infections. Open Forum Infect Dis 7(8):ofaa302

BEVA (2018) British Equine Veterinary Association: "Protect ME – the award-winning Antibiotic use toolkit from BEVA". https://www.beva.org.uk/protectme. Accessed 28 May 2023

Blondeau JM, Fitch SD (2019) Mutant prevention and minimum inhibitory concentration drug values for enrofloxacin, ceftiofur, florfenicol, tilmicosin and tulathromycin tested against swine pathogens *Actinobacillus pleuropneumoniae, Pasteurella multocida* and *Streptococcus suis*. PLoS One 14(1):e0210154

Blondeau JM, Shebelski SD (2016) Comparative in vitro killing of canine strains of *Staphylococcus pseudintermedius* and *Escherichia coli* by cefovecin, cefazolin, doxycycline and pradofloxacin. Vet Dermatol 27(4):267

Blondeau JM, Zhao X, Hansen G, Drlica K (2001) Mutant prevention concentrations of fluoroquinolones for clinical isolates of *Streptococcus pneumoniae*. Antimicrob Agents Chemother 45(2):433–438

Bokma J, Boone R, Deprez P, Pardon B (2019) Risk factors for antimicrobial use in veal calves and the association with mortality. J Dairy Sci 102(2019):607–618

Bon C, Toutain PL, Concordet D, Gehring R, Martin-Jimenez T, Smith J, Pelligand L, Martinez M, Whittem T, Riviere JE, Mochel JP (2018) Mathematical modelling and simulation in animal health. Part III: using nonlinear mixed-effects to characterize and quantify variability in drug pharmacokinetics. J Vet Pharmacol Ther 41(2):171–183

Böttcher T, Ren H, Goiny M, Gerber J, Lykkesfeldt J, Kuhnt U, Lotz M, Bunkowski S, Werner C, Schau I, Spreer A, Christen S, Nau R (2004) Clindamycin is neuroprotective in experimental *Streptococcus pneumoniae* meningitis compared with ceftriaxone. J Neurochem 91(6):1450–1460

Brown SA (1992) Pharmacology in neonatal calves, American Association of Bovine Practitioners, The Bovine Proceedings No 24: 96–100

Burch DGS (2012) Examination of the pharmacokinetic/pharmacodynamic (PK/PD) relationships of orally administered antimicrobials and their correlation with the therapy of various bacterial and mycoplasmal infections in pigs. Fellowship thesis. Royal College of Veterinary Surgeons, London, UK

Burch DGS, Hammer JM (2013) Managing Lawsonia and Brachyspira infections using pharmacokinetic and Pharmacodynamic principles. American Association of Swine Veterinarians, San Diego, CA

Burch DGS, Sperling D (2018) Amoxicillin – current use in swine medicine. Journal of veterinary pharmacology and therapeutics. J Vet Pharmacol Ther 41:356–368

Cantón R, Morosini MI (2011) Emergence and spread of antibiotic resistance following exposure to antibiotics. FEMS Microbiol Rev 35(5):977–991

Carryn S, Chanteux H, Seral C, Mingeot-Leclercq MP, Van Bambeke F, Tulkens PM (2003) Intracellular pharmacodynamics of antibiotics. Infect Dis Clin N Am 17:615–634

Chai MG, Cotta MO, Abdul-Aziz MH, Roberts JA (2020) What are the current approaches to optimising antimicrobial dosing in the intensive care unit? Pharmaceutics 12(7):638

Choe S, Lee D (2017) Parameter estimation for sigmoid Emax models in exposure-response relationship. Transl Clin Pharmacol 25(2):74–84

Chou WC, Lin Z (2023) Machine learning and artificial intelligence in physiologically based pharmacokinetic modelling. Toxicol Sci 191(1):1–14

Chou WC, Tell LA, Baynes RE, Davis JL, Maunsell FP, Riviere JE, Lin Z (2022) An interactive generic physiologically based pharmacokinetic (igPBPK) modelling platform to predict drug withdrawal intervals in cattle and swine: a case study on Flunixin, Florfenicol, and Penicillin G. Toxicol Sci 188(2):180–197

Cocco A, Vernacchio L, Horst M, Anderson A (2010) Management of acute otitis media after publication of the 2004 AAP and AAFP clinical practice guideline. Pediatrics 125(2):214–220

Coelho L (2009) C-reactive protein and procalcitonin in the assessment of infection response to antibiotic therapy. https://acutecaretesting.org/en/articles/creactive-protein-and-procalcitonin-in-the-assessment-of-infection-response-to-antibiotic-therapy. Accessed 28 May 2023

Constable P (2017) In: Constable P, Hinchcliff KW, Done S, Gruenberg W (eds) Practical antimicrobial therapeutics in book: veterinary medicine: a textbook of diseases of cattle, horses, sheep, pigs and goats, 11th edn. Saunders, pp 153–174

Cox SR, McLaughlin C, Fielder AE, Yancey MF, Bowersock TL, Garcia-Tapia D, Bryson L, Lucas MJ, Robinson JA, Nanjiani I, Brown SA (2010) Rapid and prolonged distribution of tulathromycin into lung homogenate and pulmonary epithelial lining fluid of Holstein calves following a single subcutaneous administration of 2.5 mg/kg body weight. Int J Appl Res Vet Med 8:129–137

Craig WA (1998) Pharmacokinetic/pharmacodynamic parameters: rationale for antibacterial dosing of mice and men. Clin Infect Dis 26(1):1-10; quiz 11-2

Craig WA (2003) Basic pharmacodynamics of antibacterials with clinical applications to the use of beta-lactams, glycopeptides, and linezolid. Infect Dis Clin N Am 17:479–501

Craig WA (2011) Optimizing aminoglycoside use. Crit Care Clin 27(1):107–121

Cui J, Liu Y, Wang R, Tong W, Drlica K, Zhao X (2006) The mutant selection window in rabbits infected with Staphylococcus aureus. J Infect Dis 194:1601–1608

Da Silva RG, Reyes FG, Sartori JR, Rath S (2006) Enrofloxacin assay validation and pharmacokinetics following a single oral dose in chickens. J Vet Pharmacol Ther 29(5):365–372

De Briyne N, Atkinson J, Pokludová L, Borriello SP, Price S (2013) Factors influencing antibiotic prescribing habits and use of sensitivity testing among veterinarians in Europe. Vet Rec 173:475

De Velde F (2020) Dosing optimization of Beta-lactam antibiotics using parametric and nonparametric population pharmacokinetic models. Dissertation Thesis – Erasmus University Rotterdam, GVO Drukkers & Vormgevers, EdePrinting

De Velde F, Mouton JW, de Winter BCM, van Gelder T, BCP K (2018) Clinical applications of population pharmacokinetic models of antibiotics: challenges and perspectives. Pharmacol Res 134:280–288

Drlica K (2003) The mutant selection window and antimicrobial resistance. J Antimicrob Chemother 52(1):11–77

Duquette SC, Fischer CD, Williams AC, Sajedy S, Feener TD, Bhargava A, Reti KL, Muench GP, Morck DW, Allison J, Lucas MJ, Buret AG (2015) Immunomodulatory effects of tulathromycin on apoptosis, efferocytosis, and proinflammatory leukotriene B4 production in leukocytes from Actinobacillus pleuropneumoniae-or zymosan-challenged pigs. Am J Vet Res 76(6):507–519

EMA (2011) European Medicines Agency: Reflection paper on the use of macrolides, lincosamides and streptogramins (MLS) in food-producing animals in the European Union: development of resistance and impact on human and animal health. https://www.ema.europa.eu/en/documents/scientific-guideline/reflection-paper-use-macrolides-lincosamides-streptogramins-mls-food-producing-animals-european_en-0.pdf. Accessed 28 May 2023

EMA (2015a) European Medicines Agency: Gentamicin-referral article 35 veterinary medicinal products containing gentamicin presented as solution for injection to be administered to horses. https://www.ema.europa.eu/en/documents/referral/gentamicin-article-35-veterinary-medicinal-products-containing-gentamicin-presented-solutions_en.pdf. Accessed 28 May 2023

EMA (2015b) European Medicines Agency: Article 35 of Directive 2001/82/EC for all veterinary medicinal products containing colistin as a sole active substance to be administered orally. https://www.ema.europa.eu/en/documents/referral/colistin-oral-article-35-referral-annex-i-ii-iii_en.pdf. Accessed 28 May 2023

EMA (2016a) European Medicines Agency: Guideline on Demonstration of efficacy for veterinary medicinal products containing antimicrobial substances, EMA/CVMP/627/2001-Rev1. https://www.ema.europa.eu/en/demonstration-efficacy-veterinary-medicinal-products-containing-antimicrobial-substances. Accessed 28 May 2023

EMA (2016b) European Medicines Agency: Guideline on the use of pharmacokinetics and pharmacodynamics in the development of antimicrobial medicinal products, EMA/CHMP/594085/2015. https://www.ema.europa.eu/en/documents/scientific-guideline/guideline-use-pharmacokinetics-pharmacodynamics-development-antimicrobial-medicinal-products_en.pdf. Accessed 15 May 2023

EMA (2017) European Medicines Agency: Guideline on the conduct of efficacy studies for intramammary products for use in cattle, EMA/CVMP/344/1999-Rev.2. https://www.ema.europa.eu/en/documents/scientific-guideline/guideline-conduct-efficacy-studies-intramammary-products-use-cattle_en.pdf. Accessed 28 May 2023

EMA (2018a) European Medicines Agency: Reflection paper on dose optimisation of established veterinary antibiotics in the context of SPC harmonisation, EMA/CVMP/849775/2017. https://www.ema.europa.eu/en/documents/scientific-guideline/draft-reflection-paper-dose-optimisation-established-veterinary-antibiotics-context-summary-product_en.pdf. Accessed 28 May 2023

EMA (2018b) Reflection paper on use of aminoglycosides in animals in the European Union: development of resistance and impact on human and animal health, EMA/CVMP/AWP/721118/2014. https://www.ema.europa.eu/en/documents/scientific-guideline/reflection-paper-use-aminoglycosides-animals-european-union-development-resistance-impact-human_en.pdf. Accessed 28 May 2023

EMA (2018c) Guideline on the conduct of bioequivalence studies for veterinary medicinal products, EMA/CVMP/016/2000-Rev.3-corr*. Guideline on the conduct of bioequivalence studies for veterinary medicinal products (europa.eu). Accessed 19 June 2023

EMA (2019) Categorisation of antibiotics in the European Union: Answer to the request from the European Commission for updating the scientific advice on the impact on public health and animal health of the use of antibiotic in animals, EMA/CVMP/CHMP/682198/2017. https://www.ema.europa.eu/en/documents/report/categorisation-antibiotics-european-union-answer-request-european-commission-updating-scientific_en.pdf Accessed 19 June 2023

EMA – DRAXXIN (2020) European Medicines Agency: Product information DRAXXIN EPAR. https://www.ema.europa.eu/en/documents/product-information/draxxin-epar-product-information_en.pdf. Accessed 28 May 2023

EMA – NAXCEL (2020) European Medicines Agency: Product information Naxcel (cattle) EPAR. https://www.ema.europa.eu/en/documents/product--information/naxcel-epar-product-information_en.pdf. Accessed 28 May 2023

EMA – ZACTRAN (2020) European Medicines Agency: Product information Zactran EPAR. https://www.ema.europa.eu/en/documents/product-information/zactran-epar-product-information_en.pdf. Accessed 28 May 2023

EMA – ZUPREVO (2016) European Medicines Agency: Product information Zuprevo EPAR. https://www.ema.europa.eu/en/documents/product-information/zuprevo-epar-product-information_en.pdf. Accessed 28 May 2023

EMEA (2006) Scientific Discussion: Summary of the Dossier, Convenia, EMEA/CVMP/215997/2006. https://www.ema.europa.eu/en/documents/scientific--discussion/convenia-epar-scientific-discussion_en.pdf. Accessed 19 June 2023

EP/EC (2001) European Parliament and European Council: Directive 2001/82/EC on the community code relating to veterinary medicinal products. http://eur-lex.europa.eu/LexUriServ/LexUriServ.do?uri=CELEX:32001L0082:en:NOT. Accessed 28 May 2021

EP/EC (2019) Regulation (EU) 2019/6 of the European Parliament and of the Council of 11 December 2018 on veterinary medicinal products and repealing Directive 2001/82/EC. https://eur-lex.europa.eu/legal-content/en/TXT/?uri=CELEX:32019R0006. Accessed 28 May 2023

Erskine RJ, Bartlett PC, VanLente JL, Phipps CR (2002) Efficacy of systemic Ceftiofur as a therapy for severe clinical mastitis in dairy cattle. J Dairy Sci 85:2571–2575

Ersoy SC, Heithoff DM, Barnes LV, Tripp GK, House JK, Marth JD, Smith JW, Mahan MJ (2017) Correcting a fundamental flaw in the paradigm for antimicrobial susceptibility testing. EBioMed 20:173–181

Ersoy SC, Abdelhady W, Li L, Chambers HF, Xiong YQ, Bayer AS (2019) Bicarbonate resensitization of methicillin-resistant *Staphylococcus aureus* to β-lactam antibiotics. Antimicrob Agents Chemother 63:496–419

FDA (2018) Highlights of prescribing information: ZEMDRI (plazomicin) injection, for intravenous use Initial U.S. Approval: 2018. https://www.accessdata.fda.gov/drugsatfda_docs/label/2018/210303orig1s000lbl.pdf. Accessed 28 Mar 2023

Fidler M, Wilkins JJ, Hooijmaijers R, Post TM, Schoemaker R, Trame MN, Xiong Y, Wang W (2019) Nonlinear mixed-effects model development and simulation using nlmixr and related R open-source packages. CPT Pharmacometrics Syst Pharmacol 8(9):621–633

Fink-Gremmels J (2008) Implications of hepatic cytochrome P450-related biotransformation processes in veterinary sciences. Eur J Pharmacol 585:502–509

Firth EC, Nouws JF, Driessens F, Schmaetz P, Peperkamp K, Klein WR (1986) Effect of the injection site on the pharmacokinetics of procaine penicillin G in horses. Am J Vet Res 47(11):2380–2384

Fischer CD, Beatty JK, Stephanie C, Morck DW, Lucas MJ, Buret AG (2013) Direct and indirect anti-inflammatory effects of tulathromycin in bovine macrophages: inhibition of CXCL-8 secretion, induction of apoptosis, and promotion of efferocytosis. Antimicrob Agents Chemother 57(3):1385–1393

Fissell WH (2013) Antimicrobial dosing in acute renal replacement. Adv Chronic Kidney Dis 20(1):85–93

Fogsgaard KK (2015) Recovery from mastitis in dairy cows – development of behaviour, milk production and inflammatory markers in the weeks during and after naturally occurring clinical mastitis. Ph.D. thesis AARHUS University Press, 113 p

Foster DM, Martin LG, Papich MG (2016) Comparison of active drug concentrations in the pulmonary epithelial lining fluid and interstitial fluid of calves injected with Enrofloxacin, Florfenicol, Ceftiofur, or Tulathromycin. PLoS One 11(2):e0149100. Erratum in: PLoS One 11(7):e0159219

Giguère S, Tessman RK (2011) Rational dosing of antimicrobial agents for bovine respiratory disease: the use of plasma versus tissue concentrations in predicting efficacy. Int J Appl Res Vet Med 9(4):342–355

Giguère S, Prescott JF, Dowling PM (2013) Antimicrobial therapy in veterinary medicine, 5th edn. Wiley

Goto Y, Fukunari K, Suzuki T (2023) Multiplex RT-qPCR application in early detection of bovine respiratory disease in healthy calves. Viruses 15(3):669

Gu M, Zhang N, Zhang L, Xiong M, Yang Y, Gu X, Shen X, Ding H (2015) Response of a clinical Escherichia coli strain to repeated cefquinome exposure in a piglet tissue-cage model. BMC Vet Res 11:169

Guyonnet J, Manco B, Baduel L, Kaltsatos V, Aliabadi MH, Lees P (2010) Determination of a dosage regimen of colistin by pharmacokinetic/pharmacodynamic integration and modelling for treatment of G.I.T. Disease in pigs. Res Vet Sci 88:307–314

Hansen E, Karslake J, Woods RJ, Read AF, Wood KB (2020) Antibiotics can be used to contain drug-resistant bacteria by maintaining sufficiently large sensitive populations. PLoS Biol 18(5):e3000713

Hémonic A, Hugues L, Corrégé I (2016) Medication via drinking water by a dosing pump: motivations, restraints and use practices. J Rech Porc 48:357–358

Horgen L, Jerome A, Rastogi N (1998) Pulsed-exposure and postantibiotic leukocyte enhancement effects of amikacin, clarithromycin, clofazimine, and rifampin against intracellular Mycobacterium avium. Antimicrob Agents Chemother 42(11):3006–3008

Huang L, Zhang H, Li M, Ahmad I, Wang Y, Yuan Z (2018) Pharmacokinetic-pharmacodynamic modelling of tylosin against *Streptococcus suis* in pigs. BMC Vet Res 14:319

Huang Z, Hu Z, Zheng H, Xia X, Gu X, Shen X, Yang H, Ding H (2020) The PK/PD integration and resistance of tilmicosin against *Mycoplasma hyopneumoniae*. Pathogens 9(6):487

Hudson W (2019) Shorter antibiotic courses, but why? Aust Prescr 42(4):120

Hurley JC, Guidet B, Offenstadt G, Maury E (2012) Endotoxemia and mortality prediction in ICU and other settings: underlying risk and co-detection of gram-negative bacteraemia are confounders. Crit Care 16(4):148

Idowu OR, Peggins JO, Cullison R, Bredow J (2010) Comparative pharmacokinetics of enrofloxacin and ciprofloxacin in lactating dairy cows and beef steers following intravenous administration of enrofloxacin. Res Vet Sci 89(2):230–235

Isgren CM (2021) Improving clinical outcomes via responsible antimicrobial use in horses. Equine Veterinary Education, Wiley on behalf of EVJ Ltd., pp 1–11

Javed A, Slingerland CJ, Wood TM, Martin NI, Broere F, Weingarth MH, Veldhuizen EJA (2023) Chimeric peptidomimetic antibiotic efficiently neutralizes lipopolysaccharides (LPS) and bacteria-induced activation of RAW macrophages. ACS Infect Dis 9(3):518–526

Jerzsele A (2012) Comparative Veterinary Pharmacokinetics in Readings in Advanced Pharmacokinetics – Theory, Methods and Applications. https://www.intechopen.com/books/readings-in-advanced-pharmacokinetics-theory-methods-and-applications/veterinary-pharmacokinetics. Accessed 29 May 2023

Jones C (2014) The Pollyanna phenomenon and non-inferiority: how our experience (and research) can lead to poor treatment choices, science-based medicine, Feb 14. https://sciencebasedmedicine.org/the-pollyanna-phenomenon-and-non-inferiority-how-our-experience-and-research-can-lead-to-poor-treatment-choices. Accessed 29 May 2023

Kaartinen L, Pyörälä S, Moilanen M, Räisänen S (1997) Pharmacokinetics of enrofloxacin in newborn and one-week-old calves. J Vet Pharmacol Ther. 20(6):479–482

Kang J, Hossain MA, Park HC, Kim Y, Lee KJ, Park SW (2019) Pharmacokinetic and pharmacodynamic integration of enrofloxacin against *Salmonella* Enteritidis after administering to broiler chicken by per-oral and intravenous routes. J Vet Sci 20(2):15

Karachalios G, Charalabopoulos K (2003) Biliary excretion of antimicrobials drugs. Chemotherapy 48(6):280–297

Kashuba AD, Nafziger AN, Drusano GL, Bertino JS Jr (1999) Optimizing aminoglycoside therapy for nosocomial pneumonia caused by gram-negative bacteria. Antimicrob Agents Chemother 43(3):623–629

Keshawani A, Keshawani A, Mishra MK, Maqbool MA, Pathak S, Biswal PK (2017) Various approaches towards taste masking: an overview. Int J Pharmaceut Res Appl 2(4):1–9 https://www.researchgate.net/publication/323995560_Various_approaches_towards_taste_masking_An_overview. Accessed 28 May 2023

Krause KM, Serio AW, Kane TR, Connolly LE (2016) Aminoglycosides: an overview. Cold Spring Harb Perspect Med 6(6):a027029

Kristoffersson AN, David-Pierson P, Parrott NJ, Kuhlmann O, Lave T, Friberg LE, Nielsen EI (2016) Simulation-based evaluation of PK/PD indices for

meropenem across patient groups and experimental designs. Pharm Res 33(5):1115–1125
Kumaraswamy M, Lin L, Olson J, Sun CF, Nonejuie P, Corriden R, Döhrmann S, Ali SR, Amaro D, Rohde M, Pogliano J, Sakoulas G, Nizet V (2016) Standard susceptibility testing overlooks potent azithromycin activity and cationic peptide synergy against MDR Stenotrophomonas maltophilia. J Antimicrob Chemother 71(5):1264–9126
Kuti JL (2016) Optimizing antimicrobial pharmacodynamics: a guide for your stewardship program. Rev Med Clin Condes 5:615–624
Ledesma C, Rosario C, Gracia-Mora J, Tapia G, Gutiérrez L, Sumano H (2018) Antibacterial activity of amoxicillin in vitro and its oral bioavailability in broiler chickens under the influence of 3 water sanitizers. Poult Sci 97(7):2391–2399
Lees P (2004) Veterinary advances in PK/PD modelling. J Vet Pharmacol Ther 27(6):395
Lees P (2010) PK-PD integration and PK-PD modelling: alternatives to dose titration studies for selecting optimal dosage schedules of antimicrobial drugs. https://www.ema.europa.eu/en/documents/other/paper--information-pk-pd-modelling-prof-p-lees_en.pdf. Accessed 28 May 2023
Lees P, Concordet D, Aliabadi FS, Toutain PL (2006) Drug selection and optimization of dosage schedules to minimise antimicrobial resistance. In Aarestrup et al. (eds) antimicrobial resistance in bacteria of animal origin. ASM Press, American Society for Microbiology, pp 53
Lees P, Svendsen O, Wiuff C (2008) Strategies to minimise the impact of antimicrobial treatment on the selection of resistant bacteria. In Guardabassi L, Jensen LB, Kruse H (eds) Guide to antimicrobial use in animals. Blackwell Publishing, pp 77–101
Lewis RE, Klepser ME, Ernst EJ, Lund BC, Biedenbach DJ, Jones RN (1999) Evaluation of low-dose, extended-interval clindamycin regimens against Staphylococcus aureus and Streptococcus pneumoniae using a dynamic in vitro model of infection. Antimicrob Agents Chemother 43(8):2005–2009
Li M, Gehring R, Tell L, Baynes R, Huang Q, Riviere JE (2014) Interspecies mixed-effect pharmacokinetic modelling of penicillin G in cattle and swine. Antimicrob Agents Chemother 58(8):4495–4503
Li M, Gehring R, Riviere JE, Lin Z (2018) Probabilistic physiologically based pharmacokinetic model for penicillin G in milk from dairy cows following Intramammary or intramuscular administrations. Toxicol Sci 164(1):85–100
Lin L, Nonejuie P, Munguia J, Hollands A, Olson J, Dam Q, Kumaraswamy M, Rivera H Jr, Corriden R, Rohde M, Hensler ME, Burkart ME, Pogliano J, Sakoulas G, Nize V (2015) Azithromycin synergizes with cationic antimicrobial peptides to exert and therapeutic activity against highly multidrug-resistant gram-negative bacterial pathogens. EBioMed 2:690–698
Lin L, Wong H (2017) Predicting oral drug absorption: mini review on physiologically-based pharmacokinetic models. Pharmaceutics 9(4):41
Lin Z, Gehring R, Mochel JP, Lavé T, Riviere JE (2016a) Mathematical modelling and simulation in animal health – part II: principles, methods, applications, and value of physiologically based pharmacokinetic modelling in veterinary medicine and food safety assessment. J Vet Pharmacol Ther 39(5):421–438
Lin Z, Cuneo M, Rowe JD, Li M, Tell LA, Allison S, Carlson J, Riviere JE, Gehring R (2016b) Estimation of tulathromycin depletion in plasma and milk after subcutaneous injection in lactating goats using a nonlinear mixed-effects pharmacokinetic modelling approach. BMC Vet Res 12(1):258
Lipcsey M, Larsson A, Eriksson MB, Sjölin J (2008) Effect of the administration rate on the biological responses to a fixed dose of endotoxin in the anesthetized pig. Shock 29(2):173–180
Liu P, Müller M, Derendorf H (2002) Rational dosing of antibiotics: the use of plasma concentrations versus tissue concentrations. Int J Antimicrob Agents 19(4):285–290
Lu C, Li A (2001) Species comparison in P450 induction: effects of dexamethasone, omeprazole, and rifampin on P450 isoforms 1A and 3A in primary cultured hepatocytes from man, Sprague-Dawley rat, minipig, and beagle dog. Chem Biol Interact 134:271–281
Luo W, Chen D, Wu M, Li Z, Tao Y, Liu Q, Pan Y, Qu W, Yuan Z, Xie S (2019) Pharmacokinetics/pharmacodynamics models of veterinary antimicrobial agents. J Vet Sci 20(5):40
Luo W, Qin H, Chen D, Wu M, Meng K, Zhang A, Pan Y, Qu W, Xie S (2020) The dose regimen formulation of tilmicosin against *Lawsonia intracellularis* in pigs by pharmacokinetic-pharmacodynamic (PK-PD) model. Microb Pathog 147:104389
Lv J, Deng S, Zhang L (2021) A review of artificial intelligence applications for antimicrobial resistance. Biosafety Health 3(1):22–31
Martinez MN, Toutain PL, Turnidge J (2013) The pharmacodynamics of antimicrobial agents. In: Giguère S, Prescott JF, Dowling PM (eds) Antimicrobial therapy in veterinary medicine, 5th edn. Wiley
Martín-Jiménez T, Riviere JE (1998) Population pharmacokinetics in veterinary medicine: potential use for therapeutic drug monitoring and prediction of tissue residues. J Vet Pharmacol Ther 21(3):167–189
Matthews SG, Miller AL, Plötz T, Kyriazakis I (2017) Automated tracking to measure behavioural changes in pigs for health and welfare monitoring. Sci Rep 7:17582
McDonald PJ, Wetherall BL, Pruul H (1981) Postantibiotic leukocyte enhancement: increased susceptibility of bacteria pretreated with antibiotics to activity of leukocytes. Rev Infectious Dis 3(1):38–44. JSTOR. www.jstor.org/stable/4452512. Accessed 20 June 2023
Mekala P, Jagadeeswaran A, Arivuchelvan A, Sethilkumar P, Nanjappan K, Krishnamurthy TRG (2015) Comparison of pharmacokinetics of Enrofloxacin

after single oral bolus and pulse dose administration in broiler chicken. Proc Nat Acad Sci India Sect B 87(3):679–683

Messenger KM, Papich MG, Blikslager AT (2011) Distribution of enrofloxacin and its active metabolite, using an in vivo ultrafiltration sampling technique after the injection of enrofloxacin to pigs. J Vet Pharmacol Ther 35(5):452–459

Mestorino N, Errecalde JO (2012) Pharmacokinetic – Pharmacodynamic considerations for bovine mastitis treatment. In A bird's-eye view of veterinary medicine. INTECH. https://www.intechopen.com/books/a-bird-s-eye-view-of-veterinary-medicine/pharmacokinetic-pharmacodynamic-considerations-for-bovine-mastitis-treatment. Accessed 29 May 2023

Mevius DJ, Breukink HJ, Guelen PJM, Jansen T, de Grève B (1990) Pharmacokinetics, metabolism and renal clearance of flumequine in veal calves. J Vet Pharmacol Therap 13:159–169

Moore RD, Lietman PS, Smith CR (1987) Clinical response to aminoglycoside therapy: importance of the ratio of peak concentration to minimal inhibitory concentration. J Infect Dis 155:93–99

Morthorst D (2002) Bioavailability of amoxicillin in weanling piglets after oral and parenteral application by feed and water under different conditions (Bioverfügbarkeit von Amoxicillin bei Absetzferkeln nach peroraler und parenteraler Applikation über Futter und Trinkwasser unter verschiedenen Bedingungen (In German)). https://elib.tiho-hannover.de/receive/etd_mods_00002692. Accessed 15 May 2023

Mouton JW, Dudley MN, Cars O, Derendorf H, Drusano GL (2005) Standardization of pharmacokinetic/pharmacodynamic (PK/PD) terminology for anti-infective drugs: an update. J Antimicrob Chemother 55(5):601–607

Muller EA (2017) Aminoglycosides: revision of clinical breakpoints and implications for clinical practice, 27th ESCMID, Vienna 2017; 1–42: 24

Nakamura M, Imaoka S, Miura K, Tanaka E, Misawa S, Funae Y (1994) Induction of cytochrome P-450 isozymes in rat renal microsomes by cyclosporin A. Biochem Pharmacol 48:1743–1746

Nebbia C, Ceppa L, Dacasto M, Carletti M, Nachtmann C (1999) Oxidative metabolism of monensin in rat liver microsomes and interactions with tiamulin and other chemotherapeutic agents: evidence for the involvement of cytochrome P-450 3A subfamily. Drug Metab Dispos 27:1039–1044

Nedbalcova K, Zouharova M, Sperling D (2019) Post-antibiotic effect of marbofloxacin, enrofloxacin and amoxicillin against selected respiratory pathogens of pigs. Veterinarni Medicina 64:67–77

Neethirajan S (2020) Transforming the adaptation physiology of farm animals through sensors. Animals 10(9):1512

Nemeth J, Oesch G, Kuster SP (2015) Bacteriostatic versus bactericidal antibiotics for patients with serious bacterial infections: systematic review and meta-analysis. J Antimicrob Chemother 70(2):382–395

Nielsen EI, Cars O, Friberg LE (2011) Pharmacokinetic/pharmacodynamic (PK/PD) indices of antibiotics predicted by a semimechanistic PKPD model: a step toward model-based dose optimization. Antimicrob Agents Chemother 55(10):4619–4630

Nussbaumer-Pröll AK, Knotzer S, Eberl S, Reiter B, Stimpfl T, Jäger W, Poschner S, Zeitlinger M (2019) Impact of erythrocytes on bacterial growth and antimicrobial activity of selected antibiotics. Eur J Clin Microbiol Infect Dis 38(3):485–495

Onakpoya IJ, Walker AS, Tan PS, Spencer EA, Gbinigie OA, Cook J, Llewelyn MJ, Butler CC (2018) Overview of systematic reviews assessing the evidence for shorter versus longer duration antibiotic treatment for bacterial infections in secondary care. PLoS One 13(3):e0194858

Onufrak NJ, Forrest A, Gonzalez D (2016) Pharmacokinetic and pharmacodynamic principles of anti-infective dosing. Clin Ther 38(9):1930–1947

Ouyang X, Hoeksma J, Lubbers RJM (2022) Classification of antimicrobial mechanism of action using dynamic bacterial morphology imaging. Sci Rep 12:11162

Ovando G, Gorla H, Luders C, Poloni G, Errecalde C, Prieto G, Puelles I (1999) Comparative pharmacokinetics of enrofloxacin and ciprofloxacin in chickens. J Vet Pharmacol Ther 22(3):209–212

Owens RC Jr, Shorr AF (2009) Rational dosing of antimicrobial agents: pharmacokinetic and pharmacodynamic strategies. Am J Health Syst Pharm 66(12) (Suppl 4):23–30

Owens WE, Xiang ZY, Ray CH (1990) Determination of milk and mammary tissue concentrations of ceftiofur after intramammary and intramuscular therapy. J Dairy Sci 73:3449–3456

Pardon B, Buczinski S (2020) Bovine respiratory disease diagnosis: what progress has been made in infectious diagnosis? Vet Clin North Am Food Anim Pract 36(2):425–444

Pelligand L, Lees P, Sidhu PK, Toutain PL (2019) Semi-mechanistic modelling of florfenicol time-kill curves and in silico dose fractionation for calf respiratory pathogens. Front Microbiol 10:1237

Petel D, Winters N, Gore GC, Papenburg J, Beltempo M, Lacroix J, Fontela PS (2018) Use of C-reactive protein to tailor antibiotic use: a systematic review and meta-analysis. BMJ Open 8(12):e022133

Plumb DC (2018) Plumb's veterinary drug handbook, 9th edn. PharmaVet; Blackwell Pub, Stockholm

Post C, Rietz C, Büscher W, Müller U (2020) Using sensor data to detect lameness and mastitis treatment events in dairy cows: a comparison of classification models. Sensors (Basel) 20(14):3863

Post C, Rietz C, Büscher W, Müller U (2021) The importance of low daily risk for the prediction of treatment events of individual dairy cows with sensor systems. Sensors (Basel) 21(4):1389

Potter T, Illambas J, Pelligand L, Rycroft A, Lees P (2013) Pharmacokinetic and pharmacodynamic integration and modelling of marbofloxacin in calves for

Mannheimia haemolytica and Pasteurella multocida. Vet J 195:53–58
Pouwels KB, Hopkins S, Llewelyn MJ, Walker AS, McNulty CA, Robotham JV (2019) Duration of antibiotic treatment for common infections in English primary care: cross sectional analysis and comparison with guidelines. BMJ 364:l440
Rázquin-Olazarán I, Shahrour H, Martínez-de-Tejada G (2020) A synthetic peptide sensitizes multi-drug resistant Pseudomonas aeruginosa to antibiotics for more than two hours and permeabilizes its envelope for twenty hours. J Biomed Sci 27(1):85
Rea RS, Capitano B (2007) Optimizing use of aminoglycosides in the critically ill. Semin Respir Crit Care Med 28:596–603
Reeves PT, Roesch C, Raghnaill MN (2015) Routes of administration and dosage forms. https://www.msdvetmanual.com/pharmacology/pharmacology--introduction/routes-of-administration-and-dosage-forms. Accessed 19 Jun 2023
Reyns T, De Boever S, Baert K, Croubels S, Schauvliege S, Gasthuys F, De Backer P (2007) Disposition and oral bioavailability of amoxicillin and clavulanic acid in pigs. J Vet Pharmacol Ther 30(6):550–555
Riviere JE, Gabrielsson J, Fink M, Mochel J (2015) Mathematical modeling and simulation in animal health. Part I: moving beyond pharmacokinetics. J Vet Pharmacol Therap 39:213–223
Roggeveen LF, Guo T, Driessen RH, Fleuren LM, Thoral P, van der Voort PHJ, Girbes ARJ, Bosman RJ, Elbers P (2020) Right dose, right now: development of autoKinetics for real time model informed precision antibiotic dosing decision support at the bedside of critically ill patients. Front Pharmacol 11:646
Roggeveen LF, Guo T, Fleuren LM, Driessen R, Pl T, van Hest RM, Mathot RAA, Swart EL, de Grooth HJ, van den Bogaard B, Girbes ARJ, Bosman RJ, Elbers Roggeveen PWG (2022) Right dose, right now: bedside, real-time, data-driven, and personalised antibiotic dosing in critically ill patients with sepsis or septic shock – a two-centre randomised clinical trial. Crit Care 26:265
Roth WJ, Kissinger CB, McCain RR, Cooper BR, Marchant-Forde JN, Vreeman RC, Hannou S, Knipp GT (2013) Assessment of juvenile pigs to serve as human pediatric surrogates for preclinical formulation pharmacokinetic testing. AAPS J 15:763–774
Ruiz-Romero RA, Vargas-Bello-Pérez E (2023) Non-aureus staphylococci and mammaliicocci as a cause of mastitis in domestic ruminants: current knowledge, advances, biomedical applications, and future perspectives – a systematic review. Vet Res Commun. https://doi.org/10.1007/s11259-023-10090-5. Accessed 19 Jun 2023
Salas JR, Jaberi-Douraki M, Wen X, Volkova VV (2020) Mathematical modelling of the 'inoculum effect': six applicable models and the MIC advancement point concept. FEMS Microbiol Lett 367(5):fnaa012
Sandberg A, Hessler JH, Skov RL et al (2009) Intracellular activity of antibiotics against Staphylococcus aureus in a mouse peritonitis model. Antimicrob Agents Chemother 53:1874–1883
Sang K, Hao H, Huang L, Wang X, Yuan Z (2016) Pharmacokinetic-pharmacodynamic modelling of enrofloxacin against Escherichia coli in broilers. Front Vet Sci 2:80
Shimizu T, Harada K, Kataoka Y (2013) Mutant prevention concentration of orbifloxacin: comparison between Escherichia coli, Pseudomonas aeruginosa, and staphylococcus pseudintermedius of canine origin. Acta Vet Scand 55(1):37
Sidhu P, Landoni M, Aliabadi F, Lees P (2010a) Pharmacokinetic and pharmacodynamic modelling of marbofloxacin administered alone and in combination with tolfenamic acid in goats. Vet J 184:219–229
Sidhu P, Landoni M, Aliabadi F, Lees P (2010b) PK–PD integration and modelling of marbofloxacin in sheep. Res Vet Sci 88:134–141
Sidhu P, Landoni M, Aliabadi F, Toutain PL, Lees P (2011) Pharmacokinetic and pharmacodynamic modelling of marbofloxacin administered alone and in combination with tolfenamic acid in calves. J Vet Pharmacol Ther 34:376–387
Sidhu P, Rassouli A, Illambas J, Potter T, Pelligand L, Rycroft A, Lees P (2014) Pharmacokinetic -pharmacodynamic integration and modelling of florfenicol in calves. J Vet Pharmacol Ther 37:231–242
Siegel TW, Earley DL, Smothers CD, Sun F, Ricketts AP (2004) Cellular uptake of the triamilide tulathromycin by bovine and porcine phagocytic cells in vitro. Poult Sci 82:186
Sjölund M, Ekstrand C, Wallgren P, Bondesson U, Pringle M, Bengtsson B (2020) Exposure to benzylpenicillin after different dosage regimens in growing pigs. Acta Vet Scand 62(1):55
Soliman GA (2000) Tissue distribution and disposition kinetics of enrofloxacin in healthy and *E. coli* infected broilers. Dtsch Tierarztl Wochenschr 107(1):23–27
Soraci AL, Amanto F, Tapia MO, de la Torre E, Toutain PL (2014) Exposure variability of fosfomycin administered to pigs in food or water: impact of social rank. Res Vet Sci 96(1):153–159
Szucs G, Tamási V, Laczay P, Monostory K (2004) Biochemical background of toxic interactions between tiamulin and monensin. Chemico Biol Interact 147:151–161
Tanigawa M, Sawada T (2002) Effects of subminimal inhibitory concentrations of amoxicillin against Actinobacillus pleuropneumoniae. J Vet Med B Infect Dis Vet Public Health 49(10):513–518
Thabit AK, Fatani DF, Bamakhrama MS, Barnawi OA, Basudan LO, Alhejaili SF (2019) Antibiotic penetration into bone and joints: an updated review. Int J Infect Dis 81:128–136
Toutain PL, del Castillo JRE, Bousquet-Mélou A (2002) The pharmacokinetic-pharmacodynamic approach to a rational dosage regimen for antibiotics. Res Vet Sci 73(2):105–114

Toutain PL, Ferran AA, Bousquet-Mélou A, Pelligand L, Lees P (2016) Veterinary medicine needs new green antimicrobial drugs. Front Microbiol 7:1196

Toutain PL, Bousquet-Mélou A, Damborg P, Ferran A, Mevius D, Pelligand L, Veldman KT, Lees P (2017a) En route towards European clinical breakpoints for veterinary antimicrobial susceptibility testing: a position paper explaining the VetCAST approach. Front Microbiol 8:2344

Toutain PL, Potter T, Pelligand L, Lacroix M, Illambas J, Lees P (2017b) Standard PK/PD concepts can be applied to determine a dosage regimen for a macrolide: the case of tulathromycin in the calf. J Vet Pharmacol Ther 40:16–27

Toutain PL, Sidhu PK, Lees P, Rassouli A, Pelligand L (2019) VetCAST method for determination of the pharmacokinetic-pharmacodynamic cut-off values of a long-acting formulation of florfenicol to support clinical breakpoints for florfenicol antimicrobial susceptibility testing in cattle. Front Microbiol 10:1310

Toutain PL, Pelligand L, Lees P, Bousquet-Mélou A, Ferran AA, Turnidge JD (2021) The pharmacokinetic/pharmacodynamic paradigm for antimicrobial drugs in veterinary medicine: recent advances and critical appraisal. J Vet Pharmacol Ther 44(2):172–200

Trouchon T, Lefebvre S (2016) A review of enrofloxacin for veterinary use. Open J Vet Med 6(2):40–58

Turnidge J (2003) Pharmacodynamics and dosing of aminoglycosides. Infect Dis Clin N Am 17(3):503–528

Van Bambeke F, Tulkens PM (2011) Optimizing aminoglycoside dose based on PK/PD, PK-PD course. https://www.facm.ucl.ac.be/cooperation/Vietnam/WBI-Vietnam-October-2011/Course-on-PK-PD/5A--Aminoglycosides.pdf. Accessed 28 May 2023

Van Bambeke F, Barcia-Macay M, Lemaire S et al (2006) Cellular pharmacodynamics and pharmacokinetics of antibiotics: current views and perspectives. Curr Opin Drug Discov Devel 9:218–230

Vandael F, Filippitzi M, Dewulf J, Daeseleire E, Eeckhout M, Devreese M, Croubels S (2019) Oral group medication in pig production: characterising medicated feed and drinking water systems. Vet Rec 185(13):40

Vermeulen B, De Backer P, Remon JP (2002) Drug administration to poultry. Adv Drug Deliv Rev 54(6):795–803. https://doi.org/10.1016/s0169-409x(02)00069-8

Villarino N, Brown SA, Martin-Jimenez T (2013) The role of the macrolide tulathromycin in veterinary medicine. Vet J 198:352–357

Vinks AA, Derendorf H, Mouton JW (2014) Fundamentals of antimicrobial pharmacokinetics and pharmacodynamics. Springer, Heidelberg, pp 223–255

Wang K, Kestenbaum B (2018) Proximal tubular secretory clearance: a neglected partner of kidney function. Clin J Am Soc Nephrol x 13(8):1291–1296

Wang J, Hao H, Huang L, Liu Z, Chen D, Yuan Z (2016) Pharmacokinetic and pharmacodynamic integration and modelling of enrofloxacin in swine for *Escherichia coli*. Front Microbiol 7:36

Wei G, Yun H (2022) Antibacterial and antibiofilm activities of novel cyclic peptides against methicillin-resistant Staphylococcus aureus. Int J Mol Sci 23(14):8029

Wen X, Gehring R, Stallbaumer A, Riviere JE, Volkova VV (2016) Limitations of MIC as sole metric of pharmacodynamic response across the range of antimicrobial susceptibilities within a single bacterial species. Sci Rep 6:37907

Wilson HL, Daveson K, Del Mar CB (2019) Optimal antimicrobial duration for common bacterial infections. Aust Prescr 42(1):5–9

Wong KF, Luk JM (2009) Endotoxin-neutralizing peptides as gram-negative sepsis therapeutics. Protein Pept Lett 16(5):539–542

Wright WE, Line VD (1980) Biliary excretion of cephalosporins in rats: influence of molecular weight. Antimicrob Agents Chemother 17(5):842–846

Wu Y, Huang J, Quan S, Yang Y (2022) Light regimen on health and growth of broilers: an update review. Poult Sci 101(1):101545. https://doi.org/10.1016/j.psj.2021.101545

Xiao X, Jiang L, Lan W, Jiang Y, Wang Z (2018) In vivo pharmacokinetic/Pharmacodynamic modelling of Enrofloxacin against *Escherichia coli* in broiler chickens. BMC Vet Res 14(1):374

Xiong M, Wu X, Ye X, Zhang L, Zeng S, Huang Z, Wu Y, Sun J, Ding H (2016) Relationship between cefquinome PK/PD parameters and emergence of resistance of *Staphylococcus aureus* in rabbit tissue-cage infection model. Front Microbiol 7

Yancey RJ Jr, Evans RA, Kratzer DD, Paulissen JB, Carmer SG (1990) Efficacy of ceftiofur hydrochloride for treatment of experimentally induced colibacillosis in neonatal swine. Am J Vet Res 51(3):349–353

Zangaro C (2018) Use of antibiotics administered in water to pigs in the Post-VFD era, Michigan State University, Pork extension. Available online at: https://www.canr.msu.edu/news/use-of-antibiotics-administered-in--water-to-pigs-in-the-post-vfd-era. Accessed 15 Mar 2023

Zelenitsky SA, Harding GKM, Sun S, Ubhi K, Ariano RE (2003) Treatment and outcome of *Pseudomonas aeruginosa* bacteraemia: an antibiotic pharmacodynamic analysis. J Antimicrob Chemother 52(4):668–674

Zeng QL, Mei X, Su J, Li XH, Xiong WG, Lu Y, Zeng ZL (2017) Integrated pharmacokinetic–pharmacodynamic (PK/PD) model to evaluate the in vivo antimicrobial activity of Marbofloxacin against *Pasteurella multocida* in piglets. BMC Vet Res 13:178

Zeng D, Sun M, Lin Z, Li M, Gehring R, Zeng Z (2018) Pharmacokinetics and pharmacodynamics of tildipirosin against *Pasteurella multocida* in a murine lung infection model. Front Microbiol 18(9):1038

Zhao DH, Zhang CY, Zhang Z, Liu ZC, Liu BT, Yu JJ, Guo JP, Deng H, Liu YH (2013) Population pharmacokinetics of cefquinome in pigs. J Vet Pharmacol Ther 36(4):313–319

Zhao DH, Zhang Z, Zhang CY, Liu ZC, Deng H, Yu JJ, Guo JP, Liu YH (2014) Population pharmacokinetics of valnemulin in swine. J Vet Pharmacol Ther 37(1):59–65

Zhao Y, Guo L-L, Fang B, Liu B (2018) Pharmacokinetic/pharmacodynamic (PK/PD) evaluation of tulathromycin against *Haemophilus parasuis* in an experimental neutropenic Guinea pig model. PLoS One 13(12):0209177

Zhou Q, Zhang G, Wang Q, Liu W, Huang Y, Yu P, Li Y, Ding H, Fang B (2017) Pharmacokinetic/pharmacodynamic modelling of tulathromycin against *Pasteurella multocida* in a porcine tissue cage model. Front Pharmacol 8:392

Zimmermann P, Ziesenitz VC, Curtis N, Ritz N (2018) The immunomodulatory effects of macrolides-a systematic review of the underlying mechanisms. Front Immunol 9:302

Ziv G, Rasmussen F (1975) Distribution of labelled antibiotics in different components of milk following intramammary and intramuscular administrations. J Dairy Sci 58(6):938–946

Swine

Daniel Sperling and David G. S. Burch

Abstract

Pork remains the most widely consumed meat worldwide. The changes in swine production over the years have resulted in increased disease challenges and pathogen transmission due to larger units, increased population density and throughput, due to the significant intensification of production. Antimicrobials have been widely used in swine production over several decades, and management of the rational use of the antimicrobials for control and treatment of diseases is an important responsibility requiring specific veterinary knowledge and understanding. At the same time, great effort has been made in terms of improvement of biosecurity, management practices, and improvement of genetics and health status coming from Specific Pathogen-Free (SPF) farms, in an attempt to control and reduce antimicrobial use. Besides this fact, many bacterial infections require targeted antimicrobial treatment. The three main reasons for antimicrobial administration nowadays in pigs are the treatment of respiratory infections, systemic infections characterised by septicaemia and digestive infections, especially in piglets and finishing pigs. This chapter's aim is to provide an overview of the major consideration to antimicrobial therapy in swine with a specific focus on selected bacterial diseases, the minimum inhibitory concentrations (MICs) of specific pathogens, the optimal use of antimicrobials utilising their pharmacokinetics, possible alternative tools and biosecurity in swine medicine, considering current developments as well as new veterinary regulations.

Keywords

Antimicrobials in swine medicine · Pigs · Susceptibility patterns · Pharmacokinetics and pharmacodynamics · Treatment of swine infectious diseases

1 Introduction

With the increase in worldwide demand for meat, fast-growing species like pigs with efficient feed conversion will have a major share in the growth in the livestock sector, and pig production is expected to continue to grow. The increase is not spread evenly: Asia is leading, whereas pig numbers in North America and Europe are increasing more slowly or are stationary. In some of the EU countries, a decrease of production was more pronounced during the last few years, with interyear increase by 2% noted in the period

D. Sperling (✉)
Ceva Sante Animale, France, Libourne

D. G. S. Burch (Deceased)
Berkshire, UK

L. Pokludová (ed.), *Antimicrobials in Livestock 2: Choices of Prevention and Treatment*,
https://doi.org/10.1007/978-3-031-60807-0_2

2020–2021, followed again by a decrease by 3.7–6.1% in different production categories of pigs (data for 27 EU Member States, December 2021–December 2022). Spain (23%) and Germany (20%) were the biggest producers within European Union in 2021, while about 39% is covered by five countries (France, Poland, Denmark, the Netherlands and Italy), remaining 18% covers production of further EU countries (European Commission, DG Dashboard: Pigmeat 2023). Intensification of production has resulted in a steady decline of small farms (backyard farms) and an increase in the larger holdings, particularly in Russia and the Asian countries.

Only two countries and EU production cover two-third (76%) of the total world production of pigmeat, while China's production prevails (45%), followed by European Union (20%) and USA (11%). Russian Federation, together with Brazil, were the fourth biggest world producers, each contributed 4% to world production of pig meat. Rapid intensification of production has occurred mainly in China, influenced dramatically by the African swine fever (ASF) outbreak, which started in 2018 (Yan et al. 2023). About 20 years ago, it was estimated that 74% of swine production was in the form of backyard farms (annual production 1–49 heads), and commercial farms which produced more than 1000 heads were estimated only 5%. By 2010, commercial farms accounted for 12% and only 37% for backyards. (Schneider and Sharma 2014; Krishnasamy et al. 2015; Agribusiness research and Advisory, Rabobank International 2012). Current forecasts of Chinese swine production predict that 65% of all production is likely to come from specialised farms, 30% from commercial farms and only 5% from backyard farms, by 2025. Integrated farms, involving several subsystems, including crops and other farm animal production are able to use the benefit from synergistic interactions. Another factor, which plays an important role, is the decrease in people engaged in agriculture. Changes in the epidemiological situation in some regions, particularly the spread of some important diseases like African Swine Fever (ASF) or Porcine Epidemic Diarrhoea (PED), are having a huge impact, in some cases positive, due to improvement of biosecurity measures (Zhou et al. 2018). There is an increase in the number of large farms leading to an increased concentration of ownership of pigs especially, in the case of Russia, where the top five biggest holdings produce 33.2% of the country's total pork production in 2019 (NSSRF 2020).

These changes have resulted in increased disease challenges and pathogen transmission due to larger units, increased population density and throughput and, in some cases, a reduction in the quality of stockmanship. As a result, they have frequently contributed to increased use of antimicrobials in pork production to compensate and optimise production. The countries with the greatest expected increases in food demand and animal antibiotic use often have the least efficient farming systems (Krishnasamy et al. 2015).

At the same time, in other parts of the world, great efforts were made in terms of improvement of biosecurity, management practices, like batch management systems and "all-in-all-out" (AIAO) procedures and improvement of genetics and health status coming mainly from Specific Pathogen-Free (SPF) farms, which have enabled the reduction in antimicrobial use, especially in Europe.

In 2020, the top five world consumers of antimicrobials, for which 58% of global antimicrobial use was estimated, were China, Brazil, India, USA and Australia. Despite the need for the regular checking of consumption data and their changes, still they were also predicted to remain the top five in the 2030 (Mulchandani et al. 2023).

The top three classes by global sales for animal use in 2009 were assumed to be macrolides ($600 million), penicillins ($600 million) and tetracyclines ($500 million) (Center for Disease Dynamics, Economics & Policy 2015; State of the World's Antibiotics, 2015. CDDEP: Washington, D.C.). In comparison, situation in the EU/EEA countries differs as for macrolides. In 2015, three classes, accounting for 69.6% of total sales in the 30 countries, were tetracyclines (32.8%), penicillins (25.0%) and sulfonamides (11.8%) expressed in mg/PCU (EMA/ESVAC 2017). During the following years, started in 2019, penicillins have become the highest selling antibiotic class accounting for 31.2% of overall sales, followed by tetracyclines (25.8%) and sulfonamides (9.9%), and these three classes accounted for 66.9% of total sales in 31 countries in 2021, probably reflecting recent proposed EMA categorisation (EMA/ESVAC 2022). The European Surveillance of Veterinary Antimicrobial Consumption (ESVAC) project was launched by the EMA in 2009, developing and harmonising the approach to the collection and reporting of data on the sales of antimicrobial agents in animals, including the swine sector, from the Member States, starting with the first report from nine European countries (EMA/ESVAC 2011) but gaining finally overall sales data from 31 European countries (EMA/ESVAC 2022). For some countries participating in the pilot project of stratification of the sales data (Austria, Belgium, the Czech Republic, Denmark, France and the Netherlands), the total amount of antimicrobials allocated as used in the pig sector is available, according to a defined protocol (EMA 2018). Pigs were reported to have the highest consumption of antibiotics (32%) followed by cattle (31%) and poultry (14%) in Europe (EMA 2017). Within the network AACTING dealing with quantification of use of antimicrobials in animals at herd level, 38 different herd-level antimicrobial use collection systems, either private or governmental, collecting the data from the end users in the pig sector in 16 European countries were noted in 2020 (Sanders et al. 2020) and number of countries is contnuously increasing (AACTING 2024).

From the European perspective, specifically for the pig sector, some recent data can be reached from different countries, where system for the use data per animal species has already been established. In Denmark, the data from 2021 indicated that mostly macrolides, followed by pleuromutilins, tetracyclines and penicillins sensitive to beta-lactamases were used when aggregating the data for all production categories. While tetracycline use decreased since 2004, increase has been particularly determined for aminoglycosides (neomycin), macrolides and penicillins sensitive to beta-lactamase (DANMAP 2022). Data for the pig sector are available also from the Netherlands, where tetracyclines accounted for biggest proportion of classes used, followed by aminopenicillins and combination of sulphonamide/trimethoprim. Despite 23.1% reduction of the use of colistin in 2021, the pig farming sector still ranked highest in the amount of colistin used among the other sectors. Of the total amount of colistin used in the Dutch pig farming sector, 90.2% could be attributed to use in weaner pigs, 6.7% to use in suckling piglets and 3.2% to use in fattening pigs (The Netherlands Veterinary Medicines Institute 2022).

Based on the legal stipulation given by the Regulation (EC) 2019/6 since January 2023, the obligation for the EU Member States to collect the data on antimicrobial use in the major food producing species, including pigs, has been started. It can be therefore expected that detailed data will be available across EU region in some years.

It should be also noted that total overall use of antimicrobials in Europe has been reduced, following the ban on the use of Antimicrobial Growth Promoters (AGPs) in 2006. There was a transient increase in the use of therapeutic antimicrobials, especially in weaned and grower pigs, mainly due to enteric infections. Subsequently, there has been an improvement in the situation after implementation of changes in management practices, such as later weaning. The recent report (EMA/ESVAC 2022) confirmed an overall decline in sales of antibiotics, expressed in mg/kg Population Correction Unit (PCU) of 47% for 25 countries livestock productions, between the years 2011 and 2021. The majority of the sales of pharmaceutical forms are still used for group treatment, which is a characteristic mainly for pigs and poultry, but some countries, like the Netherlands, have tried to stop the use of in-feed medication. Despite this great success, especially in the pig sector, some challenges are still there. One of them is the ban of the zinc oxide–containing VMPs, which were used via in-feed medication, where fears exist on possible increase of use of certain antimicrobial classes for the treatment or metaphylaxis of enteric infections, like post-weaning diarrhoea and oedema disease. As summarised by Wang et al. (2020), many countries have approved the withdrawal of colistin as a feed additive in animals, including Brazil (November, 2016), Thailand (February, 2017), China (April, 2017), Japan (July, 2018), Malaysia (January, 2019), Argentina (February, 2019) and India (July, 2019).

Recent activities of the European Commission (EC) include guidelines regarding the prudent use of antimicrobials in veterinary medicine (EC 2015), ideally to stop the use of antibiotics for routine prevention and only use them for metaphylaxis or treatment. This has now passed into legislation, which came into effect in 2022. They are also endeavouring to restrict the use of critically important antibiotics (CIAs) to situations where the veterinarian has assessed, on the basis of antimicrobial susceptibility testing and relevant epidemiological data that there is no non-CIA alternative available, to contribute to the trend of reduction of use in swine as well (Burch and Sperling 2018). Targets for colistin use, now considered a last resort antibiotic in human medicine, have also been set.

The USA has followed a different approach and has put antimicrobial use in feed under veterinary control for those compounds considered of importance in human health and has stopped their use for growth promotion by implementing the Veterinary Feed Directive in 2017 (Sneeringer et al. 2019). Preventive use is still permitted, however. Similar activities can be seen in Asian countries, such as better control of feed medication and the restriction of certain antibiotics (Kirchhelle 2018).

The most important bacterial pathogens and diseases requiring antimicrobial use in swine are summarised in Table 1, and their disease patterns are highlighted (enteric diseases), (respiratory diseases) and (septicaemic diseases).

Any use of antimicrobial agents leads to selection pressure and the potential development of bacterial resistance. Resistance development in swine pathogenic bacteria complicates treatment of infections and, therefore, has to be regarded as primarily an animal health problem and potentially an economical burden. An example of serious consequences can be seen by the spread of multi-drug-resistant strains of *Brachyspira hyodysenteriae*, which frequently led to complete depopulation or even the end of production on affected farms, due to the lack of effective antibiotics for treatment and control of clinical swine dysentery (Sperling et al. 2011). In addition, pigs are often colonised by bacteria capable of transferring to and causing infections in humans, such as *Salmonella* and *Campylobacter* spp. Thus, the use of antimicrobials in pigs can also lead to the selection of resistance in these zoonotic bacteria and thereby, may potentially complicate treatment of human infections. The recently described occurrence of Extended-Spectrum β-Lactamase (ESBL) *Escherichia coli* strains being carried in pigs, mainly associated with the use of cephalosporins, as well as a transmissible, plasmid-mediated resistance to colistin with a possible risk for humans, will lead to new regulations and potentially to a reduction of the panel of effective antibiotics allowed to be used in swine medicine, especially in Europe (Dohmen et al. 2017; El Garch et al. 2018; Liu et al. 2016). This aspect has to be taken into consideration when choosing antimicrobials for treatment of bacterial infections in pigs.

The purpose of this chapter is to describe the current use of antimicrobial agents in swine production and to suggest possible strategies to reduce their overall use and to use them more effectively, prudently and responsibly.

2 Main Pathogens and Diseases in Swine

The major bacterial diseases in pigs (Table 1), where antimicrobial treatment plays an important role, can be divided into three major categories (Taylor 2013). Table 2 summarises, considering the EMA-AMEG categorisation of antimicrobials, the first ("D"), the second ("C") and the last ("B") treatment choices of the main bacterial diseases of pigs.

Table 1 Major bacterial pathogens, diseases and ages groups most commonly affected in swine

Bacterial species	Disease	Age
Respiratory		
Pasteurella multocida (D toxin)	Atrophic rhinitis	1–8 weeks
Bordetella bronchiseptica		Nasal distortion lasts for life
Mycoplasma hyorhinis	Pneumonia, polyserositis and arthritis	3–10 weeks of age
Mycoplasma hyopneumoniae	Enzootic pneumonia	Grower and finisher pig
Pasteurella multocida	Mycoplasma-induced respiratory disease (MIRD)	Grower and finisher—Usually secondary invader
Actinobacillus pleuropneumoniae	Pleuropneumonia	Grower and finisher—MDA last for 10 weeks
Actinobacillus suis	Septicaemia	5–28 days
	Pleuropneumonia	Weaning to slaughter
Septicaemia/bacteraemia		
Streptococcus suis	Meningitis, endocarditis, arthritis, peritonitis	2–10 weeks
Glaesserella (formerly Haemophilus) parasuis	Glässer's disease (arthritis, pericarditis, peritonitis)	2–10 weeks
Mycoplasma hyosynoviae	Mycoplasmal arthritis	16 weeks plus (usually breeding animals: gilts)
Erysipelothrix rhusiopathiae	Erysipelas (dermatitis, arthritis, endocarditis)	Growers, finishers and sows
Mycoplasma suis	Infectious anaemia- Eperythrozoonosis	Weaners and growers, breeding animals
Enteric		
Escherichia coli	Neonatal diarrhoea (ND)	1–7 days
	Pre-weaning diarrhoea	Less frequent compared to ND
	Post-weaning diarrhoea and Oedema disease	5–14 days after weaning, nursery period
Enterococcus spp.	Neonatal porcine diarrhoea	Piglets 2–20 days old
Clostridium perfringens (Cp)	Type C—Necrotic enteritis	1–7 days
	Type A—Diarrhoea	Neonatal diarrhoea till weaning
	HBS-haemorrhagic bowel syndrome	Mainly finishing pigs, role of CpA remain still unclear
Clostridioides (formerly Clostridium) difficile	Diarrhoea, ill thrift, typhlocolitis	3–7 days
Salmonella spp.	Typhimurium—Occasional diarrhoea, septicaemia, death	Grower pigs 6–16 weeks
	Derby—Occasional diarrhoea	Grower pigs 6–16 weeks
	Choleraesuis—Septicaemia diarrhoea, PRDC, death	Finishing pigs 12–16 weeks
Lawsonia intracellularis	Porcine proliferative enteropathy (ileitis)	Grower pigs (nursery, early finishing pigs)
	Regional/necrotic ileitis	Grower pigs
	Porcine haemorrhagic enteropathy	Finishing pigs and young adults (gilts) 16–40 weeks
Brachyspira hyodysenteriae *Brachyspira hampsonii* *Brachyspira suanatina*	Swine dysentery	Growers and finishers, 6–26 weeks
		All ages in primary breakdown
Brachyspira pilosicoli *Brachyspira murdochii* *Brachyspira intermedia*	Intestinal spirochaetosis 'colitis' Porcine Colonic Spirochaetosis	Grower pigs

Key: PRDC—porcine respiratory disease complex; MDA—maternally derived antibodies

Table 2 The first, second and last treatment of main bacterial diseases of pigs

Infection/disease	First choice (Cat. D 'Prudence')	Second choice (Cat. C 'Caution')	Last resort (Cat. B 'Restrict')	Comments
Respiratory				
M. hyopneumoniae:				
Enzootic Pneumonia	Tetracyclines (Inj, IW, IF)	Pleuromutilins: Tiamulin (Inj, IW, IF) Valnemulin (IF)	Marbofloxacin[a] Enrofloxacin[a]	Effective piglets and gilt vaccination commercially available
		Macrolides: Tylosin (Inj, IW, IF) Tilmicosin (IW, IF) Tylvalosin (IW, IF) Tulathromycin (Inj)		
		Lincomycin (Inj, IW, IF) Florfenicol (Inj, IW, IF)		
P. multocida				
Pneumonia	Penicillin (Inj, IW, IF)	Florfenicol (Inj)	Marbofloxacin[a] Enrofloxacin[a]	Common *M. hyopneumoniae* involvement
	Amoxicillin (Inj, IW, IF) Penethamate (Inj)	Macrolides: Tulathromycin (Inj) Tildipirosin (Inj) Gamithromycin (Inj)		
Rhinitis atrophicans (dermonecrotoxine positive)	Tetracyclines (Inj, IW, IF) Trimethoprim/S[b] (Inj, IW, IF)			Vaccination available
A. pleuropneumoniae				
Pleuropneumonia	Penicillin (Inj, IW, IF) Amoxicillin, (Inj, IW, IF) Penethamate (Inj)	Cephalosporins 1 & 2G (Inj)	Cephalosporins 3 and 4G[a]	Effective pig vaccination available together with zootechnical interventions (microclimate etc.)
	Tetracycline (Inj, IW, IF) Trimethoprim/S[b] (Inj, IW, IF)	Florfenicol, (Inj, IW, IF)	Marbofloxacin[a] Enrofloxacin[a]	
		Tiamulin (Inj, IW)		
		Macrolides: Tilmicosin (IW, IF) Tulathromycin (Inj) Tildipirosin (Inj) Gamithromycin (Inj)		

Septicaemia/bacteraemia				
G. parasuis				
Glässer's disease	Penicillin, (Inj, IW, IF) Amoxicillin (Inj, IW, IF)	Florfenicol (Inj, IW, IF)	Cephalosporins 3 and 4G[a] Marbofloxacin[a] Enrofloxacin[a]	Pig vaccination difficult due to the serotype variability
	Tetracycline (Inj, IW, IF) Trimethoprim/S* (Inj, IW, IF)	Amoxicillin/clavulanate (Inj, IW) Cephalosporins 1 & 2G (Inj)		
S. suis				
Meningitis, Septicaemia, Arthritis	Penicillin (Inj, IW, IF) Amoxicillin (Inj, IW, IF)	Florfenicol (Inj, IW, IF)	Cephalosporins 3 and 4G[a]	Pig vaccination difficult due to the serotype variability
	Trimethoprim/S[b] (Inj, IW, IF)	Cephalosporins 1 and 2G (Inj)		Autogenous vaccines available
M. hyosynoviae and M. hyorhinis				
Arthritis, Polyserositis	Tetracyclines (Inj, IW, IF)	Tiamulin (Inj, IW, IF) Lincomycin (Inj, IW, IF)		No commercial vaccines available; some tetracycline and macrolide resistance
		Macrolides (Inj, IW, IF)		
Enteric				
E. coli				
Neonatal Diarrhoea	Trimethoprim/S[b] (OD,[c] Inj)	Aminoglycosides: Neomycin, Apramycin (OD)	Cephalosporins 3 and 4G[a] Enrofloxacin (Inj, OD)[a] Colistin (OD)[a]	Effective sow vaccination available
		Amoxicillin/clavulanate (Inj, IW)		Colostrum management
Pre-weaning Diarrhoea	Trimethoprim/S[b] (Inj, IF, IW)	Aminoglycosides (Neomycin, Apramycin (OD)	*Zinc oxide (IF)*[d] Fluoroquinolones (Enrofloxacin (Inj, OD)[a]	Colostrum/milk management
Post-weaning diarrhoea	Trimethoprim/S[b] (Inj, IF, IW)	Aminoglycosides: Neomycin (IW, IF) Apramycin (IW, IF) Paromomycin (IW)	*Zinc oxide (IF)*[d] Fluoroquinolones (Enrofloxacin (Inj, OD)[a]	Piglet vaccination available
		Amoxicillin/clavulanate (Inj/IW)		Susceptibility testing required
MMA syndrome				
E. coli—dominant pathogen	Trimethoprim/S[b] (Inj) Amoxicillin (Inj)	Aminoglycosides (Inj) Amoxicillin/clavulanate (Inj)	Fluoroquinolones (Enrofloxacin (Inj) Marbofloxacin (Inj)[a]	Complex approach including management and feeding regime is important
Salmonella spp.				
Diarrhoea	Trimethoprim/S[b]	Aminoglycosides: Neomycin, Apramycin (IW, IF) Spectinomycin (IW, IF) Amoxicillin/clavulanate (Inj, IW)	Colistin (IW, IF)[a] Fluoroquinolones[a]	Complex control—acidification of water or feed Vaccination available

(continued)

Table 2 (continued)

Infection/disease	First choice (Cat. D 'Prudence')	Second choice (Cat. C 'Caution')	Last resort (Cat. B 'Restrict')	Comments
Septicaemia	Trimethoprim/S[b] (Inj)	Aminoglycosides (Inj) Amoxycillin/clavulanate (Inj, IW)	Fluoroquinolones[a]	
C. perfringens				
Necrotic enteritis and CpA enteritis	Penicillin (Inj, IW, IF) Aminopenicillins (Inj, IW)	Tylosin (Inj, IW, IF) Tiamulin (Inj, IW, IF) Valnemulin (IF) Amoxicillin/clavulanate (Inj, IW)	Cephalosporins 3 and 4G[a]	Effective pre-farrowing sow vaccination
L. intracellularis				
Ileitis	Tetracyclines (Inj, IW, IF)	Pleuromutilins (Inj, IW, IF) Macrolides (Inj, IW, IF) Lincomycin (Inj, IW, IF)		Pig vaccination
PHE	Tetracycline (Inj)	Tiamulin (Inj) Tylosin (Inj)		
B. hyodysenteriae				
Swine dysentery	Tetracyclines (Doxycycline (IW, IF)	Pleuromutilins: Tiamulin (Inj, IW, IF) Valnemulin (IF)		Wide-spread resistance; MIC determination needed especially for Macrolides
		Lincomycin (Inj, IW, IF) Macrolides (Inj, IW, IF)		
B. pilosicoli				
Colitis	Tetracyclines (Doxycycline (IW, IF)	Pleuromutilins: Tiamulin (Inj, IW, IF) Valnemulin (IF)		Less wide-spread resistance; MIC determination needed especially for Macrolides
		Lincomycin (Inj, IW, IF) Macrolides (Inj, IW, IF)		
Skin				
S. hyicus				
Exudative Epidermitis	Penicillin, (Inj, IW, IF) Amoxicillin (Inj, IW, IF) Trimethoprim/S* (Inj, IW, IF)	Florfenicol (Inj, IW, IF) Cephalosporins 1 and 2G (Inj)	Cephalosporins 3 and 4G[a] Enrofloxacin[a]	Frequently reported resistance to several groups of ATB including first choice

Adapted Burch et al. 2008 (Guidelines for Antimicrobial Use in Swine) with update to EU Categorisation of Antimicrobials Guideline 2019

[a] Antimicrobials from the last resort category according to the Answer to the request from the European Commission

[b] Sulphonamide

[c] Inj—injectable; OD—oral dosing; IW—in water; IF—in feed for updating the scientific Advice on the impact on public health and animal health of the use of antibiotics in animals—Categorisation of antimicrobials EMA/CVMP/CHMP/682198/2017

[d] Therapeutic use (as veterinary medicinal product) banned in EU since June 2022

2.1 Respiratory Pathogens

2.1.1 *Actinobacillus pleuropneumoniae*

One of the most important respiratory pathogens causing highly contagious, potentially fatal respiratory infections of mainly finishing pigs, is *A. pleuropneumoniae* (App). There are currently 19 serotypes based on sequencing and serological identification (Stringer et al. 2021). Among them, serovars 1, 5, 9, 11 and 16 are considered the most virulent due to their expression of ApxI and ApxII toxins (Sassu et al. 2018). Approaches to controlling the disease include good husbandry, biosecurity and vaccination which is mainly based on toxoid vaccines providing multiple serotype cross protection. Concerning App, but other respiratory diseases as well, data provided and evaluated using the lung score programs are a valuable means of monitoring herd lung health and asses control programmes in order to reduce ATB use on farms (Sipos et al. 2021). In the acute stage, the disease is characterised by haemorrhagic necrotising pneumonia and fibrinous pleuritis and often requires targeted antibiotic treatment preferably by injection, to limit severity, mortality and spread. In many herds, there is a bout of coughing in growing pigs and recovery but leads to about 30% of pigs having lesions of pleurisy at slaughter causing significant economic impact. Reports from many countries suggest that 30–50% of all pigs are infected (Taylor 2013). A variety of antibiotics are commonly used for both treatment and metaphylaxis, including mainly tetracyclines, penicillins (amoxicillin), trimethoprim/sulphonamides, macrolides and the fluoroquinolones. *Actinobacillus pleuropneumoniae* is considered a pathogen with the potential to develop antimicrobial resistance, with possible geographically distinct patterns, especially in case of tetracyclines and penicillins (Vanni et al. 2012; Burch 2013). Variability in antimicrobial susceptibility of *A. pleuropneumoniae* isolates originating from a single farm has been described as well (Dayao et al. 2015). Variable minimum inhibitory concentration (MIC) susceptibility patterns were also reported for tiamulin and tulathromycin, which could be also associated with different diagnostic procedures and culture methods (Pridmore et al. 2011; Casals et al. 1990; Aarestrup and Jensen 1999; Godinho et al. 2005; Matter et al. 2007). Despite that, overview of MICs and susceptibility threshold (linked to clinical breakpoints as set by CLSI vet S01, 2023) and ECOFFs can be helpful for the set of antimicrobials used in clinical practice (Table 3). To get insight into the proportion of

Table 3 Antimicrobial susceptibility of *A. pleuropneumoniae* in 158 EU isolates

Antimicrobial	MIC_{50} (µg/ml)	MIC_{90} (µg/ml)	Range (µg/ml)	Resistance (%)	S[1]/(T)ECOFF[2]
Amoxicillin	0.25	32	0.12–128	11.4[a]	≤0.5/1
Amoxicillin + clavulanic acid	0.25	0.5	0.12–2.0	0	NA/(1)
Cefquinome (4G)	0.015	0.03	0.004–0.5	0[b]	NA/(0.06)
Ceftiofur (3G)	0.015	0.03	0.008–2	0	≤2/0.06
Marbofloxacin	0.03	0.06	0.015–2	1.3[c]	NA/NA
Enrofloxacin	0.03	0.06	0.015–2.0	1.3[c]	≤0.25/0.125
Florfenicol	0.25	0.5	0.12–16	0.6	≤2/1
Trimethoprim+ sulphonamide	0.06	0.12	0.008–32	1.9	NA/0.25
Tetracycline	0.5	16	0.25–128	23.6	≤0.5/2
Tilmicosin	16	16	2.0–32	0.6	≤16/(32)
Tiamulin	8	16	2.0–16	0	≤16/(32)
Tulathromycin	32	32	4.0–64	0	≤64/64

Adapted from El Garch et al. (2016), last available clinical breakpoints (CLSI 2023) and (T)ECOFF (MIC EUCAST 2023)
[a] CLSI VET08, 4th ed. (2018) breakpoint for ampicillin and swine respiratory disease
[b] CLSI VET08, 4th ed. (2018) breakpoint for ceftiofur and swine respiratory disease
[c] CLSI VET08, 4th ed. (2018) breakpoint for enrofloxacin and swine respiratory disease
S[1] Susceptible, according to the CLSI (VET 01S, ed. 6: 2023)
(T) ECOFF[2], according EUCAST, 2023, numbers in brackets are tentative ECOFFs (gained from 3 to 4 distributions)
NA—not available till 2023 in CLSI/EUCAST

the European isolates resistance, based on the selection of the results of 10 recent European region studies and MICs detected in isolates of *A. pleuropneumoniae*, selected and commented results of analysis performed by team led by EFSA AHAW (2021) are mentioned. According to the weighted arithmetic proportion of resistance, the tetracyclines are reported as the group with the highest resistance proportion 20.1% (but at the same time with weighted standard deviation (21.6) calculated, indicating heterogeneity among studies involved). Much lover patterns were indicated for combination of sulphonamide/trimethoprim (8.3%) and for penicillins (7.8%). It should be noted that for 10 studies, where penicillin was tested, weighted standard deviation was 15.7, indicating heterogeneity among them (EFSA 2021). Despite some limitations, these results correspond to the fact that these groups of antimicrobials belong to the top three highest consumption classes during the last decade in Europe (ESVAC 2021). Practically all isolates were susceptible to ceftiofur and tulathromycin.

2.1.2 *Pasteurella multocida*

Pasteurella multocida, the causative agent of pneumonic pasteurellosis and progressive atrophic rhinitis, is considered an important agent able to cause respiratory infections usually in older pigs, following viral or especially *Mycoplasma hyopneumoniae* infections, resulting in an increase in their severity, secondary bacterial invasion and long-term colonisation. Because of the common mycoplasma involvement, the tetracyclines are the most commonly used antibiotic followed by the macrolides, pleuromutilins and finally, fluoroquinolones by injection. Susceptibility to penicillins (amoxicillin) remains high and with a stable profile. With the exception of tetracycline where 20.4% of isolates of *P. multocida* were resistant, the resistance to the other antibiotics was zero to low in a large-scale pan-European antibiotic susceptibility monitoring programme (Table 4) (El Garch et al. 2016). Geographical differences might play an important role in selection of effective antibiotic treatment. Antimicrobials tested in seven studies within European region, which were involved in the dataset analysed by experts in EFSA AHAW (2021) panel for presence of the resistant isolates, are commented. Starting with comments on beta-lactams, very low proportions of resistant isolates were detected across studies for ceftiofur and combi-

Table 4 Antimicrobial susceptibility of *P. multocida* in 152 EU isolates

Antimicrobial	MIC $_{50}$ (μg/ml)	MIC $_{90}$ (μg/ml)	Range (μg/ml)	Resistance (%)	S[1]/(T)ECOFF[2]
Amoxicillin	0.25	0.5	0.12–128	2.6[a]	≤0.5/0.5
Amoxicillin + clavulanic acid	0.25	0.5	0.12–1.0	0	NA/
Cefquinome (4G)	0.03	0.06	0.008–0.5	0[b]	NA/0.125
Ceftiofur (3G)	0.004	0.004	0.002–0.06	0	≤2/0.06
Marbofloxacin	0.03	0.06	0.015–0.5	0[c]	NA/NA
Enrofloxacin	0.008	0.03	0.08–0.25	0[c]	≤0.25/0.06
Florfenicol	0.5	0.5	0.25–32	0.7	≤2/1
Trimethoprim+ sulphonamide	0.06	0.5	0.03–32	5.3	NA/0.125
Tetracycline	0.5	2	0.12–128	20.4	≤0.5/2
Tilmicosin	1	32	2.0–32	2	≤16/32
Tiamulin	32	64	4.0–128	nd	NA/64
Tulathromycin	2	4	1.0–8	0	≤16/8

Adapted from El Garch et al. (2016), last available clinical breakpoints (CLSI 2023) and (T)ECOFF (2023)
[a] CLSI VET08, 4th ed. (2018) breakpoint for ampicillin and swine respiratory disease
[b] CLSI VET08, 4th ed. (2018) breakpoint for ceftiofur and swine respiratory disease
[c] CLSI VET08, 4th ed. (2018) breakpoint for enrofloxacin and swine respiratory disease
S[1] Susceptible, according to the CLSI (VET 01S, ed. 6: 2023)
(T) ECOFF[2], according EUCAST, 2023, numbers in brackets are tentative ECOFFs
NA—not available till 2023 in CLSI/EUCAST

nation of amoxicillin and clavulanic acid, while in penicillin weighted arithmetic proportion of resistance 30.7% was calculated. In the case of amoxicillin, analysis led to 10.4% weighted arithmetic proportion of resistance, but this result should be interpreted with caution, due to the fact that five studies involved are heterogeneous (weighted standard deviation calculated 14.5). The highest prevalence of resistance (weighted arithmetic mean proportion of resistance 28.3%) was detected in tetracyclines, but again in seven studies involved weighted standards deviation was 16.7. For sulphonamide trimethoprim, results vary among regions (studies), with highest reported proportion of resistant isolates from Spain reported in study by Petrocchi-Rilo et al. (2018). Resistance was low in fluoroquinolones, florfenicol and mostly to tulathromycin as well, with exception of a Spanish study (Cid et al. 2019) reporting 11.6% of 69 isolates resistant to that drug. For more exact interpreting of the results from such analysis EFSA AHAW (2021), it is important to take into consideration geographical differences of MIC pattern and consequently, the frequency of resistance as well as possible differences in studies as for methodology of examination.

2.1.3 *Mycoplasma hyopneumoniae*

Mycoplasma hyopneumoniae (Mhp) causes enzootic pneumonia, a chronic respiratory disease characterised clinically by coughing and resulting in considerable economic losses. Several vaccines are available to reduce the consequences of infection and clinical disease, but medication with antimicrobials in feed or water is still a common practice (Vicca et al. 2004). Successful eradication of *M. hyopneumoniae* based on targeted medication by effective antibiotics (e.g. pleuromutilins, macrolides), together with specific management measures applied at a farm level, has been described (Burch et al. 2016). The antimicrobial susceptibility testing of *M. hyopneumoniae* field isolates is not a routine practice in daily diagnostic practice, due to the fact that *M. hyopneumoniae* is a fastidious and slowly growing microorganism. The interpretation of results of laboratory testing can be difficult as well, as there are no specific laboratory standards and clinical breakpoints set. Limited numbers of susceptibility studies with field strains are available for the same reasons. The most frequently used groups of antibiotics for treatment in the field are: the tetracyclines, the macrolides, the pleuromutilins, the amphenicols and the fluoroquinolones. Due to the specific cellular structure of *Mycoplasma* spp. characterised by the absence of a cell wall, it is refractory to all antimicrobials that target the cell wall (e.g. penicillins, cephalosporins) (Lysnyansky and Ayling 2016).

Based on published pan-European antimicrobial susceptibility monitoring programme involving 50 field isolates of *M. hyopneumoniae* from Belgium, Spain and the United Kingdom (period 2010–2012, Mycopath I), the most effective antimicrobials were the macrolides (tulathromycin) and the pleuromutilins (tiamulin, valnemulin) with a narrow range of MICs and variability in between countries (Klein et al. 2017). The next 2015–2016 MycoPath II pan-European antimicrobial susceptibility monitoring survey, where 147 porcine *M. hyopneumoniae* from Belgium, France, Germany, Great Britain, Hungary, Italy and Spain were tested, brought similar $MIC_{50/90}$ results. Majority of them were within ± two dilution steps, except for the MIC_{50} of oxytetracycline with more than two dilution steps higher. MIC_{50}/MIC_{90} (mg/L) values were: for macrolides: tulathromycin $\leq$0.001/0.004; tylosin 0.016/0.063; spiramycin 0.063/0.25; tilmicosin 0.063/0.5; gamithromycin 0.063/0.5; for tetracyclines: doxycycline 0.25/1; oxytetracycline 0.25/2; for pleuromutilins: tiamulin 0.016/0.063 and valnemulin $\leq$0.001/0.004; for fluoroquinolones: enrofloxacin 0.063/1; marbofloxacin 0.063/2 and finally for florfenicol 0.5/1 and lincomycin 0.063/0.25. Values of MICs differ for the fluoroquinolones, tulathromycin and tylosin for individual countries, but inter-country comparison limits sample size per country (de Jong et al. 2021).

2.2 Septicaemic Diseases

2.2.1 *Streptococcus suis*

The disease caused by virulent serotypes of *S. suis* is typical for piglets before weaning (10–14 days) and during the nursery period. It is characterised by septicaemia with localisation in joints, meninges and heart valves (Taylor 2013). Infection can be picked up in piglets from the sow as early as 5 days by aerosol, ingestion or contact and invasion may take place via the tonsil and consequently control, and its eradication is very difficult. At present, the prevention and control of the disease are, besides antimicrobial use, based on the use of autogenous vaccines, which can provide passive immunity to piglets or active immunity in growing pigs, but are often effective only against the homogeneous strain.

The penam penicillins, including amoxicillin, are probably the most widely used antibiotics for the treatment and control of systemic/septicaemic infections in pigs, such as streptococcal meningitis, which affects 20–30% of British herds, usually in the post-weaning period (Burch and Sperling 2018). Penicillins are considered as drugs of choice and recommended for empiric treatment of *S. suis* infection not only in pigs but also in human medicine (Marie et al. 2002). Classically, clinically sick pigs are injected with penicillins, and the remainder of the group treated via the drinking water. This commonly leads to metaphylactic medication mainly delivered in the feed, during the anticipated period of risk, usually after weaning. Worldwide, the most common infection is caused by *S. suis* type 2, but there are 35 different serotypes. However, serotype 9 has become the most prevalent serotype among invasive isolates in some European countries with a large pig industry such as the Netherlands and Spain (Rieckmann et al. 2020). Long-acting forms of ceftiofur (third-generation cephalosporin), given at the time of processing (castration, teeth clipping, tail docking) in the first week of life, were particularly popular for prevention of infections in some countries, but its use is coming under more scrutiny, as the last resort option due to the associated production of ESBL resistance genes and, therefore, has been voluntarily banned in some European Union (EU) Member States, like Denmark, the Netherlands and France, as well due to the EMA categorisation of ATB. What is considered of need to pay attention as well is situation with long-acting injectable macrolides, especially tulathromycin. The experience gained in some studies (the Czech Republic) or use data in pig sector (the Netherlands) show increasing use of LA macrolide (tulathromycin) mainly during the piglets processing in farrowing house, which can be considered in relation to increasing level of resistance to macrolides as well as multidrug resistance of the serotypes concerned (Matiašovic et al. 2021; The Netherlands Veterinary Medicines Institute 2022). Macrolides should not be used for the treatment and metaphylaxis of septicaemic infections in piglets in general, due to the PK profile and fast distribution to organs (e.g. lungs). High-weighted arithmetic proportion of resistance in *S. suis* (59.5%, 3 studies/542 isolates, weighted standard deviation 7.9, European region) was detected in tylosin (EFSA AHAW 2021). Moreover, evidence for cross-antimicrobial resistance genes selection (combination of macrolide, lincosamide and lincomycin/spectinomycin use versus macrolide resistance) has been determined in the resistome analysing data (Van Gompel et al. 2019) and similarly in data by the team led by Matiašovic (2021). This is not surprising, as mechanisms behind macrolide and lincosamide resistance are common to both classes.

Very low levels of resistance to commonly used antimicrobials except to tetracycline were reported in a pan-European study including isolates from several important swine producing countries (El Garch et al. 2016). Recently, results from European studies indicate very low levels of resistance to beta-lactams (lowest detected in aminopenicillins and ceftiofur, slightly higher for penicillin beta-lactamase sensitive), while very high levels of resistance were reported for the class of tetracyclines (EFSA AHAW 2021). High proportion of isolates susceptible to specifically penicillins and aminopenicillins remains, despite the worldwide use of beta-lactams in pigs for over 50 years (Segura et al. 2020). There should be noted that despite situation in availability of interpretive criteria for *S. suis* has improved with time and according to current CLSI standards (6th ed. 2023) following MIC breakpoints (μg/ml) are available: for ampicillin, enrofloxacin and tetracyclin—susceptible ≤0.5; intermediate 1; resistant ≥2; for ceftiofur and florfenicol—susceptible <2; intermediate 4; resistant ≥8 and for penicillin G susceptible ≤0.25; intermediate 0.5; resistant ≥1 (CLSI 2023). Still some veterinary specific breakpoints and by international standards set interpretive criteria are missing (e.g. for macrolides, where (tentative ECOFF only is available 0.5 for erythromycin).

The difference in serotype susceptibility was also described, especially in the case of the virulent serotype 2 (Marie et al. 2002).

2.2.2 *Glaesserella* (Formerly *Haemophilus*) *parasuis*

Glaesserella (formerly *Haemophilus*) *parasuis* (Gps) may be the cause of great economic losses for nursery piglets and breeders as well, due to the cost of antibiotic therapy and piglet death in acute forms of the disease (Oliveira et al. 2001). Gps is the causative agent of Glässer's disease and has many similarities with *S. suis*, including colonisation of the respiratory tract and invasion causing septicaemia and then lodging in joints, occasionally brain but more frequently serosal surfaces, usually in young pigs 3–6 weeks of age and may also be associated with coughing and dyspnoea. It is widespread in herds (up to 30% colonisation level) but did not always cause major problems (Taylor 2013). With the advent of porcine reproductive and respiratory syndrome virus (PRRSV) and other immuno-suppressive agent's affecting macrophages, the situation seemed to get worse. A high antigenic heterogeneity exists among *G. parasuis* strains. According to the currently worldwide accepted classification, 15 serovars of *G. parasuis* (1–15) have been defined, and serovar 5 is considered as the most virulent one. The serovar 5 vaccine is available on the EU market. However, it is necessary to say that a large number of non-typeable *H. parasuis* isolates also exist, which makes vaccination covering the spectrum of serotypes difficult (Kielstein and Rapp-Gabrielson 1992).

Antibiotic treatment is one of the commonly used measures for the control of *G. parasuis* infections. Treatment is recommended especially for individual pigs and to control severe and acute outbreaks due to the Gps systemic infection (Aragon et al. 2012). Effective treatment options belong mainly to the penicillin group, amoxicillin), tetracyclines and trimethoprim/sulphonamides but also to the cephalosporins, florfenicol and the fluoroquinolones. Results of susceptibility testing of European isolates confirmed high susceptibility to selected antibiotics, applying the breakpoints for *A. pleuropneumoniae* and/or also for *P. multocida),* with low levels of resistance/decreased susceptibility to trimethoprim/sulfamethoxazole and ciprofloxacin (El Garch et al. 2016; CLSI VET08, 4th ed., 2018; Nedbalcova et al. 2017). Antibiotic susceptibility profiles can be variable in different countries even between EU countries, based on selection of preferred treatment options. Czech, Danish and British isolates were reported susceptible to most antibiotics, whereas high rates of resistance are reported from China and Spain (Aarestrup et al. 2004; De la Fuente et al. 2007; Zhou et al. 2010; Nedbalcova et al. 2017).

2.3 Enteric Infections

2.3.1 *Salmonella* spp. Infections

Salmonella enterica spp., especially *S.* Typhimurium and *S.* Derby, can occur in pigs causing clinical disease with diarrhoea and depressed growth, as well as deaths. Other *S. enterica* serovars can cause infection and colonisation of pigs like: Derby, Saintpaul, Infantis, Heidelberg, Typhisuis. *Salmonella* Choleraesuis is primarily adapted to swine (host-specific) and is rather rarely reported in EU countries, but is still frequently reported from North America and Asia, causing septicaemia and is an important part of the porcine respiratory disease complex (PRDC) (Wales and Davies 2017). Definitive (phage) Type (DT) 104 of serovar Typhimurium and the recent monophasic variant 4.5.12i are multi-antibiotic resistant. Generally, the susceptibility patterns of salmonella isolates are similar to *E. coli* with slightly less frequent reported resistance (Burch 2013).

In clinical cases, colistin or antimicrobials from the groups of aminoglycosides or fluoroquinolones can be the antibiotic of choice, with low-level resistance reported. Despite this, susceptibility testing is recommended. For an acute severe outbreak, the injectable route of administration should be preferred to oral one due to the necessity to reach systemic availability of the drug, as well as due to the lowered appetite and intake of the feed and possibly water by animals, especially those with a high fever. Amoxicillin in combination with clavulanic acid administered via drinking water has also proven successful. Resistance to ampicillin, streptomycin and oxytetracycline was described in JVARM (2016) on a total 143 isolates from pigs: Typhimurium (58) and Choleraesuis (40).

A high prevalence of MDR (multidrug-resistant) *Enterobacteriaceae* including *Salmonella* spp. and high resistance-gene diversity in livestock manure from pigs was recently reported (Amador et al. 2019; Liu et al. 2015).

Generally, *Salmonella* spp. control via acidification of the drinking water or occasionally liquid feed together with vaccination has been successfully achieved, if there are major clinical problems. Especially, vaccination would be promising tool in order to reduce ATB use on farms. Long-term vaccination with a live-attenuated S. Typhimurium vaccine can reduce the level of S. Typhimurium in the herd environment and decrease the use of ATB for treatment (van der Wolf et al. 2021).

2.3.2 Clostridial Infections

The main causative agents of clostridial enteric infections of piglets are C. perfringens type A, type C and Clostridium difficile. Clostridium perfringens nomenclature is based on different toxin gene presence and production (major toxin A-F). Clostridium perfringens type A (CpA) is an anaerobe, commonly isolated from the intestines of piglets and is considered a normal inhabitant of the piglet's microbiome. *C. perfringens* type A belongs to the most prevalent pathogens associated with cases of neonatal diarrhoea (ND) together with *E. coli* and RVA (rotavirus type A) and was the main pathogen detected in faeces of neonatal diarrheic piglets in Spain and Germany, just recently (Mesonero-Escuredo et al. 2018; Mertens et al. 2022). Current evidence suggests its involvement in diarrhoea in nursing piglets, especially those less than 4 days of age. Pathogenic strains of CpA have to present not only alpha toxin also but beta-2 toxin, *which is important for* development of clinical diarrhoea (Van Asten et al. 2010). CpA-caused enteritis is characterised by a non-haemorrhagic mucoid diarrhoea in suckling and feeder pigs (Barker et al. 1993; Songer and Uzal 2005). *Clostridium perfringens* type C (CpC) infection is characterised by haemorrhagic, often fatal, necrotic enteritis in young piglets, with deep mucosal necrosis in the small intestine, sometimes affecting the caecum and proximal portion of the colon. Prevention (against CpA and CpC) is based on effective vaccination of sows before farrowing with consequent passive immunisation of their offspring via toxin neutralising antibodies in colostrum. Due to the character of clinical disease caused by CpC, antibiotic treatment has little value and prevention via mass vaccination of breeding animals is far more effective. A well-known interaction between CpA and *Cystoisospora suis* will result in an increase in clinical disease and mortality (Mengel et al. 2012). *Clostridium perfringens* susceptibility data from in vitro phenotype testing as investigated and published during the last different European countries and Brasil are provided in Table 5.

Clostridioides (formerly *Clostridium difficile*) is commonly found in the colon of clinically normal animals, but establishment and toxin production in proximity to the mucosal surface lead to lesion development, affecting mainly colonic mucosa with a certain level of age resistance (Arruda et al. 2013). *Clostridioides difficile* is commonly associated with a spectrum of diseases in humans and other animals and use of antimicrobials, especially penicillin-related compounds, is considered a risk factor for development of disease in humans. Almost all antimicrobials have been implicated in the devel-

Table 5 Antimicrobial susceptibility to various antibiotics against *C. perfringens*

Antimicrobial	MIC_{50} (μg/ml)	MIC_{90} (μg/ml)	Range (μg/ml)
C. perfringens—Belgium (58 isolates) Dutte & Devriese, 1980			
Tiamulin	–	–	0.25–4.0
Lincomycin	2	256	0.12–>512
Penicillin G	0.12	0.5	0.06–1.0
Tetracycline	16	32	0.06–>64
C. perfringens—Belgium (95 isolates) Devriese et al., 1993			
Tylosin	≤0.12	0.25	≤0.12–>64
C. perfringens—Czech Republic (123 isolates) Masarikova, 2006			
Tylosin	1	8	<0.125–>16
Tiamulin	8	16	<0.5–>32
C. perfringens—Czech Republic (46 isolates) Masarikova et al., 2012			
Ampicillin	0.125	2	0.03–4
Amoxicillin	0.03	2	≤0.01–4
Amoxicillin-clavulanic acid (ratio 4:1)	0.03	1	≤0.01–4
C. perfringens—Italy (30 isolates) Agnoletti et al., 2010			
Tiamulin	4	64	0.125–128
Valnemulin	0.125	8	0.063–32
C. perfringens—Denmark (38 isolates) Agnoletti et al., 2010			
Tiamulin	2	4	0.25–8
Valnemulin	0.063	0.125	0.016–0.25
C. perfringens—Brazil (50 isolates) Salvarani et al., 2012			
Amoxicillin	0.5	1	0.25–8
Ceftiofur	0.5	8	0.25–8
Florfenicol	2	64	0.5–64
Josamycin	16	>256	2–>256
Lincomycin	16	>256	2–>256
Tetracycline	16	256	2–256
Tylosin	0.5	16	0.25–32

opment of CDI, especially when given in combination (Owens et al. 2008). Immature intestinal flora commonly observed in neonatal animals including piglets is also considered a major risk factor to *C. difficile* infection.

In piglets, antibiotic administration did not necessarily appear to influence the presence and severity of lesions and possibly the role of antimicrobials remains unclear (Arruda et al. 2013; Arruda 2014). The *Clostridioides difficile*–associated disease (CdAD) has emerged as a cause of enteritis in neonatal pigs, especially in the USA. Several reports described clinical outbreaks caused by *C. difficile,* either as an uncomplicated or mixed form of infection with an impact on production, i.e. a decreased weaning weight 10%–15% (Songer and Uzal 2005).

Due to the presence and colonisation by clostridia in clinically healthy animals, only isolation of these organisms from intestinal content and tissues of animals cannot be considered as proof of diagnosis and specific interpretive criteria (i.e., numbers of organisms, clinical manifestation of diarrhoea), and other tests (gene or toxin detection) are required to reliably establish a final diagnosis (Songer and Uzal 2005).

As an interim therapy prior to immunity being established, effective antibiotics, which could be used especially for treatment of acute outbreaks of CpA infection, belong mainly to the penicillin group, with very low MICs being detected for aminopenicillins and amoxicillin in combination with clavulanic acid (Dutta and Devriese 1980; Masarikova et al. 2012; Salvarani et al. 2012). Variable susceptibility to tiamulin and tylosin was detected with possible geographical differences reflecting different antibiotic policy and standards (Masarikova 2006; Agnoletti et al. 2010; Salvarani et al. 2012). Extensive resistance is described to the tetracyclines and macrolides, generally (Rood et al. 1978, 1985).

As a result of the complexity and cost of testing, *C. difficile* is rarely tested for antimicrobial susceptibility. Contrasting susceptibility patterns to both pleuromutilins (tiamulin, valnemulin) were shown by isolates of Danish and Italian origin (Agnoletti et al. 2010). Recently published results of a study carried out on 94 swine isolates of *C. difficile* indicated high resistance to clindamycin (56.4%), levofloxacin (100%) and erythromycin (23.4%). A lower rate of resistance was described against ampicillin, linezolid and rifampicin (14.9%; 8.5%; 2.1%) and for amoxicillin in fixed combination with clavulanic acid, vancomycin, and metronidazole, no resistance was detected (Thitaram et al. 2016). Despite the set of antibiotics tested, which were selected according to the relevance for human medicine, it confirms good in vitro susceptibility to amoxicillin alone or in fixed combination with clavulanic acid.

2.3.3 *E. coli* Infections

Escherichia coli can cause gastrointestinal tract infections, which may lead to neonatal diarrhoea (ND), septicaemia and death in very young piglets, usually during the first week of age of piglets, further diarrhoea and uneven growth can be seen at 7–10 days of age and up to weaning, but the most common time is after weaning in 90% of all herds. The enterotoxigenic *Escherichia coli* pathotypes (ETEC) are characteristic by specific fimbriae presence, strains causing neonatal colibacillosis mostly carry the fimbriae F4 (K88), F5 (K99), F6 (987P) or F41, while the ETEC of post-weaning diarrhoea carry the fimbriae F4 (K88) and F18 (Luppi 2017). A number of factors contribute to the severity of the disease, which can result in death, diarrhoea and poor growth performance. Following groups of ETEC toxins are produced, namely the two major classes of enterotoxins, heat-stable toxin (ST) and heat-labile toxin (LT), as well as the enteroaggregative heat-stable toxin 1 (EAST1). Oedema disease, which occurs usually shortly after weaning, characterised by enterotoxemia with nervous signs and caused by Shigatoxigenic *E. coli* strains (STEC) producing shigatoxin (Stx2e) has recently started to be preferably controlled by effective vaccination based on toxoid vaccines (Kim et al. 2022).

A number of management factors, additives (zinc oxide), vaccines and also antibiotics can be used to mitigate the problem. The therapeutic use of zinc, which is frequently used mainly in premix form for medication of feed after weaning, was debated, and finally, the Committee for Medicinal Products for Veterinary Use (CVMP) of the European Medicines Agency (EMA) recommended as indicated in the document EMA/CVMP/147249/2017 (CVMP 2017) the withdrawal of existing marketing authorisations for veterinary medicinal products containing zinc oxide. In 2017, the European Commission adopted a Commission Implementing Decision to withdraw, by 26 June 2022, the marketing authorisations of veterinary medicines containing zinc oxide for oral administration to food-producing animals. Due to the huge expected impact of the zinc oxide ban within the period of the high pressure on reduction of use of antimicrobials, studies for managing *E. coli* infections, including investigations of anti-enteric colibacillosis agents in piglets, have started to be prioritised. Except already known concepts, like prebiotics, probiotics, synbiotics, organic acids, phytogenic substances, unfortunately none of them able to fully and easily replace zinc oxide in pig farming, bacteriophages, spray-dried healthy animals specifically processed plasma, encapsulated IgY immunoglobulins, porcine lactoferrin, antisense oligonucleotides and aptamers were tested within different studies and projects (Castro et al. 2022), and some of them seem experimentally promising, but a long way still remains to become part of the farm practices. Currently, therefore, from veterinary medicinal products authorised, vaccines seem to be the only potent and available tool, but limited by specificity for exact spectrum of *E. coli* and costs significantly differing with zinc oxide—containing VMPs' use.

Vaccination with products based on specific toxoid Stx2e has been shown to be the most successful method for preventing clinical and subclinical form of OD, together with a positive effect by reducing the usage of antibiotics (ATB) and zinc oxide (ZnO) on farms with enzootic OD (Lillie-Jaschinski et al. 2013; Sperling et al. 2022). Particularly, the frequent use of colistin was significantly reduced by vaccination against OD, when compared to a former period without vaccination (Lillie-Jaschinski et al. 2013). On the opposite, there is in general lack of commercially available vaccines for PWEC, limited to oral vaccination of pigs against both F4 and F18-ETEC (Nadeau et al. 2017). General limitation is mainly interference with the lactogenic immunity of piglets, potential absence of cross-protection between strains (adhesins variants) and as well optimal timing of the vaccination (Rhouma et al. 2017). Maternal vaccination, immunisation of

Table 6 Minimum and maximum proportion of resistance (%R or %R + I*), weighted arithmetic mean and weighted standard deviation (WSD) gained from the studies of *E. coli*[a] in European isolates

Antimicrobial	No of Studies/No of isolates tested	Weighted arithmetic proportion of R	Min resistance	Max resistance	Weighted standard deviation
Aminopenicillins	13/8554	63.9	26	98.5	12.7
Amoxicillin + clavulanic acid	6/3786	5.7	2	29.6	10.7
Ceftiofur (3G)	12/8842	4.2	0	15.5	2.8
Colistin	8/5150	9.7	0	76.9	13.8
Fluoroquinolones	14/8934	8.5	0.1	56.5	12.3
Amphenicols* (AT and DK)	2/694 and 90	–	16.7	18.5	–
Apramycin	6/6915	11.5	5	73	16.3
Gentamicin	12/8216	11.7	0	70	16.2
Neomycin	6/6654	15.7	3.8	20	3.8
Spectinomycin	5/6262	35.7	30.3	51	7.7
Sulfonamides	4/1495	65.4	35.2	75	9.7
Trimethoprim+ sulphonamide	9/4309	51.1	26.1	79.1	10.7
Doxycycline* (ES and AT)	2/161 and 694	–	62.7	67.7	–
Tetracycline	13/8503	71.5	25	96.7	11.4

Adapted from EFSA AHAW (2021) and Barros et al. (2023)*
[a]Note: Despite the *E. coli* isolates tested coming from clinical samples, there cannot be ensured that all isolates involved in all studies for which results were summarised are ethiological agens causing or being directly associated with diseases/pathologies. To ensure that samples represent ethological agens of the disease mostly sampling inner organs or confirm in isolates that toxin, fimbriae or other virulence factors are present (some laboratories performs test on haemolytic *E. coli* isolates from clinically diseased animals only)

sows during pregnancy against ETEC infections, is on the contrary very effective way to control ND *E. coli* infections, implemented in the field (Matias et al. 2017).

As for antimicrobials (please refer to Table 6 summary, European region data only), aminopenicillins, sulfonamides and tetracycline were proven to have very high (63.9; 65.4 and 71.5%) weighted arithmetic proportion of resistance, and weighted standard deviation among studies involved was 12.7; 9.7 and 11.4). Being aware of certain limitations, results therefore indicated that in European countries these relatively old antibiotic classes may have very limited efficacy against porcine *E. coli* infections (EFSA AHAW 2021). On the contrary, cephalosporins of the third generation and amoxicillin/clavulanic acid combination both have very low resistance detected within studies involved in EFSA AHAW (2021) summary. Fluoroquinolones, the aminoglycosides (neomycin, apramycin, aminosidine) and colistin are antibiotics with still low level of resistance and often employed, either via oral treatment or by injection during an acute outbreak. The administration route is usually via medicated feed with medication of an entire group of pigs at risk (metaphylaxis). A drop of the use of antibiotics in medicated feed (mainly premixes) has been reported and an increased trend to use water medication instead (ESVAC 2022). There is also a big pressure to minimising the use of both fluoroquinolones and colistin that are involved in category B = "Restrict" in a new, four-group categorisation, which corresponds to Category 2 in the first AMEG report. Categorisation was published as document (EMA/CVMP/CHMP/682198/2017) elaborated by the AMEG group and officially published in 2020 (EMA 2020). Much attention has been paid to colistin recently due to studies performed in China and consequently as well in EU countries, describing a transmissible, plasmid-mediated resistance gene to colistin in commensal *E. coli*, (mostly due to *mcr*-1, *mcr*-4 and *mcr*-5, but until now *mcr* genes 1–10 have been described) and this fact led to the re-assessment of the low risk from use of colistin in animals and to the likely reclassification of colistin to Category B of the proposed European classification system, which includes medicines reserved for treating infections in humans, for which no effective alternative treatments exist (El Garch et al. 2018; Liu et al. 2016; Chen et al. 2018; EMA 2020).

2.3.4 Strongly β-Haemolytic *Brachyspira* spp.

Swine dysentery (SD) is a severe enteric disease of grower and finisher pigs associated with bloody to mucoid diarrhoea with significant adverse effect on growth performance and increased mortality of affected animals. Historically, *B. hyodysenteriae* was considered as the causative agent of SD but recently, since 2007, there has been a reported increase of diagnosis of SD outbreaks in North America along with the detection of a novel pathogenic species—*Brachyspira hampsonii* (Chander et al. 2012). Classic SD has been experimentally reproduced following this strain inoculation in pigs (Burrough et al. 2012). Another strongly β-haemolytic spirochaete, *Brachyspira suanatina*, was isolated from pigs with dysentery-like disease in Sweden and Denmark (Råsbäck et al. 2007).

For the therapy and control of SD, various antimicrobials were frequently used, as the development of efficacious vaccines for protection was challenging, and commercial products are not available at present (Burrough 2017).

Pleuromutilins (tiamulin and valnemulin) are the most widely used antibiotics and are considered as drugs of first choice and have been successfully used in the eradication of SD. Recently, pleuromutilin resistance has been increasingly reported from several countries with a chronic, endemic presence and clonal spread of multiresistant isolates within a country but also in between different countries (Rohde et al. 2004; Rugna et al. 2015; Sperling et al. 2011; Joerling et al. 2018; Swedres-Svarm 2019). Recently, 39.1% isolates of *B. hyodysenteriae* were resistant to both substances (Tia, Val) isolated in the period 1990–2016 in Germany (Joerling et al. 2018), in EFSA AHAW (2021) report data signalise even worse status (Tia 48.8% and Val 54.3%) Isolates of *B. hyodysenteriae* and *B. hampsonii* originating from the USA are generally susceptible to pleuromutilins, with heterogeneous susceptibility to doxycycline and low susceptibility to lincomycin and tylosin (Clothier et al. 2011; Mirajkar et al. 2016). High levels of resistance were reported to macrolides (especially tylosin) and lincosamides in *B. hyodysenteriae* worldwide, which is not surprising, as both groups of antimicrobials were widely used for both preventive and therapeutic medication against SD on pig farms before antibiotic growth promoters were banned (Prasek et al. 2014; Hommez et al. 1998; Franklin et al. 2006; Kirchgässner et al. 2016; Mahu et al. 2017; EFSA AHAW 2021). Increases of the MIC of tylvalosin (second-generation macrolide) authorised for the treatment of SD suggest that the mutation causing tylosin resistance seems to be associated (Karlsson et al. 2004; Prasek et al. 2014).

As *B. hyodysenteriae* and *E. coli* were agreed with >66% certainty by EFSA AHAW (2021) being two most relevant antimicrobial resistant bacteria in swine for the EU (with *E. coli*), it should be of high importance to focus on tools that could help to minimise/eradicate this disease from the farm, rather than be treated with antimicrobials. Total depopulation or at least partial depopulation together with effective medication based on MIC testing, necessarily combined with implementing strict biosecurity measures, seems to be the possible tools for eradication of *B. hyodysenteriae* from clinically infected pig farms (Neirynck et al. 2020).

2.3.5 *Lawsonia intracellularis*

Porcine proliferative enteropathy (PE) is one of the most economically important, intestinal infectious diseases, which has been recognised in all major swine industries throughout the world. Classical clinical manifestations of the disease are recorded post weaning, during the late nursery phase and finishing period. Ileitis is characterised by the thickening of the small intestinal mucosa with certain extension of lesions to the caecum and proximal colon due to enterocyte proliferation, associated with the presence of the

Gram-negative, intracellular bacterium, *Lawsonia intracellularis* (LI) (Lawson et al. 1993). Proliferative enteropathy has been reported in several different animal species (Cooper and Gebhart 1998). In pigs, four main pathologies were described: proliferative enteropathy, necrotic enteritis, regional ileitis and proliferative haemorrhagic enteropathy. The persistence on farms in carrier pigs (growers and breeding stock), as well as different vectors present on farms, makes the disease difficult to eradicate (Friedman et al. 2008).

Effective antimicrobial treatments, as well as commercially available vaccines, are both commonly administered, and thus, the disease is mainly controlled by antibiotic treatment and vaccination at a herd level (Jacobson et al. 2010). Metaphylactic treatment on highly affected farms, including entire batches of pigs, was found to be most efficient in reducing both high-level LI shedding and diarrhoea when compared to the treatment of diarrhoeic pens or individual diarrhoeic pigs (Larsen et al. 2016).

Knowledge of the antimicrobial susceptibility of *L. intracellularis* is important for selection of an optimal treatment option, but the published data on its in-vitro antimicrobial susceptibility are very limited. Three major studies have been carried out so far, evaluating the susceptibility of LI: McOrist et al. (1995) included three isolates of *L. intracellularis* in a European study, Wattanaphansak et al. (2009) tested six isolates from North America and four from Europe in a US study, and Yeh et al. (2011) tested two isolates from Asia (South Korea). A regular update of standard diagnostics and susceptibility service in the field (strain isolation and MIC testing) is not an option as only a few laboratories in the world can perform standard in vitro antimicrobial susceptibility testing, because of difficult isolation and maintenance of strains in cell culture (Wattanaphansak et al. 2009).

The above-mentioned studies reveal no pattern of increased resistance over time among the North American and European strains of *L. intracellularis,* which is not the case of Asian strains tested, which generally showed higher MICs to selected antimicrobials. Recently, MIC susceptibility testing results for isolates from Brazil and Thailand were published confirming pleuromutilins as the most active against *L. intracellularis* strains tested. In contrast, resistance to lincomycin, gentamicin, trimethoprim, colistin and bacitracin in in-vitro conditions was described (Wattanaphansak et al. 2019).

Based on these studies, a range of effective antimicrobials have been recommended for the treatment of pigs with *L. intracellularis*. Effective treatment options belong mainly to the pleuromutilins (tiamulin, valnemulin), macrolides (tylosin, tylvalosin), lincosamides and tetracyclines (chlortetracycline). Those antimicrobials groups have been well evaluated in clinical papers focused on control and treatment of LI infection in the field (McOrist et al. 1996, 2000; Marsteller et al. 2001; Larsen et al. 2016).

Commercial live-attenuated and inactivated bacterin-based *L. intracellularis* vaccines are available for prophylactic use and are used in the field (Karuppannan and Opriessnig 2018).

2.4 Skin Infections

2.4.1 *Staphylococcus hyicus*: Exudative Epidermitis (Greasy Pig Disease)

Exudative epidermitis (EE) is one of the most common skin diseases of pigs caused by virulent strains of *Staphylococcus hyicus,* which produce exfoliative toxins. So far, a total of six toxins have been discovered, namely ExhA, ExhB, ExhC and ExhD and SHETA and SHETB (Andresen 1998; Andresen and Ahrens 2004; Sato et al. 2000). Infection can be manifested by a generalised or local form typically in younger animals (until 5–7 weeks of age). Strains of *S. chromogenes* producing exfoliative toxin type B, ExhB, were identified by the use of a multiplex PCR, specific for the exfoliative toxins from *S. hyicus,* and were capable of inducing general infections in pigs as well (Andresen et al. 2005). From cases of EE, *S. aureus* was also frequently isolated (van Duijkeren et al. 2007; Park et al.

2013a). Lesions are characterised by excess sebaceous exudation, inflammation, which then develops into epidermal erosions and crusts and consequently with a dark and greasy texture, usually without pruritus (Taylor 2013; Gourreau et al. 2015).

Individual antimicrobial treatment by injection is most frequently combined with topical treatment based on antibiotic, antiseptic and topical oil administration in the field (Park et al. 2013b). The most frequently used topical antibiotics found in a survey carried out in Canada were a mixture of procaine penicillin G and novobiocin (69%), penicillin G (18.7%) and cephapirin benzathine (Park et al. 2013b).The most frequently used and proposed groups of antibiotic for therapy of EE are penicillins (penicillin G, aminopenicillins), gentamicin, florfenicol, trimethoprim in combination with sulphonamide and from the last resort choices, cephalosporins and quinolones (Taylor 2013; Frana 2012; Aarestrup et al. 2008).

Previously, numerous antimicrobial agents have been used for the treatment of *S. hyicus*. Resistance to several antimicrobials is quite common, as well as sharing resistance genes, as staphylococci do not exist in genetic isolation (Schwarz et al. 2018). The development of resistance against many antimicrobial agents, such as erythromycin, penicillin, amoxicillin, lincomycin, streptomycin, sulfonamides, tetracycline and trimethoprim, has been observed and reported in swine industries in Europe, Canada and Asia (Aarestrup and Jensen 2002; Aarestrup and Schwarz 2006; Aarestrup et al. 2008; Park et al. 2013a, 2018). It is difficult to predict the susceptibility of *S. hyicus,* and treatment has to be performed according to the knowledge of the specific farms and routine antimicrobial susceptibility testing of the pathogen. Proper isolation/identification of the causative pathogen with an evaluation of the MIC susceptibility profile is needed (Aarestrup et al. 2008), what was done also in the Czech Republic National Monitoring of Target Pathogens, where MICs of *S. hyicus* isolates from diseased pigs sent by veterinarians were evaluated (n = 51, over the period 2017–2021). Susceptibility patterns determined are as follows: penicillin 82.37%, erythromycin 15.2%, clindamycin 35.95%, tetracycline 57.33% and florfenicol 7.7%, ceftiofur and gentamicin with full susceptibility.

Another management procedure that can be applied and thereby reduce the use of antimicrobial treatment is the autogenous vaccination of sows with a vaccine based on specific, toxin positive *S. hyicus* isolates. This was shown to reduce the metaphylactic treatment with antimicrobials, morbidity and mortality rates in weaned pigs (Arsenakis et al. 2018). Hygiene with segregation of infected animals may also reduce infection pressure.

2.5 Postpartum Dysgalactia Syndrome (PPDS)

Postpartum dysgalactia syndrome (PPDS) is a common health problem in sows, which can affect sow life longevity leading to substantial economic losses (Niemi et al. 2017). Alteration in both milk yield and composition from affected sows has a strong impact on weight gain and growth rate of piglets, consequently causing high pre-weaning mortality. Postpartum dysgalactia syndrome also covers other syndromes and nomenclature as the MMA complex or syndrome (mastitis-metritis-agalactia), coliform mastitis (CM), fulfilling specific diagnostic criteria such as presence of one or more of the following clinical signs: mastitis, metritis, vaginal pH > 8.0, lack of appetite and a rectal temperature >39.4 °C. Other typical manifestations in cases of systemic infection are anorexia, constipation and depression.

Mastitis is one of the dominant clinical signs with a key role played by *E. coli* and endotoxin from Gram-negatives in pathogenesis was confirmed experimentally, while metritis is found only occasionally (Gerjets and Kemper 2009; Ross et al. 1981; Gerjets et al. 2008; Bertschinger et al. 1990; Waldmann and Wendt 2001).

The syndrome is present worldwide in all important pig production areas, and the incidence is estimated between 0.5 and 60% at a farm level, with an average incidence of about 13% (Hirsch et al. 2004; Kemper and Gerjets 2009). It is important to emphasise that PPDS are multifactorial and complex including factors like management and hygiene, feeding, water supply and animal specific factors such as body condition and age of the sows (Pendl et al. 2017; Martineau et al. 2013).

The pathogens most frequently isolated from mastitis-affected sows belong to the class of coliforms, represented by *Escherichia coli, Enterobacter* spp., *Citrobacter* spp., and *Klebsiella* spp. *Streptococcus* spp., *Staphylococcus* spp., *Pseudomonas* spp. and *Corynebacterium* spp. were isolated as well (Gerjets and Kemper 2009; Kemper and Gerjets 2009).

Susceptibility testing of the porcine subset of ß-haemolytic streptococci (54) from the urinary/genital tract as well as MMA syndrome (18 isolates) confirmed no resistance detected to penicillins, cephalosporins and chloramphenicol contrary to high resistance observed to sulfamethoxazole (20%), gentamicin (35%) and tetracycline (93%) (Schwarz et al. 2007). Among the 87 *E. coli* isolated from infections of the urogenital tract including MMA, the highest percentage of resistance was detected for sulfamethoxazole (59%), tetracycline (54%), potentiated sulphonamides (41%) and ampicillin (38%). Resistance to amoxicillin/clavulanic acid, cephalosporins and gentamicin was rare (Grobbel et al. 2007).

Treatment strategy and recommendation of antibiotic selection should be made according to results of susceptibility testing, considering the fact that administration of antibiotic is only part of a complex approach, including administration of non-steroidal anti-inflammatory drugs (NSAIDs) and repeated administration of oxytocin, together with management focused on feed and water intake and supportive treatment to piglets (Martineau et al. 2013; Pendl et al. 2017; Farmer et al. 2019).

Postpartum administration of prostaglandins may have a beneficial effect and prevent clinical endometritis due to the positive effect on involution of uterus and improved litter size at next farrowing on problematic animals (higher parity sows) and farms (Waldmann and Heide 1996; Vanderhaeghe et al. 2008).

The most frequently used therapy are broad-spectrum antibiotics belonging to aminopenicillins, tetracyclines, trimethoprim sulphonamides, macrolides or fluoroquinolones administered parenterally (Taylor 2013).

The selection of suitable antibiotics should be made according to susceptibility and MIC testing and the ability of the antibiotic to reach sufficient concentrations in colostrum/milk above the MIC. Systemic availability needs to be considered as well. Important PK considerations are lipid solubility, degree of ionisation and extent of binding to serum and milk/udder proteins. High lipid solubility, low degree of ionisation and less plasma protein binding contribute to a better transfer into milk (Mestorino and Errecalde 2012). Good distribution into the mammary gland of sows following intramuscular administration was recorded by these antibiotics: sulphonamides macrolides, aminoglycosides, lincomycin, trimethoprim, tiamulin and from antibiotics of last resort, the quinolones (Gruet et al. 2001; Fraile 2013).

3 Pharmacokinetics and Pharmacodynamics (PK/ PD)

3.1 Route of Administration

The route of administration of an antibiotic to a treated animal can directly influence drug absorption, its bioavailability and consequently, its actual efficacy. Absorption is best described by the drug's bioavailability, which is defined as the percentage of a drug's dose that reaches the systemic circulation. For example, when the entire dose is administered by the intravenous (IV) route, which is not a typical way for pigs, except in experimental PK/PD studies, theoretically 100% of that dose is bioavailable and is the comparator for other routes. In contrast, absorption and bioavailability after oral administration (water, feed) can be variable depending on the type of antibiotic, the formulation, as well as on the health status of the treated animal (e.g. effect of food on absorption in an enteric infection) and, in some antimicrobials, can differ also in fed or fasted animals. Only a few antibacterial agents have excellent bioavailability after oral administration. The following routes of administration of antibiotics are frequently used in swine medicine.

3.1.1 Injection

For individual administration, injection is primarily used for the treatment of clinically affected animals with decreased consumption of feed and water. Injection treatment is usually the preferred option for initial treatment when a more rapid onset of treatment effect is required and expected. Subcutaneous (SC) or intramuscular (IM) routes of administration have usually good absorption from the injection site, depending on the formulation, which provide significant benefit, especially with antibiotics with low oral bioavailability, like the aminoglycosides. On the other side, injection of pigs, other than baby piglets, is laborious and is primarily to treat clinically ill pigs either with acute respiratory infections such as *Actinobacillus pleuropneumoniae*, systemic infections caused by *Streptococcus suis* and *Glaesserella parasuis* or enteric infections such as swine dysentery caused by *Brachyspira hyodysenteriae*, which may be too sick to eat or drink. It is a highly effective route of administration to an individual animal, but many antibiotics require repeated injections on a daily basis or combination with consecutive oral medication. There is also an issue of the high volumes to be administered to adult animals (e.g. sows) that may complicate the situation (the need to divide injection volume either due to irritation or due to residues). The development of long-acting formulations has improved the issue regarding compliance, to complete the course of treatment and reduce the stress from repeated administrations. A long-acting antibiotic formulation is important mainly for time-dependent bactericidal and bacteriostatic antibiotics due to the pattern of their antimicrobial activity. Single injections of a long-acting antibiotic provide a sustained action because of the maintenance of high plasma concentrations of the antibiotic after administration of one single injection (Sun et al. 2004). This is particularly important for time-dependent acting antibiotics such as amoxicillin (Burch and Sperling 2018). There has been an increase in the prophylactic use of third-generation cephalosporins (ceftiofur) in baby piglets to prevent a variety of wound infections post processing (tail docking and castration) and also frequent regular treatment programmes based on cephalosporins to control early infections with *Streptococcus suis* and *Glaesserella parasuis*. This may have triggered the selection of MRSA clones (specifically CC398), which has become widespread on the continent of Europe (Guardabassi et al. 2007; Armand-Lefevre et al. 2005) and also in North America (Khanna et al. 2008) and the selection of extended-spectrum β-lactamase (ESBL) producing *Escherichia coli* with the possible transfer between animals and humans (Hammerum et al. 2014). It was also described that human ESBL carriage was associated with average number of hours working on the farm per week and presence of ESBLs in pigs (Dohmen et al. 2015). Prophylactic use of cephalosporins of the third and fourth generations is considered as the main risk for selection of ESBL type of resistance. It has also probably led to the very high level of resistance (41.8%) to ceftiofur which has been reported in isolates of porcine *E. coli* from clinical cases in the USA (Frana et al. 2012). People in the farming community with direct contact with a reservoir of pigs are at risk from the exchange of ESBL genes and vice versa; the diversity in ESBL among people from the general and clinical population suggesting rather numerous different sources (Dorado-Garcia et al. 2018). Similarly, in the last decade, the long-acting injectable macrolides (tulathromycin) use has become more frequent. Tulathromycin can effectively cover swine respiratory diseases caused by *Actinobacillus pleuropneumoniae, Pasteurella multocida, Glaesserella parasuis, Mycoplasma hyopneumoniae* and *Bordetella bronchiseptica* via high distribution to lung and slow elimination following a single dose of tulathromycin, with proven significant accumulation in neutrophils and alveolar macrophages

(Nutsch et al. 2005; EMA-Documents-Product Information-Draxxin 2008). It should be noted that prudent use of this antimicrobial should be considered, not only due to the fact that macrolides belong among CIAs in WHO classification but also due to the zoonotic potential of several Gram-positive pathogens of pigs, like *S. suis*. Those bacteria that can be affected by transfer of macrolide and lincosamide resistance due to the presence of *erm*(T) gene causing resistance to both groups. Although the low detection rate of *erm*(T) gene among the *S. suis* population due to the fitness cost of the *erm*(T)-carrying plasmid and ICE (integrative and conjugative elements), the presence of *erm*(T) in *S. suis* and its potential transmission to other Gram-positive pathogens will be of important significance (Yu et al. 2022). Gen erm (T) has been already identified in *Lactobacillus*, *Streptococcus*, *Staphylococcus*, *Enterococcus* and *Erysipelothrix* and also in Gram-negative organisms, such as *Glaesserella parasuis*.

In contrast, the antibiotic route of administration, such as by injection, may decrease the raising of antimicrobial resistance in gut microbiota due to the generally higher drug doses which are needed for oral therapy in order to achieve the same effective antibiotic concentration in target tissues and direct exposure of gut microbiota after oral medication, which has been described for tetracyclines and aminopenicillins. In the experiments with tetracycline or ampicillin administered either orally (in feed), or via injection (IV), oral administration resulted in rapid rise of resistance genes as detected in mice faeces. Much lower and slower resistance spread was detected after intravenous route of administration. Despite these results, the type of antimicrobial, including its metabolisation and route of excretion should be considered when assessed the effects on the gut microbiome (Zhang et al. 2013). There is a reduction of the gut microbiome's exposure in antimicrobials with renal excretion. The early treatment strategy, reducing the exposure of digestive microbiota by lowering the doses due to the fact of low pathogenic inocula, while having a minimal impact on the large gut microbiota, was as well described based on pig model assessment, comparing bactericidal activity at the gut vs. the systemic level for marbofloxacin (Ferran et al. 2013).

An alternative to classical needle injection has been adopted mainly by US swine producers and some countries in EU (Netherlands), which is needle-free injection (NF) of vaccines, iron, toltrazuril but also antibiotics (Gerlach et al. 2012; Imeah et al. 2020). It is suggested that needle-free administration may provide reductions in lateral transmission of diseases by needle contamination, reduce the risk of broken needles in meat products and increase worker safety (Reinbold et al. 2010). There is not much data published confirming potential bioequivalence of this route of administration, and more data is needed. A study comparing the pharmacokinetics of ampicillin after needle-free injection with conventional intramuscular administration demonstrated the feasibility of NF injection but bioavailability of IM injection was superior (Apley et al. 2007).

3.1.2 Oral Administration

Oral administration of antibiotics is the most common route of administration in swine medicine. Aggregated across the 31 countries (ESVAC 2019), the sales (mg/PCU, data 2017, aggregated for 31 countries) of pharmaceutical forms for group treatment accounted for 89.4% of the total sales—premixes accounted for 28.8%, oral powders for 9.9% and oral solutions for 50.7%.

Piglets may be treated individually by oral dosers, which is typical mainly for administration of toltrazuril in order to control *Cystoisospora suis* (coccidiosis) infections in piglets, with new injectable toltrazuril iron combo products authorised in 2019. The oral administration of antibiotics via dosers has proved most useful for the control of neonatal diarrhoea, usually associated with *E. coli* and *C. perfringens* type A, where gut active and systemically active formulations of enrofloxacin, trimethoprim/sulphonamide and amoxicillin have proved very effective. Oral treatment still commands great importance and is used in situations where it is necessary to anticipate predictable disease outbreaks with possible serious consequences (mortality and serious welfare issues) at the start of an outbreak of a disease in a group/herd to avoid in-contact spread of the infection (metaphylaxis). Oral use of antimicrobials within the metaphylactic regimen plays an important role for control of post-weaning *E. coli* (PWEC) infections and some acute septicaemic diseases like *S. suis* meningitis. Later on, when the *E. coli* infection is primarily in the gut, gut-active antibiotics, such as the aminoglycosides (neomycin, apramycin), aminocyclitol (spectinomycin) and the colistin, are mostly used. The situation as for the extent of the use of orally administered antimicrobials will probably start to change after the regulation on veterinary medicinal product together with delegated act on oral medication will come into action, altogether with the stricter conditions for prophylaxis and metaphylaxis and pressure on minimising/phasing out of use of colistin—the last resort antimicrobial.

Water medication is also widely used and becoming more popular in some countries, due to the introduction of effective automatic dosing machines and general improvement of technology on farms. Especially in Europe, the policy of restriction and minimising use of premixes and in-feed medication led to the voluntary ban of feed medication in some countries and attempts to reduce it in others; consequently, there is a concurrent increase in the use of water medications. In the past, it was limited to pen troughs or individual pen tanks. Larger header tanks allowed whole sheds/barns to be treated, but in some cases, difficulties may appear depending on the size of tank as well as frequently required 2–3 administrations of a drug each day to ensure an adequate drug intake and duration of activity. Modern farms with a high density of animals use mainly automatic water proportioner machines nowadays. The antibiotic is dissolved in a concentrate or is available already in liquid form, which flows into the main water system at a set target rate (usually 1–2%, but available up to 5%) providing a benefit for antibiotic products which have lower solubility. However, the antibiotic in general needs to be sufficiently soluble in order to avoid problems with the administration and obstruction of the water system.

The quality of water (pK, pH, water temperature) needs to be considered as well ensurance of proper dissolution and stability of the active substance. For example, hard water and high content of calcium may limit some of the antibiotics used (tetracyclines, enrofloxacin) due to poor solubility and even precipitation. It is also extremely important to be aware of incompatibility of certain materials (e.g. certain metals) from which especially older delivery systems are made with certain antimicrobials (e.g. doxycycline). Also, it should be borne in mind that systems should be regularly cleaned, as algae and biofilms can influence both distribution of antimicrobials and can accumulate molecules of antimicrobials and produce resistant bacteria, that can be released into the drinking water (e.g. from biofilm bacterial populations) (Martelli et al. 2017; Lumb et al. 2017). When using water-based medication systems, it is important to take into account the pig's water intake per day, ambient temperature and drinking behaviour (influenced also by health status). In experiments, an optimal timing of administration of products was identified according to the observation of specific drinking behaviour during morning and during a period of about 6 h (from 7:00 to 13:00) (Jackson 2007). In many cases, the advantages of prompt treatment, controlled dose rates and duration of treatment have helped the replacement of medication via the feed for treatment purposes, and the use of water medication instead of feed premixes could be considered as a step forward to rational use of antibiotics on pig farms.

In 2017, feed medication was in at least seven countries (from 31 countries, data 2017, ESVAC 2019) the main route of antimicrobial administration in the swine industry. But the situation has started to be changed dramatically, especially after regulation (EU) 2019/4 on medicated feed was published in the Official Journal, where the ban of preventive use of antimicrobials in the medicated feed was stipulated. Despite coming in force since January 2022, many countries started to change their usage patterns before this date. Data 2021, published by ESVAC (2022) and available also in interactive database on EMA web page, indicated that only three European countries have proportion of antimicrobials administered via medicated premixes more or near to 50% of the total consumption (considering mg/PCU data sets). There are significant differences in between countries within the EU with regard to the frequency of use of medicated feed (premixes). For example, Cyprus, Greece, Portugal and Malta followed by Switzerland and Bulgaria were (in 2021) still very high users of antimicrobials via premixes incorporated into medicated feed, exceeding 40% proportion of overall use based on mg/PCU dataset (ESVAC 2022). For treatment of disease, it can be argued that it is not the most efficient route of administration, as it may take some days for the feed to be manufactured, delivered and work through the storage bin system to get to the pigs. At the same time, pigs with clinical disease usually have a significant decrease in appetite and consumption of feed. For ease of administration, it is the simplest route, but also should be considered are the storage conditions and risk of cross-contamination of further batches of non-medicated feed, which can cause contamination of subtherapeutic levels of antimicrobials administered to subsequent groups of animals. For the early treatment (metaphylaxis) and in some specific cases for the prophylaxis of certain diseases before its ban in EU (REGULATION (EU) 2019/4), it was considered in previous decades as ideal, as it can be planned that the pigs receive the medicated feed on arrival or transfer from one shed/barn to another, especially when they are known to come from an infected source or when groups of animals are mixed and moved in a production cycle. Nowadays, prophylaxis via medicated feed is banned, and metaphylaxis is coming under scrutiny, and other preventive tools (e.g. vaccination, immunomodulators, improved biosecurity measures, probiotics and prebiotics) should be used whenever feasible and appropriate. According to an EU RONAFA report (EMA 2017), some specific occasions can justify antibiotic use for prevention of specific contagious bacterial infections characterised by rapid progression to an epidemic with mortality as a major outcome, but for such cases only other means of oral or injectable antimicrobials administration are recently allowed in certain conditions, as specified by legal provisions (regulation on VMP (EU) 2019/6). Other factors play a role and should be applied in a decision tree, for instance, the absence of effective vaccines or other means to establish herd immunity. A typical example might be *Streptococcus suis* or other pathogens causing septicaemia with high mortality or swine dysentery caused by *B. hyodysenteriae*.

The aim of metaphylaxis is to eliminate or reduce as much as possible the infectious agent in the early stage of disease and avoid horizontal transmission, so that it does not cause disease in the next growing phase and potentially reduce the contamination of the environment of the barn. According to EMEA/CVMP/627/2001 document, there should be set up a threshold for the initiation of metaphylactic treatment (e.g. proportion of clinically diseased animals at a certain time point within the group and the severity of clinical signs). The drug is administered at a therapeutic dose in an attempt to eliminate the infection, whether it is *B. hyodysenteriae*, *S. suis* or typically PWEC (*E. coli* infection). Low concentrations of bacteria (say 10^2) respond better to lower levels of antibiotic and are less likely to mutate and develop resistance than high concentrations of organisms ($>10^6$), which are found in clinical infections. Several in-vitro but as well in-vivo studies showed that early antimicrobial treatments targeting low pathogen inoculums are efficacious at doses much lower than those required to eradicate high inoculums. This may well help to avoid the selection and amplification of ESBL—carrying *Enterobacterales* in the gut microbiota (Vasseur et al. 2014, 2017). This actually supports early treatment (animals in prepatent period of infection) rather than waiting for high levels of disease before treatment. Antibiotics are often used at lower levels to prevent infection or re-infection from a contaminated environment, especially in the case of swine dysentery, basically programmes are combined and e.g. an initial step is treatment with full dose followed by prolonged medication at a lower level, due to the survivability of the organism in the environment. The concentrations in feed are generally lower than the treatment level but are effectively inhibitory, preventing the multiplication of the organism, so to be effective drug concentrations in the gut contents must exceed the minimum inhibitory concentration (MIC) of the bacterium. Those programmes should not replace improved biosecurity, especially correct cleaning and disinfection and dose of active substance should, whenever possible, respect the current SPC of a product used for medication of pigs.

Liquid (wet) feeding is becoming a very popular way to improve performance and FCR on farms in the finishing phase of production but is also becoming available as well in farrowing house for sows nowadays. Liquid feeding involves the use of a diet prepared either from a mixture of liquid food industry residues (often whey or skimmed milk) and conventional dry food, or from dry raw materials mixed with water mixed at a central point on farm (Brooks and Beal 2001). Medication via liquid feed is used for metaphylactic group treatment, as medicating the pigs using dry pellets or meal is not very practical in this system or medication of the drinking water. There is limited data available evaluating suitability of antibiotics containing veterinary medicinal products for liquid feed medication. In general premixes, oral powder forms and water-soluble formulations including concentrates/liquids are used in the field. Products used should be stable and homogeneous in mixture and an exact specific recommendation for each specific type of product needs to be followed. Medicated liquid feed should not be prepared and stored or fermented. Introduction of antibiotics should take place at the beginning of the mixing procedure and thoroughly mixed, preferably pre-dissolved, if a water-soluble form is used. The final medicated feed should be fed to pigs immediately and be consumed very quickly.

The spread of antimicrobial residues to soil via manure after oral treatment is nowadays frequently discussed as well. Recently, quantitative risk assessments of estimated levels of antimicrobial residues were described with the highest estimated level in manure after storage in pits for doxycycline due to its high stability in manure. In contrast, antimicrobials like penicillin and amoxicillin pose the least risk due to their direct degradation (Filippitzi et al. 2019).

3.2 Pharmacokinetic/Pharmacodynamic (PK/PD) Considerations

3.2.1 Introduction

The pharmacokinetics of an antibiotic describes its kinetics after its administration via various routes and is based on the following processes: absorption, distribution, metabolism and excretion. The main goal of the treatment of pigs is to achieve a sufficient unbound concentration of a specific, effective antibiotic at the site of infection, e.g. blood/plasma for septicaemic infections; lungs/bronchial fluids/leucocytes for respiratory pathogens and gut contents for enteric infections, to obtain a positive therapeutic effect. The pharmacokinetics (PK) of antibiotics closely relate to their pharmacodynamics (PD) and to the time course of the antimicrobial effects at the specific site of the infection. Their PD effects can be defined, e.g. rate and extent of bacteriostatic or bactericidal activity by the minimal inhibitory concentrations (MICs) and minimal bactericidal concentrations (MBCs) and possibly post-antibiotic effect (PAE) and post-antimicrobial leucocyte enhancement effect (PALE). Both PAE and PALE would be important parameters to asses in future for establishing optimal treatment intervals. For example, PALE is described for certain antibiotics and is defined as the degree of sensitisation of the pathogen to phagocytosis and intracellular killing mechanisms of leucocytes (McDonald et al. 1981; Nedbalcova et al. 2019). For bactericidal compounds, the MBC: MIC ratios are approximately 1–2, but for bacteriostatic antibiotics, the ratio can be much higher, up to 50 even. The PK and PD relationships for bactericidal antimicrobial compounds to achieve a good bacterial kill and clinical response have usually been determined. The objective of this chapter is not to describe in detail all the pharmacokinetic parameters and the microbiological techniques evaluating MIC, which will be found in earlier parts of this book (VOL II, chapter

"Choice of the Antimicrobial Drug and Dosing Schedules Optimisation Using PK/PD Characteristics"), but to explain basic parameters to enable readers to consider PK/PD concepts in the therapy of systemic, respiratory and enteric infections of pigs.

Plasma concentrations are considered the primary pharmacokinetic parameter for use and the maximum concentration (C_{max}) in plasma of antibiotics after administration is an important parameter for the prediction of efficiency of an antibiotic treatment, especially for concentration-dependent antibiotics (CDAs) (Burch 2010), such as the fluoroquinolones and aminoglycosides. In the case of PK evaluations in respiratory infections, for CDAs this generally applies although bronchial secretions may also be helpful for respiratory infections and gut contents concentrations also apply for enteric infections. A ratio of the C_{max} in plasma divided by the minimum inhibitory concentration (MIC) for classical CDAs should be approximately 10–12 (Toutain 2003).

The area under the curve (AUC) is defined by plasma concentration of antibiotics against time and basically represents the absorption of the antibiotic—the systemically available amount of antibiotic and is measured over a set period of time (e.g. 0–24 h) (Martinez et al. 2013). The $AUC_{0\text{-}24h}$ can be divided by the MIC of the antibiotic against the organism, and a ratio of 100–120 will usually give a good bactericidal effect for bactericidal antibiotics but may not for bacteriostatic antibiotics, where a higher ratio is usually required (Burch 2012). An additional important consideration must be applied and that is the parameter AUC must be related to the antibiotic's free concentration, unbound to plasma proteins.

The AUC/MIC can also apply to the aminoglycosides, fluoroquinolones and importantly the penicillins, aminopenicillins and cephalosporins, which are both concentration- and time-dependent in their bacterial killing effect. Time above the MIC (T > MIC) is important for the penicillins and cephalosporins, because of their time-dependent nature. For Gram-positive organisms, like *S. suis* and also some Gram-negative bacteria, such as *G. parasuis,* a bactericidal effect can be achieved if the plasma concentration T > MIC lasts for 40% of the time during a 24 h period. For enteric bacteria, like *E. coli* and *Salmonella* spp., the Time > MIC must be 100%, however. Therefore, the unbound plasma concentration of such an antibiotic must be kept for a certain period of time (≥40%/≥9.6 h above the MIC) is critical, especially for the treatment of systemic pathogens causing septicaemia in pigs such as *S. suis* or *G. parasuis*.

The plasma concentration of antibiotics is frequently influenced by route of administration. Typically, for medication via feed, the C_{max} parameter is comparatively low in comparison by injection and the concentration in plasma may fluctuate depending on the feeding pattern of the animals. For example, ad libitum feeding during the day gives a lower more prolonged plasma concentration than once a day feeding or oral dosing. The oral bioavailability in pigs varies depending, not only on time of feeding but also by physico-chemical characteristics of the antimicrobial, for example, enrofloxacin, sulfadiazine with trimethoprim have a high bioavailability (>80%), whereas in comparison, lincomycin varies from 73% in unfed pigs to 41% in fed animals and spiramycin 60%–24%, respectively (Baggot and Giguère 2013; Nielsen 1997) and the aminoglycosides, colistin and spectinomycin have almost no bioavailability when given orally, <10%, due to very limited absorption from the gastrointestinal tract.

The systemic availability of an antibiotic after IM injection can vary as well, but usually provides a higher bioavailability, C_{max} and AUC in comparison with oral treatment at the same dose (see Fig. 1). Systemic availability is also influenced by formulation, e.g. prolonged release in long-acting formulations, tissue irritation, volume administered (Baggot and Giguère 2013).

These basic PK/PD relationships work well when the MIC is similar to the minimum bactericidal concentration (MBC), but when the MBC/MIC ratio is much higher, e.g. for bacteriostatic drugs such as tetracyclines, macrolides, lincosamides and pleuromutilins, then the classic PK/PD relationships can be markedly changed.

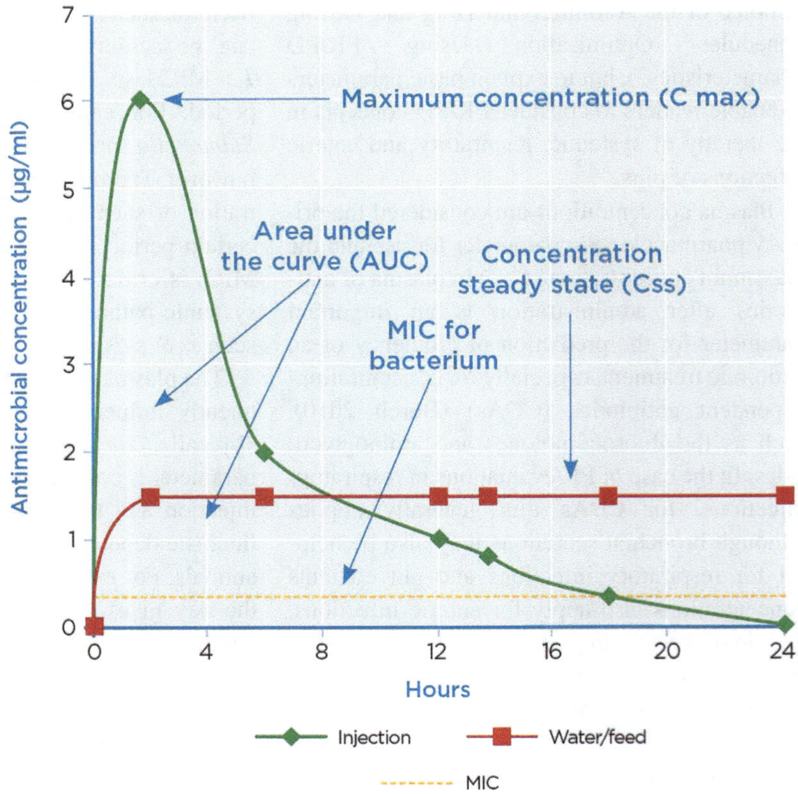

Fig. 1 Basic pharmacokinetic parameters used following different route of administration of antimicrobials (adopted from Burch 2013)

However, they can be restored by the use of the MBC in the calculations. This was shown to be the case for various antimicrobials against *Mycoplasma hyopneumoniae* (Burch 2004).

3.2.2 Therapy of Septicaemic Infections in Pigs

In general, sepsis/septicaemia is described as a systemic infection associated with the presence of a pathogen (microorganism) and/or their toxins in the bloodstream, including the systemic reaction of the immune system (Fecteau et al. 2009). Common pathogens with ability to cause sepsis in swine are listed in Table 1.

The plasma concentration is the primary pharmacokinetic parameter to consider for evaluation of antibiotics for the treatment and control of septicaemic infection of pigs due to the location of the pathogen in the bloodstream, as well as possibly the intracellular location in macrophages after phagocytosis.

Affected pigs should be treated individually, as early as possible, preferably with injectable antibiotics. Systemically absorbed antibiotics are sometimes used for a few days in feed or water as metaphylaxis to groups experiencing high morbidity in combination with an initial injectable treatment based on the previous history of the clinical manifestation of the disease in positive herds. Successful sepsis treatment requires a fast initiation of antimicrobial therapy, preferably with bactericidal antibiotics active against the causative agent (Pardon and Deprez 2018). Due to the necessity for the rapid onset of treatment of sepsis, usually, empiric antibiotic treatment is employed, based on the history of the disease/pathogen and its susceptibility, which is circulating on the affected farm.

In septicaemia-affected animals, the pharmacokinetics might be altered and need to be considered. Significant changes were described, for example, with ceftiofur in *S. suis*/PRRSV co-infected pigs, where they demonstrated lower values of C_{max} and AUC (Day et al. 2015; Seitz et al. 2016). These changes in pharmacokinetics might be caused by pH changes in the inflamed tissue and decreased tissue perfusion, for β-lactams and for aminoglycosides, an increased volume of distribution was described as well (Liang and Kumar 2015). In general, such data are commonly missing for specific antibiotics and pathogen in swine medicine, as PK studies are normally carried out in healthy animals.

One of the most frequently occurring septicaemic infection in pigs is caused by *S. suis* and the most common strain is capsular type 2 (Gottschalk and Segura 2000). Amoxicillin and penicillin are considered the main treatments of choice (Burch and Sperling 2018). Below is described amoxicillin as an example for PK/PD evaluation for the treatment of septicaemia in pigs.

Short-acting (SA) and long-acting (LA) injectables formulations of amoxicillin are available for treatment and are frequently used in the field. The LA formulation had a lower C_{max}, a greater $AUC_{0-\infty}$ and mean residence time (MRT) than the SA formulation, which makes it more suitable for a time-dependent antibiotic to act (see Fig. 2).

Other routes of administration, such as the oral route, provide reduced bioavailability and consequently a reduced AUC of amoxicillin in plasma (Burch and Sperling 2018). A rapid reduction of amoxicillin plasma concentrations following oral administration by gavage was observed in comparison with injection. When amoxicillin was administered orally at 20 mg/kg bodyweight, the

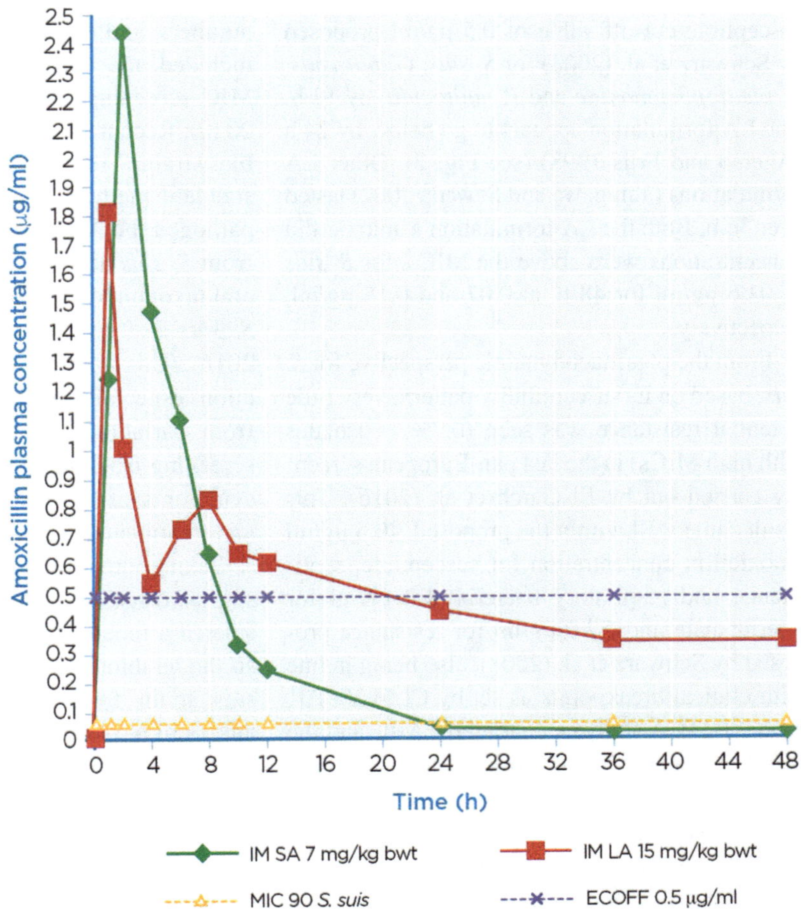

Fig. 2 Comparison of the amoxicillin pharmacokinetics of SA formulation given at the standard dose of 7 mg/kg IM bodyweight and the LA formulation given at 15 mg/kg IM bodyweight (modifying Agersø and Friis 1998) with the proposed cut off of 0.5 μg/ml systemic and respiratory infections and the MIC 90 for *S. suis* of 0.06 μg/ml (adopted from Burch and Sperling 2018)

C_{max} was 3.79 μg/ml, T_{max} 1 h, AUC_0-∞ 12.52 μg.h/ml and half-life elimination time ($T_{1/2\ el}$) was 1.68 h in a study carried out by Bes et al. (1997) and in an additional study (Ageso and Friis 1998) results at 10 mg/kg bodyweight gave a C_{max} at 1.6 μg/ml, T_{max} 1.9 h, AUC_0-∞ 6.5 μg.h/ml, both using a microbiological assay. It demonstrates the rapid reduction of amoxicillin plasma concentrations following oral administration. As a consequence of a very short elimination half-life after oral administration, normally, there is a requirement to receive multiple oral doses during the day to achieve a sufficient time above the MIC (Burch and Sperling 2018).

Prolonged plasma concentrations were seen after SA and especially LA amoxicillin use (Agerso and Friis 1998). Both the oil-based LA and SA injectable forms' plasma concentrations fall more slowly than the orally administered amoxicillin but the SA formulation, given even at double its approved dose, appears to go below the susceptible cut-off value of 0.5 μg/ml proposed by Schwarz et al. (2008) for *S. suis, G. parasuis, A. pleuropneumoniae* and *P. multocida*, by 12 h. The LA formulation appears to go below by 24 h (Agersø and Friis 1998) (see Fig. 2). Other LA formulations (Tanigawa and Sawada 2002) lasted over 36 h. Both the LA formulation's amoxicillin concentrations were above the MIC_{90} for *S. suis* of 0.06 μg/ml for 48 h, at 0.07 and 0.15 μg/ml, respectively.

From the pharmacodynamic perspective for *S. suis*, based on its susceptibility pattern, very little potential resistance was seen (0.7% of isolates with high MICs) in the last pan-European screening carried out by El Garch et al. (2016). This result came well within the proposed ≤0.5 μg/ml amoxicillin concentration for susceptible, septicaemic and respiratory bacteria, 1.0 μg/ml for intermediate and ≥2.0 μg/ml for resistance proposed by Schwarz et al. (2008), also being in line with clinical breakpoints as set by CLSI VET01 ED6 (2023). A study evaluating the MIC and the MBC of selected French isolates described only marginal bactericidal activity at the MIC for selected isolates of *S. suis* but at 2- and 4-times MIC, they were clearly bactericidal, on a time-dependent basis (Richez et al. 2012).

In general, resistance reported for *S. suis* isolates from clinically diseased animals (brain sampled) to the beta-lactams (or more specifically to the aminopenicillin group, including amoxicillin) has been demonstrated to be generally very low worldwide based on data produced primarily for ampicillin (Varela et al. 2013). In contrast, samples from upper respiratory tract screenings containing isolates of *S. suis*—usually with variable serotypes, showed a higher prevalence of resistance to beta-lactams (Segura et al. 2020). Isolates can generally be regarded as invasive as the sites used for the sampling should be considered as sterile in healthy animals (brain swabs and/or cerebrospinal fluid in case of meningitis and in case of septicaemic animal internal organs such as spleen and liver). However, this is not the case for isolates obtained from samples of the respiratory tract as *S. suis* colonises the upper respiratory tract of piglets physiologically in high numbers, and consequently lungs should not be included, the same is valid for the analysis of MIC screenings and reports (Rieckman et al. 2020). Recent studies reported several "*S. suis*-like strains" that were identified as *S. suis* by standard methods for the identification of this pathogen, but were regarded as distinct species from *S. suis* according to the standards of several taxonomic analyses, and reclassification of *S. suis*-like strains is discussed (Okura et al. 2016). This is important to take into consideration especially when susceptibility patterns from variable serotypes obtained during the screening from animals/farms without classical symptoms of *S. suis* infection are interpreted and treatment selected.

Taking into consideration its PK/PD relationship as discussed, amoxicillin is primarily considered a time-dependent bactericidal antibiotic, so the antibiotic needs to exceed the MIC of *S. suis* strain for sufficient time. The effective amoxicillin plasma concentration, taking into consideration its low plasma protein binding (PPB) of 24% for healthy pigs and its MIC90 of 0.06 μg/ml, is well above the 40% time required for both SA and LA formulations. The LA formulation at the dose of 15 mg/kg/bw should be especially effective for a 48 h period between injections, if necessary (Burch and Sperling 2018).

The differences in C_{max} and AUC parameters after different routes of administration favours the use of the LA injectable form of amoxicillin for the treatment of clinical cases of *S. suis* septicaemia and oral products for the metaphylactic treatment of groups of animals during the outbreak or on farms with a recurring *S. suis* infection. Similar results could be expected with *G. parasuis* septicaemias with the LA injectable formulation and an oral dose of 20 mg/kg bodyweight, but doubts have been expressed regarding the treatment of respiratory infections caused by *A. pleuropneumoniae* and *P. multocida* (Rey et al. 2014), as their MICs tend to be higher and an oral dose of 40 mg amoxicillin/kg is likely to be required. Clinical breakpoints for both oral (dose 20 mg/kg bodyweight/day) as well as LA injectable (15 mg/kg bodyweight/for 48 interval, if necessary) together with MIC are presented as visualisation in Fig. 3.

3.2.3 Therapy of Respiratory Infections in Pigs

Respiratory infections of pigs caused by bacterial pathogens are one of the most frequent infections in swine, usually manifested in the form of mixed infections with bacteria, mycoplasmas and viruses—termed the Porcine Respiratory Disease Complex (PRDC). Therapy is essentially targeting a lung infection/pneumonia, thus the concentration of the antimicrobial that can be achieved in lungs is crucial. Acute pneumonias are considered serious disease conditions with rapid development, potentially leading to death, especially associated with *A. pleuropneumoniae*, thus individual treatment via the injection route is recommended for the same reasons, as discussed in the septicaemia section. Plasma antimicrobial concentrations are still used as basic parameters for prediction of lung tissue and bronchial fluid concentration and C_{max} lung/plasma ratio together with AUC lung/plasma parameter are both important and will vary from one antibiotic to another. The level of penetration or concentration of the

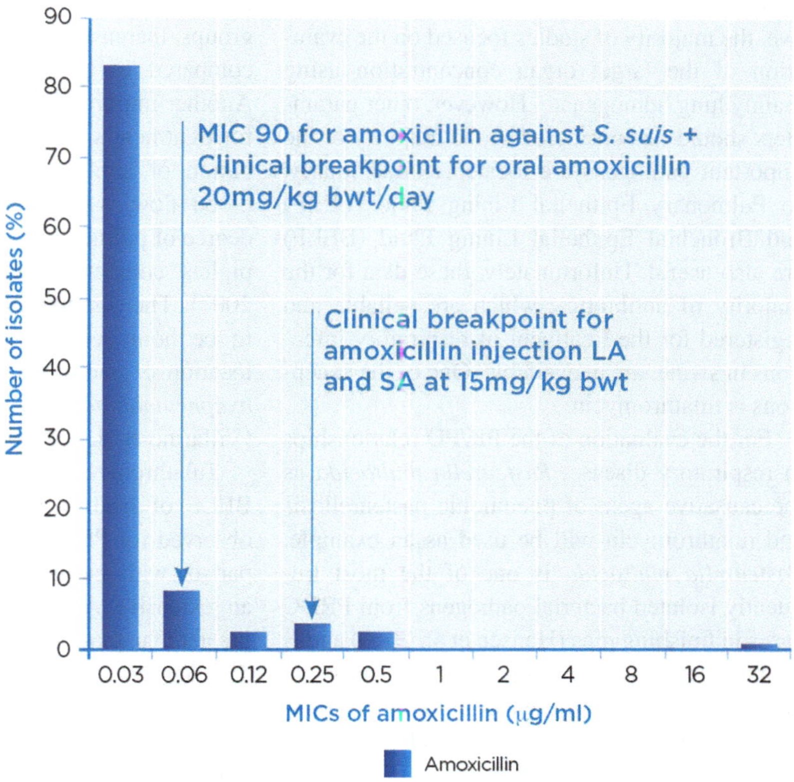

Fig. 3 MICs distribution of strains of *S. suis* isolated from EU clinical cases 2009–2012 and clinical breakpoints for oral and injectable treatment (El Garch et al. 2016)

antibiotic found in lung depends on many factors like: type of antibiotic, lipid solubility, cell membrane penetration, dissociation constants pKa, molecular weight, as well as epithelial permeability and absorptive capacity (Agerso and Friis 1998; Burch 2010).

Besides the target tissue concentration of the antibiotic, the exact location of the respiratory pathogenic bacteria within the lung is an important factor. For example, *A. pleuropneumoniae,* causing rapid, fatal pleuropneumonia, colonises the lower respiratory tract by adhesion to the bronchial epithelium and pneumocytes lining the alveoli. In later stages of infection, it can be isolated from devitalised lung lesions, which are poorly penetrated by antibiotics and with possible extracellular tissue binding to debris and fibrin (Bossé et al. 2002; Van Overbeke et al. 2002; Gottschalk 2012).

Another important pathogen is *Mycoplasma hyopneumoniae*, which colonises the cilia and airway epithelium with its main location in the interstitial fluid of the cranial and ventral parts of the lung, where the typical lesions of enzootic pneumonia are found. From a PK/PD perspective, the majority of studies focused on the evaluation of the target organ concentration using mainly lung homogenate. However, other parameters should be considered as the majority of the important pathogens are located extracellularly, so Pulmonary Epithelial Lining Fluid (PELF) and Bronchial Epithelial Lining Fluid (BELF) are also useful. Unfortunately, these data for the majority of antibiotics, which are suitable and registered for the treatment of respiratory infections in swine, are unavailable. One of the exceptions is tulathromycin.

For the evaluation of the PK/PD relationships in respiratory disease, *Pasteurella multocida* as the causative agent of pneumonic pasteurellosis and tulathromycin will be used as an example. *Pasteurella multocida* is one of the most frequently isolated bacterial pathogens from PRDC cases in finishing pigs (Hansen et al. 2010) and is characterised by its extracellular location in affected pigs and its presence in porcine respiratory tract mucus (Letellier et al. 1991).

Tulathromycin is an azalide derivative of the macrolide, erythromycin, with additional nitrogen groups, and it has been classed as a triamilide (Evans 2005). Tulathromycin has the unique ability to accumulate in lung tissue rather than plasma. In swine, tulathromycin is characterised by a rapid rate of absorption and systemic bioavailability after the IM injection of 2.5 mg/kg (Villarino et al. 2013a, b, c). The concentration in plasma is rather low with a C_{max} of 616 ng/mL, but lung tissue concentrations were described at 24–181 times those in plasma over a period of 0–360 h (Benchaoui et al. 2004; Villarino et al. 2013c). It is important to mention that it is widely accepted that lung homogenate concentration is not the exact PK biophase for lung infections, as the drug concentration is inside the cell and the bacteria are outside, thus a much more complex evaluation is needed to understand PK/PD relationships in the case of respiratory infections (Mouton et al. 2008; Villarino et al. 2013c; Toutain et al. 2017). Tulathromycin has been shown to accumulate in neutrophils and alveolar macrophages in pigs at 16.6 and 8.1 times the extracellular fluid due to the charged nitrogen groups, increasing the intracellular concentration compared to other macrolides (Evans 2005). Another important factor, which may play a role for treatment success, is the difference in distribution of antibiotics within the lungs, as the blood flow is not homogenous with a higher degree of pulmonary vascular reactivity in young piglets compared with adults (Brogan et al. 2007). The distribution of tulathromycin seems to be homogeneous, considering the different location of pathogens within the lungs (e.g. *M. hyopneumoniae,* and *A. pleuropneumoniae*) (Villarino et al. 2013b).

Tulathromycin rapidly penetrates PELF and BELF of healthy and infected pigs. The T_{max} observed for PELF ($\leq$24 h) was shorter in comparison with the BELF parameter (72 h) and with an extensive $AUC_{0–17\ days}$ > 90 and 223 times greater than plasma (Villarino et al. 2013b). These data suggest a certain level of variability of penetration into different levels of lung tissue with possible changes in different compartments. As other macrolides, tulathromycin tends to accumulate as well intracellularly, as already noted, this is the case especially in neutrophils, blood macrophages and broncho-alveolar cells (BACs) was reported as well (Siegel et al. 2004; Villarino et al. 2013c). The rate of elimination was lower in the lipopolysaccharide (LPS)-treated pigs, suggesting different kinetics in BACs in healthy and infected pigs (Villarino et al. 2013a).

Recently, the pharmacokinetic profile of tulathromycin was established in serum, inflamed exudates and non-inflamed transudates of calves at the dosage of 2.5 mg/kg. When the bacteria were grown in broth and serum, there was a reported 50 times reduction of MIC for *P. multocida* in serum (Toutain et al. 2017). Using these results, it indicated that there were sufficient serum concentrations of tulathromycin to explain the efficacy in clinical use, using classical PK/PD concepts. Consequently, classical methods of in vitro measurement of MIC (in broth/agar) performed according to recognised standards, may be misleading as a basis for estimating the in vitro potency of tulathromycin (macrolides) and rather biological fluids (serum) with optimal pH should be used for assessment (Lees et al. 2017).

Pasteurella multocida has a stable MIC profile and susceptibility pattern to tulathromycin with no resistance observed in EU isolates in the period 2009–2012, with a MIC_{50} of 2 µg/ml and a MIC_{90} of 4 µg/ml (El Garch et al. 2016). A recent study carried out on *P. multocida* originating from calves confirmed tulathromycin's potent bactericidal effect at concentrations approximately equal to the serum MIC with the $AUC_{(0-24\ h)}$/MIC of approximately 24 h. The main source of variability was the serum MIC and the actual tulathromycin plasma exposure (Toutain et al. 2017). The distribution of the MIC pattern of *P. multocida* isolates in broth was very similar to the one published by El Garch et al. (2016) for porcine isolates (visualised in Fig. 4). It is considered that similar conclusions are likely to be applicable in the case of the treatment of *P. multocida* in swine with tulathromycin.

This example demonstrates the complexity of PK/PD analysis in relation to respiratory infec-

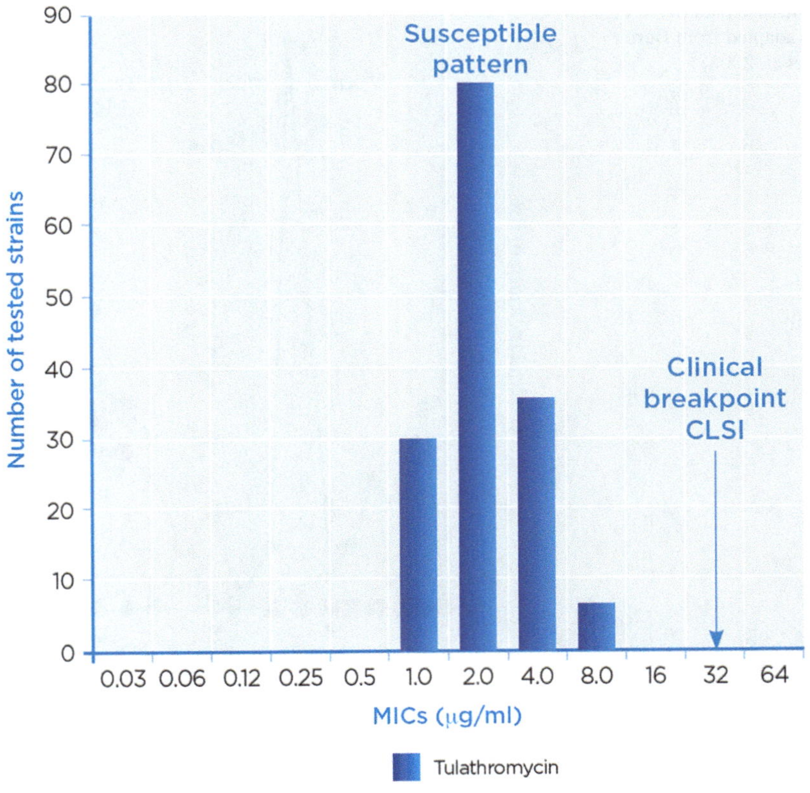

Fig. 4 Susceptibility pattern for tulathromycin against isolates *P. multocida* (El Garch et al. 2016)

tions, as many antibiotics do not concentrate in lung tissue or bronchial fluids yet still are highly effective in the treatment of pneumonia. Good clinical efficacy data, which are correlated in dose–titration studies with organisms with a known MIC, may be more helpful from a clinical perspective.

3.2.4 Therapy of Enteric Infections

For the effective therapy of enteric infections and optimal selection of antibiotics, the concentration achieved in gut contents and intestinal tissue related to the MIC or the intracellular MIC (iMIC) of the bacteria involved needs to be evaluated. Bacterial pathogens causing enteric infection have typically different sites of infection, e.g. the colon in the case of *Brachyspira* spp. infections; intracellular location with a very short extracellular life prior to cell penetration in the ileum and proximal colon in the case of *Lawsonia intracellularis*; the possible colonisation of the whole gastro-intestinal tract (GIT) including regional associated lymph nodes in the case of *Salmonella* spp. infections and in particular the small intestine with regard to pathogenic *Escherichia coli*. As indicated in Fig. 5, characteristic for bacterial enteric pathogens is typical age pattern and clinical manifestation, which is frequently driven by specific sensitivity depends on age and category of pigs.

Besides the treatment of clinically affected animals, preferably by injection, metaphylactic medication (high level treatment) via feed or water is commonly employed to treat the animals on arrival in a nursery or finishing barn with the first signs of enteric infection, to clear up any sub-clinical infection and to cover the period of risk according to the farm history of the disease. If the pigs have already been exposed to the infection and have started to build up immunity, then there should be good control from future infectious challenges, e.g. ileitis caused by *L. intracellularis* but some infections, such as swine dysentery caused by *B. hyodysenteriae* can take some time to develop full immunity.

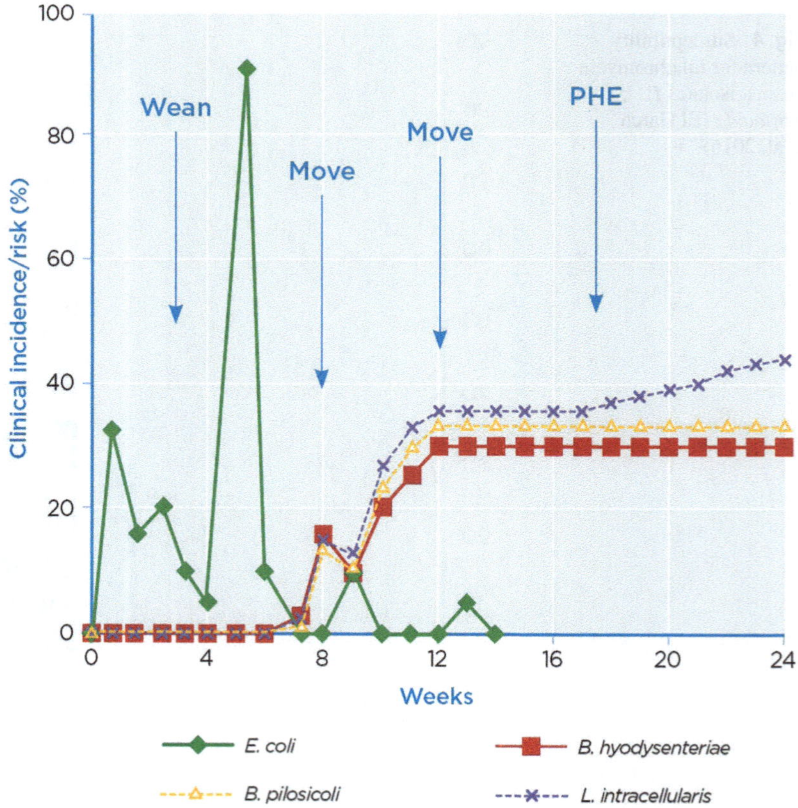

Fig. 5 Common enteric disease patterns in pigs (adapted from Burch et al. 2008)

From a pharmacokinetic perspective after feed or water medication, the plasma concentration is relatively unimportant but the concentration in the different parts of the gut are the most critical aspects. Some antibiotics are well absorbed so smaller residues remain in the colon but others are almost unabsorbed, such as the aminoglycosides, spectinomycin and colistin, so high concentrations can be achieved as they pass down the gut and concentrate in the colon, prior to being excreted in the faeces. In contrast, following injection, the excretion route is the most important in order to reach an intestinal site of infection. There are, in general, two ways how an antibiotic can reach the site of infection; by the biliary route, if the antibiotic is excreted through the liver or by active secretion to the intestinal tract from epithelial cells of the intestinal mucosa (fluoroquinolone carrier-mediated transport across the apical membrane) (Martinez et al. 2006; Griffiths et al. 1994). The presence of enterohepatic recirculation can increase the residence time of antibiotics.

Intestinal absorption and final gut concentration depend as well on other factors like the structure and integrity of the intestinal membrane and the specific transport mechanisms involved (Kim et al. 2017). For example, in the case of the fluoroquinolone, enrofloxacin, following injection, high therapeutic concentrations are achieved in the gut following excretion via the liver and bile, to treat *E. coli* infections. Similarly, injections of the pleuromutilin, tiamulin, also achieve high concentrations in the gut to treat such infections as *B. hyodysenteriae*.

Considering an antibiotic's route of administration, mechanism of excretion (urine or hepatic mechanisms of clearance) and intestinal permeability to deliver the antibiotic, either directly or indirectly, to the intestinal lumen, can help to predict whether it will have an effect on an enteric pathogen, but as well as on the microbiota (Kim et al. 2017).

There is limited data available regarding gut concentrations of antibiotics. As they flow down the gut, there are great fluctuations in concentration. Particularly, in the small intestine, the drug concentration depends on a number of factors such as gastric emptying, the presence of liquid only, which passes more quickly, or the presence of food and its particle size, as well as possible breakdown of the drug as it passes into the intestine (Burch 2012).

Regarding concentrations in the ileum and colon, the major effects are the absorption of the compound and its stability. There is a concentrating effect in the colon due to the food and fluids being absorbed of about 1.6 times the concentration of the drug in the feed, providing that it passes directly through unchanged. Often the concentration of the drug especially in the colon reaches a steady state, and this can be used in PK calculations to determine the AUC_{24h} by multiplying the concentration by 24 h, and this figure can be used in relation to the PD of the specific bacterium. Frequently, the AUC_{24h}/MIC at 24 h indicates an inhibitory effect and at 120 h a bactericidal effect (Burch 2012), which correlates to 4–5 times the concentration of the antibiotic in the gut. Other factors, like faecal (protein) binding and pH, might have a negative influence on the actual effective concentration of the antibiotic, as reported regarding plasma protein binding and possibly the presence of some minerals may have an effect on bioavailability, such as the presence of calcium or other components/additives like kaolin on the tetracyclines.

For the therapy of enteric infections in pigs, frequently bacteriostatic groups of antibiotics are used, e.g. pleuromutilins, macrolides, lincosamides and tetracyclines, the first three groups for swine dysentery caused by *B. hyodysenteriae* and all four groups for the treatment of *L. intracellularis*, the cause of ileitis. Other groups such as the aminoglycosides (apramycin and neomycin), spectinomycin and colistin are used for treating *E. coli* and *Salmonella* spp. infections.

For an example of a PK/PD evaluation, *Lawsonia intracellularis* and tiamulin will be used. Tiamulin is primarily a bacteriostatic antibiotic belonging to the pleuromutilin class and is considered a very effective option for the treatment and control of enteric infections caused by *L. intracellularis* and *Brachyspira* spp. The concentrations of tiamulin in the colon contents (CCC) were published after administration by

Table 7 Tiamulin colon and ileum contents concentration following oral administration

Tiamulin concentration (ppm)	Dosage rate (mg/kg bwt)	Colon contents concentration (μg/ml)	Ileum contents concentration (μg/ml)
In feed (14 days)			
38.5	1.98	<1.98 (LOQ) 0.99 (E)	0.29
110	6.6	2.84	0.82
220	13.2	8.05	2.33
In water (5 days)			
60	6.16	2.16	0.63
120	13.2	5.59	1.62
180	20.9	18.58	5.39

feed and drinking water suggesting accumulation mainly in CCC (Anderson et al. 1994; Burch 2005). Ileum content concentration (ICC) was estimated to be approximately 29% of the CCC (Burch 2005). Feed medication achieved higher concentrations in the gut than water medication at the same dose, possibly due to the feed impairing the absorption and bioavailability of tiamulin (Burch and Hammer 2013). Tiamulin colon and ileum contents concentrations following oral administration (in feed for 14 days) or in water (for 5 days) and at certain dosages are presented in Table 7.

Pleuromutilins, in general, attain relatively low concentrations in the ileum in comparison with other antibiotics used for treatment, like the macrolides and lincomycin, because of their high level of absorption (approximately 90%) (Burch 2005).

From a pharmacodynamic perspective, *L. intracellularis* is a rather difficult pathogen to grow in vitro, as it requires a cell culture, and there are only a few published papers on how drugs worked against this infectious agent (McOrist et al. 1996; Wattanaphansak et al. 2009, 2019; Yeh et al. 2011). A limited range of studies are available evaluating the intracellular and extracellular MICs, but they demonstrate the high efficacy of tiamulin, when the PK of the ileal contents is compared with the intracellular MIC. Lawsonia penetrate the intestinal cells rapidly (within 2 h) leaving little time for exposure to an antibiotic extracellularly, consequently the ability of the antibiotic to accumulate within the cells is crucial and the iMIC more relevant.

The intracellular minimum bactericidal concentration (iMBC) for tiamulin has been determined (McOrist and Gebhart 1995) against a single strain of *L. intracellularis,* and the tiamulin concentration required was similar to the iMICs from different strains, so a precise ratio between iMIC and iMBC is not known (Burch and Hammer 2013). However, clinical trials showed an inhibitory effect and lesion reduction following feed medication at 38.5 ppm for 28 days to already infected pigs (Walter et al. 2001); 50 ppm tiamulin in feed given prior to infection completely prevented lesion development at 3 times the iMIC and at 72 h for the $AUC_{0\text{-}24h}$/iMIC and administration of tiamulin in feed at 150 ppm, 7 days after infection for 14 days, completely eliminated infection demonstrating a more bactericidal effect at ICC concentration 12.7 times the iMIC and 305 h for the $AUC_{0\text{-}24h}$/iMIC (McOrist et al. 1996; Burch and Hammer 2013) see Fig. 6.

Walter et al. (2001) demonstrated that tiamulin given in the drinking water at 60 ppm for 5 days to pigs previously infected with *L. intracellularis* 7 days before, also substantially reduced gross and microscopic lesions of ileitis and *L. intracellularis* shedding. The ICC/iMBC ratio was 2.5, and the $AUC_{0\text{-}24h}$/iMBC was 60 h.

3.2.5 Alternatives and Attempts to Reduce Antibiotic Use in Swine

Antibiotics and targeted antimicrobial treatment of bacterial infections represent an important tool to ensure health, welfare and productivity in livestock production, especially in swine. There is the increasing concern about the associated

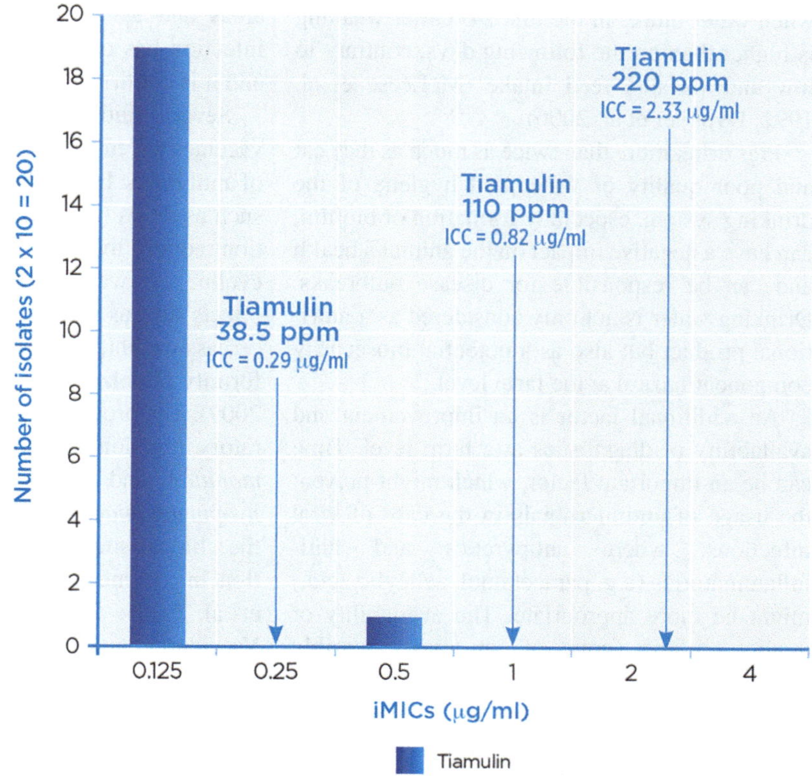

Fig. 6 Concentration of tiamulin in the ileum contents in relation to the iMIC susceptibility pattern of *L. intracellularis* (Anderson et al. 1994; Wattanaphansak et al. 2009)

development of antimicrobial resistance for certain groups of antibiotics used in veterinary medicine, which are at the same time of major importance for human: CIAs—Critically Important Antimicrobials (WHO 2019). The aim of this section is to provide current information about possibilities how to reduce/optimise the use of antibiotics in swine farms by examining possible alternatives, as well as by improvement of external management like biosecurity measures (external and internal biosecurity) and the organisation of production in the form of batch management. Such efforts should not compromise herd health, the welfare of animals or the economics of production and the competitiveness of the industry. Currently, there are several documents and reviews discussing and proposing efficient ways how to reduce the use of antibiotics on swine production. These have been mainly prepared on a national level, but also centralised activities in Europe have been led by the EMA and its specific body, the Committee for Medicinal Products for Veterinary use (CVMP).

The profile and characteristics of farms, which are generally using less antibiotics, were identified by several recent studies (Collineau et al. 2017). Top farms, according to the survey, had mainly improved external and internal biosecurity and improved disease status, especially concerning gastrointestinal and respiratory health (Collineau et al. 2017). There have been described many alternatives to antibiotics in the swine sector in the literature including antimicrobial peptides, feed additives like pre and probiotics, organic acids and enzymes, immunostimulants and vaccines (Cheng et al. 2014; Postma et al. 2016). Not only dietary modifications and specific additives including metals like copper and zinc can reduce pathogen load by direct antimicrobial effect and by improving the microbiome of the gut, but also acidification and regular disinfection of drinking water can have a positive effect. Reducing the pH of drinking water to nursery pigs resulted in the lower faecal *E. coli* excretion (De Busser et al. 2011). This strategy would be important mainly just after weaning, when water intake in the first 24 h after weaning is higher than on the following days, contrary to low and variable feed intake (McLeese et al. 1992; Dybkjær et al. 2006).

Pigs drink more than twice as much as they eat and poor quality of water and hygiene of the drinking system, especially formation of biofilm, can have a negative impact on the animal's health and can be responsible for disease outbreaks. Drinking water is not only considered as a nutritional product but also as a potential biosecurity component/hazard at the farm level.

An additional factor is an improvement and availability of diagnostics at a farm level. Time can be an important factor, which might prevent the usage of antimicrobials in the case of viral infections, where antipyretics and anti-inflammatories (e.g. paracetamol, salicylic acid,) might be more appropriate. The availability of quality on-farm (point of care) tests possibly including tools covering the susceptibility testing and profile of a bacterial pathogen might improve/correct the treatment decision tree and minimise the empiric approach. Frequently, the antimicrobials are used in the field before results of diagnostics/susceptibility testing are available and treatment needs to be initiated. First-line treatment is currently based mainly on a veterinarian's experience and history of infection (pathogen susceptibility profile) on farm, and it is a response often to syndromic indications only, e.g. diarrhoea and/or coughing, which might be of a more complex aetiology with an initial role of a viral disease. Future implementation of the detection of all common virulence factors and most resistance phenotypes with help of molecular genetic methods, like sequencing, would improve the diagnostics and leading time for the results, like proposed for *E.coli* infections of piglets (Vereecke et al. 2023).

Vaccination is one of the frequently discussed ways of how to reduce general consumption of antibiotics at a farm level. Vaccines are considered as the highest feasible approach to routine antimicrobial treatment, with a high degree of efficacy (Hoelzer et al. 2018; Postma et al. 2016). Vaccination might be used as a strategic approach to prevent the incursion of disease from epidemic areas and as an emergency vaccination when infection has occurred in an area or on a farm, and it is confirmed.

Several studies demonstrated effectiveness of vaccines to reduce significantly the consumption of antibiotics, like in the case of enteric infections such as ileitis (*L. intracellularis*) where vaccination reduced the usage of tetracyclines (oxytetracycline) as well as maintaining other positive effects such as average daily gain (ADG), overall carcass weight, feed conversion ratio and pig uniformity (Hoelzer et al. 2018; Bak and Rathkjen 2009). The proven effect was confirmed in respiratory infections control like App (*A. pleuropneumoniae*) and in the case of Mhp (*M. hyopneumoniae*), where vaccination improved the clinical situation and outperformed medication in the mean range of lung lesions (Kruse et al. 2015; Le Grand and Kobisch 1996). Vaccination, using a toxoid-based vaccine against App on an endemically affected farm, reduced individual injectable treatment during the outbreak and consequently, the management cost, providing a benefit of 0.43 euro/pig (Catelli et al. 2017).

The offer of effective solutions is essential especially for nursery enteric and systemic bacterial disease problems, like *E. coli* and *Salmonella* infections. Effective tool to reduce colistin sulphate for the management of clinical Salmonella infection represents vaccination. After the implementation of the live attenuated *Salmonella* Typhimurium vaccine, in combination with standard hygienic precautions, the clinical form of Salmonellosis was reduced to a minimum and hardly any antimicrobial use was necessary (measured by Belgian antimicrobial treatment index (BD100, AMCRA) (van der Wolf et al. 2021). The similar effect on reduction of use of antimicrobials under the field conditions was described in case of the vaccination against oedema disease (OD) in the field trials. After the implementation of the early vaccination with inactivated toxoid-based vaccine, the metaphylactic use of colistin sulphate during the nursery period was stopped on farm with the history of OD, because no deaths due to OD had occurred anymore. The defined daily dose per animal per year was therefore relevantly reduced from 12.6 in 2012 to 2.6 in 2013 (Fricke et al. 2015). Similarly, *E. coli* F4/F18 vaccination against PWD has led to similar technical performance parameters and mortality, in combination with a significant reduction in use of antimicrobials via medicated feeding stuffs (Vangroenweghe 2021).

As antibiotic treatment is still the most widely used therapeutic approach to improve the health condition of pigs affected with PWD and OD, colistin being one of the main antimicrobials used before the description of *mcr-1* gene, vaccination can be relevant option especially for control of OD (Mesonero-Escuredo et al. 2018).

Similarly, vaccination against systemic viruses like porcine reproductive and respiratory syndrome (PRRS) virus and PCV-2 (together with *L. intracellularis*) led to a considerable reduction in antibiotic consumption together with an improvement in ADG and in the case of PRRS vaccination, a reduction of more than 50% in consumption (Bak and Havn 2011; van Looveren et al. 2015). In general, enteric and respiratory infections are known to be major indications for increased antimicrobial consumption and their effective control by increased vaccination rate makes sense, where possible. The vaccination rate and amount of vaccine used as a factor influencing reduction of antibiotic use on particular farms strongly depends on herd health status, which is obvious when comparing SPF type of production with conventional ones. However, farms, which were ranked as top farms (defined by high technical performance and low antimicrobial usage) were able to obtain such status in densely populated pig areas, by increasing focus on internal biosecurity and increased vaccination (Collineau et al. 2017). At the same time, it described a positive association between number of pathogens vaccinated against and antimicrobial use, suggesting that many farms trying to control the disease situation used a combination of both approaches in the field (Postma et al. 2016). This was especially the case, at the beginning of the health programme intervention (start of vaccination), where a strong effect of immunity after vaccination could not be expected due to the time taken for the onset of immunity and the number of vaccinated/ protected animals present on the whole enterprise.

A limitation of many current vaccines used in swine production is that usually they do not prevent colonisation of mucosal surfaces (e.g. the intestine, urogenital tract and respiratory tract) which are the most important portals of entry of pathogens, so eradication via mass vaccination is rather difficult, as well as a certain level of poor cross-serotype protection and necessary match with field strains, which exists especially in some of the pathogens like *S. suis* (Hoelzer et al. 2018). Currently, several attempts are discussed on how to improve current vaccines and defined diseases for which vaccines could reduce antimicrobial use in animals, including pigs (AHG 2015). Identified diseases, where new or improved vaccine would significantly reduce the need for antibiotic use with vaccine research priority are respiratory ones and *E. coli*, *B. hyodysenteriae* and Rotavirus from enteric pathogens of swine (AHG 2015).

Besides certain level of limitations, vaccines probably represent the single most cost-effective medical countermeasure that can be used to fight against antibiotic resistance (AHG 2015).

Many antibiotic alternatives have already been proposed and adopted by the food animal industry, including swine. Alternative products can enhance animal immunity and productivity and prevent infection at the same time, which make them attractive for commercial operations, but to date, there are fewer options available for real treatment during a disease outbreak. The aim of this part is not to list all possible alternative described in the literature, but provide some examples, which already have regular presence in the field. More detailed lists of possible alternatives with future implementation in swine medicine were discussed in several reviews (Thacker 2013; PEW CT 2017; Liu et al. 2018; Xiong et al. 2019).

One of the examples currently used is different types of probiotics. Probiotics are live cultures of microorganisms (e.g., yeast, fungi, and bacteria) that are added to the diet to improve the balance of microbial communities in the gastrointestinal tract and thereby gut health, which offer improvements in performance. Such products use the mechanism of competitive exclusion, especially if used soon after birth or during the critical periods (e.g. after weaning). This helps the animals establish a community of beneficial bacteria in the gut before pathogens can colonise the intestine. The main benefit is usually seen in young animals. Other possible modes of actions were proposed, like modulation of host immune responses, diarrhoea reduction and antitoxic effects and modulation of nutrient digestibility (Liao and Nyachoti 2017). Probiotics might be given to animals that have been treated with antibiotics, to re-colonise a gut depopulated by therapeutic treatment (Hughes and Heritage 2002; Pamer 2016). Reviews by FAO, the European Medicines Agency (EMA) and EFSA have concluded that probiotics are effective growth promoters in swine, with the potential to prevent diarrhoea and reduce mortality due to infections with *E. coli* in piglets (FAO 2016). Improvements in weight gain of over 7% in piglets after weaning, significant increases in feed efficiency in sows and efficacy in preventing post-weaning diarrhoea in young piglets have been reported (Mallo et al. 2010; Böhmer et al. 2006; Taras et al. 2006). There are as well several reports evaluating a series of experiments, where results were not consistent in response to probiotics use, especially in controlled experiments (Nousiainen and Setala 1993).

In a similar category we might find prebiotics, which are usually organic compounds such as certain sugars that, when added to the diet, are indigestible by animals but are broken down by certain beneficial microorganisms in the gut and selectively stimulate these and another microorganisms' growth (Vyas and Ranganathan 2012). Prebiotics can favour the presence of beneficial microorganisms in the intestine and stabilise the gut microbiome.

Zinc and other trace minerals are frequently being used due to the antimicrobial properties when added in diets at levels higher than the standard nutritional requirement of pigs. One of the most widely used forms is zinc oxide, an inorganic compound with relatively poor absorption from the GIT. It is used mainly for the prevention of post-weaning diarrhoea caused by *E. coli*, usually at a dosage of 160 mg ZnO/kg body weight for 14 consecutive days, which equates to 2500 ppm zinc in feed. Beside zinc, different forms of copper were used (200–250 ppm) such as sulphate, carbonate and chloride salts and found to be efficacious (Wallace 1967; Cromwell et al. 1998). Several antimicrobial mechanisms of action by zinc oxide were proposed, such as production of hydrogen peroxide generated from the surface of zinc oxide causing inhibition of the growth of the cells, the affinity between zinc oxide and bacterial cells, as well as increased gene expression of antibacterial peptides in the small intestine and a positive effect on the stability and diversity of the microbiota particularly in respect to coliforms (Luppi 2017; Pluske et al. 2018). Other investigators showed that zinc oxide reduced bacterial adherence of ETEC F4 with invasion and modulating cytokine gene expression (Roselli et al. 2003).

Actually, ZnO was together with colistin probably one of the most frequently used compounds during the post-weaning period of piglets, in order to control diarrhoea caused mainly by *E. coli*. Zinc oxide's therapeutic use in future has been recently determined from an EU perspective, and has been banned since June 2022, following concerns over environmental contamination issues. Several studies have also focused attention on heavy metals including Zn used in animal farming and possible mechanisms that could promote the spread of antibiotic resistance via co-selection of antimicrobial resistance. Zinc was associated with co-selection of methicillin-resistant *Staphylococcus aureus* (MRSA) CC398 in Denmark and an increased proportion of multiresistant *E. coli* in gut microbiota (Aarestrup et al. 2008; Yazdankhah et al. 2014).

Acidifiers, in the form of organic acids such as formic acid, fumaric acid, citric acid and their combinations, have been shown to have benefit in the post-weaning period, playing a role in improved growth promotion and disease prevention especially against *E. coli* diarrhoea and are frequently used. The mode of action is mainly based on maintenance/establishment of a proper gastric pH, because the acid secretion in young pigs does not reach an appreciable level until 3–4 weeks post weaning (Cranwell and Moughan 1989). Proper gastric (intestinal) pH is a strongly connected function to several enzymatic and biochemical systems, like pepsin activity, efficiency of protein digestion, stimulation of secretion of pancreatic proteolytic enzymes and bicarbonate system as well (Ravindran and Kornegay 1993). Acidification of the stomach environment is the likely mode of action exerting direct bactericidal properties for certain microorganisms, in particularly for the coliforms. Several investigations have shown a strong direct bactericidal effect of organic acid by the penetration of the bacterial cells and disrupting their normal physiology (Lambert and Stratford 1999; Suiryanrayna and Ramana 2015).

The review of trials evaluating an effect of organic acids (fumaric, citric and formic acids) reported in general improvement of growth rate, weight gain and feed efficiency with certain levels of variability and mixed results (Ravindran and Kornegay 1993). It was reported that regardless of the organic acids used in the feed, the incidence and severity of diarrhoea in pigs were reduced, with an improved performance of the treated group compared to control animals. Organic acids seem to be one of the practical options for management of PWEC complex (Tsiloyiannis et al. 2001). The described differences in efficiency might be dependent on inclusion rate, type of acid used and differences in feed formulas. Organic acids can be both bacteriostatic and bactericidal, and these actions depend on the levels of their inclusion (Lückstädt and Mellor 2011).

Organic acids are useful and also have an effect on the maintenance of the hygiene status on feeding systems on farms, especially using liquid feed, with reported effects on the prevention of development of Haemorrhagic Bowel Syndrome (HBS), as well as acting as preservatives of high-moisture grains and as mould inhibitors in feed (Crenshaw et al. 1986; Grahofer et al. 2017).

New trends, which become very important and play the role in rational use of antibiotic in swine medicine as well, are the application of the principles and techniques of process engineering to livestock farming to monitor, model, and manage animal production is called precision livestock farming (PLF) (Vranken and Berckmans 2017). Different systems and technologies enable to monitor and improve animal health, welfare and consequently can improve (reduce) medication by antimicrobials. One of practical example is pig cough monitoring, based on what early respiratory problems can be detected and action can be taken. Early treatment in this case provide benefit of reduction of antibiotics. Tracking systems for recording of individual animals treated by antimicrobials provide as well better evidence, secure against possible antibiotic residues in meat and provide possibility to produce antibiotic-free pigs.

Other methods to reduce antibiotic consumption on farms, which are not based on feed additives, have been described as an effective way using different management tools, including biosecurity. Biosecurity is basically the implementation of specific measures along the production chain, in order to minimise the risk of introduction of infectious diseases into a farm (external biosecurity), as well as their spread and circulation within the farm (internal biosecurity). As a consequence of its role in disease prevention and control, biosecurity has been shown to have a positive impact in reducing the volume of antimicrobials used by implementing specific measures, including improvement of external and internal biosecurity (Laanen et al. 2013; Postma et al. 2016; Collineau et al. 2017). An expert survey conducted among 111 European pig experts, identified reinforced internal and external biosecurity, as well as improved zootechnical conditions (e.g. climate of the stable) as the most promising alternatives in terms of perceived effectiveness, feasibility and return on investment (Postma et al. 2015).

The importance of biosecurity is very frequently discussed due to the spread of African Swine Fever (ASF) in the EU but as well recently the epidemic in China affecting several important production provinces and a number of neighbouring countries. There are several previous examples of spread of diseases due to insufficient implementation of biosecurity measures, such as the porcine epidemic diarrhoea in USA and Mexico (PED) (Scott et al. 2016), porcine reproductive and respiratory syndrome (HP-PRRS) (Brookes et al. 2015) the foot-and-mouth disease (FMD) epidemic in the United Kingdom in 2001 (Ellis-Iversen et al. 2011) and PCV2 infection (Madec et al. 2008). The most important emerging and re-emerging diseases of pigs over the last decades have been highly host-specific viruses, rather than multi-host pathogens, and some bacterial pathogens (e.g. *S. suis*, *G. parasuis*) have increased importance in modern production systems, which needs to be taken into consideration as well (Davis 2012).

There are different levels of biosecurity implementation in between producing countries/areas described, even between EU countries (Filippitzi et al. 2017). There are currently available tools evaluating the level of external and internal biosecurity providing qualitative and quantitative evaluation and analysis of implemented biosecurity, like the Biocheck UGent™ questionnaire (Filippitzi et al. 2017).

External biosecurity is primarily defined as ways of preventing the introduction of disease into the farm and even country and it is strongly influenced by location, size of the farm and type of production (farrow to finish × multi-site production) and then consequently by purchasing policy with isolation/quarantine of newly introduced animals (gilt acclimatisation and management) and semen purchase rather than boars. External biosecurity also includes control of contact by visitors, wildlife, neighbouring livestock and pets to the operation and dead carcass management and transport.

Internal biosecurity measurements within farm operation representing mainly separating groups of animals based on risk (age category) using especially the batch farrowing system with the possibility to manage an "all-in-all-out" (AIAO) system. Internal management includes specific organisation of workload (working lines) with proper cleaning and disinfection, clothing and tool restrictions and a specific biosecurity action plan with training of personnel.

Sow batch management system, which is practiced on the majority of commercial farms nowadays, has proven an effective way how to improve the health situation. Applying an AIAO system will reduce antibiotic consumption on a farm level, due to the health improvement. There are different types of batch management applied, like 1 week batching on majority of large farm enterprises up to a three to five-week system, using frequently gilt synchronization and providing the biggest improvement in health, beside some difficulties as a multi-week system is not always suitable for large enterprises. The main benefit is due to the strict separation of consecutive groups of animals, the predicted flow of animals on the farm and time for proper cleaning and disinfection due to the empty rooms, especially in case of the multi-week systems, resulting in postponing seroconversion and reducing the prevalence of Mhp, App and *L. intracellularis* (Vangroenweghe et al. 2012). Mixing pigs of different age in farms, where strict AIAO is not practiced, is one of the commonly agreed important risk factors for increasing cross-infection and playing a role in the incidence of PMWS, enzootic pneumonia, and other infectious diseases. It may be responsible for the increase of post-weaning mortality over 2%, which is normally the acceptable figure, and this results in a higher antibiotic consumption (Rose et al. 2003; Madec 1998; Tucker 2006).

The change either from continuous or weekly management to 4–5 weeks batch farrowing system resulted in the significant increase of the performance and reduced medication costs due to the improvement of the general health status of the weaned and fattening pigs, applying an all-in/all-out policy and the separation of age groups in farms (Wolter and Ellis 2001; Rose et al. 2003).

4 Conclusions

Over the last 50 years, there have been many changes in the way we have used antibiotics in pig production. In 1969, in the United Kingdom, Swann postulated that antibiotics of importance in human medicine should not be used for growth promotion. This led to a switch from penicillins, tetracyclines and sulphonamides but not the macrolides to being used as therapeutics only, rather than as AGPs and they then were supplied under a veterinary prescription or control. Antibiotic growth promoters were supplied without prescription primarily for inclusion in feed. As new compounds became available, the narrower spectrum activity antibiotics with Gram-positive activity (e.g. avilamycin, avoparcin, virginiamycin and bacitracin) became growth promoters and the broader spectrum antibiotics with Gram-negative activity became therapeutics. In the EU, the more successful AGPs often could be used for prevention of certain diseases at low concentrations in feed; hence, their popularity and enhanced efficacy on farms that had certain enteric diseases such as ileitis (*L. intracellularis*) and swine dysentery (*B. hyodysenteriae*) and even *E. coli*. Examples of such growth promoters were tylosin, carbadox and olaquindox, respectively. The therapeutic antibiotics often had claims and could be used for low level preventive use and also therapeutic or treatment indications. In the USA, many in-feed antibiotics had all three claims and the levels used were controlled by the feed mills and only over recent years new antibiotics could be classed that they must have a veterinary directive or prescription for use. In many parts of the world, antimicrobial use, especially in feed, was uncontrolled and based on cost-effectiveness evaluation on the farms that wanted to use them. Generic production of antibiotics frequently reduced the price of the antibiotic and increased the cost/benefit effect and encouraged further use.

In the EU, major changes occurred when Denmark and subsequently the EU decided to ban certain growth promoters, primarily avoparcin, which had a related antibiotic product in human medicine, vancomycin, which was primarily used for MRSA treatment in man and also enterococci infections. This was extended to virginiamycin, as streptogramins were used in man also. This led the EU to ban all growth promoters, under the 'precautionary principle', as there was very limited data that proved it was a major issue or risk to human medicine. From 2006 in the EU, the use of antibiotics as growth promoters has stopped. The rest of the world continued to use them until 2017 when the USA also decided to stop using so called "medically important" antibiotics as AGPs, which were considered of importance in human health (Sneeringer et al. 2019). In addition, in-feed antibiotics should come under veterinary direction and control (FDA 2012—GFI 209). Despite coming to stricter rules, according to the FDA, some antimicrobials are still considered as NMI (not medically important) and are broadly marketed, including among these were the ionophores as well as e.g. pleuromutilins and bacitracin—differently from WHO list, where bacitracin and a representative of pleuromutilin—retapamulin were considered as important (FDA 2018; WHO 2019). This concept is being slowly adopted in other countries around the world as part of the consideration for the responsible use of antibiotics in animals under the 'one health' initiatives that are being promoted by the WHO, FAO and WOAH (previously OIE) and recently by UNEP as well. These Quadripartite members, the Food and Agriculture Organization of the United Nations (FAO), the United Nations Environment Programme (UNEP), and the World Organisation for Animal Health (WOAH), launched, in October 2022, new One Health Joint Plan of Action.

The latest changes proposed in the EU are pushing this even further by stopping the use of antibiotics for the prevention of disease, except in individual animals (EC regulations 4/2019 and 6/2019). If infection and disease are starting to develop in a herd, then early treatment or metaphylaxis may be used. This has been already imposed and regulated by Member States since 2022 and has initiated major changes in the use of medication in pig medicine. The use of medicated feed is expected to continue diminishing dramatically. Prevention claims have started to be removed from product licences of medicated premixes, only therapeutic and metaphylactic claims will be allowed. This is already starting to happen in some Member States. It is expected that the use of water medication will increase, as if disease is present, treatment should be initiated promptly to stop it spreading and causing further losses. Pig farms are starting to change their medication systems by the introduction of automatic water proportioners already. Injectable use may increase also, especially to treat the initial cases of disease before pen/house medication can be introduced. Liquid feed systems may be less affected as they can introduce medication quite rapidly also.

Whether these changes will roll out around the world, we will have to wait and see. The USA is reluctant to make further changes, i.e. banning preventive use but further lobbying may overturn this. Will the rest of the world introducing these changes also remains to be seen? These restrictions may have an impact on the cost of production. In less regulated countries, there may be an effect on international trade to the EU and competition in general. The EU has put in the legislation (EC, 2019/6 Article 118) that imported animals or animal products must be produced to the same regulations as they have and not use designated antimicrobials reserved for the treatment of infections in humans.

Another change, which may alter antibiotic use in pigs in the future, is the possibility for the EU authorities to remove compounds that are considered critical for use in human medicine. Despite the fact, that colistin, the third- and fourth-generation cephalosporins and fluoroquinolones were not withdrawn from animal use as they are not included in the list as stipulated by the Commission Implementing Regulation (EU) 2022/1255, pressure of minimising of their use continue. Moreover, already defined list of antimicrobials or groups of antimicrobials to be reserved for treatment of certain infections in humans, as provided for in this Regulation should be kept under continual review in the light of new scientific evidence or emerging information and therefore can be broadened. So apart from changes in use, there may be further restrictions on the availability of some compounds. The USA have faced this in the past, as oral medication with fluoroquinolones was stopped in 2005, which primarily affected the poultry industry but they did continue to allow injectable forms in pigs and cattle. Chloramphenicol was also banned for use in veterinary medicine in the UK and EU in the 1980s and yet florfenicol is now commonly used. We will have to see if the EU will follow up on this but it may act as a disincentive for the pharmaceutical industry to develop new compounds for veterinary use as most new developments have historically come from the medical development of antibiotics. Minimalisation of use of certain antimicrobials could, in some EU Member States, especially those belonging among "small markets" endanger treatment possibilities by non-availability of certain veterinary medicinal products containing antimicrobials.

Moving from an EU/world context to the local farm level context, there should be mentioned the new behavioural models that might have increasing importance. The changes of management and treatment practices replacing and reducing the need for the use of antimicrobials should be supported by behavioural change of veterinarians and farmers (Speksnijder and Wagenaar 2018). Veterinarians ideally should change from reactive and curative antimicrobial prescribers towards a more proactive role as animal health consultants and experts for farmers, without relying on prescribing antimicrobials. These are changes, which can be seen on different countries nowadays.

Apart from the regulatory and political developments over antibiotic use in animals, there have been some great steps forward regarding our understanding of how antibiotics work in relation to treatment of pig diseases. The use of pharmacokinetics, pharmacodynamics, susceptibility patterns and resistance development has helped us to understand how to use antibiotics to achieve the best therapeutic benefit from their use. Improving likely dosing success rates by using PK/PD determinations may impact future dose rates, which will have to be re-assessed for the older compounds, and this is under discussion currently, in the EU, through VETCAST.

Overall, it is considered that antibiotics will remain an important part of pig production around the world, to treat sick animals, as we have no other such potent tools as antibiotics that enable us to treat effectively, clinically sick animals.

From a veterinary perspective, it is our aim to produce healthy animals, under good welfare systems of production, to provide safe and good quality meat products for human consumption and rational and responsible antibiotic use will continue to play an integral part in achieving this.

References

AACTING (2024) Overview of farm-level AMU monitoring systems. https://aacting.org/monitoring-systems/. Accessed 21 June 2024

Aarestrup FM, Jensen NE (1999) Susceptibility testing of *Actinobacillus pleuropneumoniae* in Denmark. Evaluation of three different media of MIC determinations and tablet diffusion tests. Vet Microbiol 64:299–305

Aarestrup FM, Jensen LB (2002) Trends in antimicrobial susceptibility in relation to antimicrobial usage and presence of resistance genes in *Staphylococcus hyicus* isolated from exudative epidermidis in pigs. Vet Microbiol 89:83–94

Aarestrup FM, Schwarz S (2006) Staphylococci and streptococci. In: Aarestrup FM (ed) Antimicrobial resistance in bacteria of animal origin. ASM Press, Washington DC, pp 187–206

Aarestrup F, Seyfarth A, Angen O (2004) Antimicrobial susceptibility of *Haemophilus parasuis* and *Histophilus somni* from pigs and cattle in Denmark. Vet Microbiol 101:143–146

Aarestrup FM, Duran CO, Burch DGS (2008) Antimicrobial resistance in swine production. Anim Health Res Rew 9(2):135–148

Agersø H, Friis C (1998) Bioavailability of amoxycillin in pigs. J Vet Pharmacol Ther 21:41–46

Agnoletti F, Drigo I, Bano L, Klein U, Szancer J (2010) Pleuromutilin susceptibility of *Clostridium perfringens* and *Clostridium difficile* isolates from pigs in Italy and Denmark. Proceedings of the 21st international pig veterinary society congress, Vancouver, Canada, 1: 83

Agribusiness Research and Advisory (2012) Rabobank International; September (2012) Industrialization of China's Pork Supply Chain. Industry Note #329. Food & Agribusiness Research and Advisory. https://www.farminguk.com/content/knowledge/20120919_China%20Pork__Revised%20FINAL%20Approved_WS.PDF. Accessed 12 Apr 2022

AHG (2015) On prioritisation of diseases for which vaccines could reduce antimicrobial use in animals: report of the meeting of the OIE ad hoc group on prioritisation of diseases for which vaccines could reduce antimicrobial use in animals. World Organisation for Animal Health, Paris. https://www.oie.int/fileadmin/SST/adhocreports/Diseases%20for%20which%20Vaccines%20could%20reduce%20Antimicrobial%20Use/AN/AHG_AMUR_Vaccines_Apr2015.pdf. Accessed 1 Jun 2023

Amador P, Fernandes R, Prudêncio C, Duarte I (2019) Prevalence of antibiotic resistance genes in multidrug-resistant *Enterobacteriaceae* on Portuguese livestock manure. Antibiotics 8:23

Anderson MD, Stroh SL, Rogers S (1994) Tiamulin (Denagard®) activity in certain swine tissues following oral and intramuscular. Proceedings AASP meeting, Chicago, pp 115–118

Andresen LO (1998) Differentiation and distribution of three types of exfoliative toxin produced by *Staphylococcus hyicus* from pigs with exudative epidermitis. FEMS Immunol Med Microbiol 20:301–310

Andresen LO, Ahrens P (2004) A multiplex PCR for detection of genes encoding exfoliative toxins from *Staphylococcus hyicus*. J Appl Microbiol 96:1265–1270

Andresen LO, Ahrens P, Daugaard L, Bille-Hansen V (2005) Exudative epidermitis in pigs caused by toxigenic *Staphylococcus chromogenes*. Vet Microbiol 105(3–4):291–300

Apley MD, Coetzee JF, Imerman PM, Karriker LA, Gehring R (2007) Ampicillin pharmacokinetics in swine following needle-free, intramuscular and intravenous administration. J Vet Pharmacol Ther 30:417–421

Aragon V, Segales J, Oliveira S (2012) Glässer's disease. In: Straw B, Zimmerman J, Karriker L, Ramirez A, Schwartz K, Stevenson G (eds) Diseases of swine. Wiley Blackwell, West Sussex, pp 760–769

Armand-Lefevre L, Ruimy R, Andremont A (2005) Links clonal comparison of *Staphylococcus aureus* isolates from healthy pig farmers, human controls and pigs. Emerg Infect Dis 11(5):711–714

Arruda PHE (2014) *Clostridium difficile* infection in neonatal piglets: pathogenesis, risk factors, and prevention. Graduate theses and dissertations. Iowa State University, 14096

Arruda PH, Madson DM, Ramirez A, Rowe E, Lizer JT, Songer JG (2013) Effect of age, dose and antibiotic therapy on the development of *Clostridium difficile* infection in neonatal piglets. Anaerobe 104–110

Arsenakis I, Boyen F, Haesebrouck F, Maes DGD (2018) Autogenous vaccination reduces antimicrobial usage and mortality rates in a herd facing severe exudative epidermitis outbreaks in weaned pigs. Vet Rec 182(26):744

Baggot JD, Giguère S (2013) Principles of antimicrobial drug bioavailability and disposition. In: Giguère S, Prescott JF, Dowling PM (eds) Antimicrobial therapy in veterinary medicine, 5th edn. Wiley, pp 41–77

Bak H, Havn K (2011) Significantly reduced use of antimicrobials with PCV2 and ileitis vaccination in a Danish herd. In: Safe pork conference proceedings, Netherlands, 2011

Bak H, Rathkjen PH (2009) Reduced use of antimicrobials after vaccination of pigs against porcine proliferative enteropathy in a Danish SPF herd. Acta Vet Scand 51:1

Barker IK, Van Dreumel AA, Palmer N (1993) The alimentary system. In: Jubb KF, Kennedy PC, Palmer N (eds) Pathology of domestic animals, vol 2, 4th edn. Academic Press, San Diego, CA, pp 1–318
Barros MM, Castro J, Araújo D, Campos AM, Oliveira R, Silva S, Outor-Monteiro D, Almeida C (2023) Swine Colibacillosis: global epidemiologic and antimicrobial scenario. Antibiotics 12:682
Benchaoui HA, Nowakowski M, Sherington J, Rowan TG, Sunderland SJ (2004) Pharmacokinetics and lung tissue concentrations of tulathromycin in swine. J Vet Pharmacol Ther 27:203–210
Bertschinger HU, Bürgi E, Eng V, Wegmann P (1990) Lowering of the incidence of puerperal mastitis in the sow by protection of the mammae from contamination. Schweiz Arch Tierheilkd 132:557–566
Bes N, Baleri T, March R, Riera P (1997) Pharmacokinetics of amoxicillin following oral administration in swine. J Vet Pharmacol Ther 20(Suppl 1):47–48
Böhmer BM, Kramer W, Roth-Maier DA (2006) Dietary probiotic supplementation and resulting effects on performance, health status, and microbial characteristics of primiparous sows. J Anim Physiol Anim Nutr (Berl) 90(7–8):309–315
Bossé JT, Janson H, Sheehan BJ, Beddek AJ, Rycroft AN, Kroll JS, Langford PR (2002) *Actinobacillus pleuropneumoniae*: pathobiology and pathogenesis of infection. Microbes Infect 4:225–235
Brogan TV, Mellema JD, Martin LD, Krueger M, Redding GJ, Glenny RW (2007) Spatial and temporal heterogeneity of regional pulmonary blood flow in piglets. Pediatr Res 62:434–439
Brookes VJ, Hernández JM, Holyoake P, Ward MP (2015) Industry opinion on the likely routes of introduction of highly pathogenic porcine reproductive and respiratory syndrome into Australia from south-East Asia. Aust Vet J 93(1–2):13–19
Brooks PH, Beal J (2001) Liquid feeding of pigs: potential for reducing environmental impact and for improving productivity and food safety. Recent Adv Anim Nutr Austr 13
Burch DGS (2004) The comparative efficacy of antimicrobials for the prevention and treatment of enzootic pneumonia and some of their pharmacokinetic and pharmacodynamic relationships. Pig J 53:8–27
Burch DGS (2005) Pharmacokinetic, pharmacodynamic and clinical correlations relating to the therapy of colonic infections in the pig and breakpoints determinations. Pig J 56:8–24
Burch DGS (2010) Antimicrobial concentrations in plasma and lung and their relationships to bacterial respiratory infections. Pig J 63:34–49
Burch DGS (2012) Examination of the pharmacokinetic/pharmacodynamic (PK/PD) relationships of orally administered antimicrobials and their correlation with the therapy of various bacterial and mycoplasmal infections in pigs. Fellowship thesis, Royal College of Veterinary Surgeons, London, p 76
Burch DGS (2013) Antimicrobial drug use in swine. In: Giguère S, Prescott J, Dowling P (eds) Antimicrobial therapy in veterinary medicine. Wiley, Hoboken, pp 553–568
Burch DGS, Hammer JM (2013) Managing Lawsonia and Brachyspira infections using pharmacokinetic and pharmacodynamic principles. AASV Annual Meeting: Purpose-Inspired Practice, pp 231–235
Burch DGS, Sperling D (2018) Amoxicillin-current use in swine medicine. J Vet Pharmacol Ther 41(3):356–368
Burch DGS, Duran CO, Aarestrup FM (2008) Guidelines for antimicrobial use in swine. In: Guardabassi L, Jensen LB, Kruse H (eds) Guide to antimicrobial use in animals. Blackwell Publishing, Oxford
Burch DGS, Grogan R, O'Neill C, Klein U (2016) Successful attempt to eradicate PRRS virus and enzootic pneumonia from a breeder/finisher herd. In: Proceedings of the 24th international pig veterinary society, Dublin, p 331
Burrough ER (2017) Swine dysentery: Ethiopathogenesis and diagnosis of a reemerging disease. Vet Pathol 54(1):22–31
Burrough ER, Strait EL, Kinyon JM, Bower LP, Madson DM, Wilberts BL, Schwartz KJ, Frana TS, Songer JG (2012) Comparative virulence of clinical *Brachyspira* spp. isolates in inoculated pigs. J Vet Diagn Invest 24:1025–1034
Casals JB, Nielsen R, Szancer J (1990) Standardisation of tiamulin for routine sensitivity of *Actinobacillus (Haemophilus) pleuropneumoniae* (App). In: Proceedings of the 11th international pig veterinary society congress, Lausanne, p 43
Castro J, Barros MM, Araújo D, Campos AM, Oliveira R, Silva S, Almeida C (2022) Swine enteric colibacillosis: current treatment avenues and future directions. Front Vet Sci 9:981207
Catelli E, Catella A, Canelli E, Luppi A, Caleffi A, Arioli E, Borghetti P, Martelli P (2017) Evaluation of the efficacy of a vaccine against *A. pleuropneumoniae* Clinical outcome, mortality and pleuritis lesions at slaughterhouse. In: Proceeding of 9th ESPHM conference, Prague
Center for Disease Dynamics, Economics & Policy: State of the World's Antibiotics (2015) CDDEP, Washington, DC
Chander Y, Primus A, Oliveira S, Gebhart CJ (2012) Phenotypic and molecular characterization of a novel strongly hemolytic Brachyspira species, provisionally designated "*Brachyspira hampsonii*". J Vet Diagn Invest 24(5):903–910
Chen L, Zhang J, Wang J, Butaye P, Kelly P, Li M, Yang F, Gong J, Yassin AK, Guo W, Li J, Song C, Wang C (2018) Newly identified colistin resistance genes, mcr-4 and mcr-5, from upper and lower alimentary tract of pigs and poultry in China. PLoS One 13(3)
Cheng G, Ha H, Xie S, Wang X, Dai M, Huang L, Yuan Z (2014) Antibiotic alternatives: the substitution of antibiotics in animal husbandry? Front Microbiol 5:Article 217/I
Cid D, Fernandez-Garayzabal JF, Pinto C, Dominguez L, Vela AI (2019) Antimicrobial susceptibility of *Pasteurella multocida* isolated from sheep and pigs

in Spain - short communication. Acta Vet Hung 67:489–498

Clothier KA, Kinyon JM, Frana TS, Naberhaus N, Bower L, Strait EL, Schwartz K (2011) Species characterization and minimum inhibitory concentration patterns of *Brachyspira* species isolates from swine with clinical disease. J Vet Diagn Invest 23(6):1140–1145

CLSI VET01S (2023) Performance Standards for Antimicrobial Disk and Dilution Susceptibility Tests for Bacteria Isolated From Animals. 6th ed. CLSI supplement VET01S. Clinical and Laboratory Standards Institute; 2023

CLSI VET08 (2018) Clinical and laboratory standards institute (CLSI). Performance standards for antimicrobial disk and dilution susceptibility tests for bacteria isolated from animals, 4th ed. CLSI supplement VET08. Clinical and Laboratory Standards Institute, Wayne, PA

Collineau L, Rojo-Gimeno C, Léger A, Backhans A, Loesken S, Nielsen EO, Postma M, Emanuelson U, Beilage EG, Sjölund M, Wauters E, Stärk KDC, Dewulf J, Belloc C, Krebs S (2017) Herd-specific interventions to reduce antimicrobial usage in pig production without jeopardising technical and economic performance. Prev Vet Med 144:167–178

Cooper DM, Gebhart CJ (1998) Comparative aspects of proliferative enteritis. J Am Vet Med A 212:1446–1451

Cranwell PD, Moughan PJ (1989) Biological limitations imposed by the digestive system to the growth performance of weaned pigs. In: Barnett JL, Hennesy DP (eds) Manipulating pig production II. Australasian Pig Science Association, Werribee, pp 140–159

Crenshaw JD, Peo ER, Lewis AJ, Schneider NR (1986) The effects of sorbic acid in high moisture sorghum grain diets on performance of weanling swine. J Anim Sci 63:831–837

Cromwell GL, Lindemann MD, Monengue HJ, Hall DD, Orr DE Jr (1998) Tribasic cooper chloride and copper sulfate as copper sources for weanling pigs. J Anim Sci 76:118–123

CVMP (2017) Committee for Medicinal Products for veterinary use: meeting of 14–16 March 2017 EMA/CVMP/147249/2017. European Medicines Agency. Press release. https://www.ema.europa.eu/en/news/committee-medicinal-products-veterinary-use-cvmp-meeting-14-16-march-2017. Accessed 29 Aug 2022

DANMAP (2022) DANMAP 2021 - Use of antimicrobial agents and occurrence of antimicrobial resistance in bacteria from food animals, food and humans in Denmark. https://www.danmap.org/reports/2021. Accessed 12 June 2023

Davis PR (2012) One world, one health: the threat of emerging swine diseases. A North American perspective. Transbound Emerg Dis 59(Suppl. 1):18–26

Day DN, Sparks JW, Karriker LA, Stalder KJ, Wulf LW, Zhang J, Kinyon JM, Stock ML, Gehring R, Wang C, Ellingson J, Coetzee JF (2015) Impact of an experimental PRRSV and *Streptococcus suis* coinfection on the pharmacokinetics of ceftiofur hydrochloride after intramuscular injection in pigs. J Vet Pharmacol Ther 38(5):475–481

Dayao DAE, Dawson S, Kienzle MJ-P, Gibson JS, Blackall PJ, Turni C (2015) Variation in the antimicrobial susceptibility of *Actinobacillus pleuropneumoniae* isolates in a pig, within a batch of pigs, and among batches of pigs from one farm. Microb Drug Resist 21(4):491–496

De Busser EV, Dewulf J, De Zutter L, Haesebrouck F, Callens J, Meyns T, Maes W, Maes D (2011) Effect of administration of organic acids in drinking water on faecal shedding of E. coli, performance parameters and health in nursery pigs. Vet J 188:184–188

de Jong A, Youala M, Klein U, El Garch F, Moyaert H, Simjee S, Maes D, Gyuranecz M, Pridmore A, Thomson JR, Ayling RD (2021) MycoPath study group. Antimicrobial susceptibility monitoring of mycoplasma hyopneumoniae isolated from seven European countries during 2015-2016. Vet Microbiol 253:108973

De la Funete AJ, Tucker AW, Navas J, Blanco M, Morfia SJ, Gutiérrez-Martín CB (2007) Antimicrobial susceptibility patterns of *Haemophilus parasuis* from pigs in the United Kingdom and Spain. Vet Microbiol 120:184–191

Dohmen W, Bonten MJM, Bos MEH, van Marm S, Scharringa J, Wagenaar JA, Heederik DJJ (2015) Carriage of extended-spectrum ß-lactamases in pig farmers is associated with occurrence in pigs. Clin Microbiol Infect 21:917–923

Dohmen W, Dorado-Garcıa A, Bonten MJM, Wagenaar JA, Mevius D, Heederik DJJ (2017) Risk factors for ESBL-producing Escherichia coli on pig farms: a longitudinal study in the context of reduced use of antimicrobials. PLoS One 12(3):e0174094

Dorado-Garcia A, Smid JH, Van Pelt W, Bonten MJM, Fluit AC, Van Den Bunt G, Wagenaar JA, Hordijk J, Dierikx CM, Veldman KT, de Koeijer A, Dohmen W, Schmitt H, Liakopoulos A, Pacholewicz E, Lam TJGM, Velthuis AG, Heuvelink A, Gonggrijp MA, van Duijkeren E, van Hoek AHAM, de Roda Husman AM, Blaak H, Havelaar AH, Mevius DJ, Heederik DJJ (2018) Molecular relatedness of ESBL/AmpC-producing *Escherichia coli* from humans, animals, food and the environment: a pooled analysis. J Antimicrob Chemother 73(2):339–347

Dutta GN, Devriese LA (1980) Susceptibility of *Clostridium perfringens* of animal origin to fifteen antimicrobial agents. J Vet Pharmacol Ther 3:227–236

Dybkjær L, Jacobsen AP, Togersen FA, Poulsen HD (2006) Eating and drinking activity of newly weaned piglets: effects of individual characteristics, social mixing, and addition of extra zinc to the feed. J Anim Sci 84:702–711

EFSA AHAW Panel (EFSA Panel on Animal Health and Welfare), Nielsen SS, Bicout DJ, Calistri P, Canali E, Drewe JA, Garin-Bastuji B, Gonzales Rojas JL, Gortazar Schmidt C, Herskin M, Michel V, Miranda Chueca MA, Padalino B, Pasquali P, Roberts HC, Sihvonen LH, Spoolder H, Stahl K, Velarde A, Viltrop

A, Winckler C, Dewulf J, Guardabassi L, Hilbert F, Mader R, Baldinelli F, Alvarez J (2021) Scientific opinion on the assessment of animal diseases caused by bacteria resistant to antimicrobials: swine. EFSA J 19(12):7113, 66 pp

El Garch F, de Jong A, Simjee S, Moyaert H, Klein U, Ludwig C, Marion H, Haag-Diergarten S, Richard-Mazet A, Thomas V, Siegwart E (2016) Monitoring of antimicrobial susceptibility of respiratory tract pathogens isolated from diseased cattle and pigs across Europe, 2009-2012. Vet Microbiol 194:11–22

El Garch F, de Jong A, Bertrand X, Hocquet D, Sauget M (2018) *mcr*-1-like detection in commensal *Escherichia coli* and *Salmonella* spp. from food-producing animals at slaughter in Europe. Vet Microbiol 213:42–46

Ellis-Iversen J, Smith RP, Gibbens JC, Sharpe CE, Dominguez M, Cook AJC (2011) Risk factors for transmission of foot-and-mouth disease during an outbreak in southern England in 2007. Vet Rec 168(5):128–128

EMA (2017) RONAFA report scientific opinion—EMA and EFSA joint scientific opinion on measures to reduce the need to use antimicrobial agents in animal husbandry in the European Union, and the resulting impacts on food safety (RONAFA). https://doi.org/10.2903/j.efsa.2017.4666

EMA (2018) European Medicines Agency: Stratification of sales data of antimicrobials by species Data collection protocol 2017 (EMA/284404/2018)

EMA (2020) European Medicines Agency: Answer to the request from the European Commission for updating the scientific advice on the impact on public health and animal health of the use of antibiotics in animals - Categorisation of antimicrobials (EMA/CVMP/CHMP/682198/2017). https://www.ema.europa.eu/en/documents/report/categorisation-antibiotics-european-union-answer-request-european-commission-updating-scientific_en.pdf. Accessed 29 Aug 2022

EMA/ESVAC (2011) European Medicines Agency, European Surveillance of Veterinary Antimicrobial Consumption (ESVAC). 'Trends in the sales of veterinary antimicrobial agents in nine European countries (2005–2009)' (EMA/238630/2011)

EMA/ESVAC (2017) European Medicines Agency, European Surveillance of Veterinary Antimicrobial Consumption (ESVAC). 'Sales of veterinary antimicrobial agents in 30 European countries in 2015' (EMA/184855/2017)

EMA/ESVAC (2019) European Medicines Agency, European Surveillance of Veterinary Antimicrobial Consumption (ESVAC). 'Sales of veterinary antimicrobial agents in 31 European countries in 2017' (EMA/275982/2018)

EMA/ESVAC (2022) European Medicines Agency, European Surveillance of Veterinary Antimicrobial Consumption (ESVAC). 'Sales of veterinary antimicrobial agents in 31 European countries in 2021 (EMA/795956/2022)

EMA-Documents-Product Information-Draxxin (2008) European Public Assessment report Product information. https://www.ema.europa.eu/en/documents/product-information/draxxin-epar-product-information_en.pdf. Accessed 23 June 2023

European Comission, DG Dashboard: Pigmeant, 2023, regularly updated: Pigmeat_Dashboard.pdf (europa.eu). Accessed 24 June 2023

European Commission (EC) (2015) Commission notice—guidelines for the prudent use of antimicrobials in veterinary medicine (2015/C 299/04) 11.9.2015, Official Journal of the European Union, C299/7-26

Evans NA (2005) Tulathromycin: an overview of a new triamilide antimicrobial for livestock respiratory disease. Vet Ther 6:83–95

FAO (2016) Probiotics in animal nutrition—Production, impact and regulation by Yadav S. Bajagai, Athol V. Klieve, Peter J. Dart and Wayne L. Bryden. Editor Harinder P.S. Makkar. FAO Animal Production and Health Paper No. 179. Rome. http://www.fao.org/3/a--i5933e.pdf. Accessed 29 Aug 2022

Farmer Ch, Maes D, Peltoniemi O (2019) Mammary system. In: Zimmerman JJ, Karriker LA, Ramirez A, Schwartz KJ, Stevenson GW, Zhang J (eds) Diseases of swine, 11th edn. Wiley, pp 313–338

FDA (2012) Food and Drug Administration: CVM GFI # 209 – The Judicious use of medically important antimicrobial drugs in food-producing animals. https://www.fda.gov/regulatory-information/search-fda-guidance-documents/cvm-gfi-209-judicious-use-medically-important-antimicrobial-drugs-food-producing-animals. Accessed 29 Aug 2022

FDA (2018) Food and Drug Administration: Summary Report on Antimicrobials Sold or Distributed for Use in Food-Producing Animals 2017. https://www.fda.gov/media/119332/download. Accessed 29 Aug 2022

Fecteau G, Smith BP, George LW (2009) Septicaemia and meningitis in the newborn calf. Vet Clin N Am Food Anim Pract 25:195–208

Ferran AA, Bibbal D, Pellet T, Laurentie M, Gicquel-Bruneau M, Sanders P, Schneider M, Toutain PL, Bousquet-Melou A (2013) Pharmacokinetic/pharmacodynamic assessment of the effects of parenteral administration of a fluoroquinolone on the intestinal microbiota: comparison of bactericidal activity at the gut versus the systemic level in a pig model. Int J Antimicrob Agents 42(5):429–435

Filippitzi ME, Brinch Kruse A, Postma M, Sarrazin S, Maes D, Alban L, Nielesen LR, Dewulf J (2017) Review of transmission routes of 24 infectious diseases preventable by biosecurity measures and comparison of the implementation of these measures in pig herds in six European countries. Transbound Emerg Dis 65:381–398

Filippitzi ME, Devreese M, Broekaert K, Rasschaert G, Daeseleire E, Meirlaen J, Dewulf J (2019) Quantitative risk model to estimate the level of antimicrobial residues that can be transferred to soil via manure, due to oral treatments of pigs. Prev Vet Med 167:90–100

Fraile L (2013) Treatment of mastitis metritis Agalactia syndrome in sows. In: Fraile L (ed) Antimicrobial therapy in swine: a practical approach. Grupo Asis Biomedia, S.L., Zaragoza-Spain

Frana TS (2012) Staphylococcosis. In: Zimmerman JJ, Karriker LA, Ramírez A, Schwartz KJ, Stevenson GW (eds) Diseases of swine. Wiley-Blackwell Publishing, pp 3059–82

Frana T, Boyes N, Madson D (2012) Antimicrobial susceptibility patterns associated with virulence factors in swine *Escherichia coli* isolates. In: Proceedings of the 43rd meeting of the American Association of Swine Veterinarians, Denver, Colorado, pp 57–60

Franklin A, Pringle M, Hampson DJ (2006) Antimicrobial resistance in *Clostridium* and *Brachyspira* spp. and other anaerobes. In: Aarestrup FM (ed) Antimicrobial resistance in bacteria of animal origin. ASM Press, Washington, pp 127–144

Fricke R, Bastert O, Gotter V, Brons N, Kamp J, Selbitz HJ (2015) Implementation of a vaccine against Shigatoxin 2e in a piglet producing farm with problems of Oedema disease: case study. Porcine Health Manage 1:6

Friedman M, Bednar V, Klimes J, Smola J, Mrlík V, Literák I (2008) *Lawsonia intracellularis* in rodents from pig farms with the occurrence of porcine proliferative enteropathy. Lett App Microbiol 47:117–121

Gerjets I, Kemper N (2009) Coliform mastitis in sows: a review. J Swine Health Prod 17:97–105

Gerjets I, Kruse S, Krieter J, Kemper N (2008) Diagnosis of MMA affected sows: bacteriological differentiation, temperature measurement and water intake. In: 20th International pig veterinary society congress, Durban, South Africa, p 236

Gerlach BM, Houser TA, Hollis LC, Tokach MD, Nietfeld JC, Higgins JJ, Anderson GA, Goehring BL (2012) Incidence and severity of *Arcanobacterium pyogenes* injection site abscesses with needle or needle-free injection. Meat Sci 92:805–807

Godinho KS, Keane SG, Nanjiani IA, Benchaoui HA, Sunderland SJ, Jones MA, Weatherley AJ, Gootz TD, Rowan TG (2005) Minimum concentrations of tulathromycin against respiratory bacterial pathogens isolated from clinical cases in European cattle and swine and variability arising from changes in in-vitro methodology. Vet Therapeut 6:113–121

Gottschalk M (2012) Actinobacillosis. In Straw, BE, Zimmerman JJ, D'Allaire S, Taylor DJ (eds) Blackwell Publishing, Ames, 10th edn. pp 653–659

Gottschalk M, Segura M (2000) The pathogenesis of the meningitis caused by *Streptococcus suis*: the unresolved questions. Vet Microbiol 76(3):259–272

Gourreau JM, Drolet R, Martineau GP, Morvan H, Pastoret PP, Pin D, Scott DW (2015) Greasy pig disease and staphylococcal skin disease. In: Gourreau JM et al (eds) Atlas of porcine dermatology. OIE, pp 39–54

Grahofer A, Gurtner C, Nathues H (2017) Haemorrhagic bowel syndrome in fattening pigs. Porcine Health Managem 3:27

Griffiths NM, Hirst BH, Simmons NL (1994) Active intestinal secretion of the fluoroquinolone antibacterials ciprofloxacin, norfloxacin and pefloxacin; a common secretory pathway? J Pharmacol Exp Ther 269:496–502

Grobbel M, Lubke-Becker A, Alesik E, Schwarz S, Wallmann J, Werckenthin C, Wieler LH (2007) Antimicrobial susceptibility of *Escherichia coli* from swine, horses, dogs and cats as determined in the BfT-GermVet monitoring program 2004-2006. Berl Münch Tierärztl Wochenschr 120:391–401

Gruet P, Maincent P, Berthelot X, Kaltsatos V (2001) Bovine mastitis and intramammary drug delivery: review and perspectives. Adv Drug Deliv Rev 50(3):245–259

Guardabassi L, Stegger M, Skov R (2007) Retrospective detection of methicillin resistant and susceptible *Staphylococcus aureus* ST398 in Danish slaughter pigs. Vet Microbiol 122(3–4):384–386

Hammerum AM, Larsen J, Andersen VD, Lester CH, Skovgaard Skytte TS, Hansen F, Olsen SS, Mordhorst H, Skov RL, Aarestrup FM, Agersø Y (2014) Characterization of extended-spectrum b-lactamase (ESBL)-producing *Escherichia coli* obtained from Danish pigs, pig farmers and their families from farms with high or no consumption of third- or fourth-generation cephalosporins. J Antimicrob Chemother 69(10):2650

Hansen MS, Pors SE, Jensen HE, Hansen VB, Bisgaard M, Flachs EM, Nielsen OL (2010) An investigation of the pathology and pathogens associated with porcine respiratory disease complex in Denmark. J Comp Pathol 143:120–131

Hirsch AC, Philipp H, Kleemannm R (2004) Efficacy of meloxicam for treatment of the mastitis-metritis-agalactia syndrome in sows. Prakt Tierarzt 85:842–848

Hoelzer K, Bielke L, Blake DP, Cox E, Cutting SM, Devriendt B, Erlacher-Vindel E, Goossens E, Karaca K, Lemiere S, Metzner M, Raicek M, Collell Suriñach M, Wong NM, Gay C, Van Immerseel F (2018) Vaccines as alternatives to antibiotics for food producing animals. Part 1: challenges and needs Vet Res 49:64

Hommez J, Castryck F, Miry C, Lein A, Devriese LA, Haesebrouck F (1998) Susceptibility of different *Serpulina* species in pigs to antimicrobial agents. Vlaams Diergeneeskd Tijdschr 67:32–35

Hughes P, Heritage J (2002) Antibiotic growth-promoters in food animals. Food and Agriculture Organization of the United Nations, Rome

Imeah B, Penz E, Rana M, Trask C (2020) Needle-less injector study team. Economic analysis of new workplace technology including productivity and injury: the case of needle-less injection in swine. PLoS One 15(6):e0233599

Jackson CJ (2007) Drinking behavior in nursery aged pigs. Retrospective Theses and Dissertations 14664. https://lib.dr.iastate.edu/rtd/14664. Accessed 29 Aug 2022

Jacobson M, Fellström C, Jensen-Waern M (2010) Porcine proliferative enteropathy: an important disease with questions remaining to be solved. Vet J 184:264–268

Joerling J, Barth SA, Schlez K, Willems H, Herbst W, Ewers C (2018) Phylogenetic diversity, antimicrobial susceptibility and virulence gene profiles of *Brachyspira hyodysenteriae* isolates from pigs in Germany. PLoS One 13(1):e0190928

JVARM (2016) Ministry of Agriculture forestry and fisheries of Japan, National veterinary Assay Laboratory: Report on the Japanese Veterinary Antimicrobial Resistance Monitoring System, 2012–2013

Karlsson M, Fellstrom C, Johansson K-E, Franklin A (2004) Antimicrobial resistance in *Brachyspira pilosicoli* with special reference to point mutation in the 23S rRNA gene associated with macrolide and lincosamide resistance. Microb Drug Resist 10:204–208

Karuppannan AK, Opriessnig T (2018) *Lawsonia intracellularis*: revisiting the disease ecology and control of this fastidious pathogen in pigs. Front Vet Sci 5:181

Kemper N, Gerjets I (2009) Bacteria in milk from anterior and posterior mammary glands in sows affected and unaffected by postpartum dysgalactia syndrome (PPDS). Acta Vet Scand 51:26

Khanna T, Friendship R, Dewey C, Weese JS (2008) Methicillin resistant *Staphylococcus aureus* colonization in pigs and pig farmers. Vet Microbiol 128(3–4):298–303

Kielstein P, Rapp-Gabrielson VJ (1992) Designation of 15 serovars of *Haemophilus parasuis* based immunodiffusion using heat stable antigen extracts. J Clin Microbiol 30:862–865

Kim S, Covington A, Pamer EG (2017) The intestinal microbiota: antibiotics, colonization resistance, and enteric pathogens. Immunol Rev 279(1):90–105

Kim K, Song M, Liu Y, Ji P (2022) Enterotoxigenic *Escherichia coli* infection of weaned pigs: intestinal challenges and nutritional intervention to enhance disease resistance. Front Immunol 13:885253

Kirchgassner C, Schmitt S, Borgstrom A, Wittenbrink MM (2016) Antimicrobial susceptibility of *Brachyspira hyodysenteriae* in Switzerland. Schweiz Arch Tierheilkd 158(6):405–410

Kirchhelle (2018) Pharming animals: a global history of antibiotics in food production (1935–2017). Palgrave Commun 4:96. https://doi.org/10.1057/s41599-018-0152-2

Klein U, de Jong A, Moyaert H (2017) Antimicrobial susceptibility monitoring of *Mycoplasma hyopneumoniae* and *Mycoplasma bovis* isolated in Europe. Vet Microbiol 204:188–193

Krishnasamy V, Joachim O, Silbergeld E (2015) Antimicrobial use in Chinese swine and broiler poultry production. Antimicrob Resist Infect Control 4:17

Kruse AB, Nielsen LR, Alban L (2015) Vaccination against *Actinobacillus pleuropneumoniae* as an alternative strategy to antimicrobial use in Danish pig herds. In: Safe pork conference proceedings, Porto, Portugal, 2015

Laanen M, Persoons D, Ribbens S, de Jong E, Callens B, Strubbe M, Maes D, Dewulf J (2013) Relationship between biosecurity and production/antimicrobial treatment characteristics in pig herds. Vet J 198(2):508–512

Lambert RJ, Stratford M (1999) Weak-acid preservatives: modelling microbial inhibition and response. J Appl Microbiol 86:157–164

Larsen I, Nielsen SS, Olsen JE, Nielsen JP (2016) The efficacy of oxytetracycline treatment at batch, pen and individual level on *Lawsonia intracellularis* infection in nursery pigs in a randomised clinical trial. Prev Vet Med 124:25–33

Lawson GHK, McOrist S, Jasni S, Mackie RA (1993) Intracellular bacteria of porcine proliferative enteropathy: cultivation and maintenance in vitro. J Clin Microbiol 31:1136–1142

Le Grand A, Kobisch M (1996) Comparison of the use of a vaccine and sequential antibiotic treatment in a herd infected with *Mycoplasma hyopneumoniae*. Vet Res 27(3):241–253

Lees P, Illambas J, Potter TJ, Pelligand L, Rycroft A, Toutain PL (2017) A large potentiation effect of serum on the in vitro potency of tulathromycin against *Mannheimia haemolytica* and *Pasteurella multocida*. J Vet Pharmacol Ther 2017 Oct;40(5):419–428

Letellier A, Dubreuil D, Roy G, Fairbrother JM, Jacques M (1991) Determination of affinity of *Pasteurella multocida* isolates for porcine respiratory tract mucus, and partial characterization of the receptors. Am J Vet Res 52(1):34–39

Liang SY, Kumar A (2015) Empiric antimicrobial therapy in severe sepsis and septic shock: optimizing pathogen clearance. Curr Infect Dis Rep 17(7):493

Liao SF, Nyachoti M (2017) Using probiotics to improve swine gut health and nutrient utilization. Anim Nutr 3(4):331–343

Lillie-Jaschniski K, Köchling M, Lindner T (2013) Erste Erfahrungen mit dem Einsatz von ECOPORC SHIGA, der neuen Vakzine gegen die Ödemkrankheit—Auswertungen aus Feldversuchen in Deutschland. Tierärztl Umsch 68:377–382

Liu Z, Zhang Z, Yan H, Li J, Shi L (2015) Isolation and molecular characterization of multidrug-resistant Enterobacteriaceae strains from pork and environmental samples in Xiamen. China J Food Prot 78(1):78–88

Liu YY, Wang Y, Walsh TR, Yi LX, Zhang R, Spencer J, Doi Y, Tian G, Dong B, Huang X, Yu LF, Gu D, Ren H, Chen X, Lv L, He D, Zhou H, Liang Z, Liu JH, Shen J (2016) Emergence of plasmid-mediated colistin resistance mechanism MCR-1 in animals and human beings in China: a microbiological and molecular biological study. Lancet Infect Dis 16(2):161–168

Liu Y, Espinosa CD, Abelilla JJ, Casas GA, Lagos LV, Lee SA, Stein HH (2018) Non-antibiotic feed additives in diets for pigs: a review. Anim Nutr (Zhongguo xu mu shou yi xue hui) 4(2):113–125

Lückstädt C, Mellor S (2011) The use of organic acids in animal nutrition, with special focus on dietary potassium diformate under European and Austral-Asian conditions. In: Recent advances in animal nutrition-Australia, vol 18, pp 123–130

Lumb KR, Robertson J, Scott HE, Woolfenden NJ (2017) Optimising the use of antimicrobials: preparing the industry for in-water delivery in the short term and improving hygiene and more effectively targeting medication in the longer term. Final Report prepared by RAFT Solutions Pty Ltd for Agriculture and Horticulture Development Board. https://pork.ahdb.org.uk/media/274247/51510014_ft014_raft--solutions-ltd-waterreport_approved_september-2017.pdf. Accessed 29 Aug 2022

Luppi A (2017) Swine enteric colibacillosis: diagnosis, therapy and antimicrobial resistance. Porcine Health Manage 3:16

Lysnyansky I, Ayling RD (2016) *Mycoplasma bovis*: mechanisms of resistance and trends in antimicrobial susceptibility. Front Microbiol 27:595. (eCollection)

Madec F, Bridoux N, Bounaix S, Jestin A (1998) Measurement of digestive disorders in the piglet at weaning and related risk factors. Prev Vet Med 35:53–72

Madec F, Rose N, Grasland B, Cariolet R, Jestin A (2008) Post-weaning multisystemic wasting syndrome and other PCV2-related problems in pigs: a 12-year experience. Transbound Emerg Dis 55(7):273–283

Mahu M, Pasmans F, Vranckx K, de Pauw N, Vande Maele L, Vyt P, Vandersmissen T, Martel A, Haesebrouck F, Boyen F (2017) Presence and mechanisms of acquired antimicrobial resistance in Belgian *Brachyspira hyodysenteriae* isolates belonging to different clonal complexes. Vet Microbiol 207:125–132

Mallo JJ, Rioperezb J, Honrubiaa P (2010) The addition of *enterococcus faecium* to diet improves Piglet's intestinal microbiota and performance. Livest Sci 133(1):176–178

Marie J, Morvan H, Berthelot-Hérault F, Sanders P, Kempf I, Gautier-Bouchardon AV, Jouy E, Kobisch M (2002) Antimicrobial susceptibility of *Streptococcus suis* isolated from swine in France and from humans in different countries between 1996 and 2000. J Antimicrob Ch 50:201–209

Marsteller T, Winkelman N, Gebhart C, Armbruster G, Weldon W, Muller R, Weatherford J, Symanowski J (2001) Efficacy of intramuscular tylosin for the treatment and control of porcine proliferative enteropathy caused by *Lawsonia intracellularis*. Vet Ther 2(1)

Martelli F, Lambert M, Butt P, Cheney T, Tatone FA, Callaby R, Rabie A, Gosling RJ, Fordon S, Crocker G, Davies RH, Smith RP (2017) Evaluation of an enhanced cleaning and disinfection protocol in *Salmonella* contaminated pig holdings in the United Kingdom. PLoS One 12(6):e0178897

Martineau GP, Yannig Le Treut D, Guillou D, WaretSzkuta A (2013) Postpartum dysgalactia syndrome: a simple change in homeorhesis? J Swine Health Prod 21

Martinez M, McDermott P, Walker R (2006) Pharmacology of the fluoroquinolones: a perspective for the use in domestic animals. Vet J 172:10–28

Martinez M, Toutain PL, Turnidge J (2013) The pharmacodynamics of antimicrobial agents. In: Giguère S, Prescott JF, Dowling PM (eds) Antimicrobial therapy in veterinary medicine, 5th edn. Wiley, pp 79–105

Masarikova M (2006) In vitro susceptibility to tylosin and tiamulin in Czech isolates of *cpb2+ Clostridium perfringens* type A. In: Proceedings of the 19th international pig veterinary society congress; Copenhagen, Denmark, vol 2, p 458

Masarikova M, Cizek A, Smola J, Sperling D (2012) In vitro susceptibility of Czech porcine isolates of *Clostridium perfringens* type a *cpb2+* to amoxicillin, ampicillin and amoxicillin-clavulanic acid. In: Proceedings of 22nd international pig veterinary society congress; Jeju, Korea, vol 2, p 639

Matías J, Berzosa M, Pastor Y, Irache JM, Gamazo C (2017) Maternal vaccination. Immunization of sows during pregnancy against ETEC infections. Vaccines (Basel) 5(4):48

Matiašovic J, Nedbalcová K, Žižlavský M, Fleischer P, Pokludová L, Kellnerová D, Nechvátalová K, Šimek B, Czanderlová L, Zouharová M, Bernardy J, Králová N, Šlosárková S (2021) *Streptococcus suis* isolates-serotypes and susceptibility to antimicrobials in terms of their use on selected repopulated Czech pig farms. Pathogens 10(10):1314

Matter D, Rossano A, Limat S (2007) Antimicrobial resistance profile of *Actinobacillus pleuropneumoniae* and *Actinobacillus porcitonsillarum*. Vet Microbiol 122:146–156

McDonald PJ, Wetherall BL, Pruul H (1981) Post-antibiotic leukocyte enhancement: increased susceptibility of bacteria pretreated with antibiotics to activity of leukocytes. Rev Infect Dis 3:38–44

McLeese JM, Tremblay ML, Patience JF, Christison GI (1992) Water intake patterns in the weanling pig: effect of water quality, antibiotics and probiotics. Anim Prod 54:135–142

McOrist S, Gebhart CJ (1995) *In vitro* testing of antimicrobial agents for proliferative enteropathy (ileitis). Swine Health Prod 3(4):146–149

McOrist S, Mackie RA, Lawson GH (1995) Antimicrobial susceptibility of ileal symbiont intracellularis isolated from pigs with proliferative enteropathy. J Clin Microbiol 33:1314–1317

McOrist S, Smith SH, Shearn MF, Carr MM, Miller DJ (1996) Treatment and prevention of porcine proliferative enteropathy with oral tiamulin. Vet Rec 139:615–618

McOrist S, Muller WA, Kratzer D, Sjosten CG (2000) Therapeutic efficacy of water-soluble lincomycin-spectinomycin powder against porcine proliferation enteropathy in a European field study. Vet Rec 146:61–65

Mengel H, Krüger M, Krüger MU, Westphal B, Swidsinski A, Schwarz S, Mundt HC, Dittmar K, Daugschies A (2012) Necrotic enteritis due to simultaneous infection with *Isospora suis* and clostridia in newborn piglets and its prevention by early treatment with toltrazuril. Parasitol Res 110(4):1347–1355

Mertens N, Theuß T, Köchling M, Dohmann K, Lillie-Jaschniski K (2022) Pathogens detected in 205 German farms with porcine neonatal diarrhea in 2017. Vet Sci 9(2):44

Mesonero-Escuredo S, Strutzberg-Minder K, Casanovas C, Segalés J (2018) Viral and bacterial investigations on the aetiology of recurrent pig neonatal diarrhoea cases in Spain. Porcine Health Manag 4:5

Mestorino N, Errecalde JO (2012) Pharmacokinetic-pharmacodynamic considerations for bovine mastitis treatment. IntechOpen, Paris

MIC EUCAST (2023) Antimicrobial wild type distributions of microorganisms, MIC distributions for *Actinobacillus pleuropneumoniae*. https://mic.eucast.org/search/?search%5Bmethod%5D=mic&search%5Bantibiotic%5D=-1&search%5Bspecies%5D=24&search%5Bdisk_content%5D=-1&search%5Blimit%5D=50. Accessed 24 June 2023

Mirajkar NS, Davies PR, Gebhart CJ (2016) Antimicrobial susceptibility patterns of *Brachyspira* species isolated from swine herds in the United States. J Clin Microbiol 54(8):2109–2119

Mouton JW, Theuretzbacher U, Craig WA, Tulkens PM, Derendorf H, Cars O (2008) Tissue concentrations: do we ever learn? J Antimicrob Chemother 61:235–237

Mulchandani R, Wang Y, Gilbert M, Van Boeckel TP (2023) Global trends in antimicrobial use in food-producing animals: 2020 to 2030. PLOS Glob Public Health 3(2):e0001305

Nadeau É, Fairbrother JM, Zentek J, Bélanger L, Tremblay D, Tremblay CL, Röhe I, Vahjen W, Brunelle M, Hellmann K, Cvejić D, Brunner B, Schneider C, Bauer K, Wolf R, Hidalgo Á (2017) Efficacy of a single oral dose of a live bivalent *E. coli* vaccine against post-weaning diarrhea due to F4 and F18-positive enterotoxigenic *E. coli*. Vet J 226:32–39

Nedbalcová K, Zouharová M, Šperling D (2017) The determination of minimum inhibitory concentrations of selected antimicrobials for porcine *Haemophilus parasuis* isolates from The Czech Republic. Acta Vet Brno 86:175–181

Nedbalcova K, Zouharova M, Sperling D (2019) Post-antibiotic effect of marbofloxacin, enrofloxacin and amoxicillin against selected respiratory pathogens of pigs. Veterinarni Med 64:67–77

Neirynck W, Boyen F, Chantziaras I, Vandersmissen T, Vyt P, Haesebrouck F, Dewulf J, Maes D (2020) Implementation and evaluation of different eradication strategies for *Brachyspira hyodysenteriae*. Porcine Health Manag 6:27

Nielsen P (1997) The influence of feed on the oral bioavailability of antibiotics/chemotherapeutics in pigs. J Vet Pharmacol Therapy (Suppl 1):30–31

Niemi JK, Bergman P, Ovaska S, Sevón-Aimonen M-L, Heinonen M (2017) Modeling the costs of postpartum dysgalactia syndrome and locomotory disorders on sow productivity and replacement. Front Vet Sci 4:181

Nousiainen J, Setala J (1993) Lactic acid bacteria as animal probiotics. In: Salminen S, Wright A (eds) Lactic acid bacteria. Marcel Dekker, New York, NY, pp 315–356

NSSRF (2020) National Union of Swine Breeders of Russia: Rating of the largest pork producers in the Russian Federation by the end of 2019 (in Russian). http://nssrf.ru/images/statistics/243917_810.pdf. Accessed 1 Sept 2020

Nutsch RG, Hart FJ, Rooney KA, Weigel DJ, Kilgore WR, Skogerboe TL (2005) Efficacy of tulathromycin injectable solution for the treatment of naturally occurring Swine respiratory disease. Vet Ther 6(2):214–224

Okura M, Osaki M, Nomoto R, Arai S, Osawa R, Sekizaki T, Takamatsu D (2016) Current taxonomical situation of *Streptococcus suis*. Pathogens 5:E45

Oliveira S, Galina L, Pijoan C (2001) Development of a PCR test to diagnose *Haemophilus parasuis* infections. J Vet Diagn Invest 13:495–501

Owens RC Jr, Donskey CJ, Gaynes RP, Loo VG, Muto CA (2008) Antimicrobial-associated risk factors for *Clostridium difficile* infection. Clin Infect Dis 46(Suppl 1):19–31

Pamer EG (2016) Resurrecting the intestinal microbiota to combat antibiotic-resistant pathogens. Science 352:535–538

Pardon B, Deprez P (2018) Rational antimicrobial therapy for sepsis in cattle in face of the new legislation on critically important antimicrobials. Vlaams Diergeneeskundig Tijdschrift 87(1)

Park J, Friendship RM, Poljak Z, Weese JS, Dewey CE (2013a) An investigation of exudative epidermitis (greasy pig disease) and antimicrobial resistance patterns of *staphylococcus hyicus* and *Staphylococcus aureus* isolated from clinical cases. Can Vet J 54(2):139–144

Park J, Friendship RM, Weese JS, Poljak Z, Cate E, Dewey CE (2013b) An investigation of resistance to β-lactam antimicrobials among staphylococci isolated from pigs with exudative epidermitis. BMC Vet Res 9:211

Park JE, Easwaran M, Park JW, Shin HJ (2018) Antimicrobial susceptibility of *Staphylococcus hyicus* isolated from Korean pigs with exudative epidermitis. J Prev Vet Med 42(1):41–45

Pendl W, Jenny B, Torgerson PR, Spring P, Kümmerlen D, Sidler X (2017) Effect of herd health management on the prevalence of postpartum dysgalaktie syndrome (PPDS) and the treatment incidence. Schweiz Arch Tierheilkd 159:109

Petrocchi-Rilo M, Gutiérrez-Martín CB, Méndez-Hernández JI, Rodríguez-Ferri EF, Martínez-Martínez S (2018) Antimicrobial resistance of *Pasteurella multocida* isolates recovered from swine pneumonia in Spain throughout 2017 and 2018. Vet Anim Sci 7:100044

PEW CT (2017) The Pew Charitable Trusts: Alternatives to Antibiotics In Animal Agriculture (2017). Available online at: https://www.pewtrusts.org/en/research-and-analysis/reports/2017/07/alternatives-to-antibiotics-in-animal-agriculture. Accessed 29 Aug 2022

Pluske JR, Turpin DL, Kim JC (2018) Gastrointestinal tract (gut) health in the young pig. Anim Nutr 4:187–196. https://doi.org/10.1016/j.aninu.2017.12.004

Postma M, Stärk KDC, Sjölund M, Backhans A, Beilage EG, Lösken S, Belloc C, Collineau L, Iten D, Visschers V, Nielsen EO, Dewulf J (2015) Alternatives to the use of antimicrobial agents in pig production: a multi-country expert-ranking of perceived effectiveness, feasibility and return on investment. Prev Vet Med 118(4):457–466

Postma M, Backhans A, Collineau L, Loesken S, Sjölund M, Belloc C, Emanuelson U, Grosse Beilage E, Stärk KDC, Dewulf J, MINAPIG consortium (2016) The biosecurity status and its associations with production and management characteristics in farrow-to-finish pig herds. Animal 10:478–489

Prasek J, Sperling D, Lobová D, Smola J, Cizek A (2014) Antibiotic susceptibility of *Brachyspira hyodysenteriae* isolates from Czech swine farms: a 10-year follow-up study. Acta Vet Brno 83:003–007

Pridmore A, Burch D, Lees P (2011) Determination of minimum inhibitory and minimum bactericidal concentrations of tiamulin against field isolates of *Actinobacillus pleuropneumoniae*. Vet Microbiol 151(3–4):409–412

Råsbäck TD, Jansson S, Johansson KE, Fellström C (2007) A novel enteropathogenic, strongly haemolytic spirochaete isolated from pig and mallard, provisionally designated "*Brachyspira suanatina*" sp. nov. Environ Microbiol 983–991

Ravindran V, Kornegay ET (1993) Acidification of weaner pig diets: a review. J Sci Food Agric 62:313–322

Reinbold JB, Coetzee JF, Hollis LC, Nickell JS, Riegel CM, Christopher JA, Ganta RR (2010) Comparison of *Anaplasma marginale* disease transmission with needle-free versus needle injection. Am J Vet Res 71(10):1178–1188

Rey JF, Laffont CM, Croubels S, de Backer P, Zemerline C, Bousquet E, Guyonnet J, Ferran AA, Bousquet-Melou A, Toutain PL (2014) Use of Monte Carlo simulation to determine pharmacodynamic cutoffs of amoxicillin to establish a breakpoint for antimicrobial susceptibility in pigs. Am J Vet Res 75:124–131

Rhouma M, Fairbrother JM, Beaudry F, Letellier A (2017) Post weaning diarrhoea in pigs: risk factors and non-colistin-based control strategies. Acta Vet Scand 59(1):31

Richez P, Keck N, Burren X (2012) *Streptococcus suis*: bactericidal profile of amoxicillin, susceptibility and resistance in porcine clinical strains. In: Proceedings of the 17th international congress of the European Association for Veterinary Pharmacology and Toxicology Noordwijkerhout, The Netherlands, pp 188–189

Rieckmann K, Pendzialek SM, Vahlenkamp T, Baums CG (2020) A critical review speculating on the protective efficacies of autogenous *Streptococcus suis* bacterins as used in Europe. Porcine Health Manage 6:12. https://doi.org/10.1186/s40813-020-00150-6

Rohde J, Kessler M, Baums CG, Amtsberg G (2004) Comparison of methods for antimicrobial susceptibility testing and MIC values for pleuromutilin drugs for *Brachyspira hyodysenteriae* isolated in Germany. Vet Microbiol 102:25–32

Rood JI, Maher EA, Somers EB, Campos E, Duncan CL (1978) Isolation and characterization of multiply antibiotic-resistant *Clostridium perfringens* strains from porcine faeces. Antimicrob Agens Chemother 13:871–880

Rood JI, Buddle JR, Wales AJ, Sidhu R (1985) The occurrence of antibiotic resistance in *Clostridium perfringens* from pigs. Aust Vet J 62:276–279

Rose N, Larour G, Le Diguerher G, Eveno E, Jolly JP, Blanchard P, Oger A, Le Dimna M, Jestin A, Madec F (2003) Risk factors for porcine post-weaning multisystemic wasting syndrome (PMWS) in 149 French farrow to-finish herds. Prev Vet Med 61:209–225

Roselli M, Finamore A, Garaguso I, Britti MS, Mengheri E (2003) Zinc oxide protects cultured enterocytes from the damage induced by *Escherichia coli*. J Nutr 133(12):4077–4082

Ross RF, Orning AP, Woods RD, Zimmermann BJ, Cox DF, Harris DL (1981) Bacteriologic study of sow agalactia. Am J Vet Res 42:949–955

Rugna G, Bonilauri P, Carra E (2015) Sequence types and pleuromutilin susceptibility of *Brachyspira hyodysenteriae* isolates from Italian pigs with swine dysentery: 2003–2012. Vet J 203:115–119

Salvarani FM, Silva ROS, Pires PS, Cruz Junior ECC, Albefaro IS, Guedes RMC, Lobato FCF (2012) Antimicrobial susceptibility of *Clostridium perfringens* isolated from piglets with or without diarrhea in Brazil. Braz J Microbiol 1030–1033

Sanders P, Vanderhaeghen W, Fertner M, Fuchs K, Obritzhauser W, Agunos A, Carson C, Borck Høg B, Dalhoff Andersen V, Chauvin C, Hémonic A, Käsbohrer A, Merle R, Alborali GL, Scali F, Stärk KDC, Muentener C, van Geijlswijk I, Broadfoot F, Pokludová L, Firth CL, Carmo LP, Manzanilla EG, Jensen L, Sjölund M, Pinto Ferreira J, Brown S, Heederik D, Dewulf J (2020) Monitoring of farm-level antimicrobial use to guide stewardship: overview of existing systems and analysis of key components and processes. Front Vet Sci 7:540

Sassu EL, Bossé JT, Tobias TJ, Gottschalk M, Langford PR, Henning-Pauka I (2018) Update on *Actinobacillus pleuropneumoniae*: knowledge, gaps and challenges. Transbound Emerg Dis 65:72–90

Sato H, Watanabe T, Higuchi K, Teruya K, Ohtake A, Murata Y, Saito H, Aizawa C, Danbara H, Maehara N (2000) Chromosomal and extrachromosomal synthesis of exfoliative toxin from *Staphylococcus hyicus*. J Bacteriol 182:4096–4100

Schneider M, Sharma S (2014) China's Pork miracle? Institute for Agriculture and Trade Policy. http://www.iatp.org/blog/201402/bracing-for-impacts-as-china-enters-industrial-meat-complex. Accessed 29 Aug 2022

Schwarz S, Alesik E, Grobbel M, Lubke-Becker A, Werckenthin C, Wieler LH, Wallmann J (2007) Antimicrobial susceptibility of streptococci from various indications of swine, horses, dogs and cats as determined in the BfT-GermVet monitoring program 2004-2006. Berl Münch Tierärztl Wochenschr 120:380–390

Schwarz S, Böttner A, Goossens L, Hafez HM, Hartmann K, Kaske M, Kehrenberg C, Kietzmann M, Klarmann D, Klein G, Krabisch P, Luhofer G, Richter A, Schulz B, Sigge C, Waldmann KH, Wallmann J, Werckenthin C (2008) A proposal of clinical breakpoints for amoxicillin applicable to porcine respiratory tract pathogens. Vet Microbiol 126(1–3):178–188

Schwarz S, Feßler AT, Loncaric I, Wu C, Kadlec K, Wang Y, Shen J (2018) Antimicrobial resistance among staphylococci of animal origin. In: Aarestrup AA, Schwarz S, Shen J, Cavaco L (eds) Antimicrobial resistance in bacteria from livestock and companion animals. ASM Press, Washington, pp 127–157

Scott A, McCluskey B, Brown-Reid M, Grear D, Pitcher P, Ramos D, Spencer D, Singrey S (2016) Porcine epidemic diarrhea virus introduction into the United States: root cause investigation. Prevent Vet Med 123

Segura M, Aragon V, Brockmeier SL, Gebhart C, Greeff A, Kerdsin A, O'Dea MA, Okura M, Saléry M, Schultsz C, Valentin-Weigand P, Weinert LA, Wells JM, Gottschalk M (2020) Update on *Streptococcus suis* research and prevention in the era of antimicrobial restriction: 4th international workshop on *S. suis*. Pathogens 9(5):374

Seitz M, Valentin-Weigand P, Willenborg J (2016) Use of antibiotics and antimicrobial resistance in veterinary medicine as exemplified by the swine pathogen *Streptococcus suis*. Curr Top Microbiol Immunol 398:103–121

Siegel TW, Early DL, Smothers CD, Sun F, Rickets AP (2004) Cellular uptake of the triamilide tulathromycin by bovine and porcine phagocytic cells in vitro. J Anim Sci 82(Suppl. 1):186

Sipos W, Cvjetković V, Dobrokes B, Sipos S (2021) Evaluation of the efficacy of a vaccination program against *Actinobacillus pleuropneumoniae* based on lung-scoring at slaughter. Animals (Basel) 11(10):2778

Sneeringer S, Bowman M, Clancy M (2019) The U.S. and EU Animal Pharmaceutical Industries in the age of antibiotic resistance, ERR-264, U.S. Department of Agriculture, Economic Research Service. The U.S. and EU Animal Pharmaceutical Industries in the Age of Antibiotic Resistance (usda.gov). Accessed 24 Jun 2023

Songer JG, Uzal FA (2005) Clostridial enteric infections in pigs. J Vet Diagn Invest 17:528–536

Speksnijder DC, Wagenaar JA (2018) Reducing antimicrobial use in farm animals: how to support behavioural change of veterinarians and farmers. Anim Front 8:4–9

Sperling D, Smola J, Cizek A (2011) Characterisation of multiresistant *Brachyspira hyodysenteriae* isolates from Czech pig farms. Vet Record 168(8):215–215

Sperling D, Isaka N, Karembe H, Vanhara J, Vinduska J, Strakova N, Kalova A, Kolackova I, Karpiskova R (2022) Effect of the vaccination against Shiga toxin 2e in a farm with history of oedema disease, caused by atypical *Escherichia coli* producing Shiga toxin (STEC). Vet Med-Czech 67(10):510–518

Stringer OW, Bossé JT, Lacouture S, Gottschalk M, Fodor L, Angen Ø, Velazquez E, Penny P, Lei L, Langford PR, Li Y (2021) Proposal of *Actinobacillus pleuropneumoniae* serovar 19, and reformulation of previous multiplex PCRs for capsule-specific typing of all known serovars. Vet Microbiol 255:109021

Suiryanrayna MV, Ramana JV (2015) A review of the effects of dietary organic acids fed to swine. J Anim Sci Biotechnol 6:45

Sun Y, Scruggs DW, Peng Y, Johnson JR, Shukla AJ (2004) Issues and challenges in developing long-acting veterinary antibiotic formulations. Adv Drug Deliv Rev 56(10):1481–1496

Swann report (1969) Report by the Joint Committee on the use of Antibiotics in Animal Husbandry and Veterinary Medicine (Cmnd 4190)

Swedres-Svarm (2019) Sales of antibiotics and occurrence of resistance in Sweden. Solna/Uppsala. ISSN: 1650-6332. https://www.folkhalsomyndigheten.se/contentassets/fb80663bc7c94d678be785e3360917d1/swedres-svarm-2019.pdf. Accessed 29 Aug 2022

Tanigawa M, Sawada T (2002) Effects of subminimal inhibitory concentrations of amoxicillin against *Actinobacillus pleuropneumoniae*. J Veterinary Med Ser B 49:513–518

Taras D, Vahjen W, Macha M, Simon O (2006) Performance, diarrhea incidence, and occurrence of *Escherichia coli* virulence genes during long-term administration of a probiotic *Enterococcus faecium* strain to sows and piglets. J Anim Sci 84(3):608–617

Taylor D (2013) Bacterial diseases. In: Pig diseases, 9th edn. D.J. Taylor, Glasgow, pp 117–261

Thacker PA (2013) Alternatives to antibiotics as growth promoters for use in swine production: a review. J Anim Sci Biotechnol 4(1):35

The Netherlands Veterinary Medicines Institute (2022) Usage of Antibiotics in Agricultural Livestock in the Netherlands in 2021. Trends and benchmarking of livestock farms and veterinarians (SDa/1156/2021). https://cdn.i-pulse.nl/autoriteitdiergeneesmiddelen/userfiles/sda%20jaarrapporten%20ab-gebruik/ab-rapport-2020/uk-sda-report-usage-of-antibiotics-in-agricultural-livestock-in-nl-2020-def-rev.-oct-21.pdf. Accessed 20 Jun 2023

Thitaram S, Frank J, Siragusa G, Bailey JS, Dargatz DA, Lombard J, Haley C, Lyon SA, Fedorka-Cray P (2016) Antimicrobial susceptibility of *Clostridium difficile* isolated from food animals on farms. Int J Food Microbiol 227:1–5

Toutain PL (2003) Pharmacokinetics /pharmacodynamics integration in dosage regimen optimization for veterinary medicine. J Vet Pharmacol Therapy 26(Suppl 1):1–8

Toutain PL, Potter T, Pelligand L, Lacroix M, Illambas J, Lees P (2017) Standard PK/PD concepts can be applied to determine a dosage regimen for a macrolide: the case of tulathromycin in the calf. J Vet Pharmacol Ther 40:16–27

Tsiloyiannis VK, Kyriakis SC, Vlemmas J, Sarris K (2001) The effect of organic acids on the control of porcine post-weaning diarrhoea. Res Vet Sci 70(3):287–293

Tucker AW (2006) Porcine multi-systemic wasting syndrome (PMWS): a review. The Pig Site. Porcine multi-systemic wasting syndrome (PMWS): a review | The Pig Site. Accessed 24 Jun 2023

Van Asten AJAM, Nikolaou GN, Gröne A (2010) The occurrence of cpb2-toxigenic *Clostridium perfringens* and the possible role of the β2-toxin in enteric disease of domestic animals, wild animals and humans. Vet J 183(2):135–140

Van der Wolf P, Meijerink M, Libbrecht E, Tacken G, Gijsen E, Jaschniski KL, Schüller V (2021) *Salmonella typhimurium* environmental reduction in a farrow-to-finish pig herd using a live attenuated *Salmonella typhimurium* vaccine. Porcine Health Manag 7(1):43

Van Duijkeren E, Jansen MD, Flemming SC, de Neeling H, Wagenaar JA, Schoormans AHW, van Nes A, Fluit C (2007) Methicillin- resistant *Staphylococcus aureus* in pigs with exudative Epidermitis. Emerg Infect Dis 13(9):1408–1410

Van Gompel L, Luiken REC, Sarrazin S, Munk P, Knudsen BE, Hansen RB, Bossers A, Aarestrup FM, Dewulf J, Wagenaar JA, Mevius DJ, Schmitt H, Heederik DJJ, Dorado-García A, Smit LAM, EFFORT consortium (2019) The antimicrobial resistome in relation to antimicrobial use and biosecurity in pig farming, a metagenome-wide association study in nine European countries. J Antimicrob Chemother 74(4):865–876

Van Looveren F, De Jonghe E, Maass P, De Backer P (2015) Reduction of antibiotic use after implementation of Ingelvac® PRRS MLV piglet vaccination in a Belgian wean to finish farm. In: Safe pork conference proceedings, Porto

Van Overbeke I, Chiers K, Charlier G, Vandenberghe I, Van Beemuen J, Ducatelle R, Haesebrouck F (2002) Characterization of the in vitro adhesion of *Actinobacillus pleuropneumoniae* to swine alveolar epithelial cells. Vet Microbiol 88:59–74

Vanderhaeghe C, Dewulf J, Daems A, Van Soom A, de Kruif A, Maes D (2008) Influence of postpartum cloprostenol treatment in sows on subsequent reproductive performance under field conditions. Reprod Domest Anim 43(4):484–489

Vangroenweghe FA (2021) 237 stable performance with reduced antibiotic use in piglets vaccinated with an *E. coli* F4/F18 vaccine for the prevention of F18-ETEC post-weaning diarrhea. J Anim Sci 99(Suppl 1):139–140

Vangroenweghe F, Suls L, Van Driessche E, Maes D, De Graef E (2012) Health advantages of transition to batch management system in farrow-to-finish pig herds. Vet Med 57:83–91

Vanni M, Merenda M, Barigazzi G, Garbarino C, Luppi A, Tognetti R, Intorre L (2012) Antimicrobial resistance of *Actinobacillus pleuropneumoniae* isolated from swine. Vet Microbiol 156:172–177

Varela NP, Gadbois P, Thibault C, Gottschalk M, Dick P, Wilson J (2013) Antimicrobial resistance and prudent drug use for *Streptococcus suis*. Anim Health Res Rev 14(1):68–77

Vasseur MV, Laurentie M, Rolland JG, Perrin-Guyomard A, Henri J, Ferran AA, Toutain PL, Bousquet-Mélou A (2014) Low or high doses of cefquinome targeting low or high bacterial inocula cure *Klebsiella pneumoniae* lung infections but differentially impact the levels of antibiotic resistance in fecal flora. Antimicrob Agents Chemother 58:1744–1748

Vasseur MV, Lacroix MZ, Toutain P-L, Bousquet-Melou A, Ferran AA (2017) Infection-stage adjusted dose of beta-lactams for parsimonious and efficient antibiotic treatments: a *Pasteurella multocida* experimental pneumonia in mice. PLoS One 12(8):e0182863

Vereecke N, Van Hoorde S, Sperling D, Theuns S, Devriendt B, Cox E (2023) Virotyping and genetic antimicrobial susceptibility testing of porcine ETEC/STEC strains and associated plasmid types. Front Microbiol 4:1139312

Vicca J, Stakenborg T, Maes D, Butaye P, Peeters J, de Kruif A, Haesebrouck F (2004) In-vitro susceptibilities of *mycoplasma hyopneumoniae* field isolates. Antimicrob Agents Ch 48:4470–4472

Villarino N, Lesman S, Fielder A, García-Tapia D, Cox S, Lucas M, Robinson J, Brown SA, Martín-Jiménez T (2013a) Pulmonary pharmacokinetics of tulathromycin in swine. Part I: lung homogenate in healthy pigs and pigs challenged intratracheally with lipopolysaccharide of *Escherichia coli*. J Vet Pharmacol Ther 36:329–339

Villarino N, Lesman S, Fielder A, García-Tapia D, Cox S, Lucas M, Robinson J, Brown SA, Martín-Jiménez T (2013b) Pulmonary pharmacokinetics of tulathromycin in swine. Part 2: intra-airway compartments. J Vet Pharmacol Ther 36:340–349

Villarino N, Brown SA, Martín-Jiménez T (2013c) Understanding the pharmacokinetics of tulathromycin: a pulmonary perspective. J Vet Pharmacol Ther 37(3):211–221

Vranken E, Berckmans D (2017) Precision livestock farming for pigs. Anim Front 7(1):32–37

Vyas U, Ranganathan N (2012) Probiotics, prebiotics, and synbiotics: gut and beyond. Gastroenterol Res Pract 872716:1

Waldmann KH, Heide J (1996) Investigation on uterine contraction post partum and effects of different uterotonicia in puerperal sows. Proc Int Vet Soc Congr 14:614

Waldmann KH, Wendt M (2001) Lehrbuch der Schweinekrankheiten, 4th edn. Parey Verlag, Stuttgart

Wales AD, Davies RH (2017) Salmonella vaccination in pigs: a review. Zoonoses Public Health 64(1):1–13

Wallace HD (1967) High level cooper in swine feeding. Int Copper Res Assoc, New York

Walter D, Knittel J, Schwarz K, Kroll J, Roof M (2001) Treatment and control of porcine proliferative enteropathy using different tiamulin delivery methods. J Swine Health Prod 9(3):109–115

Wang Y, Xu C, Zhang R, Chen Y, Shen Y, Hu F, Shen J (2020) Changes in colistin resistance and *mcr*-1 abundance in *Escherichia coli* of animal and human origins following the ban of colistin-positive additives in China: an epidemiological comparative study. Lancet Infectious Dis. Published Online June 4, 2020. https://doi.org/10.1016/S1473-3099(20)30149-3

Wattanaphansak S, Singer RS, Gebhart CJ (2009) *In vitro* antimicrobial activity against 10 North American and European *Lawsonia intracellularis* isolates. Vet Microbiol 134:305–310

Wattanaphansak S, Pereira CER, Kaenson W, Assavacheep P, Tantilertcharoen R, Resende TP, Barrera-Zarate JA, Oliveira-Lee JSV, Klein U, Gebhart C, Guedes RMC (2019) Isolation and in vitro antimicrobial susceptibility of porcine *Lawsonia intracellularis* from Brazil and Thailand. BMC Microbiol 19:27

WHO (2019) Critically important antimicrobials for human medicine, 6th revision. World Health Organization, Geneva. Licence: CC BY-NC-SA 3.0 IGO. https://apps.who.int/iris/bitstream/handle/10665/312266/9789241515528-eng.pdf?ua=1. Accessed 29 Aug 2022

Wolter BF, Ellis M (2001) The effects of weaning weight and rate of growth immediately after weaning on subsequent pig growth performance and carcass characteristics. J Anim Sci 81:363–369

Xiong X, Tan B, Song M, Ji P, Kim K, Yin Y, Liu Y (2019) Nutritional intervention for the intestinal development and health of weaned pigs. Front Vet Sci 6:46

Yan Z, Wang M, Li X, Jiang H (2023) Impact of African swine fever epidemic on the cost intensity of pork production in China. Agriculture 13(2):497

Yazdankhah S, Rudi K, Bernhoft A (2014) Zinc and copper in animal feed - development of resistance and co-resistance to antimicrobial agents in bacteria of animal origin.Microb Ecol Health Dis. 2014 Sep 26;25

Yeh JY, Lee JH, Yeh HR, Kim A, Lee JY, Hwang JM, Kang BK, Kim JM, Choi IS, Lee JB (2011) Antimicrobial susceptibility testing of two *Lawsonia intracellularis* isolates associated with proliferative hemorrhagic enteropathy and porcine intestinal Adenomatosis in South Korea. Antimicrob Agents Chemother 55(9):4451–4453

Yu R, Xu Y, Schwarz S, Shang Y, Yuan X, Zhang Y (2022) *erm*(T)-mediated macrolide-Lincosamide resistance in Streptococcus suis. ASM J Microbiol Spectrum 10(1)

Zhang L, Huang Y, Zhou Y, Buckley T, Wang HH (2013) Antibiotic administration routes significantly influence the levels of antibiotic resistance in gut microbiota. Antimicrob Agents Chemother 57:3659–3666

Zhou X, Xu X, Zhao Y, Chen P, Zhang X, Chen H, Cai X (2010) Distribution of antimicrobial resistance among different serovars of *Haemophilus parasuis* isolates. Vet Microbiol 141:168–173

Zhou X, Li N, Luo Y, Liu Y, Miao F, Chen T, Zhang S, Cao P, Li X, Tian K, Qiu H, Hu R (2018) Emergence of African Swine fever in China. Transbound Emerg Dis 65:1482–1484

Yazdankhah S, Rudi K, Bernhoft A (2014) Zinc and copper in animal feed - development of resistance and co-resistance to antimicrobial agents in bacteria of animal origin. Microb Ecol Health Dis. 2014 Sep 26;25.

Poultry

Lucie Pokludová

Abstract

Poultry represents the second most commonly farmed animals, producing two important food commodities meat and eggs and having big perspective in further development. New breeds and hybrids as products of genetic selection, with fast fattening cycles, but suffer by the diseases, from which in Europe, but worldwide as well, most frequently treated by antimicrobials are coccidiosis and necrotic enteritis, followed by colibacillosis. Within the European Union antibiotics are authorised as veterinary medicinal products belonging among prescription-only medicines, while anticoccidial substances (namely ionophore antibiotics as well as those chemically synthetised) belong to coccidiostats and being approved as feed additives approaching farms without prescription of vets. Intensive farming of big flocks counting thousands to tens of thousands of birds exist, practically all antimicrobials are administered via medicated drinking water and therefore requiring functional and clean water delivery systems. Individual treatment is hardly possible and if used (e.g. vaccines injections) it is at early life stage. Despite pressure from market and customers as well as risk of antimicrobial resistance led currently to enforcement of prudent use of antimicrobials in many countries worldwide, there are still countries using antimicrobials as growth promoters. On the other hand, trend of responsible use or more and more no use of antimicrobials during life span of birds is coming. To succeed with farming without antimicrobials, complex, holistic approach should be accepted and implemented into practice using alternative tools to improve animal health, working on biosecurity and animal welfare and step by step minimise the need of antimicrobials use. To this improvement can help also precisely and timely performed vaccination programmes tailored for the conditions of grandparents, parents and finally food-producing birds—broilers for meat production or laying hens.

Keywords

Antimicrobials · Poultry medicine · Bacterial pathogens of poultry · Susceptibility · Pharmacokinetics and pharmacodynamics · Alternatives to antimicrobials

L. Pokludová (✉)
Institute for State Control of Veterinary Biologicals and Medicines, Brno, Czech Republic
e-mail: pokludova@uskvbl.cz

L. Pokludová (ed.), *Antimicrobials in Livestock 2: Choices of Prevention and Treatment*,
https://doi.org/10.1007/978-3-031-60807-0_3

1 Introduction

Due to their short reproductive cycle and their worldwide popularity as a food, poultry represent the most highly selected livestock, primarily directed at economic traits which have reduced costs of production. Chickens for fattening (broiler chicks) have been subjected to intense genetic selection. In the past 50 years, broiler growth rates have increased by over 300% (from 25 g per day to 100 g per day) (Knowles et al. 2008). EU (European Union) demand for poultry meat has been rising consistently for many years. The main reason is market demand—except prices, the fact of consumers' perception of poultry meat being healthy are major drivers (EC 2022). The question is the meaning of the word "healthy". Does this perception come from the fact that the poultry meat is low fat, easily digestive—or is it somehow linked with other factors? It seems that consumers do not take into account the fact of risks of zoonosis (two most frequent coming in majority cases from poultry sector products are Campylobacteriosis and Salmonelosis), especially when hygienic rules in the meat/eggs processing are not properly followed. It also seems that it is not taken into account the antimicrobials (including antibiotics as well as anticoccidials) used in the poultry sector production, because still "zero antimicrobial" production at all stages of life including parent flocks is not that common worldwide as well as in the EU. Despite the fact that during last 10 years in 25 European countries that have provided sales data continuously between 2011 and 2021, aggregated consumption of antimicrobials (mg/PCU) declined by 46.5% (ESVAC 2022), it has been and is assumed cross-link among veterinary antimicrobials consumption (in poultry sector fluoroquinolones and quinolones) and human health. For example, within the report (JIACRA III 2021 in the case of *Salmonella*, a significant association in univariate analysis was found between consumption of fluoroquinolones and other quinolones in poultry and resistance to fluoroquinolones in *Salmonella* from poultry. Multivariate analysis showed a direct effect of both fluoroquinolone consumption and resistance in poultry on the occurrence of fluoroquinolone resistance in *C. jejuni* from humans (JIACRA III 2021). Notwithstanding these information and hesitations that may come from above-mentioned risks, COVID-19 lockdown measures strengthened the importance of affordability and convenience of poultry meat for home cooking. The EU per capita consumption of poultry meat, supported by imports of breast meat, reached average 23.4 kg per capita in 2020–2022 (and is expected to continue with annual growth 0.3%, i.e. be approximately 24.7 kg per capita in 2032), the consumption of eggs reported was on average 13.6 kg per capita in 2020–2022 (and is expected to raise up to 14.4 kg per capita in 2032) as documented in the report (EC 2022).

1.1 Characterisation of the Poultry Sector

Poultry is one of the most widespread food industries worldwide. Chicken is the second most commonly farmed species. In 2021, almost138 million tonnes poultry meat (as carcass weight equivalent) were produced (FAO 2023). As for stratification of the production more than 46% of the chicken population lives and more than 35% (tonnes poultry meat) is produced in Asia, and similarly in South and North America together. In 2020 global population was 33 billion birds, commercial layers can lay over 300 eggs per year, while indigenous hens' productivity is 40–60 eggs only. Worldwide is recognised more than 1600 chicken breeds (FAO 2023).

European production was 13.3 million tonnes in 2021 (EU-DG Agri Dashboard 2023). Poultry meat production will be dominating more than half of the growth of all the additional meat produced by 2025. The European Union is counted amongst the world's largest poultry meat producers and exporters. The European Union import (mainly breast meat and poultry preparations) comes from Brazil, Ukraine and Thailand. At the same time, however, the European Union belongs among three biggest exporters of poultry products. Within the EU exist six main poultry producing countries covering 73% of the total

EU production: Poland (21%), Spain (13%), Germany (12%), France (11%), Italy (9%) and the Netherlands (7%). As for the stratification of the poultry species meat in EU production, broiler (84.4%), turkey (near 12.7%) and duck (near 2.6%) sectors create near to hundred per cent of the EU production, 0.3% is allocated to other poultry species (EU-DG Agri Dashboard 2023).

The whole poultry sector can be divided into two main areas producing meat and eggs for human consumption which include: commercial breeding (mostly specific hybrid breed raised in the EU specifically for meat or for egg production, as, e.g. Ross, Cobb, ISA Brown, ISA Sussex, ISA Dual, Decalb White, Bowans Brown, Bowans Sperwer, or some locally bred as Moravian BSL and Moravia Barred), with parent and grandparent flocks of importance and village/backyard breeding (small-scale flocks for both eggs and meat production). Geneticists design special breeding programmes to select birds with the best characteristics for egg or meat production. There are two main types of commercial chicken (*Gallus gallus*) breeds: layers and meat (broiler) chickens.

A large diversity of antimicrobials are used to raise poultry in most countries worldwide (Mehdi et al. 2018). As of January 2017, the United States Food and Drug Administration (FDA) fully implemented Guidance for Industry #209 and #213. It means removal of all growth promotion claims for medically important antimicrobials. Additionally, all medically important antimicrobials administered via the water (claims treatment, prevention, or control of disease) became veterinary prescription only. Those administered via the feed became Veterinary Feed Directive (VFD) drugs, a category tantamount to a veterinary prescription. All antimicrobials currently approved in the United States for poultry except bacitracin, bambermycins, ionophores and avilamycin are classified as medically important (Smith et al. 2020). It should be also noted the withdrawal of the use of fluoroquinolones in the USA in 2005 (Smith et al. 2020). In the European Union, there is no difference among antimicrobials, all are considered as prescription only (including, e.g. bacitracin). Bambermycins as well as avilamycin are not currently (2023) used in the EU. Ionophores are non-prescription and subject of the feed additive legislation, supervised by EFSA (European Food Safety Authority) in the EU.

Considering of more and more prudent (rational, responsible) and finally minimum or zero use of antimicrobials, so called “alternatives to antimicrobials” should not be forgotten. Despite the fact that this term is broadly used, it should be used with great caution, keeping in mind the reality of everyday practice. In fact, no equally potent substance (comparing to antimicrobials) has been available yet, especially for an acute outbreak of the disease and considering also economic aspect—the similar price. Usually there is the need for complexity of measures and products “altering the use of antimicrobials”. The aim of these “alternative products” is to maintain a low mortality rate, a good level of animal yield while preserving environment, user and consumer health. Among these, the most popular have started to be probiotics, prebiotics, symbiotic, enzymes, organic acids, immunostimulants, bacteriocins, bacteriophages, phytogenic feed additives, phytoncides, nanoparticles and essential oils (Mehdi et al. 2018). Despite from this list it could seem that the era of antimicrobials is over, it is not the case, it is a long way of improvement of the conditions of farming, care on the animals, improving welfare and biosecurity, testing balance of environmental factors in the housing facilities (considering intensity of poultry farming), genetics of the animals, properly designed and constantly as high quality as possible feeding stuffs (e.g. without mycotoxins) supplemented by the “alternative products” with proven (it means evidence based) positive effects and minimum negative impacts.

A lot of discussions have been and is held how to achieve antibiotic-/antimicrobial-free production. Definition what is meant by this term “antibiotic-free” is one of the first critical issues. In the United States for example non-therapeutic uses of medically important antibiotics is restricted and the era of more veterinary supervision has been started since 2017 (Smith et al.

2020). On the other hand, there is still allowed to use broadly some molecules, which are listed as non-medically important (NMI)—for example bacitracin, bambermycin, avilamycin as well as ionophore anticoccidials. Previously also pleuromutilins were classified as NMI as molecules exclusively used in veterinary medicine only, what has changed with authorisation of retapamulin and lefamulin in human medicine for treatment of MRSA (methicillin-resistant *Staphylococcus aureus*) (U.S. Food and Drug Administration 2020a; U.S. Food and Drug Administration 2020b). Moreover, chemically synthetised anticoccidials are still used routinely, as prevention of coccidiosis and together with NMI also in control of necrotic enteritis. The above examples belong among the restrictions demanded by law. But marketing demands have started to be main driver, rather than health (minimising resistance) aspects (Singer et al. 2019). Sale the "production with additional value", which can bring to producers also more money from customers became more and more popular. It has started with "certified organic production" and is recently more targeted with respect of the limiting of use of antimicrobials to marks as RWA "Raised Without Antimicrobials" or NAE "No Antibiotic Ever" or AF "Antibiotic Free" (Feeks 2019). It should be noted that "Antibiotic Free", "Residues Free" or "Chemicals Free" are generally disallowed in the United States, as they can belong to false, misleading advertisement (Smith 2019) and hardly to be proven (especially considering that for those tissues for which Maximum Residue Limits (MRLs) are established so called "chemical safety" is well established for all food production both in the USA and in the EU). The aim of RWA or NAE is not only to provide "antibiotic free" meat, offal and eggs from gained from poultry, but also to minimise selective pressure for antimicrobial resistance. It should be noted that for declaration of RWA or NAE, any *antibiotics* (medically important as well as non-medically important) and any ionophore anticoccidials should not be used during the whole life of the birds (including no administration at hatch or in ovo) (USDA 2016). Therefore, vaccination and chemically synthetised ionophores were allowed only (the USA), but in Canada only vaccination or non-pharmaceutical alternative tools were used to control coccidiosis in the RWA/NAE flocks till 2016, when it was reconsidered and chemically synthetise anticoccidials have started to be allowed again. If the flock gets sick, it cannot be marked as RWA or NAE. The possibility of the success of RWA or NAE is based on combination and complexity of measures starting at grandparents and parents flocks (well-designed vaccination tailored for expected epidemiological threats and pressure expected in broilers for fattening as well as for the laying birds flocks, biosecurity measures in the whole chain, well balances feed, good quality of litter, good quality of drinking water as well as altogether environmental parameters, no overstocking and high quality care on animals). In December 2015 RWA or NAE created 13.0% of birds grown in the USA comparing to February 2017 with 38.7% and in 2018 with more than 50% (Donahue 2017; Feeks 2019). The most common weights of RWA/NAE birds were among 2.36–3.18 kg, with 2.9% increase in caloric conversion and 0.68% higher mortality (Donahue 2017). Example coming from Canada (Cardwell 2019) speaks about 2.31 kg at 38 days age. According to Agri-Stats data in 2018, the mortality rate in "Antibiotic free" comparing to conventionally raised birds were reported to be approximately 4.2% and 2.9%, respectively (Ritter et al. 2019). Some alarming facts arrived from survey, where 565 completed responses were received, 442 self-reported as practicing veterinarians or producers, with just over 50% of respondents reported having past or current experience with RWA programmes. Many of respondents felt that RWA production has negative impacts on animal health and welfare (Singer et al. 2019). It should be avoided as much as possible situation described by Karavolias et al. (2018), "Policies aimed at eliminating or restricting the use of antibiotics in broiler production may come with potentially negative consequences with respect to good animal welfare." Series of measures that help to manage RWA or NAE programmes are listed and briefly commented in Table 1.

Table 1 Ten key measures important for farming of broilers without any antimicrobials in systems RWA or NAE

Nutrition, high quality feed, additives	High quality, well balanced, clean feed. Optimisation of nutrition strategies, including using enzymes, medium chain fatty acids, vitamins
Cleanliness	Housing cleanliness, with imperative on clean drinking water
Healthy gut	Immunostatus is very dependent on healthy gut
Environmental conditions	Managament of environmental conditions
Management of the early birds mortality	No eggs from dirty floor, good hatchery conditions + healthy, homogeneous weight one-day-old chicks from vaccinated parents
No overstocking	Avoid overstocking as it is linked with more discomfort and stress leading to higher disease, morbidity and mortality rates
Vaccination programme	Timely and proper use of vaccines (starting at parents´ flocks)
Managemenet of the most critical diseases	Necrotic enteritis and coccidiosis as well as some viral diseases that weaken birds´ immunity and health and make them more prone for bacterial diseases
Proper care on animals + trained and motivated staff	Management of care on animals, including motivation of staff
Patience + stepwise improvement + sustainability	Not all things go smoothly, benefit not arrives immediately, importance of building sustainable and vital system. Do not forget to "value and sale the quality"

2 Diseases Mostly Treated in Poultry

Main pathogens/disorders which were indicated by veterinarians (De Bryine et al. 2020) and where antimicrobials are most frequently used in broilers of *Gallus gallus* intended for fattening include: gastrointestinal disorders (coccidiosis, necrotic enteritis, dysbacteriosis), respiratory diseases (secondary bacterial infections caused by, e.g. *E. coli* with background of infectious bronchitis, Newcastle disease, infectious laryngotracheitis), locomotion-related diseases (bacterial arthritis—due to, e.g. *E. coli*, *Staphylococcus aureus* or *Enterococcus* spp., and secondary bacterial infections connected with tenosynovitis, necrosis of the femur head) and septicaemia or omphalitis. In laying hens, gastrointestinal disorders (mainly enteritis caused by *E. coli*, avian intestinal spirochaetosis and *Clostridium perfringens* as well as coccidial infestations by *Eimeria* spp., followed by respiratory and locomotion-related diseases (*E. coli* and *Mycoplasma* spp.) and secondary bacterial infections concomitant with red mite infestation. In free-range production systemic taeniosis can be an issue.

Authors of the study performed in Belgium in the period 2013–2017, collected data from 295 different flocks under veterinary intervention, originating from 50 different layer pullet houses. Collected data included farm and housing conditions, flock characteristics (date of chickens´ placement, feather colour of the pullets, number of pullets placed and hatchery of origin) and veterinary interventions (age of the birds at intervention, diagnosis and treatment). Most therapeutic interventions were for intestinal disorders: coccidiosis (92/295 flocks), necrotic enteritis (77/295), enteritis (52/295) and dysbacteriosis (36/295), with less interventions for non-gut-related problems: for example, omphalitis (19/295), colibacillosis (12/295), red mites (11/295), respiratory problems (11/295), joint problems—arthritis, osteomyelitis and osteoporosis (11/295) and vaccine adverse reactions (13/295). A minority of the veterinary interventions were related to fever (4/295), heat stress (4/295), hypoglycaemia (2/295), hypovitaminosis (2/295), dehydration (1/295) or septicaemia (1/295) (Goossens et al. 2020a). Critical point for initiation of therapy as well as for whole management of the flocks including vaccination strategy is the proper diagnosis of the diseases or even identification of carriers or sources of infections (even in the batches of arriving chicks). New methods of detection, including different polymerase chain reaction (PCR) methods can be of significant help. Quantitative nanofluidic technology-based PCR with 96 pathogen-specific PCR primers can in one run make the screening of 15 avian respiratory pathogens, including viruses, bacteria and fungi (Croville et al. 2018).

For selected pathogens, mainly acting as ethiological agents of diseases of poultry, with predominant targeting on *Gallus gallus*. Table 2 brings the linking to systemic infections as well as pathologies of respiratory, gastrointestinal and genitary tract together with drawing the attention to pathogens causing omphalitis, locomotory and skin diseases and commenting on the mainly affected production/age categories.

Table 2 Selected bacterial species involved in diseases of poultry with production/age category

Bacterial species	Disease	Production/age category mainly affected
RESPIRATORY		
Avibacterium paragallinarum	Infectious coryza	Mainly older birds, generally all ages can be affected, in juveniles less severe
Escherichia coli (APEC)	Airsacculitis	More frequent in young growing birds
Gallibacterium anatis	Respiratory tract infection	Can be isolated at all ages, but mostly isolated from breeders and commercial laying hens around peak of eggs production, in 50% as sole pathogen, in the rest concomitant infections with *E. coli*
Mycoplasma synoviae (MS)	Infectious synovitis	Mostly in at age 4–12 weeks, either with clinical signs or without clinically apparent signs
M. meleagridis (MM)	Airsacculitis	In turkeys, extension of the infection to air sac lesion mostly at age 3–5 weeks, if present in one-day-old birds then at 3–5 weeks lesions in air sac
Mycoplasma spp., *Mycoplasma gallisepticum* (MG)	Chronic respiratory disease	All ages can be affected, but very young birds seldom affected, most outbreaks after 4 weeks of age
Ornithobacterium rhinotracheale	Respiratory tract infection	All age groups. Mostly not a sole pathogen, but occurring concomitantly with other causative agents both bacteria and viruses
GASTROINTESTINAL		
Clostridiaceae (multifactorial also other bacterial species contribute)	Bacterial enteritis, Dysbacteriosis	Typically from age 10 to 14 days
Clostridium perfringens (NetB+)	Necrotic enteritis	Most commonly laying hens/parents flocks at period of lay, can be detected also in animals approximately 14 days of age (broilers 2–5 weeks, turkeys 7–12 weeks old)
SYSTEMIC INFECTIONS		
Escherichia coli (APEC)	Colibacillosis	All age groups: day-old chicks, broiler chickens and adult
Enterococcus spp.	Infections caused by enterococci as:	Starting at age approximately 14 days in broiler chicks. Acute infection course can be during the weeks 1–3. Mortality rate 5–10% around day 14, due to sepsis can occur in *E. cecorum* infection of broilers

(continued)

Table 2 (continued)

Bacterial species	Disease	Production/age category mainly affected
Salmonella spp. (invasive serotypes)	Salmonelosis	High mortality descibed in newly hatched birds and in broilers of approximately 3–7 days of age. In *S.* Pullorum also older birds can be affected fatally (2–5 weeks). Salmonella invasivity to internal organs higher at 1 day than at 7 (*S.* Enteritidis and *S.* Typhimurium)
Pasteurella multocida	Fowl cholera	Acute (mostly in turkeys) or chronic (mostly in chickens) disease
Erysipelothrix rhusiopathiae	Erysipelas	Very rarely in young broilers and pullets, sensitivity increase with age was not confirmed after oral infection
Clostridium spp.	Gangrenous dermatitis	Broiler flocks at 4–8 weeks of age, also in breeders and commercial layers
	Botulism (intoxication and/or infection)	All ages can be affected, mature hens might be of less susceptibility to clostridia producing botulinum neurotoxins
Pseudomonas aeruginosa	Septicaemia	Especially young and growing poultry, invades also fertile eggs and can cause embryo mortality. Concurrent infection with other bacteria and virus occur
SKIN ANDLOCOMOTORY		
Staphylococcus aureus	Arthritis, periarthritis, joints infection, chondronecrosis, osteomyelitis, femoral head necrosis, hock joints	All ages can be affected, usually older higher score of lesions
Enterococcus	Spondylitis, osteomyelitis, chondronecrosis	
Escherichia coli (APEC)	Arthritis	
OMPHALITIS		
Pseudomonas spp., *S. aureus, E. coli, Salmonella* spp., *Proteus* spp.	Omphalitis and egg yolk sac infections	Newly hatched and early stages
GENITARY INFECTIONS		
Gallibacterium anatis	Salpingitis	Mostly adult
Riemerella anatipestifer		
Mycoplasma gallisepticum		
Escherichia coli, Salmonella spp.		

3 Groups of Antimicrobials and Their Purposes for Use in Poultry

3.1 Historical and Current Use of Antimicrobial Growth Promoters in Poultry Sector

History of the use of antimicrobial growth promoters (AGP) in poultry sector has gone to the years after World War II, when they have started to be widely used to improve performance of broiler chickens mainly in order to promote faster growth and improve conversion rates as well as to protect against and control diseases such as necrotic enteritis caused by *Clostridium perfringens* (Castanon 2007). As for antimicrobials used in poultry sector following substances (Table 3) has been used and still are actively used in some parts of the world, as indicated by the OIE (newly WOAH) report (OIE 2020).

Thoughts about phasing out AGPs have started in 1969 with Swann report, where was indicated, that antimicrobials used for therapeutic purposes should not be used as growth promoters. Starting in mid-1990s, with concomitant improvement of housing, biosecurity, disinfections, hygiene and other factors, several EU countries banned use of AGPs, since 1997 with avoparcin ban started the era of EU ban of antimicrobial growth promoters, followed by virginiamycin, bacitracin, tylosin and spiramycin ban as AGP in 1999 and finally complete ban of all antimicrobials' use as growth promoter since January 2006. Unfortunately, this EU policy was not followed by third countries, including big players in poultry sector as the USA. For a long time, discussion has also been held on the terminology. For example, the USA divided antimicrobials to medically and non-medically important (bacitracin, ionophores and previously also pleuromutilins), with limiting the use of those with medical importance. Other countries paid more attention to critically important antibiotics only (e.g. colistin) or antibacterials as a group. Several antimicrobials (in a broader sense of meaning, i.e. antiprotozoals—mainly coccidiostats) are however still allowed to be used as AGP (worldwide), or as feed additives (EU).

Table 3 Antimicrobials used as AGP in poultry sector during the time course (ref Vol I: updated)

Importance	Classes (individual antimicrobials used) in poultry sector as AGP
Critically important antimicrobials: Highest priority antimicrobials	(a) Fluoroquinolones and quinolones (b) Polymyxins (bacitracin, colistin) (c) Macrolides (erythromycin, tilmicosin, spiramycin, josamycin, kitasamycin, tylosin)
Critically important antimicrobials: High priority antimicrobials	(a) Penicillins (amoxicillin, benzylpenicillin) (b) Aminoglycosides (neomycin)
Highly important antimicrobials	(a) Streprogramins (virginiamycin) (b) Lincosamides (lincomycine) (c) Tetracyclines (oxytetracycline, chlortetracycline, tetracycline) (d) Sulphonamides (sulfachlorpyridazine) (e) Amphenicols (florfenicol)
Important antimicrobials	(a) Pleuromutilins (tiamulin)
Not classified yet (at EU level)	(a) Ionophores (anticoccidial substances) (b) Glycophospholipids (flavofosfolipol) (c) Orthosomycines (avilamycin) (d) Other (enramycin, fosfomycin, bicozamycin)

3.2 Antimicrobials Used for the Treatment of Poultry

Most antimicrobials (speaking especially about antibiotics) in this sector is authorised for use in broilers of *Gallus gallus* for either more general or specific indications. There are several factors limiting the use either in other categories or in other commercially produced poultry (turkeys, ducks, geese, quails etc.). First, factor limiting the appropriate use is unknown (robust, evidence based) dosing schedule for each specific poultry species. Mostly the dosing schedules were tested in broiler chickens (*Gallus gallus*), less in hens laying eggs for human consumption (or parent flocks) and even less for turkeys. Second, extremely important factor limiting the use, is consumer safety of the meat and offal as well as eggs produced by the birds after the treatment. Safe withdrawal period (WP) should be established for each individual veterinary medicinal product with valid (full dossier) authorisation for any poultry species, which can limit the use in minor species (as mostly studies are performed for broiler chickens, less for laying hens, turkeys and rarely for other poultry species). Necessary pre-requisite for establishment of WP is firstly setting of Acceptable Daily Intake (ADI), including mostly microbiological ADI for antimicrobials. Based on ADI, Maximum Residue Limits for meat and offal and skin and fat in natural proportions should be established for each individual antimicrobial and/or its metabolite. Having these two key variables (ADI and MRL) and having proper residue studies, withdrawal period is set for each individual veterinary medicinal product and its conditions of use (route of administration, dosing schedule, target species/production category). In broilers in the late stage of fattening as well as in hens laying eggs for human consumption due to the established withdrawal period for eggs (longer) duration of withdrawal period is limiting factor in selection of antimicrobial (exact VMP) for treatment and metaphylaxis. For "off-label use" (= use outside of the conditions of the marketing authorisation), what is many times the case for poultry species other than *Gallus gallus*, withdrawal period can complicate the situation.

Withdrawal period valid according to the Regulation 2019/6/EC (in force since January 2022), after use outside of the terms of authorisation (so called cascade or off-label use) there should be established withdrawal period for poultry meat and eggs.

Meat and *Offal* withdrawal period should **not be less than**:

(i) The longest withdrawal period provided in its summary of the product characteristics for meat and offal multiplied by **factor 1.5**.
(ii) **Twenty-eight days** if the medicinal product is not authorised for food-producing animals.
(iii) **One day**, if the medicinal product has a zero withdrawal period and is used in a different taxonomic family than the target species authorised.

Eggs withdrawal period should **not be less than**:

(i) The longest withdrawal period for eggs provided in the summary of the product characteristics for any animal species multiplied by **factor 1.5**.
(ii) **Ten days**, if the product is not authorised for animals producing eggs for human consumption.

Article 115/2 stipulate, if the calculation of the withdrawal period according to above-defined rules leads to a fraction of days, the withdrawal period shall be rounded up to the nearest number of days.

In 2014 and 2015 European Medicines Agency (EMA 2015) in cooperation with nine Member States performed the project targeted on setting of the Defined Daily Doses (DDDvet) and Defined Course Doses (DCDvet) for three major food-producing species (pigs, cattle and broilers). The purpose was to have the units of measurement of consumption of antimicrobials in these three, in the Europe, most frequently produced animal species. It gives also the background for the list of antimicrobials authorised in the EU. Table 4 brings the overview of the antimicrobials authorised in these nine Member States, as updated by 2019 data from 31 EU/EEA countries for sector of broiler chickens for fattening, indicating substances with MRLs for eggs (so they are/were authorised also for laying hens). Table 5 lists $\text{DDDvet}_{\text{broilers}}$, which were collected and set based on the Principles (EMA 2015) for the poultry (primarily broilers). Despite the fact that DDDvet are technical units, they reflect exact authorised doses in SPCs (summary of product characteristics) and therefore can be considered, that very close/similar doses were used in the treatment in this period and with high probability they stay valid, as investment to variation of marketing authorisations of antimicrobials is hardly expected.

Despite the fact that certain antimicrobials can cover a spectrum of bacteria (susceptible strains

Table 4 Antimicrobials authorised for poultry within the EU countries (mostly for broiler of *Gallus gallus*), those with MRLs can be also authorised for laying hens

Antimicrobial class	Active substance	Combination with	MRL for eggs	Comment ref uses in poultry by the class	Spectrum of bacterial species where reasonable to perform antimicrobial susceptibility testing
Aminoglycosides	Apramycin		No	Aminoglycosides are of importance in the following poultry diseases colibacillosis, coryza, necrotic enteritis, gangrenous dermatitis, prevention of infection with *Histomonas* spp. Note: After oral administration, limited to negligible proportion absorbed from gastroinstestinal tract	*Avibacterium paragallinarum, Escherichia coli, Staphylococcus aureus*
	Neomycin		**Yes**		
	Neomycin	Erythromycin and tylosin	**Yes**		
		Amoxicillin and oxytetracycline	No[1]		
		Oxytetracycline	**Yes**		
	Paromomycin		No		
Aminocyclitols	Spectinomycin	Lincomycin	No[2]	Used in combination with lincomycin for colibacillosis (multisystemic syndromes; omphalitis, airsacculitis, fowl cholera).	Avibacterium paragallinarum, *Escherichia coli, Pasteurella multocida,* Pseudomonas *spp. and other*
Amphenicols	Florfenicol		No	The wide range class represents a useful alternative in respiratory infections of poultry.	*Avibacterium paragallinarum, Enterococcus* spp.*, Escherichia coli, Mycoplasma* spp.*, Ornithobacterium rhinotracheale, Pasteurella multocida, Gram-negative non-fermenting rods other than* Pseudomonas spp.*, Staphylococcus aureus, Streptococcus* spp.*, selected anaerobes*
	Thiamfenicol		No		
Beta- lactams	Amoxicillin		No	Beta- lactams are of importance in poultry for broad spectra of diseases:*E. coli* infections (local and systemic infections, airsacculitis, arthritis), *Clostridium* spp. infections (necrotic enteritis, botulism, ulcerative enteritis) and respiratory diseases such as fowl cholera, coryza, *Riemerella anatipestifer* infection (RA), and *Ornithobacterium rhinotracheale* infection (ORT).	*Avibacterium paragallinarum, Clostridium spp., Enterococcus spp., Erysipelothrix rhusiopathiae, Gallibacterium* spp.*, Escherichia coli, Mycoplasma* spp.*, Ornithobacterium rhinotracheale, Pasteurella multocida, Staphylococcus aureus (non MRSA, non beta-lactamase) Streptococcus spp.*
	Amoxicillin	Flumequin	No[1]		
		Lincomycin	No[1]		
		Neomycin and Oxytetracycline	No[1]		
	Ampicillin		No		
	Phenoxymethyl penicillin		No		Similarly as above aminopenicillins, except Enterobacterales and Gram-negative non-fermenting rods
Antimicrobial class	Active substance	Combination with	MRL for eggs	Comment ref uses in poultry by the class	Spectrum of bacterial species where reasonable to perform antimicrobial susceptibility testing
Fluoroquinolones	Danofloxacin		No	Quinolones and fluoroquinolones are used for treatment of *E. coli* infections (colibacillosis (local and systemic infections), airsacculitis, arthritis), *Mycoplasma* spp. infections (chronic respiratory disease) and *Pasteurella* spp. infections (fowl cholera).	*Avibacterium paragallinarum, Bordetella spp., Enterococcus spp., Erysipelothrix rhusiopathiae, Gallibacterium* spp.*, Escherichia coli, Mycoplasma* spp.*, Ornithobacterium rhinotracheale, Pasteurella multocida, Pseudomonas* spp.*,* Gram-negative non-fermenting rods*, Staphylococcus* spp.*, Streptococcus* spp.
	Difloxacin		No		
	Enrofloxacin		No		
Other quinolones	Flumequin		No		
	Oxolinic acid		No		
Lincosamides	Lincomycin		**Yes**	Used in combination with spectinomycin for colibacillosis (multisystemic syndromes; omphalitis, airsacculitis, fowl cholera).	*Clostridium* spp.*, Escherichia coli, Enterococcus* spp.*, Erysipelothrix rhusiopathiae, Mycoplasma* spp.*, Ornithobacterium rhinotracheale, Pasteurella multocida, Staphylococcus*
	Lincomycin	Spectinomycin	No[2]		
Macrolides	Erythromycin		**Yes**	Macrolides are used to treat *Mycoplasma* spp. infections in poultry (chronic respiratory disease (CRD), arthritis), fowl cholera, *Ornithobacterium rhinotracheale* infection (ORT), necrotic enteritis, avian intestinal spirochetosis, ulcerative enteritis (UE), gangrenous dermatitis.	*Avibacterium paragallinarum, Brachyspira spp., Clostridium* spp.*, Mycoplasma* spp.*, Ornithobacterium rhinotracheale, Pasteurella multocida, Staphylococcus aureus, Streptococcus* spp.
	Erythromycin	Sulafamonomethoxin	**No[3]**		
		Neomycin and tylosin	**Yes**		
	Spiramycin		No		
	Spiramycin	Oxytetracycline			
	Tilmicosin		No		
	Tylosin		**Yes**		
	Tylosin	Neomycn and Erythromycin	**Yes**		
	Tylvalosin		No		
Pleuromutilins	Tiamulin		**Yes**	Pleuromutillins are used against respiratory infections, avian intestinal spirochetosis and ulcerative enteritis.	*Brachyspira* spp.*, Clostridium* spp.*, Mycoplasma* spp.*, Staphylococcus* spp.*, Streptococcus* spp.
Antimicrobial class	Active substance	Combination with	MRL for eggs	Comment ref uses in poultry by the class	Spectrum of bacterial species where reasonable to perform antimicrobial susceptibility testing
Polymyxins	Colistin		**Yes**	Colistin (after oral administration) is used for non invasive *E.coli* infections Note: after oral administration, negligible proportion absorbed from gastroinstestinal tract	*Escherichia coli,* Pseudomonas *spp. and other Gramnegative non-fermenting rods*
Streptogramines	Virginiamycine		No	Virginiamycin is used in prevention of necrotic eneteritis caused by *Clostridium perfringens* .	*Clostridium* spp.
Sulphonamides	Sulfaclozine		No	Sulphonamides are important in the treatment of a wide range of diseases (bacterial and coccidial infections).	*Avibacterium paragallinarum, Escherichia coli, Eimeria* spp. (protozoal)*, Pasteurella multicida, Gram-negative non-fermenting rods other than* Pseudomonas spp.*, Staphylococcus* spp.*, Enterococcus* spp.*, Streptococcus* spp.
	Sulfadimethoxine		No		
	Sulfadimidine		No		
	Sulfachlorpyrdazine		No		
	Sulfaguanidine		No		
	Sulfamonomethoxine		No		
	Sulfaquinoxaline		No		

Table 4 (continued)

Antimicrobial class	Active substance	Combination with	MRL for eggs	Comment ref uses in poultry by the class	Spectrum of bacterial species where reasonable to perform antimicrobial susceptibility testing
Sulphonamides in fixed combination with trimethoprim				Sulphonamides in combination with trimethoprime are important in the treatment of a wide range of diseases (bacterial and coccidial infections).	*Avibacterium paragallinarum*, Clostridium *spp.*, *Escherichia coli*, *Eimeria* spp. (protozoal), *Erysipelothrix rhusiopathiae*, *Ornithobacterium rhinotracheale*, *Pasteurella multicida*, *Staphylococcus aureus*
	Sulfachlorpyridazine	Trimethoprim	No[4]		
	Sulfadiazine	Trimethoprim	No[4]		
	Sulfamonomethoxine	Trimethoprim	No[4]		
	Sulfadimethoxine	Trimethoprim	No[4]		
	Sulfadimidine	Trimethoprim	No[4]		
	Sulfamethoxazole	Trimethoprim	No[4]		
	Sulfaquinoxaline	Trimethoprim	No[4]		
	Sulfamethoxypyridazine	Trimethoprim	No[4]		
Antimicrobial class	Active substance	Combination with	MRL for eggs	Comment ref uses in poultry by the class	Spectrum of bacterial species where reasonable to perform antimicrobial susceptibility testing
Sulphonamides in combinations with other antimicrobials				The need for thorough justification of the combinations	
	Sulfadimidine	Neomycin, Oxytetracyclin, Sulfaguanidin	No[5]		
	Sulfadimidine	Trimethoprim, Oxytetracycline	No[5]		
	Sulfadimidine	Trimethoprim, Sulfadimethoxine	No[5]		
	Sulfadimethoxine	Oxytetracycline	No[5]		
	Sulfadimethoxine	Tylosin	No[5]		
Sulphonamides in combination				Sulphonamides are important in the treatment of a wide range of diseases (bacterial and coccidial infections).	As sulphonamides
	Sulfadimidine	Sulfaquinoxaline	No		
	Sulfadimidine	Sulfadimethoxine	No		
	Sulfadimidine	Sulfamerazine, Sulfathiazole	No		
Tetracyclines	Chlortetracycline		**Yes**	Tetracyclines are important in the treatment of a wide range of diseases (bacterial and coccidial infections).	*Avibacterium paragallinarum*, *Bordetella spp.*, *Brachyspira spp.*, *Chlamydia spp.*, *Clostridium spp.*, *Escherichia coli*, *Eimeria spp. (protozoal)*, *Enterococcus spp.*, *Erysipelothrix rhusiopathiae*, *Mycoplasma* spp., *Ornithobacterium rhinotracheale*, *Pasteurella multocida*, *Staphylococcus aureus*, *Streptococcus*
	Doxycycline		No		
	Doxycycline	Sulfaguanidine	No		
	Oxytetracycline		**Yes**		
	Tetracycline		**Yes**		

[1]is limited by no MRL for amoxicillin
[2]is limited by no MRL for spectinomycin
[3]is limited by no MRL for sulfamonomethoxin
[4]no MRL for both sulphonamide and trimthoprim
[5] is limited by no MRL for sulphonamide

Note 1: Despite the broad spectrum of possible efficacy of some antimicrobials towards certain bacterial species, there should be taken into consideration susceptibility, which may differ across time as well as geographically.
Note 2: The fact that microorganism is susceptible in vitro does not preclude fully in vivo susceptibility (as commented e.g. for aminoglycosides or colistin, which after oral administration, are only negligibly absorbed from the gastrointestinal tract, what makes them inappropriate for sytemic infections treatment).

of certain bacterial species), not only pharmacodynamic properties are of importance but pharmacokinetic of the antimicrobial ensuring achieving sufficient concentration at the site of infection either. Thinking pragmatically, the issue of availability in the pharmaceutical form suitable for administration in the flocks of poultry is of importance. One Health aspect should be kept in mind as it covers the perspective of efficacy and safety of the treatment in target species of animals, but as well the aspect of possible selection and co-selection of resistance to antimicrobials and impacts and possible consequences on public health. Risks for the environment has been also recognised, especially in areas with extensive farming. Considering both human health and animals need for the treatment there was elaborated by the AMEG group working under umbrella of the European Medicines Agency the categorisation of antibiotics. Categorisation was finalised at the end of 2019 and published at early 2020 (EMA 2019). From the perspective of the treatment of the poultry diseases, considering also this categorisation but as well as more strict approach to some antimicrobials (amoxicillin) as proposed, e.g. by AMCRA (2020) treatment recommendation for the selected poultry diseases (antimicrobials of the first, second and third choice). Table 6 is proposed specifying combination bacterial species—disease and first, second, third choice antimicrobials.

Table 5 Defined daily dose for animals (DDDvet) for antimicrobial agents for use in broilers (**$DDDvet_{broilers}$** have been assigned according to the "Principles on assignment of defined daily dose for animals (DDDvet) and defined course dose for animals (DCDvet) (EMA/710019/2014)" and were generally based on SPC information on dosing authorised in VMPs collected from nine European Union Member States at 2016). These doses should not necessarily be assumed to reflect the *exact* daily doses recommended or prescribed, but represent technical unit for monitoring of consumption. Based on the data from EMA (2016)

Antimicrobial	Oral route DDDvet in mg/kg/day
Amoxicillin	16
Apramycin	81
Chlortetracycline	30
Colistin	5.1
Difloxacin	10
Doxycycline	15
Enrofloxacin	10
Erythromycin	20
Florfenicol	30
Flumequin	14
Lincomycin (sole)	8.6
Lincomycin + spectinomycin[a]	22
Neomycin	24
Oxolinic acid	20
Oxytetracycline	39
Paromomycin	23
Phenoxymethylpenicillin	17
Spectinomycin (sole)	124
Spectinomycin + lincomycin[b]	38
Spiramycin	73
Sulfachlorpyridazine + trim[c]	30
Sulfaclozine	70
Sulfadiazine + trim[c]	34
Sulfadimethoxine	65
Sulfadimethoxine + trim[c]	31
Sulfadimidine	182
Sulfadimidine + trim[c]	25
Sulfamethoxazole + trim[c]	27
Sulfamethoxypyridazine + trim[c]	23
Sulfaquinoxaline	60
Tetracycline	71
Thiamfenicol	55
Tiamulin	23
Tilmicosin	18
Trimethoprim + sulfa	6.4
Tylosin	81
Tylvalosin	25

[a]When in combination with spectinomycin; because the dose of lincomycin is substantially lower in single substance products than in combination with spectinomycin
[b]When in combination with lincomycin; because the dose of spectinomycin is substantially higher in single substance products than in combination with lincomycin
[c]When in combination with trimethoprim; due to synergistic effect

3.3 Use of Coccidiostats in Poultry Sector

Coccidiostats belong among antimicrobials with antiprotozoal effect. The use of coccidiostats has increased in line with the increase in poultry production, and originally among these substances were counted also sulphonamides (that finally, because of the antibacterial activity as well as potential to select and co-select resistance, including this with importance for human health) were included among veterinary medicinal products. An extensive use of these ionophore coccidiostats in poultry feed in these days is of concern. Increase of their use has been noticed in some countries of Europe, e.g. UK data reports 15% increase in production between 2012 and 2016 (Kaul 2018). For years, this classification has been the subject of a recurring intense debate and controversy in opinions, mainly due to possible classification of coccidiostats as antibiotics and in opposite classify them rather as feed additives. There is a persistent rumour that coccidiostats will be phased out, despite that, it seems that they will remain on the list of feed additives. All coccidiostats (as animal-only antimicrobials) are currently not classified as veterinary medicinal products, but are listed among the feed additive falling into scope of evaluation of the EFSA (European Food Safety Authority). The EU has regulated the use of coccidiostats in Regulation No 1831/2003. The work is ongoing towards a revision of the regulation, in which the use of feed additives is outlined, as well as the use of coccidiostats. Some stakeholders, such as the Federation of Veterinarians of Europe (FVE), call for re-evaluation of the approach and classification of coccidiostats as VMPs, including listing them among the vet prescription-only veterinary medicinal products similarly to antibiotics. The main argument is that it could be of help for limiting of the resistance spread of *Eimeria* against certain types of coccidiostats. In a so-called position paper (FVE 2016), the FVE also points out that ionophore coccidiostats may be used for controlling of cancer in human medicine in the future (considering studies released for monensin with inhibiting effect on the spread of kidney cancer cells). For example, Russia has a zero-tolerance policy on coccidiostats residues in poultry production, prompting poultry farmers not to use coccidiostats at all (Poultry world 2020). Interestingly ionophores could not be used in production systems which ban the use of antibiotics such as "no antibiotics ever" (NAE) or "raised without antibiotics" (RWA) which are both popular in the USA currently (Mehdi et al. 2018).

Coccidiostats are mainly used to control coccidiosis (caused by coccidia—in poultry by seven exact species from genus *Eimeria*), to maintain intestinal integrity and alleviate pain and suffering. These animal-only antimicrobials are not used in human medicine and until these days it seems that their contribution to resistance to antibiotics is limited—but an issue is existence of significant data gaps (Wong 2019). Proven resistance appeared to be linked to individual coccidiostats vs individual antibiotics (Table 7), but still the need of further research should be of interest (especially due to some mechanisms influencing both bacteria and coccidial envelopes—e.g. thickening of the cell wall, or ABC transporters—efflux interaction) as indicated in the comments made by Wong (2019). An interesting information (VKM

Table 6 Treatment with antimicrobials for selected causal pathogens in diseases of poultry

Bacterial species	Disease	First choice	Second choice	Third choice
RESPIRATORY				
Avibacterium paragallinarum	Infectious Coryza	Sulphonamide/ trimethoprim	Doxycycline (preferred to chlortetracycline)	Amoxicillin*, enrofloxacin**
Gallibacterium anatis	Respiratory tract infections	Sulphonamide/ trimethoprim	Florfenicol	Enrofloxacin**
Mycoplasma synoviae (MS)	Infectious synovitis	Tetracycline, tiamulin, valnemulin	Macrolides	(Fluoro) quinolones**
M. meleagridis (MM)	Aerosacculitis			
Mycoplasma spp., *Mycoplasma gallisepticum* (MG)	Chronic respiratory disease			
Ornithobacterium rhinotracheale		Sulphonamide/ trimethoprim; doxycycline	Chlortetracycline	Amoxicillin*
GASTROINTESTINAL				
Clostridiaceae	Dysbacteriosis		Sulphonamide/ trimethoprim	Amoxicillin*
Clostridium perfringens (NetB+), Clostridium colinum	Necrotic enteritis	Phenoxymethylpenicillin; tylosin	Sulphonamide/ trimethoprim; doxycycline; lincomycine	Amoxicillin*
SYSTEMIC INFECTIONS				
Escherichia coli (APEC)	Colibacillosis	Sulphonamide/ trimethoprim	Apramycine; neomycine; doxycycline; chlortetracycline; spectinomycine	Amoxicllin* (fluoro)quinolones**
Enterococcus	Enterococcus spp. caused inf.	Sulphonamide/ trimethoprim; doxycycline	Chlortetracycline	Amoxicillin*
Salmonella spp. (invasive serotype	Salmonelosis			
Pasteurella multocida	Fowl cholera	Sulphonamide/ trimethoprim; doxycycline	Chlortetracycline	Amoxicllin* (fluoro)quinolones**
Erysipelothrix rhusiopathiae	Erysipelas			Amoxicillin* chlortetracycline
Clostridium spp.	Gangrenous dermatitis	Currently not recommended		
Pseudomonas spp., *S. aureus*, *E.coli*, *Salmonella* spp.	Omphalitis (and Yolk Sac Infection)	Rather prevention, if antimicrobials used it should be based on ethiological agens presented and the susceptibility		
SKIN AND LOCOMOTORY				
Staphylococcus aureus	Arthritis, periarthritis, joints affection, chondronecrosis, osteomyelitis, femoral head necrosis, hock joints	Despite some antimicrobials seems effective in in vitro susceptibility testing, the penetration to the infection site especially to joints and bones is very limited or close to zero and therefore MICs are not achieved at site of infection		Enrofloxacin**
Enterococcus	Spondylitis, osteomyelitis, chondronecrosis			Enrofloaxin**
GENITARY INFECTIONS				
Gallibacterium anatis, Riemerella anatipestifer, Mycoplasma gallisepticum, E.coli, Salmonella spp.	Salpingitis	If antimicrobials used it should be based on ethiological agens presented and the susceptibility		

*amoxicillin - ESBL and AmpC selection promotion

** enrofloxacin (fluoroquinolones)= CIAs

Table 7 Coccidiostats used as poultry feed additives and their antibacterial activity as well as potential to resistance to antibiotics

Group	Active substance	Antibacterial activity (mainly active against)	Comment to possible resistance to antibiotics
Ionophores	Lasalocid (sodium)	Gram-positives	Possible cross resistance to bacitracin (limited number ($n = 3$) of strains of *Clostridium aminophilum*), Russell and Houlihan (2003)
		No activity against Gram-negatives	
	Maduramycin (ammonium)	Gram-positives	
		No activity against Gram-negatives	
	Monensin (sodium)	Gram-positives	Monensin-resistant isolates of *Enterococcus* spp. and *Clostridium perfringens* had thicker cell walls than did unexposed controls (Simjee et al. 2012).
			Aarestrup et al. (1998) defining resistance in terms of a markedly higher MIC, found up to 6% monensin resistance in *Staphylococcus hyicus* and *Enterococcus* spp. in pigs (further information also in Butaye et al. 2000)
	Narasin	Gram-positives	Association between vancomycin resistance and reduced susceptibility to narasin among *E. faecium*
			In-feed narasin selects for *E. faecium* with reduced susceptibility to narasin (Simm et al. 2019)
	Salinomycin (sodium)	Gram-positives	Numbers of coliforms resistant to streptomycin, ampicillin, carbenicillin and cephalothin were greater ($P < 0.05$) from broiler chickens fed by salinomycin (George et al. 1982) Resistance to salinomycine (*nar*AB genes presented in isolates), correlation with presence of resistance to erythromycin, tetracyclin and ampicillin (*E. faecium* and *E. faecalis* isolates ($N = 137$)). In most isolates *nar*AB is found to be located on a plasmid carrying additional genes for other types of antibiotic resistance (Pikkemaat et al. 2022)
		No activity against Gram-negatives	
	Semduramycin (sodium)	Limited activity against Gram-negatives, minimal activity against selected Gram-positives	
Non-ionophores	Decoquinate	Most bacterial strains tested* appeared resistant to the effect of decoquinate ($c > 64$ mg/L), which is substantially higher than the concentration of decoquinate in GIT	
	Diclazuril	No substantial antibacterial activity	
	Halofuginone	No known antibacterial effect	
	Nicarbazin	No known antibacterial effect	
	Robenidine (hydrochloride)	No known antibacterial effect	

2015) come from Norway, where surveillance data (period 2002–2018) show that a high proportion of *Enterococcus faecium* isolates from poultry (25–91%) were confirmed to have reduced susceptibility to narasin (Norm/Norm-Vet, 2001–2014 in Naemi et al. 2020), moreover vancomycin-resistant *E. faecium* (VRE), all strains tested in NORMVET were proven to be narasin resistant, this indicate that narasin can co-select for vancomycin resistance. In-feed anticoccidials have been phased out since 2016 in broiler chicks based on decision of the Norwegian broiler associations. This may have contributed to a dramatic reduction in narasin-resistant isolates in broilers in 2018 (Norm/Norm-Vet 2019). In-feed use of monensin remains for turkeys. The proportion of narasin-resistant *E. faecium* isolates from turkeys remained on a high level in 2018 (Norm/Norm-Vet 2019). It was determined that narasin, salinomycin and maduramycin (but not monensin) stimulated the occurrence of vancomycin-resistant transconjugants in a concentration-dependent manner and resistance mechanism NarAB was indicated as supportive for horizontal gene transfer under selection pressure at certain concentrations/conditions, what is considered as significant risk from the perspective of VRE selection and maintenance in gut—microbiome (Naemi et al. 2020). Use of polyether ionophores as the feed additives in poultry should therefore be further revisited and in the case that sufficiently robust evidence will be given, the view as well as classification to non-medically important antibiotics may need to be re-evaluated (Wong 2019).

Coccidiostats can be grouped in two major classes: the polyether ionophores (lasalocid, maduramycin, monensin, narasin, salinomycin and semduramycin) and the non-polyether ionophores (often referred as synthetic compounds or chemicals, e.g. decoquinate, diclazuril, halofuginone, nicarbazin, robenidine). It should be also noted, that sulphonamide compounds, authorised as veterinary medicinal products have also properties and mechanism of action affecting coccidia, including those causing diseases of poultry. Polyether ionophores (also classified as ionophores antimicrobials) are produced by fermentation with several strains of *Streptomyces* spp. and *Actinomadura* spp. They have both anticoccidial and antibacterial activity, and they are also used as growth-promoting agents and as an active compound against clostridiosis (Anadón and Martínez-Larrañaga 2013). Their mechanism of action on bacteria is via lipophilic ionophore attachment to the lipid-rich cell membranes of Gram-positive bacteria. Ionophores bind Na+ and K+ and H+ and facilitate their transfer across the bacterial cell membrane, resulting in an increase in H+ concentration inside the Gram-positive cell which eventually leads to the death of the bacteria. Different ionophores can therefore inhibit the growth of poultry strains of *Clostridium perfringens* in vitro and reduce lesions in a necrotic enteritis model in vivo. Currently (September 2020), combination of polyether coccidiostats monensin with non-polyether coccidiostats nicarbazin was authorised (Huvepharma 2020).

Once coccidiosis is not controlled, the parasite can cause enteritis in birds leading to intestinal inflammation, reduced absorptive capacity, increased podo-dermatitis and increased mortality and could require the use of medically important antibiotics. Coccidiosis is common in all poultry worldwide and can compromise bird health and welfare, regardless of how they are kept, including indoor-reared, free-range, and organic. The term "coccidiosis" summarises several gut infections in poultry that are each caused by a different type of *Eimeria (Eimeria acervulina, E. brunetti, E. maxima, E. necatrix* and *E. tenella*). In addition to that, there are two more Eimeria-types (*E. mitis* and *E. praecox*) that are often not recognised as a disease, but do cause damage in poultry (Swormink 2019).

Of interest can also be results comparing pharmacokinetic of amoxicillin in healthy and coccidia- infected chickens, where healthy chickens' parameters detected were: C_{max} 3.26 ± 0.2 µg/mL reached at T_{max} of 1.03 ± 0.01 h and bioavailability was approximated 40.2, MRT was 3.99 ± 0.06 h. In chickens suffered from coccidia the absorption and distribution of amoxicillin was influenced, with increasing $T_{1/2a}$ and $T_{1/2b}$ to 1.04 and 9.14 h. Delayed T_{max} and lower C_{max} was detected. Infection with caecal coccidiosis significantly affected also AUC_{0-t}, where values of 14.96 ± 0.9 μg/mL.h and 11.6 ± 2.03 μg/mL.h were determined for non-infected and infected chickens, respectively. The above findings lead to the conclusion that the dose should be approximately 1.3-fold higher in chickens with coccidiosis (Kandeel 2015).

4 Considerations on Routes of Medication by Antimicrobials in Poultry

As commercial poultry (broilers as well as laying hens) are mostly reared in flocks, which can range from backyards small flocks counting several birds to commercial flocks, which can be very large (having 10,000 or up to 80,000 birds per hall). In the extensive farming facilities, with thousands of birds, the individual treatment after placing the animals into the hall is not possible. Majority of the treatments is therefore performed via medication of water for drinking and minimum extent (comparing, e.g. to pig sector) via feed medicated (e.g. in species of geese, ducks, turkeys and in certain regions) by antimicrobials. It is of critical importance that sick birds lose their appetite and also amount of water consumed decreases (mostly in less extent than feed consumption). This aspect should be considered, when adjusted amount of antimicrobial in water to be drunk by birds. Other aspect to be taken into consideration is that medication via water is more flexible. Readily and quickly prepared, with mostly high bioavailability of the antimicrobial active water medication is of the first choice for effective metaphylaxis (in certain part of birds has started to appear change of behaviour and vocalisation typical for certain disease or directly clinical signs of disease) or even treatment in early stages of clinical disease in most of birds. Factors playing a key role in proper administration of antimicrobials via medication of water are summarised in Table 8.

Another thorough consideration should be given to the actual water intake not only from the perspective of the bird's age and weight, but based on ambient temperature that has significant influence on amount of water consumed (increase of temperature = increase of water intake). This should be considered from the perspective of gain of sufficient concentration from the efficacy perspective (proper calculation of concentration of antimicrobial to avoid underdosing, which can be critical in antimicrobials, which effect is concentration dependent—e.g. enrofloxacin) as well from the perspective of possible toxicity (e.g. sulphonamides containing VMPs). Another factor of importance is lightening regimen. Poultry start to eat and drink, once light is turned on. This should be considered once dosed laying hens, or parent/grandparent flocks. For broilers behaviour is typical that they tend to eat and then drink in intervals (approximately 3–4 h) (Singer et al. 2020). In the case of replacement pullets, due to the regimen of control body mass gain, both feed and water is limited to several hours per day. Another fact to be considered in oral medication are possible interactions, in Table 8 interactions with acidifiers and sanitisers are commented, note should be done also to interaction with specific substances and antimicrobials (e.g. bentonite and tylosin (Devreese et al. 2012) and mycotoxin binders and tetracyclines (De Baere et al. 2018).

Table 8 Factors playing key role in the accuracy of the medication of water for drinking

Factor	Description
Animal factors	Breed, strain, age (absolute water consumption increases with age, but consumption per kg live weight decreases (Gray et al. 2021)), production category Birds activity (diseased or prostrate birds drink less)
Quality of water	Clean water Knowledge of pH and hardness of water, presence of electrolytes + checking of solubility of exact VMP (based on its formulation as well as antimicrobial contained)
Delivery device cleaning and function checking	Regular cleaning and disinfection of devices, and flushing by clean water prior new birds are put in place, biofilm formation should be avoided (can catch antimicrobial, can spread part of biofilm with microbes, can be hot spot of pathogens/resistance for flock) Drinkers/nipples obstructions should be avoided Delivery tubes with eater leaking should be repaired/changed
Delivery device setting	Drinking delivery (nipples) or bell systems properly set for comfortable approach by animals (line height adjustment, number of nipples or drinkers)
Medicated solution quality	Freshly medicated water to be prepared in intervals as stipulated in product texts of each respective VMP (based also on stability of an antimicrobial in the individual formulation)
Volume of water daily consumed	Qualified estimation or better exact measure of volume of water consumption per 24 h in the respective flock to adjust concentration of antimicrobial in medicated water
Bulk tank use	Clean bulk tank with mixing mechanism, assurance that all antimicrobial will be completely dissolved, with medicated water of proper concentration of antimicrobial dissolved in volume established to be consumed per day/12 h (period to be decided also according stability of exact VMP in water after reconstitution)
Proportioner use	Clean stock solution tank, assurance that all antimicrobial will be completely dissolved (follow instruction on maximum concentration to achieve complete solubility for preparation of stock solution), tune accurately flow rate from stock to water to deliver final proper concentration to birds treated; read the product texts—in some VMP use of proportioner is not recommended
Dosing adjustment	In authorised VMPs recommended dose linked to the indication or at least diseases of certain location/systems of the body (e.g. respiratory, gastrointestinal, locomotory) is established. Whenever is possible/appropriate such dose should be adjusted based on actual weight and water intake by animals Amount of antimicrobial in mg/kg body weight/day **x** mean actual body weight of bird **x** amount of birds in flock and considering water intake (dependent mostly on ambient temperature) per day should be used for calculation. Younger birds drink more water daily per unit of body weight than older, also laying hens consume more water daily per unit of weight than non-laying hens or roosters
Ligtening vs dosing	Pulse dose might be appropriate for certain breeding/category of animals (e.g. broiler breeder pullets) management. During dark periods birds do not usually drink and a peak of water consumption can occur just after lights are turned on (Gray et al. 2021). Care should be paid to concentration of antimicrobial/full solution/acceptance—or even possible toxicity (i.e. wide margin of safety bactericidal antimicrobials only can be used by this approach)
Influence of acidifiers	Some acidifiers (e.g. citrate, as well as natural acidifiers as grapefruit or citrus extracts) can influence pharmacokinetics of antimicrobials, as indicated study by Ledesma et al. (2018) in which citrate enhance in vitro antimicrobial activity of amoxicillin as well as improved pharmacokinetic parameters
Influence of sanitisers	Some sanitisers have been investigated to detect their effect on pharmacokinetic parameters and in vitro efficacy of amoxicillin (Ledesma et al. 2018). Sodium hypochlorite in combination with amoxicillin decrease both, the antimicrobial activity of in vitro amoxicillin and also its relative bioavailability. Also, maximum serum concentration as well as the area under the concentration-time curve and the mean residence time were lower/shorter compared to the amoxicillin administered alone (Ledesma et al. 2018)

5 Pharmacological Considerations with Respect to Use of Antimicrobials in Poultry

Two major aspects are to be considered from the pharmacological perspective. One is pharmacodynamic, involving mechanism of action of an antimicrobial, mechanism of resistance, including its transmission potential (i.e. location of genes encoding resistance on some mobile genetic element as, e.g. plasmid, transposon, integron); exact placing of the gene promoting expression of genes, possible cross- and co-resistance with other antimicrobials. For clinical practice mostly used and of almost importance is knowledge of susceptibility of the target pathogen (or from epidemiological perspective of high relevance susceptibility/resistance profile of the target pathogen population in epidemiologically related units—what means in the case of poultry profiling the parent's flocks). It should be noted the importance of correct sampling techniques, including choice of proper sampling device and transport medium, if appropriate, correct and speedy transport and further issues related to the laboratory investigation (for more detail information, please refer to the Vol. I chapter dedicated to Laboratory Investigations and Result Interpretation).

Despite the clinician having at his/her disposal results of susceptibility testing, still of great importance is compliance of them with clinical response, as in vitro testing of susceptibility, even performed by advanced techniques still represent artificial conditions of culturing, which cannot reflect in full extent the clinical conditions in birds' body affected by pathogenesis. Therefore, current science works on the "media mimicking natural conditions of affected tissues". Until such methods will be used in practice, the need for certain level of extrapolation as well as cross-linking of in vitro laboratory results, with clinical outcomes and experience is of high importance. Of importance is also the heterogeneity of population of bacteria causing certain infection (sometimes secondary infection emerging after viral infection). One of these cases is airsacculitis caused by *E. coli*, where different strains of *E. coli* can play role in the infection and its development—including becoming systemic (e.g. sepsis). Difference among the strains within such *E. coli* invasive population is determined not only by having different virulence factors but as well differences in susceptibility profile. Avian Pathogenic *E. coli* (APEC) having very large plasmids (e.g. ColV) with accumulated potential virulence genes and a parallel antibiotic resistance genes and be subject of easy horizontal transfer of these plasmids via conjugation cause serious issue (Christensen et al. 2021). Both clonal and non-clonal spread of the highly virulent strains bearing resistance to several antimicrobials is further promoted by use of antimicrobials (Argudín et al. 2017). Under experimental conditions, the presence of antibiotics in the growth medium of *E. coli* increased the expression of proteins involved in plasmid conjugation. A higher frequency of nucleic acid transfer (including genes of resistance) via conjugation to other bacterial cells was proven (Møller et al. 2017; Liu et al. 2019). The intestinal tract of poultry is considered as hot spot for the selection and horizontal transmission of antibiotic resistance genes, related to their high prevalence and diversity (Christensen et al. 2020; Lambrecht et al. 2019). Of importance for selection of antimicrobial is also immunostatus of the flock. Once the animals are immunocompetent, not weakened, e.g. by concomitant or previous viral infection, bacteriostatic antimicrobials can be administered as the mechanisms of immune defence of birds could help to recover from the disease. In other case bactericidal antimicrobial with good penetration to target site of infection are drugs of choice. Especially after infectious bursal disease (IBD), the treatment of airsacculitis caused by *E. coli* has been proven to be difficult to resolve by bacteriostatic oxytetracycline (Singer et al. 2020).

Knowledge of the pharmacokinetic behaviour of antimicrobial after oral administration mostly in water (in some cases still in feed) is key information. Some substances as, e.g. aminoglycosides or colistin are not absorbed or very poorly absorbed from gastrointestinal tract after administration via medicated feed or medicated water. In poultry diseases affecting or being located in synovial fluid or joint is critical to reach sufficient concentration, exceeding minimum inhibitory concentration of the target pathogen causing infection. More information on pharmacokinetic is listed below in the individual subchapters, especially in subchapter on colibacillosis, where especially pharmacokinetic and fate of antimicrobials in poultry body is commented.

6 Main Bacterial Pathogens and Diseases in Poultry, Where Antimicrobials Are Used

6.1 Infectious Coryza

Causative agent of infectious coryza (IC) is *Avibacterium paragallinarum.* This disease appears in pullets and layers, occasionally in broiler chickens as rhinorrhoea, affecting all ages, but mainly older animals. Chronically ill or healthy carrier birds are the reservoir of infection (Blackall and Soriano 2020). Morbidity and mortality are high, egg production is reduced in laying hens by 10–40% as well as reduced hatchability (Huevelink et al. 2018). The incubation period is 1–3 days. Typical symptoms of coryza are nasal discharge, sneezing, conjunctivitis, cerebral oedema, swelling of the intraorbital sinuses. In the mildest form of the disease, the only signs may be depression, a serous naChristensensal discharge, and occasionally slight facial swelling. Respiratory symptoms may persist for several weeks. In the case of sole infection, coryza duration is usually 2–3 weeks; in the case of concomitant disease (e.g. mycoplasmosis) duration can be longer (including swelling symptoms). Other lesions may include conjunctivitis, tracheitis, bronchitis and airsacculitis, particularly if other pathogens are involved. Transmission route is via direct contact, airborne droplets and contamination of drinking water. “All-in/all-out” management can help eradication of infectious coryza (AMCRA 2020; Soriano-Vargas 2022).

Differential diagnostics should distinguish swelling of the face and wattles from that seen in fowl cholera, but also other diseases to be considered as mycoplasmosis, laryngotracheitis, Newcastle disease, infectious bronchitis, avian influenza, ornithobacterosis, *Gallibacterium anatis* infections and vitamin A deficiency (Soriano-Vargas 2022; Crispo et al. 2019). Study by Clothier et al. (2019) reported concurrent infections with *Mycoplasma gallisepticum* (MG: 18.6%), *M. synoviae* (MS: 18.6%), infectious bronchitis virus (IBV: 12.8%) and infectious laryngotracheitis virus (ILV: 7.0%). More than 30% of birds with IC were concurrently infected with MG, MS and IBV. Interestingly, the frequency of Marek’s disease that would be expected to result in immunosuppression was not different between *A. paragallinarum* PCR-positive and negative birds (Clothier et al. 2019).

To make a diagnosis in live animals, the animals with an acute form of the disease that have not yet been treated should be selected for the sampling (samples from more than one animal is recommended). The nasal discharge can be collected with a swab from the nasal cavities or the trachea and care should be paid to collection, avoiding contamination by collecting to much material on swab. The best pure culture is obtained with exudate collected with a swab from the lung alveoli, but deep incision in the sinus post-mortem to collect the exudate with a swab can be also used. Swabs for bacteriological culturing followed by susceptibility testing should preferably be stored in a transport medium with activated charcoal to promote the chances of isolation of *Avibacterium paragallinarum.* Swabs intended for PCR will preferably be transported without transport medium. The most commonly used serotyping scheme is the Page scheme, which groups *Av paragallinarum* isolates into three serovars (A, B and C) or by Kume classification to nine serovars (A-1 to A-4; B, C1 to C4) that correlate with immunotype specificity (AMCRA 2020; Soriano-Vargas 2022). Slow-growing, fastidious, Gram-negative non-motile *A. paragallinarum* in most cases also require so called V-factor: nicotinamide adenine dinucleotide, NAD for their growth (Christensen et al.

2016). A capsule was described in some virulent strains, which has been shown to protect against bactericidal activity of normal chicken serum (Sawata et al. 1984). *A. paragallinarum* also produce AvxA toxin (a composite serine-protease RTX-toxin) with cytotoxic effect on an avian macrophage-like cell line (Küng and Frey 2013). Some preliminary findings indicated that AvxA is a common major virulence attribute of all *A. paragallinarum* serotypes and represents a possible target for future subunit or recombinant vaccine developments against IC (Frey 2019).

There is no standardised method for testing the antimicrobial susceptibility of *A. paragallinarum* (also no testing control standards). It is therefore difficult to compare studies carried out using different methods (agars/broths modifications, agar incubation in enriched CO_2 atmosphere, different interpretative criteria).

In some EU countries following antimicrobials are approved for indication of infectious coryza: ampicillin (amoxicillin), erythromycin, penicillin, sulfamethoxazole, tetracycline (doxycycline, chlortetracycline) and trimethoprim/sulfamethoxazole. Studies reporting *A. paragallinarum* in European countries are rare, one of the most recent come from the Netherlands. Huevelink et al. (2018) included in the testing also further substances or their combination (not only for therapeutic reasons): amoxicillin/clavulanic acid, cefepime, ceftiofur, clindamycin, enrofloxacin, florfenicol, neomycin, oxacillin, penicillin. Full set above-mentioned antimicrobials were tested phenotypically, but using PCR also genes of resistance to some of these antimicrobials were detected. Strains tested in the above-mentioned study (n = 44) were isolates from poultry within period 2008–2017, from total 24 Dutch farms with outbreaks of infectious coryza. Table 9 indicates MIC_{50} and MIC_{90} results. Except tetracycline and sole sulfamethoxazole, other antimicrobials indicated for infectious coryza have very low MIC_{90} (μg/mL): sulfamethoxazole/trimethoprim 0.5; ampicillin < 0.5; penicillin < 1; erythromycin < 2.0. In some countries, as a third choice, enrofloxacin is allowed to be used, for which Dutch study indicated following MIC values range: ≤0.25 to 2 μg/mL with 25.0% of isolates (MIC ≤ 0.25 μg/mL); 11.4% (MIC = 0.5 μg/mL); 36.4% (MIC = 1 μg/mL) and 27.3% (MIC = 2 μg/mL). As neomycin, florfenicol and tilmicosin are also authorised for use in poultry (not namely mentioning the coryza indication), the MICs can also be mentioned—with mostly MIC ≤ 4 μg/mL for neomycin and tilmicosin; the MIC value of 8 μg/mL was obtained for eight (18.2%) of tested isolates/neomycin and for one isolate/tilmicosin and with MIC ≤ 2 μg/mL for florfenicol.

Within treatment guidelines across Europe treatment recommendations for first and second lines treatments differ (AMCRA 2020; Brugère-Picoux et al. 2015; Boereboom et al. 2012) but mostly sulphonamide/trimethoprim combination is mentioned as first or second choice, followed by doxycycline (or other tetracyclines as chlortetracycline or oxytetracycline) and aminopenicillins (where warning include possible selection of extended spectrum beta-lactamases or AmpC beta-lactamases especially in *Enterobacterales*). Last choice mentioned by AMCRA is enrofloxacin.

Vaccination can be a solution for affected commercial operations and can provide strong

Table 9 *Avibacterium paragallinarum* (n = 44; 25 outbreaks, the Netherlands, 2008–2017) MIC_{50} and MIC_{90} (μg/mL) by broth microdilution method using CAMHB[1]

Antimicrobial	MIC_{50} (μg/mL)	MIC_{90} (μg/mL)
Ampicillin	0.25	0.25
Penicillin	0.5	1.0
Erythromycin	1.0	2.0
Sulfamethoxazole	<16	128
Tetracycline	16	>16
Trimethoprim/sulfamethoxazole	0.125	0.5

protection against strains of *A. paragallinarum* if properly sampled, isolated a processed to the autogenous vaccine. For the successful vaccination knowledge of serovars is essential (Blackall and Soriano 2020). For vaccination administration via SC and IM route, in laying hens, scheme of two injections 4 weeks apart before 20 weeks of age seem to result in better performance. Oral administration is also possible but with more concentrated inoculum in vaccine, 100 x more than in parenteral route (Blackall and Soriano 2020).

Of interest is also susceptibility to both chemical agents and physical/environmental factors. Infections exudate is inactivated in 4 h in water of ambient temperature; exudate can remain infective 24 h (at 22 °C), and 24–48 h (at 37 °C), several days (at 4 °C). In 2–10 min appear killing effect at temperatures of 45–55 °C (Blackall and Soriano 2020).

6.2 Dysbacteriosis Due to Bacterial Imbalance

Bacterial enteritis (or dysbiosis or dysbacteriosis) is one of the main indications for the use of antibacterials in poultry. Typically manifests from the age of 14 days in fattened chickens and turkeys. Changes in intestinal motility, modifications of gastric acidity, reduction in the production of bacteriostatic peptides in the pancreas, alterations in the amounts of mucus produced or in its composition, reduced IgA secretion, and focal ulcerations of mucosa result in failure of nutrient absorption, tissue necrosis, and shifts causing imbalance in gut microbiota (Oviedo-Rondón (2020). If untreated disease can lead to caked and wet litter with increased food pad dermatitis and hock burns (animal welfare problems) as mentioned by Löhren (2008). Temperature stress (heat or cold) or other environmental discomfort can also contribute dysbiosis and/or enteritis. Exposure to the stress hormones norepinephrine and epinephrine, significantly increases the proliferation of several enteropathogenic bacteria mainly: *Escherichia coli, Yersinia enterocolitica, Pseudomonas aeruginosa, Salmonella* (e.g. enteritidis, choleraesuis and typhimurium). Young animals are more prone to dysbiosis due to less stable and immature microbiota, and also the lack of digesta can change microbiota balance in periods of hours when nutrients are not present. Critical is also water quality with influence on digesta pH, sulphates, magnesium and other salts can increase passage rate or enhance water consumption (spillage water can also lead to wet litter, but different ethiology). Coccidia (*Eimeria* spp.) can also cause pathologies with impact on the patterns of fermentation in the ileum and caeca and initiation of dysbiosis.

Due to the complexity and multifactorial nature—responsibility for such enteritis cannot be attributed to any particular bacteria. Bacterial species varies as for number, ratio and strains represented. Bacterial enteritis does not cause direct mortality. It is rather an issue of welfare (stress increase, imbalanced water and feed intake). It also leads to the inconsistency of the flock as for weight gain by impacting the water/feed ratio and, hence, to the economic impacts. The quality of the bedding influences respiratory system, which is a gateway of selected pathogens. Due to the fact that intestine is location of immune defense and whole health status is influenced, dysbacteriosis can also lead to increased mortality through secondary pathologies. More cases of arthritis in broiler chickens from flocks affected by bacterial enteritis were noticed as well.

For diagnose of dysbiosis no unambiguous and easily available set of biomarkers is available, therefore autopsy of animals from the suspect herd and macroscopic intestinal scoring are recommended. A high lesion score on day 10 may lead to a higher score thereafter.

Antibiotics with a Gram-positive spectrum, mostly penicillins, e.g. water-soluble phenoxymethylpenicillin, are effective on *Clostridiaceae* and it seems they can help to improve status of the animals. Looking from another perspective the administration of therapeutic doses of antibiotics can trigger the development of dysbiosis by disrupting the gut microbiota. It is therefore preferable to use antibiotics only when the macroscopic scoring shows the presence of severe enteritis (scores exceeding 5), as recommended by AMCRA (2020).

The control of other pathologies (such as infections and infestations by coccidia), avoidance of mycotoxins and rancid fats in feed, and in complexity good feed balance and flock management are therefore essential to prevent bacterial enteritis. The most important factors to prevent dysbiosis are minimising environmental stress, supply of water with good quality (including clean tubes and drinking devices), improving feed digestibility (lowering non-starch polysaccharides as well as balancing minerals-issue of soybean meal) and avoiding antinutritional factors, mycotoxins and rancidity.

Due to the aforementioned complexity of the disease, it is not possible, based on current scientific knowledge (2023), to perform a complex bacteriological investigation, including an antibiogram or other susceptibility tests indicating unambiguously the best treatment option. Scientific notices for antibiotics therefore do not specifically mention this indication. Also searching among the approved indications of authorised VMPs with antimicrobials broad description for this indication appears as "gastrointestinal infections/disorders caused by microorganisms susceptible to" the active substance/or their combination of the exact VMP.

For poultry group treatments with drinking water are of interest, since the flock forms an epidemiological unit (AMCRA 2020). Therapeutic options are for dysbiosis indication mostly based on empirical knowledge and mostly reflecting susceptibility to antibiotics of *Clostridium* spp. Available guidelines differ in first/second choice. In Belgium (AMCRA 2020) no active substance is mentioned as a first choice. Macrolides (effective against *Clostridium* spp.) are not authorised for this indication in Belgium, but in other countries it could be different, therefore, e.g. in the Netherlands, France and the United Kingdom macrolides are authorised for broad indications, and can therefore involve also dysbiosis, with participation of *Clostridium* spp., and are recommended as second (last) choice or alternative to (phenoxymethyl) penicillin (Löhren 2008). As the second (in Belgium third) choice also aminopenicillins can be used, but are not recommended due to the warning include possible selection of extended spectrum beta-lactamases or AmpC beta-lactamases especially in Enterobacteria. As antibiograms cannot be established for this indication (full scale of germs contributing not (yet) isolable), the AMCRA (2020) gives an advice to examine the effectiveness of molecules that can be exploited via macroscopic scoring systems. Some guidelines (including AMCRA) recommend sulphonamide/trimethoprim combination also due to the concomitant impact—reduction of coccidiosis that can contribute dysbiosis.

As commented above, dysbacteriosis (or dysbiosis) due to its bacteriological characterisation and complexity, is usually diagnosed on the basis of clinical signs and an autopsy, using a scoring system with ten parameters giving better picture of gut health, than simple bacteriological investigation. Bacteria of the genus *Clostridium* (*Clostridium perfringens* and Clostridiaceae) and coccidiosis (Eimeria) may play a role in the development of dysbiosis (AMCRA 2020).

No vaccine is available, but alternative new feed additives have been introduced as probiotics, prebiotics, enzymes, organic acids, and herb extracts (phytoadditives), mycotoxin controlling systems are frequently used as a preventive measure and for less severe flare-ups. A whole set/combination of the factors that can influence GIT health conditions as well as immunity and protect against dysbacteriosis as well as some other diseases can be summarised in brief (Table 10). Some approved coccidiostats can be used in appropriate administration scheme (please refer above to the Sect. 3.3.).

Last, but of great importance both from dysbiosis perspective, but as well predisposing to several important diseases in poultry such as coccidiosis, salmonellosis, necrotic enteritis and also colibacillosis are presence of mycotoxins in feed. Mycotoxins with toxic effects affecting of inner organs (e.g. liver) as well as impairing the immune system leading to the immune suppressive conditions. Mycotoxin effect on the intestinal epithelium with final effect on nutrient absorption, cell proliferation, cytokine release,

Table 10 Factors that can influence equilibrium of the gut and minimise risk of dysbiosis

Feed ballance and composition	Minimise excess of nutrients, proper dietary fat type, importantce of maintenatnce of minimum of 1% fat in the diet to stimulate gastric function and slow release of digesta from gizzard to the small intestine
Feed quality	Quality assurance of soybean meal (Undercooked soybeans increase of antitrypsin factor concentrations, overcooked decreases protein digestibility)
	Corn genetics, moisture of corn during harvest, drying temperatures, and storage conditions can play a role in the formation of retrograded starch and reduction on salt-soluble protein which decreases digestibility
	Maximise digestion ballance and effectivity (enzymes)
	Keeping ballance of minerals
	Proper granulometry or particle size of all feed ingredients that stimulates gizzard function
	Mycotoxins in feed causing immunity impairment and predisposing for dysbiosis and diseases of intestinal tract
Good water quality, suuficient amount of water	Ensurance of good water quality from microbial as well as physicochemical standpoint Clean watter lines, nipples, drinkers, regular cleaning of the watering devices
	Minimising occurence of biofilm, endotoxins, algae and microbes´ proliferation in water plants
	Minimising water hardness, alkalinity and basic water pH, which disturb crop pH and consequently reduce the first phase of digestion
	Organic acids in feed or water create acidic conditions in the crop of broilers, but not in layers
	Water and dietary calcium levels higher than the requirement for each age group and productive phase can be deleterious to intestinal health since it may chelate nutrients, reduce enzyme activity, and promote proliferation of *Clostridium* bacteria
New feed additives	Probiotics (introduction of desirable mixture of microorganisms)
	Prebiotics promoting the growth of desirable bacteria in GIT
	Organic acids grow inhibition of bacteria
	Enzymes are used to help eliminate the anti-nutritional effects of phytate
	Water-soluble polysaccharides and/or change substrates to improve proliferation of some beneficial gut microbes
	Phytobiotics as esential oils
	Phytochemicals and natura phytoaditives
Housing environment	Optimum temperature, air velocity, and relative humidity according to the age, phase of production, size of the birds and external environmental conditions
	Environmental stress due to heat or cold, very dry or very humid air, or drafts can alter their feed intake patterns and intestinal motility thereby causing reduced digestion and the ensuing gut (and secondary also immunity) problems
	Appropriate stocking density (that can be in some extent compensated by proper ventilation and temperature regulating capacity
	Renewing bedding, after each cycle clean properly and have technological breaks, ensurance of proper humidity
	Appropriate lighting programmes
Biosecurity	Examples only: Bio-exclusion (limiting visitors and entry of vehicles and equipment that visit other poultry farms) and bio-containment, isolating houses, and controlling insects, rodents and the entry of wild birds and other animals to the internal facilities
Improvement of breederers´ flocks healh	Nutrition/feeding of broiler breeders are crucial or adequate development of their progeny. Embryo development based on egg nutrients deposited by hens. Specific nutrients such as vitamin D, trace minerals, carotenoids and fatty acids seem to play key roles in immunity and gut development. Hens also affect embryo nutrition and development by eggshell properties (determines the capacity of eggs to exchange gases and water vapour, consequently affecting embryo utilisation of yolk and nutrients). Finally, breeders can transfer good and bad intestinal microbes and immunity to their progeny, therefore intestinal breeder health and proper vaccination scheme is of importance
Hatchery conditions and early/late hatch	Elevated temperatures during the last phases of egg incubation reduce GIT tissue mass and enzymatic activity. These effects of elevated incubation temperatures have implications for the digestive capacity of the chicks and probably in the incidence of intestinal problems and resistance to parasites. Too early/too late hatching issue of lymphoid tissue development, immunity and gut development influence

immunoglobulins and finally barrier function of the epithelium can have severe impact on birds' ability to resist infections. Examples of mycotoxins—ordering from high toxicity to lower toxicity: T-2 toxin, aflatoxin, ochratoxin A, deoxynivalenol, zearalenone, fumonisin/moniliformin, influencing health status and immunostatus of poultry. Therefore, minimising mycotoxins and management strategies to achieve so, are of high importance. With respect to mitigation of toxic effects of mycotoxins in poultry, post-harvest strategies of different principles, some using also feed additives, have been introduced as well as discussed in the review of patents (Zhu et al. 2016). Not all technologies are beneficial in all aspects and therefore strategies should be subject of (commercially) independent studies. Some technologies work with inorganic (clay minerals), organic (yeast cell wall and glucomannan) feed additives as well as microorganisms and enzymes, which are incorporated to feed as additives to bind and/or inactivate mycotoxins and finally detoxify feed were reviewed by study under EFSA umbrella (Boudergue et al. 2009), more updated review focusing on adsorbents with the ability to bind to multiple mycotoxins is also available (Vila-Donat et al. 2018). Ten years later after EFSA review, Elliott et al. (2019) published data, shows that some of these substances—particularly clay minerals can have wide range of negative health effects in farm animals, including an interaction with micronutrients, vitamins and active substances in veterinary medicinal products. Considerations and comments brought in the very recent bibliography by several authors (Elliott et al. 2019; Kolawole et al. 2020; Prapapanpong et al. 2019) should be thoroughly analysed, as highlighting available data on both natural and modified adsorbents or mycotoxin binders with possible induction of cytotoxic effects including oxidative stress, reduction in cell viability, apoptosis, and DNA damage as well as reduction of feed conversion, immunosuppression and interactions as described briefly below. Concerns from the use of such substances are linked to, e.g. significant decrease of egg weight and egg yolk index, after inclusion of 2% clinoptilolite to diet contaminated with 2.5 mg AFB1/kg in laying hens (Rizzi et al. 2003). Attention should also be paid to modification of extent of the oral absorption of drugs (increase/decrease), potentially leading to therapy failure and/or higher levels of antibiotic residues in foods of animal origin. Interaction of mineral adsorbents with active substances of veterinary medicinal products has been reported for antibiotics such as tilmicosin, tylosin, paromomycin, doxycycline and coccidiostats such as monensin and salinomycin (Elliott et al. 2019). Some data (Gray et al. 1998; EFSA, 2011a, b) are reported for example for bentonite and broiler chickens (monensin, salinomycin—lowering preventive anticoccidial effect and incompatibility with robenidine). A significant alteration in pharmacokinetic profiles and oral bioavailability of oxytetracycline and amoxicillin was detected within in vivo study with broiler chickens, where additionally, significant concentration of oxytetracycline was found in the kidneys of treated birds (Osselaere et al. 2012) compared to control group. Content variable amounts of accessory minerals (quarts, non-tronite, erionite), heavy metals (lead, copper, cadmium), dioxins and trace elements can be also an issue in mineral adsorbent—that can cause toxicity induction in birds. Moreover, serum mineral profile and activities of enzymes such as glutamate dehydrogenase, aminotransferase, creatinine and glutathione peroxidase can be altered (Elliott et al. 2019). Last serious fact, reported by some teams of researchers is evidence of inefficiencies of most commercial binders or detoxifiers, as well as their affinity for only a single mycotoxin (Kolawole et al. 2020; Prapapanpong et al. 2019; Kolawole et al. 2019).

6.3 Necrotic Enteritis

Necrotic enteritis (NE) is a disease with causative agent from genus *Clostridium perfringens*. In poultry mostly types A, B and G are of importance (Mora et al. 2020). Toxin production, is the basis for a classification system of the *C. perfringens* strains that, has been recently revised to include seven toxinotypes (A, B, C, D, E, F, G), based upon the presence of genes encoding for CPA, CPB, ETX, and ITX toxins, as well as the recently included enterotoxin CPE and necrotic enteritis B-like toxin (NetB) (Rood et al. 2018). Different types produce a number of toxins, is an acute status with enterotoxemia. To date, both a haemorrhagic form of NE and the type G-associated non-haemorrhagic disease entity are commonly referred to as necrotic enteritis, but should be described separately and the first one is proposed to be newly called necro-haemorrhagic enteritis (Goossens et al. 2020).

Toxin called NetB was identified in disease causing *C. perfringens* type G isolates from broilers, which belongs to the α-haemolysin family of β-pore-forming toxins (Rood et al. 2016; Prescott et al. 2016). NetB toxin is essential to cause necrosis, together with a range of additional virulence factors that might provide an advantage when colonising the broiler intestine (Wade et al. 2020; Wade et al. 2016). NetB toxin production is positively regulated by the VirR/VirS two-component phosphorelay system and such production seems to be stimulated in concentration of *C. perfringens* exceeding 10^9 CFU/g and birds on diet with a high polysaccharides level, dysbiosis, slow food passage and low bioavailability or those being suffered by intestinal damage linked many times with presence and action of coccidia from genus *Eimeria* (Mora et al. 2020). Once the isolate contains the *net*B gene, it is located on large (85 kb) conjugative plasmid (containing also 36 additional genes, e.g. coding for two glycohydrolases, two leucocidins, chitinases, an internalin-like protein, a metalloprotease, and several adhesin-like proteins), encoding the pathogenicity loci (NELoc-1, 42 kb). It has been shown that vaccines that include NetB can protect chickens against NE. Outcomes of the specific experiments indicated that NetB will be an important antigen to include in an effective, commercially viable, necrotic enteritis vaccine (Rood et al. 2016). It also seems that possession of an additional toxin, TpeL, a member of a large clostridial glycosylating cytotoxin family found on tcp-conjugative plasmid, enhances the virulence of NetB-containing strains (Prescott et al. 2016). In strains containing TpeL disease and mortality are induced more rapidly. New information is related also to two Quorum Sensing (QS) mechanisms (LuxS and Agr-like systems), which are functional in *C. perfringens* having impact on gene expression. Namely Agr-like QS system plays a key role in virulence via regulation of the gene expression leading to release of proteins CPA, CPB, CPB2, CPE and NetB. It seems that regulation of expression of genes included in the process of adherence also exists. These genes are located on the chromosomal locus VR-10B. One of the seven described genes encoding collagen adhesion gene *cna*A. Adhesion to intestinal mucosa and tips of the villi plays role in pathogenesis of NE (Opengard and Boulianne 2020). One more factor is considered once speaking about NE: number of *C. perfringens* in the birds' gut. Enumerations of 0 to 10^5 CFU/g of *C. perfringens* have been observed in the intestine of healthy chickens, while in sick birds with NE enumerations of 10^6–10^8 CFU/g was detected as well as further necrotic enteritis predisposing factors among them bacteriocins, adhesins, proteolytic enzymes, collagenolytic enzymes, necrotic toxin enteritis B-like (NetB) and TpeL (Gharib-Naseri et al. 2019; Timbermont et al. 2011). Importance of zinc metalloproteases (especially ZmpA, but with more strong effect if together with ZmpB) for binding affinity for the mucin glycoprotein, a constituent of the mucosa of the gastrointestinal epithelium has been proved (Razmyar et al. 2017). More comprehensive information can be also found in recent publication by Mora et al. (2020), where summarisation and details on pathophysiology of the disease, including factors contributing to it as well as seriousness of the disease is described.

To date in bibliography mostly the cases of non-haemorrhagic NE has been described, where toxins produced by the bacteria cause damage to the small intestine (jejunum and ileum mostly affected), liver lesion (less frequently) and mortality. The clinical illness is usually very short, and often the only signs are a severe depression followed quickly by a sudden increase in flock mortality. There also exist subclinical forms of the G-associated non-haemorrhagic disease (Goossens et al. 2020).

Necrotic enteritis primarily affects broiler chickens (2–5 weeks old) and turkeys (7–12 weeks old) rare on litter as well as can attack layer pullets (Hargis 2014). In layer pullets it has been proven in recent study that birds kept in aviary houses were less prone to NE than flocks reared in floor systems or enriched cages (data from Belgium, study by Goossens et al. 2020a). NE was observed from 2.4 weeks of age to the end of the pullet rearing period (17 weeks), with the median age 9.8 weeks in study by Goossens et al. (2020a). The severity of clinical signs can vary with the age of the birds. Early mortality is often related to coccidiosis vaccination programmes, with *Eimeria* cycling in these flocks (Hargis 2014). Another form, necro-haemorrhagic disease is characterised by confluent mucosal necrosis, often covered by a pseudomembrane. More mild cases are characterised by multifocal, brown-grey, necro-haemorrhagic lesions (Goossens et al. 2020). Microscopically, lesions are characterised by necrosis and haemorrhage extending downwards from the tip of the villi (Hargis 2014). Lesions mainly appear in the jejunum, ileum, expanding to the caecum and duodenum, with a thin, dilated wall and with the presence of gas. Brown-grey or yellow-green mucosa can appear as well as, occasionally, lesions in liver, and kidney (Cooper et al. 2013).

Several authors cited by Goossens et al. (2020) mentioned, that necro-haemorrhagic enteritis has yet be proven being linked with specific netB-negative *C. perfringens* type A strains (experimentally was this form of diseases reproduced with netB-negative strains isolated from diseased birds (Li et al. 2018; Wilson et al. 2018; Yin et al. 2017). However, re-isolation of the inoculated *C. perfringens* strains was not reported and no studies to identify causative toxins or virulence factors have been conducted so far (Goossens et al. 2020). In order to fully understand disease pathogenesis and to develop prevention strategies, a clear distinction should be made between both types of NE mentioned above, but as for treatment options, there seems no reason for use of different antimicrobials in each of these cases.

Among important NE predisposing factors belong high dietary levels of animal by products (e.g., fishmeal), wheat, barley, oats, or rye. Disease progress is promoted by slow feed passage rate in the small intestine and, generally, by anything that promotes excessive bacterial growth and toxin production. In many cases, concurrent coccidiosis (especially *Eimeria maxima*, and *Eimeria acervulina* to a lesser extent) is associated with outbreaks in broiler chicks' flocks. Some studies supports for the fact that intestinal damage caused by *Eimeria* spp. and colonisation of birds gut causes the release of plasma proteins to the lumen of GIT, including more than 11 amino acids, growth factors, and vitamins; such factors can play a role of the growth substrate for pathogenic *C. perfringens* (Moore 2016; Timbermont et al. 2014; Van Immerseel et al. 2009). Despite this association with coccidiosis is documented, recent investigations with NetB-positive isolates have also been reported, which caused disease without predisposition from *Eimeria i*nfections. Occupancy density is also a risk factor. Fewer animals per square metre reduce the risk of the disease emerging. In layer pullets, data from Belgium (2013–2017, 295 different flocks/originating from 50 different layer pullet houses) show through multivariate analysis a significant association of NE with coccidiosis, housing system and season (Goossens et al. 2020a).

The diagnosis of necrotic enteritis can be made at autopsy by the characteristic intestinal lesions and by bacteriological examination confirming presence of *Clostridium perfringens*. Intestinal lesions samples taken immediately after death/euthanasia, by using swab, with transport medium (anaerobic conditions, preferably +/−4 °C, speed transport) and further processed by culturing and/or PCR testing. *C. perfringens* toxins can be detected with PCR or by ELISA. Specific multiplex PCR has been developed identifying genes *netB* and *tpeL relevant for NE (*Bailey et al. 2013*)*. As for ELISA serum anti-CPA (Cooper et al. 2013) or anti-NetB (Lee et al. 2011) or detection of the levels of IL-10 in serum of infected chickens and intestinal epithelial cells stimulated with *C. perfringens* by

ELISA capturing antigens (Lee et al. 2018) can be used for laboratory confirmation.

For poultry, given the rapid and very virulent nature of the disease, leading to high mortality there is of importance to dose antimicrobials as speedy as possible, mostly by using drinking water. In general, narrow spectrum penicillins and macrolides are effective against *Clostridium* spp. Once these active substances are not effective against the isolates, the second and third choice molecules can be used. Resistance to lincosamides and tetracyclines has been observed. In recent study in layer pullets in Belgium (Goossens et al. 2020a), all cases of NE were treated with antibiotics (amoxicillin the most frequently used accounting for 78%, tylosin for 13% and phenoxymethylpenicillin for 9%). Trimethoprim + sulfonamides combination can be considered as possibility in concomitant coccidiosis with possible effect on both pathologies. However, consideration should be given to the possible resistance of *Clostridium* spp., the efficacy of this trimethoprim + sulfonamides combination in vivo can be jeopardised by necrotic effect of the disease. The use of aminopenicillins should be avoided (due to the selection of organisms that produce "extended spectrum beta-lactamases" (ESBLs) or AmpC). Therefore, e.g. AMCRA (2020) guidelines in Belgium classify amoxicillin as the last, third choice.

More up to date approach is targeted on two phases—first mainly with biosecurity and sanitation, is some farms also by exact technology protocols tailored for the farm conditions. Second phase is targeted on modification of diet as well on a prevention of the disease using one of or combination of probiotics, prebiotics, synbiotics, phytogens, organic acids, and dietary modifications (Mora et al. 2020).

Examples of probiotic strains used for broiler chickens positive microbiota modulation to control necrotic enteritis caused by *Clostridium perfringens* (CP) presented by Mora et al. (2020) mentioned, e.g. *Bacillus subtilis* CGMCC 1.921: lowering CP in caeca digesta (Guo et al. 2017); *Bacillus subtilis* PB6: reduction CP, improved villi length (Jayaraman et al. 2013); *Lactobacillus acidophilus* D2/CSL CECT 4529 and *Bacillus subtilis* PB6 ATCC-PTA 6737: significant decrease of CP, *E. coli* and *Staphylococcus* spp. in caecum and ileum digesta (Forte et al. 2016); culture of *Bacillus subtilis* DSM17299, *Clostridium butyricum*, *Lactobacillus acidophilus*: decrease CP and *E. coli* in caecum, increase of strains from genus *Lactobacillus* and *Bifidobacterium* (Hossain et al. 2015); *Lactobacillus johnsonii* (significantly decrease enterobacteria and *C. perfringens* in the ileum (Olnood et al. 2015). Also, commercialised products containing mixture of "healthy microbiota" strains are available for use as prevention of NE in broilers as, e.g. Aviguard®, BROILACT®, PoultryStar® and other (Caly et al. 2015) and one product MSC™ was also evaluated as for decrease of CP enterotoxin (Craven et al. 1999).

Prebiotics conferring health benefits by improvement of the intestinal epithelium (longer villi and shallower crypts) and increase of feed conversion and efficiency with help of beneficial compounds as short-chain fatty acids (mainly acetate, propionate and butyrate), which are absorbed directly from the hindgut and used as an energy source, improving weight gain and performance (Mora et al. 2020). Most of them can act by reducing the intensity and time of histopathological conditions caused by *C. perfringens* in the jejunum and duodenum (Ricke et al. 2020). Prebiotics influence bone mineralisation (via better absorption of calcium and phosphorus), promoting symbiotic processes in gut microbiota, favouring lactic bacteria and inhibiting pathogens having also a secondary effect, decreased occurrence of toxic metabolites (ammonia, indoles, phenols, thiols)—Kheravii et al. (2018).

Effects of both probiotics (individual strains of bacteria or their mixture) and prebiotics (e.g. inulin, fructooligosaccharides and xylooligosaccharides) are combined in synbiotics. Some examples, from which several were in experiments indicated to reduce mortality, to cause decrease of *C. perfringens* in different parts of gastrointestinal tract and reduce lesions, some improve morphology of intestine (higher villi and shallow crypts), influence positively weight gain (Mora et al. 2020). Several exact examples, non-exhaustive, could be mentioned that they are produced commercially and used in practice: two well-known products, one made up of *Enterococcus faecium* and fructooligosaccharides (Biomin®IMBO) and second one containing a mixture of *Bifidobacterium animalis*, *Enterococcus faecium*, *Lactobacillus reuteri* and *Lactobacillus salivarius*, *Pediococcus acidilactici* and inulin (Poultry Star®)—as in detail also described on web page Biomin GmBH (2021). Another product is synbiotic poultry (produced by Vetafarm) containing Lactobacilli (*L. acidophilus*, *L. casei*, *L. salivarius*, *L. plantarum*, *L. rhamnosus*, *L. brevis*) and *Bifidobacteria* (*B. bifidum*, *B. lactis*) as well as *Streptococcus thermophilus* and inulin (Markowiak and Śliżewska 2018). *Saccharomyces cerevisiae*, *Enterococcus faecium* and *Bacillus* spp. are included in the product Avi-Lution®. It seems that studying the exact effect as well as concomitant factors contributing to better or worse efficacy once synbiotics are used in broilers needs to be explored in more detail in the future.

Dietary management is a promising strategy for the necrotic enteritis control in poultry. Plant extracts and herbals, organic acids and enzymes and minerals can contribute as well as different dietary restriction, modification of the content and source of macronutrients. Enzymes such as proteases, glucanase, manase, cellulase, amylase, phytase and xylanase are used as feed additives to minimise the non-starch polysaccharides negative effects and increase the digestibility, management and absorption of nutrients (Mora et al. 2020; Kiarie et al. 2013).

Herbals and plant extracts are also used for the control of *C. perfringens* in broilers. Substances from groups of essential oils: anise (*Pimpinella anisum*) and oregano (*Origanum vulgare*); phenols: thymol (*Thymus* vulgaris) and carvacrol (*Origanum vulgare*); alkaloids: protopine (e.g. *Eschscholzia californica* or *Fumaria officinalis*; benzophenanthridine (*Chelidonium majus*); piperine (*Piper nigrum*) and curcumin (*Curcuma longa*); tannins (*Castanea saliva)*, saponins (*Chenopodium quinoa*) and flavonoids (e.g. citrus origin) are currently used. The effect on intestinal development (longer villi and shallow crypts), decreased number of *C. perfringens*, intensity lesions, reduced mortality, increase serum antibody titres, induced immunogenicity and further positive effects have been reported (Mora et al. 2020).

Another option are organic acids (aliphatic or aromatic) influencing the microbiota, having an immunomodulatory effect and stimulating the intestinal mucosa. Base of the organic acid product, dose and type are key elements for the final effectivity on *C. perfringens.* Different acids influencing number of *C. perfringens* in different part of the intestine, capric-caprylic, caproic and lauric acids influencing intestinal histomorphology, except this effect hexanoic, benzoic and butyric acids influencing moreover frequency and intensity of the intestinal lesions and decrease number of CP in liver and caecum (Abdelli et al. 2020; Pham et al. 2020).

Although NE is a major cause of mortality and reduced productivity in poultry, no commercially produced vaccine is available against its causative agent *Clostridium perfringens* to this date, because authorisation of the one vaccine that was centrally approved by EMA in 2014 is no longer authorised (adjuvanted vaccine against *Clostridium perfringens* type A alpha toxoid), EMA (2021). Notwithstanding that, veterinarians in flocks suffered from NE can, except other above-mentioned antibiotics as well as the preventive and alternative tools, collect samples and go for the autogenous vaccines production. At least two autogenous vaccines producers within the EU are able to design and manufacture autogenous vaccine against this disease. Moreover, it was demonstrated that non-virulent *C. perfringens* NetB toxin positive strain with an adequate adjuvant can be used as orally administered vaccine to promote intestinal immunity against pathogenic strains (Mishra and Smyth 2017). Interestingly, attenuated *Salmonella* vaccines carrying recombinant *C. perfringens* antigens have been successfully used as a potent protection against NE as commented by Jiang et al. (2015) in study with regulated delayed lysis vaccine, as well as in newer study with *Salmonella*-vectored vaccine delivering three *Clostridium perfringens* antigens by Wilde et al. (2019). Two of the above-mentioned antigens were toxoids (α-toxin and NetB toxin), the third was fructose-1,6-bisphosphate aldolase (Fba), a metabolic enzyme. Above data indicates that exist potential for vaccine that can protect against *C. perfringens*-induced necrotic enteritis, which might be further-improved. Promising seems also possibility of immunisation of broiler chickens with five newly identified surface-exposed proteins unique to *Clostridium perfringens*, confirmed by PCR and sequencing and purified. Among them, two hypothetical proteins, two candidates were prepilin proteins (one of them P 509 with highest antibody titre) which are predicted to form the subunits of a pilus structure, and one candidate was a non-haem iron protein (Heidarpanah et al. 2023). Among new approach belong studies on combination of vaccination together with support of probiotics (Redweik et al. 2020).

6.4 Collibacillosis

Escherichia coli is a part of the common microflora of the intestinal tract of poultry as well can be found throughout upper respiratory tract and is found also on the skin and feathers of healthy birds (Nolan 2020). Within the bird's body *E. coli* is commonly found on the mucous membranes, but only some strains have specific virulence factors that can cause serious diseases. It is an avian pathogenic *E. coli* (APEC). APEC-induced colibacillosis affects all age groups of poultry. Virulence of the strain together with birds' health and immunostatus as well as stress conditions are those mainly influencing severity of the disease. The clinical signs of the disease are of high diversity in contrary to mammals, in which *E. coli* causes mainly intestinal problems. Poultry affected with APEC infection typically extraintestinal local or systemic diseases occur. In the acute stages of the disease with high mortality, we can detect mostly colisepticaemia or severe respiratory tract damage in affected individuals—airsacculitis, which later turns into a systemic infection with characteristic fibrinous lesions (perihepatitis, pericarditis). Among broad spectrum of APEC caused syndromes belong colisepticaemia, haemorrhagic septicaemia, coligranuloma (Hjarre's disease), airsacculitis (chronic respiratory disease, CRD), swollen head syndrome, polyserositis, enteritis, peritonitis, coliform cellulitis, salpingitis, orchitis, venereal colibacillosis, osteomyelitis/synovitis, panophthalmitis and omphalitis/yolk sac infection (Nolan 2020; Sargeant et al. 2019). *E. coli* strains causing extraintestinal infections are called ExPEC and most APEC can be classified into this group (Nolan 2020). Colibacillosis can develop as a primary infection, but can follow or concomitantly complicate, as secondary post-viral infection or bacterial ethiology diseases (Sargeant et al. 2019). Among viral diseases that can influence host susceptibility to further *E. coli* infections, especially in chickens belong (indexing mean availability of commercial[1] and/or autogenous vaccine[2]): adenoviral infections (type1)[2], low pathogenic avian influenza (e.g. H9; H3[2]), avian metapneumovirus[1], chicken infectious anaemia[1], infectious bronchitis[1,2], infectious bursitis[1], infectious laryngotracheitis, Marek's disease[1], Newcastle disease[1,2], reovirus[2]. Among bacterial diseases that can be present either primarily or concomitantly with *E. coli* caused pathologies belong (indexing mean availability of commercial[1] and/or autogenous vaccine[2]): *Mycoplasma*[2] *(M. gallisepticum*[1,2]*, M. meleagridis*[2]*, M. synoviae*[1,2]), *Bordetella avium*[2]*, Campylobacter jejuni, Clostridium perfringens*[2]*, Erysipelothrix rhusiopathiae*[2]*, Pasteurella multocida*[1] (USKVBL 2021; EMA 2021; Anicon 2021). There can be reasonable to consider timely administration of vaccines protecting against selected above-mentioned viral or bacterial diseases, based on consideration of the risks for the exact production type and breed in certain conditions (above was therefore indicated if vaccines are available). Another contributing factor can be presence of either protozoa (especially coccidia as *Eimeria*—e.g. *E. brunetti*, *E. tenella*) or parasites as *Ascaridia* (*A. dissimilis*, *A. galli*).

One health perspective perception is that poultry meat and eggs can be source the ExPEC as there exist a close genetic relationship among APEC and ExPEC, which can also cause urinary tract infections (as so called uropathogenic *E. coli*, UPEC) in humans (Manges 2016; Jørgensen et al. 2019; Mortensen et al. 2019) and meningitis in neonates (neonatal-meningitis *E. coli*, NMEC), and are also associated with bacteraemia, sepsis, cellulitis and other potentially fatal infections (Mellata 2013). These strains bear various virulence-associated genes (encoding adhesins and colonisation factors, invasins affecting internal organs, iron acquisition and protecting against host's immune responses). APEC, UPEC and NMEC generally form genetically distinct groups, however overlap exist especially within the B2 phylogenetic group comprising both human and chicken isolates. High virulence in the chicken infection model as well as in the neonatal rat meningitis model were proven and low or no host specificity (Zhu Ge et al. 2014; Tivendale et al. 2010; Moulin-Schouleur et al. 2007). Important and potentially zoonotic APEC were detected, as, e.g. ST117, ST354 and ST95, showing several molecular features typical for human ExPEC in recent study performed by Papouskova et al. 2020. Moreover, conjugative plasmids being able of transmission from APEC to other pathogens of human health importance were detected, some of them harbouring genes for resistance to antimicrobial, heavy metals and disinfectants (Johnson et al. 2005). Resistance to antimicrobials detected in faecal isolates of *E. coli* from broilers and turkeys' farms frequently using antimicrobials than in layers with less use (Van den Bogaard et al. 2001). Of great importance is resistance to the critically important antimicrobials—e.g. third generation cephalosporins found on plasmids isolated from human patients suffered by urinary tract infection and retail chicken meat, where human clinical isolates harboured IncK/blaCMY-2 plasmid variants highly similar to the IncK/blaCMY-2 plasmid present in the poultry (Berg et al. 2017; Tofteland et al. 2017).

Strains isolated during clinical episodes of the disease, based on the overview made by from available bibliography (references listed below) reporting data for *Gallus gallus* including laying hens from several European countries most frequently reported belong to serotypes O1, O2, O35 and O78, with more sporadic, local, occurrences of other serotypes: O111; O129; O88; O8 and O18. In layer chickens with colibacillosis and peritonitis in Italy was detected among the others also O111 serotype (Zanella et al. 2000), as well as in Greece (Koutsianos et al. 2017), moreover in Italy in laying hens was also reported O129 serotype being dominant in some flocks (Camarda et al. 2008), while another work from Northern Italy reported, among others also O88 serotype (D'Incau et al. 2006). In the United Kingdom was, as the most prevalent serotype in layers, found strains O78 (McPeake et al. 2005), while in a German study on colibacillosis in layers and broilers identified O2, O78 and O1 as the most prevalent serotypes (Ewers et al. 2004). Recent work of Czech authors primarily targeted on genomic analysis of spread of virulent APEC clones, as well as of zoonotic potential of specific poultry-associated lineages including those with multidrug resistance to antibiotics, detected O1, O8, O18 and O78 serotypes in 49 isolates, with predominant serogroups O1 (30; 32%) and O8 (13; 14%); four isolates belong to O78 and two to O18 serogroups (Papouskova et al. 2020). Additionally, serotyping of a collection of 460 isolates from chickens, turkeys and ducks identifying six serogroups (O1, O2, O5, O8, O18 and O78) accounted for 56.5% of the APEC isolates and 22.5% of the non-pathogenic isolates (France, Spain and Belgium between 1992 and 2000) was described by Schouler et al. (2012)).

Whole specific chapter could be written on virulence factors, unfortunately there is not exactly established simple set of virulence factors that ensure that *E. coli* which was collected from the sample in the poultry affected by colibacillosis is APEC. This is due to variability in size, number and virulence genes harboured on plasmids both of APEC isolates and isolates from apparently healthy birds (Doetkott et al. 1996). The potential virulence factors of APEC can be separated into the following categories: capsule, lipopolysaccharides, flagella, adhesins, iron acquisition systems, protectins, invasins and toxins (Christensen et al. 2020). Some "minimum set" of virulence genes (identified e.g. via multiplex PCR) localised on plasmid, operons or specific regions, which presence indicate APEC is shown in the Table 11 below (Skyberg et al. 2003;

Table 11 Selected genes encoding proteins and pose properties of virulence in APEC

Gen	Characteristic or protein encoded by gene
*hly*F	Avian *E. coli* haemolysin (also with *hly*D transport gene of the α-haemolysin operon)
tsh	Temperature sensitive haemagglutinin
*omp*T	Protease able to cleave colicin
iss	Outer membrane protein
*iut*A + *iut*C	Ferric aerobactin outer membrane receptor + aerobactin synthesis encoding
*iro*N	Catecholate siderophore receptor
*cnf*1 (*cnf*2)	Cytotoxic necrotising factor 1: altered cytoskeleton, necrosis
cbi	Immunity ColB operon
fim (A, H)	Fimbrial adhesins
pap (A, C, G)	Importance for adhesion, pilus function and assembly, pilus operon located

Johnson et al. 2008; Jeong et al. 2012). Compilation or even more broad list of genes, their location (chromosome/plasmid), function and source reference can be found in the compilation article by Ovi et al. (2023).

Strategies of APEC reduction in the whole production pyramid can follow the principles and procedures already established for salmonella, having acknowledged that high levels of biosecurity salmonella measures do not eliminate colibacillosis and giving a signal, that *E. coli* will need a plan on its own with specific control points (Christensen et al. 2020). Among the others from the long-term philosophy genetics (breeding for resistance to disease (Cheng et al. 2013; Rautenschlein et al. 2020) as well as pressure on the fattening speed should be re-considered. Leveraging host genetics to improve disease resistance can be an effective and sustainable way to address the APEC caused disease challenges (Monson and Lamont 2021). Selection for increased resistance in the pure lines and commercial hybrids possibilities has been already discussed (Cheng et al. 2013; Rautenschlein et al. 2020). Prospective strategies for improvement of host genetic resistance to APEC have been discussed recently by Monson and Lamont (2021), indicating that the estimated heritabilities of traits together with association with the host response and the results of genomic studies suggesting the genetic control as partial, complex and highly polygenic and that genetic improvement should be involved as a part of comprehensive disease-control programmes together with biosecurity, vaccination and immunomodulation. As touched above also slower growing breeds and well-balanced technologies linked to it, lower the pressure on disease development as indicated by several experiences. Slower-growing breeds make up nearly 40% of broiler production in the Netherlands, 24% of broiler production in France and 11% of production in the United Kingdom (Davies 2019) and those percentages have been even grown up since 2019 (Singh et al. 2021). For example, in Denmark: Danpo, the largest Danish chicken processor, had pledged to phase out broiler chickens by the end of 2021, switching from Ross genetics to the slow-growing Ranger Gold (Eurogroup for Animals 2020) or the Netherlands with similar approach going for slow-growing breeds. Development of the balanced feeding programme is of importance to allow for all body parts to grow proportionately to help keep natural behaviours and movement. For this purpose, e.g. RSPCA approved farming scheme standards for such slow growing breeds (RSPCA 2020). Another study brings the set of comparisons faster growing vs slower growing chicks (Davies 2019), amended with the data from CIWF (2020) conclusion that there exists the evidence that a combination of reducing stocking density and increasing age at slaughter (i.e. slower growing poultry) reduce the need for antibiotics.

Series of factors can influence, with might be shorter time perspective philosophy, effective control of colibacillosis. The first step is recognising the initial predisposing factor that open the doors for entrance of the infection. There have been already mentioned above in this chapter predisposing bacterial and viral diseases as well as examples of predisposing pressure due to protozoa or parasites. Also, other tools, which can lead to lower pressure from the occurrence of *E. coli* in the environment (feed, litter, water, dust etc.) are of importance.

But the strategies should be started in hatchery, because eggs contaminated with faeces/*E. coli* especially those of APEC characteristics (virulence potential) as well as antibiotic-resistant *E. coli*, including ESBL/AmpC were documented on eggshells (Mezhoud et al. 2016) are an issue. Hatching, handling with eggs as well as one-day-old chickens transfer, delivery and establishing the new environment is of critical importance (Table 14). Studies on one-day-old chicks shown them as potential reservoirs of multiresistant enterobacteria obtained vertically from breeders (Projahn et al. 2016; Projahn et al. 2018; Okorafor et al. 2019; Montoro-Dasi et al. 2020). Parent stock and hatchery as a source and eggshells as "vector" could serve as a source of multidrug-resistant bacteria and altogether resistance genes for day-old chicks (Projahn et al. 2018; Daehre et al. 2018; Osman et al. 2018). Special attention should therefore be paid to collect and discard eggs hatched on the floor, especially those with faecal contamination as well as damaged eggs (cracked shell). Even some hatcheries use the floor eggs and place them on the bottom of the hatching incubator, such eggs, especially once broken during incubation are source of infection

Table 12 Factors with proven or suspected enhancing insusceptibility to *E. coli* pressure and minimising the risk of colibacillosis outbreak (Nolan 2020; Hartady et al. 2021, modified)

Immunostatus	Supplements	Genetic/Physiologic
Immunity Passive Active	***Nutrients*** Balance of proteins	Breed (slowly raised)
Immunostimulans Phagocyte priming	***Vitamins*** A (incl β- carotene), D, E C	Good health status day-old chicks
Gut health including balanced microbiota	***Chemical elements*** (in bioavailable form) Iron- orally Selenium	Female
	Probiotics	
E. coli And balanced microbiota	***Herbal additives*** (examples): Fruit: *P. emblica* (Amla), *P. guajava* (Guava), *T. indica* (Tamarind), Leaves: *P. americana* (Avocado) *T. vulgaris* (Thyme) Mint, Onion Seeds: *P.americana* (Avocado) *Rhizome: C.longa* (Turmeric)	Mostly as bioactives inhibiting bacterial activity (*E. coli, Salmonella*) or promoting e.g. Lactobacilli in the gut. Phagocytic cell stimulation, Lactobacilli promoting, failure in gut colonisation with pathogens Inhibition of *E. coli* and *Salmonella* growth Bioactives - alkaloid, flavonoid, saponin, and tannin Bioactives - saponins, tannins, flavonoids, and terpenoid Delayed the growth of bacteria *E. coli* and *Salmonella* Inhibiting bacterial growth *E. coli* Phenolic compounds as well documented antimicrobial Lactobacilli counts increasing and inhibition of *E. coli*
	Essential oils	Antifungal activity

(through droplets, contaminating incubators, trails, crates as well as hatched chickens). Clean nesting facilities as well as egg-conveyor belts and hygiene/sanitation of hatching facilities (incubators) are of high importance. Proper airflow is of concern and hatcheries should be designed and constructed (including relevant filtering device) to eliminate transmission of both bacteria and viruses (Christensen et al. 2020).

Different technologies for sanitation used should be applied within 2 h from hatching. Formaldehyde fumigation has been broadly used, but is controversial due to human occupational exposure of this hazardous compound. Therefore, several disinfection agents and procedures have recently been investigated and compared as sanitisers with electrostatic spraying technologies, ultraviolet irradiation, low energy electron irradiation, quick burst of steam delivered at ultrasonic frequency, hydrogen peroxide, ozone and peracetic acid and plasma as the alternatives to formaldehyde fumigation (Melo et al. 2019; Tebrün et al. 2020; Christensen et al. 2020).

Attention should be undoubtedly paid also to other environmental/technological factors. Firstly, despite the fact that this is the costliest step, the thoughts should be paid to old housing facilities and critical assessment with questioning, if proper cleaning and proper setting of environmental conditions is feasible in these facilities. Whenever possible facilities should be reconstructed or newly build with consideration of all technological steps ensuring relevant biosecurity and hygiene. Everything starts with thorough mechanical cleaning and disinfection of housing facilities. Also drying off and keeping technological break among stocks' batches housed is vital for keeping environment as much as possible pathogen free. In colibacillosis, similarly as in other poultry diseases cannot be omitted—thorough balancing of temperature (especially extremes, or broad range of fluctuations are issue), avoiding of insufficient ventilation, overstocking, dry condition, poor litter quality, dusty environment, poor drinking water quality, biofilm formation of watering devices, restrictions of approach to water or feed, ammonium). Once all the factors needed for proper housing are fulfilled, day-old chickens can be placed into the housing, ideally under conditions monitoring their weight, appearance, health status, proper light regimen etc. newly housed flock should be thoroughly monitored as for behaviour, including vocalisation. Persons calmly going through the flock and watching the birds should also remove all chickens which seem weak, abnormal, with signs of sickness or dead. Such culling process can limit the possibility of horizontal spread of the disease. Christensen et al. (2020) referring to the unpublished data that have shown that systematic culling within the first week of life in broiler chickens minimise the problem of horizontal transfer and spread of colibacillosis and that "treatment with antibiotic is without effect".

Generally poor feed quality and either lacking or exceeding vitamin supplementation can negatively influence insusceptibility to pathogenic *E. coli* invasion (Vitamin A: hypervitaminosis or deficiency; Vitamin E: hypervitaminosis). Once the flock consists of normally developed birds, of good health status (including balanced intestinal microbiota), without defects and/or injuries (unhealed navel, wounds), appropriate density, low stress (including this environmental) conditions, immunocompetent (might be supported by selected types of vaccination against different primary diseases) the probability of resistance to naturally occurring *E. coli* strains, including most of them with virulence determinants, increase significantly.

Therefore, effective measures eliminating these predisposing cause(s) and support of the positive factors (see Table 12) minimise the risk of *E. coli* become a clinical infection with serious impact on birds' health, welfare as well as economic profit.

Environmental load minimisation via thermal inactivation (up to 90%) of *E. coli* in litter can be achieved, based on properly balanced temperature, duration, being more effective in low moisture and more rapid in ammonia presence (Himathongham et al. 2000). Most APEC strains are also inhibited by pH less than 4.5, or exceeding 9, but even these low or high pH values are not bactericidal. Organic acids seem more effective (e.g. citric, salicylic, tartaric acid) for significant reduction of *E. coli* in litter. Unfortunately examining plasmids confirmed that some strains of APEC can harbour genes supporting tolerance to acids or bile, as well as disinfectants (Mellata et al. 2012). It also seems that viable *E. coli* are minimised by drying after 48 h, as a 24- or 48-h drying period for faecal matter on broiler transport cage flooring significantly lower bacterial numbers (*Campylobacter* and *E. coli*) on these surfaces (Berrang and Northcutt 2005). Also 265-nm UV-C LED light can significantly reduce viable load of *Campylobacter* and *E. coli* on transport crates for chickens (Moazzami et al. 2021). Problematic is minimising the environmental load via disinfection, as some APEC strains have been proven to bear plasmids conferring resistance to disinfectants. Examples of *E. coli* mobile genetic elements that can be transferred, e.g. by conjugation are, e.g. IncHI2 (pAPEC-O1-R) conferring resistance to potassium tellurite, silver nitrate, copper sulphate, benzalkonium chloride as well as antibiotics: streptomycin, tetracycline, gentamicin, sulfisoxazole (Johnson et al. 2006), another one is IncF (pAPEC-O2-R) that has been proven to encode resistance to quarternary ammonium salts, silver, other heavy metals together with resistance to antimicrobials (Johnson et al. 2005; Yang et al. 2015).

Colibacillosis control by using of veterinary medicinal products (VMPs) has been performed via prevention and treatment regimens using different classes of antimicrobials for decades. Recently, due to antimicrobial resistance and strengthening pressure on prudent use of antimicrobials, investigation of introduction of effective alternative strategies is in the centre of both scientists and pharmaceutical and feed/water additives and probiotics/prebiotics/phytoadditives producing companies as well as companies offering tools of new technologies. Some experience moreover exists in laying hens as the regulations on maximum residue levels in eggs for human consumption limited the antimicrobial VMPs available to poultry practitioners due to the withdrawal period established for eggs what makes the use of antimicrobials more costly (Ghunaim et al. 2014). Therefore, issue of immunisation of pullets/hens against colibacillosis with live, inactivated or sub-unit vaccines based on different technologies, was investigated as well performed with different outcomes in practice. The original presumption, that effective vaccination strategy against colibacillosis will be based on vaccination by either commercial or autogenous vaccines directly against *E. coli*, was replaced by more broad view considering also biology of the disease including primary infections—predisposing factors of colibacillosis. Namely airsacculitis and other respiratory tract affecting forms started to be reconsidered together with infectious bronchitis virus (Dwars et al. 2009). Studies performed to investigate virulence of both these pathogens, bacteria and virus, as well as effect on IBV vaccination on subsequent colibacillosis were cornerstones for implementing the thought how vaccines against concomitant primary disease can help to minimise occurrence of secondary *E. coli* infection (Matthijs et al. 2005; Matthijs et al. 2009). This leads to the study on day-old chickens that were vaccinated by IBV vaccine strain H120 by coarse spray or ocularly. It was proven that IBV vaccination reduced both the number of broilers with *E. coli* airsacculitis and the severity of airsacculitis significantly after challenge with IBV-M41 (28 days after IBV vaccination) and *E. coli* 506 challenge (28 + 5 days after IBV vaccine and IBV-M41 challenge). Vaccine administered ocularly reduced systemic infection, but did not protect against airsacculitis (Matthijs et al. 2005).

Vaccines targeted directly on the immunogenicity against *E. coli*, or better APEC is difficult to produce as genetic diversity is vast, therefore early attempts were proven to be effective against homologous strains only (Fancher et al. 2020). Live attenuated, inactivated or sub-unit vaccines were tested for the immunisation in different production types (pullets or hens). Studies (outside to those involved in authorisation of the immunological veterinary medicinal product) exist that live vaccines have been repeatedly proven to contribute positively to bird protection performed by La Ragione et al. (2013) and Mombarg et al. (2014). One study report indicated moderate immunogenicity and serogroup-specific protection of O78 delta gale, delta purA and delta aroA mutants of APEC (Kariyawasam et al. 2004a).

An attenuated *E. coli* live vaccine can be used in broiler chickens and adult animals. An inactivated vaccine can be used for the prevention of colibacillosis in the offspring of breeding animals with twice administrations during the rearing period. If recurring problems on the farm occur (AMCRA-mentioned scenario is more than half of the flocks among the last six installed required treatment for colibacillosis), vaccination is preferred, both for fattening chickens and for adult animals. Van Goor et al. (2017) evaluated a recombinant antigens (rAg) vaccine containing common ExPEC surface proteins combination EtsC, OmpA, OmpT and TraT against APEC infections in chickens. Decrease of bacterial loads in heart and spleen, as well as drop down of gross colibacillosis lesion scores in air sac, heart and liver, together with immune-related mRNA expression in the spleen and bursa and elicitation of IgY was proven in this experiment (Van Goor et al. 2017). It also seems, that could be also back the concept of so-called Bacterial Ghost (BG) platform (Langemann et al. 2010), which was experimentally proven with over 90% immune protection (targeting O2 serotype) with antibody levels highest in the BG immunised group, which also had no pathological lesions associated with colibacillosis (Hu et al. 2019). Studies at commercial levels, with different serotypes should be performed to proof effectiveness in large-scale settings.

Live vaccine, commercially produced was authorised in 2012 by European Medicines Agency and since this time has been available in EU region (EMA 2013). This vaccine contains an attenuated O78 *E. coli* strain in lyophilised form with a gene (aroA) deleted. This vaccine is intended for use in chickens with administration as a single dose either by coarse spray vaccination onto the birds (which consume the vaccine when preening this feathers) from 1 day of age. Another possibility is administration via drinking water from 5 days of age. In turkeys it is given by coarse spray vaccination from 1 day of age with a second dose 3 weeks later. When adding this vaccine into drinking water a sufficient amount to supply one dose of reconstituted vaccine for every chicken should be added to as much water as is consumed by the chickens in 3 h. After vaccination via spraying is declared 8 weeks protection against colibacillosis lesions and 12 weeks protection against mortality from colibacillosis in chickens, 12 weeks protection against lesions and mortality is declared with drinking water vaccination, e.g. in commercial vaccine Poulvac *E. coli* (EMA 2013). In another vaccine approved in EU since 2014 is declared partial passive immunisation of broiler chickens during their first 7 weeks of life by vaccination of the broiler breeders. It covers postnatal colibacillosis (air sac disease and septicaemia) caused by fimbrial F11-antigen and flagellar Ft-toxin containing *E. coli*. This vaccine is administered via intramuscular or subcutaneous injection in broiler breeder hens. Vaccination scheme contains: two doses, with an interval of at least 6 weeks. First vaccination at 6–12 weeks of age, revaccination at 14–18 weeks of age (VMD-DEFRA 2020). Inactivated vaccines can also be effective against colibacillosis (Gomis et al. 2007; Gregersen et al. 2010). There is declared effective protection upon challenge with O78 wild strains. Generally, as commented by (Sadeyen et al. 2015) vaccines with killed or attenuated strain of bacteria protect against infection with the homologous strain, but less efficiency against heterologous strains cannot be excluded in them. It should be also mentioned autogenous vaccines. Those are produced from pathogenic strains isolated from within the affected flock or epidemiologically related (e.g. parent flock). The vaccine should be then used only in these flocks. Within EU exist several companies with high advanced technologies with Good Manufacturing Practice, that are able to produce such vaccines. Autogenous vaccines started to be widely used in the field for colibacillosis control (Hera and Bureš 2004; Landman and van Eck 2017). Recently study targeted on the assessment of protection rate against colibacillosis conferred by several vaccination programmes in conventional pullets were carried out (Koutsianos et al. 2020). Within this study, a live commercial vaccine and/or an inactivated autogenous vaccine were used. Assessment and comparison of two applications of an autogenous vaccine consisting of O78, O111 and O18 strains and two applications of a live commercial vaccine combined with a single application of the autogenous vaccine was performed. Significant positive effect against intratracheal challenge with either the O78 or O111 strain was confirmed. Moreover, O18 challenge results lead to the idea that combining both vaccine types heterologous protection can be extended (Koutsianos et al. 2020).

Passive immunisation via using inactivated vaccines in breeders provides passive protection against homologous challenge in progeny. It was proven that complete immunisation of chickens from these breeders last 2 weeks and partial immunisation even longer (Heller et al. 1990). Egg-yolk antibodies (purified immunoglobulins IgY) induced by immunising hens using selected *E. coli* antigens (lipopolysaccharide (LPS), type 1 pilus adhesin (FimH), P pilus adhesin (PapG), aerobactin outer membrane receptor (IutA) from APEC O78) was studied in broiler chickens as for protection potential against respiratory/septicaemic disease caused by APEC. Intramuscular administration of 100 mg purified IgY to 11-day-old broiler chickens was used and challenge 3 days later with homologous (O78) or heterologous (O1 or O2) *E. coli* by the intra-air sac route was performed. Except FimH, protectivity (90–100%) of all antigens against the homologous challenge was proven, but only anti-PapG and anti-IutA were effective against heterologous challenge. All the above serogroups of *E. coli* were covered by protection of *anti-PapG IgY.* Hence vaccination of broiler breeders to induce anti-PapG and anti-IutA antibodies may provide passive protection of progeny chicks against respiratory/septicaemic disease caused by APEC (Kariyawasam et al. 2004a, b). There has been also described immunopotentiation in the cases of co-administration of antigen together with modified *E. coli* non-toxic heat-labile enterotoxin (nLT), which once administered orally or parenterally stimulated antibody production.

Treatment with antimicrobials is complicated by resistance and in a lot of APEC isolates even multiresistance to tetracyclines, the combination trimethoprim + sulfonamides and amoxicillin frequently occur (Prátová 2020). Antimicrobial-resistance profile ranged from full susceptibility to all antimicrobials tested to multidrug resistance, with dominating resistance to β-lactams (ampicillin) and first-generation quinolones (nalidixic acid) (Papouskova et al. 2020, Sgariglia et al. 2019). Specific attention should be paid to the substance with promising effect introduced decades ago in poultry medicine—fluoroquinolone enrofloxacin. Unluckily, the resistance to this substance either in low level as well in high level, caused by chromosomal mutations as well as plasmid-mediated quinolone resistance (PMQR) are backgrounds for the issue of decreased susceptibility/increased resistance. In Czechia (Prátová 2020) in samples of *E. coli* both from inner organs and from GIT ($N = 124$) 51% were sensitive, 20.2% intermediary and 28.2% resistant to fluoroquinolones (using CBP $S < 0.25$ mg/L; $I = 0.5–1$ mg/L; $R > 2$ mg/L), another study by Nedbalcová et al. (2020) APEC (PCR confirmed virulence genes *iut*A, *hly*F, *iss*, *iro*N, *omp*T and *fim*C) shown 53.75% resistant, 28.75% intermediary and 17.5% fluoroquinolone-resistant isolates. Isolates coming from Italy (Sgariglia et al. 2019) were detected as resistant to fluoroquinolones in 18%. Approximately 21% of intermediary susceptible to fluoroquinolones and 10% of fluoroquinolone-resistant APEC isolates was determined in Flanders (2017/2018, $N = 125$-one isolate per one farm). These above isolates within work of Temmerman et al. (2020) were further analysed and it was confirmed that chromosomal mutations in DNA gyrase and topoisomerase IV was the main source of decreased susceptibility to fluoroquinolones, what de-emphasise the role of PMQR mechanisms. Point mutations in genes encoding for DNA gyrase (gyrA and gyrB) and genes encoding for DNA topoisomerase IV (parC and parE) are of importance, as both gyrase and topoisomerase play an essential role in the remodelling of DNA topology and DNA replication (Redgrave et al. 2014). The regions of DNA sequences where these target-site mutations rise are well known as the QRDR = quinolone resistance determining regions (Yoshida et al. 1990). PMQR is an issue due to gene located on plasmids with spread also horizontally. Proteins translated from these genes play role in target protection (qnr A, B, C, D, E, S), reducing fluoroquinolone efficacy (aac (60)-Ib-cr) and efflux mechanisms (oqxAB and qepA). Moreover, fluoroquinolones are ranked by different international bodies as critically important antimicrobials for human medicine (WHO). Because of association with fluoroquinolone-resistant food-borne zoonotic bacteria such as *E. coli*, *Salmonella* and *Campylobacter*, currently veterinary use is put into the thorough scrutiny (EMA 2019), Walsh and Fanning 2008). Resistance in poultry sector varies across regions, but practically worldwide significantly increased due to the long and intensive use in poultry sector in some regions. As documented in review of the fluoroquinolone resistance made by Vanni et al. (2014) rates of *E. coli* isolated from broilers around the globe (Europe, the USA and Asia). The prevalence of resistance to enrofloxacin varied greatly: from 16% in the USA to 90% in China (causes could be both local intensity of use and methodological differences between the studies). Unique study was performed by the team from University of Ghent (Temmerman et al. 2020) (N = 125, APECs, isolated November 2017–June 2018, Flanders, one strain per farm). Most prevalent serotype was O78 (19%), followed by O2 (8%) and O1 (5%), 68% APEC strains non-identifiable using the three antisera. Gradient strip and agar dilution were used for the minimum inhibitory concentration (MIC) and the mutant prevention concentration (MPC) testing. Strains were characterised by PCR, gel electrophoresis and gene sequencing. APEC strains with MIC > 0.125 μg/mL (classified as non-wild type (NWT)), created 40%, from them 18% harbour PMQR-associated genes. Comparing with clinical breakpoints, 21% of isolates were classified as clinically intermediate ($0.5 \leq$ MIC ≤ 1 μg/mL) and 10% as clinically resistant (MIC ≥ 2). One or two mutations in *gyr*A was detected in majority (92%) of the NWT strains, in 22% topoisomerase IV parC subunit was affected, but three strains only carried a mutation in *par*E. The most common MPC value for susceptible strains was 0.125 μg/mL. Despite it is more than promoted responsible and minimised use of fluoroquinolones the above results from Temmerman and colleagues (2020) study bring the thoughts that approved dosage together with monitoring of drinking behaviour ensuring intake of the sufficient dose still can keep enrofloxacin as the last resort, but very potential antimicrobial, especially considering its excellent pharmacokinetic properties. Therefore, in the case that dosing of 10 mg/kg b.w./day that was considered by the Committee for Medicinal Products for Veterinary Use (CVMP) at European Medicines Agency as possibly accelerating development of bacterial resistance to enrofloxacin, which would limit its use as treatment of last resort for colibacillosis in poultry, we should think about higher dose. More PK/PD tailored—higher dose exceeding MPC and reaching target site of infection can be still considered as useful. The CVMP conclusion that enrofloxacin should no longer be used for treatment of *E. coli* infections in chickens and turkeys (CVMP 2018) might be in the future revisited, if sufficient evidence occurs supported by responsible use PK/PD tailored regimens and prudently defined conditions limiting such use as exceptional, for thoroughly justified and limited cases.

There should be also mentioned one phenomenon, which is, especially in some countries an issue- resistance to beta-lactams as well as presence of extended spectrum beta-lactamases. Cephalosporin of third generation—ceftiofur was previously used in ovo as well as in day-old chickens in some countries in Europe. The resistance of broiler isolates of *E. coli* to cefotaxime (also third generation cephalosporin) increased since 2003 to more than 20% in 2007 in the Netherlands and sharp decrease was observed here after the ban on the use of ceftiofur in hatcheries in 2010 to reach a level of 2.9% in 2014 (Bennani et al. 2020). Despite the fact that cephalosporins are not authorised and should not be used in poultry, molecule that can select for some types of extended spectrum beta-lactamases can be amoxicillin, which is widely used in other cases co-selection by other antimicrobials, including fluoroquinolones. Resistance to beta-lactams was associated largely with TEM-type beta-lactamase production (Papouskova et al. 2020). Genes *bla*CMY-2 along with *bla*CTX-M-1 are the most common ESBL/AmpC beta-lactamase encoding genes detected in poultry *E. coli* isolates (Van Hoek et al. 2016). Co-occurrence of *bla*-CMY-2, *bla*-CMY-59 and *qnr*S-1 was observed also in isolates from Czechia (Papouskova et al. 2020). Genes of ESBLs/AmpC harboured together with other genes of resistance to antimicrobials located on plasmids, e.g. IncX3: bla_{SHV-12}, *qnr*-S1 were described in isolates of *E. coli* form chicken broilers gastrointestinal tract (Apostolakos et al. 2020). It means that despite the fact that use of some antimicrobials is not allowed in poultry sector, the resistance genes still can be kept in population via co-selection. Such strains can serve as a source of resistance determinants in horizontal gene transfer.

Somehow controversial is the study of Mathers et al. (2004) on inhibition of multiresistance-conferring plasmid pBR325 (Tet, Amp, Cp, 6.0 kb) into selected Gram-negative strains with the use of an in vitro model. Inhibition 80.6–99.9% (*P*, 0.001) of resistance transfer in *E. coli* by ionophore anticoccidials (lasalocid and salinomycin) and ionophore/antimicrobial combinations (with chlortetracycline, bacitracin, lincomycin, tylosin, virginiamycin and chemical anticoccidial roxarsone) was reported, but under in vitro conditions only—therefore need of further investigation is of the need to evaluate impact on clinical practice (also in the light of the recent studies on *E. faecium* by Naemi et al. 2020 on narasin, salinomycin and maduramycin and co-selection of vancomycin resistance).

In the cases, where no above-mentioned preventive measures succeed and whenever possible after known situation on susceptibility, especially due to the high prevalence of resistance as well as possibility of occurrence of multidrug resistance, it is necessary to start treatment or metaphylaxis using antimicrobials. For such cases is of extreme importance to ensure all factors as listed in Table 8 in this chapter to ensure maximally intake of the recommended dose via medicated water. Attention should also be paid to the following of all instructions in the respective individual veterinary medicinal product SPC (summary of product characteristics) not only to achieve concentration sufficient for the birds' treatment, but also to follow the instruction for mixing, for administration and for withdrawal period.

In exceptional cases, based on clinical examination, course of the disease in the flock and after through consideration of veterinarian, especially in antimicrobials where bacterial killing is predominantly concentration dependant, higher dose could be used, if thoroughly justified and also supported, e.g. by MIC knowledge. For such exceptional cases at the same time should be kept in mind necessity of increase of withdrawal period in line with legal requirement (see also above in this chapter). Antibiotics for the treatment of poultry disease have been reviewed by Löhren (2008), who also recommend to choose from active compounds not being critical for human medicine (currently classified as AMEG A and B categories as specified in EMA (2019).

Guidelines in some European countries (NL, BE) recommend as the first line sulphonamide/trimethoprim combination, which is bactericidal (AMCRA 2023; Boereboom et al. 2012). Both active ingredients are rapidly absorbed, steady state concentrations in plasma is available after 12 h from the start of medication onwards in poultry, it is widely distributed in tissues and have short elimination half-life 1.7 h for sulfamethoxazole and 1 h for trimethoprim. In broiler chicks sulfamethoxazole C_{max} approximately 9.0 μg/g and in trimethoprim C_{max} 0.12 μg/g were established. High concentrations are available in kidneys, liver and lungs. Concentrations of sulfamethoxazole in tissues are lower than in plasma, binding to proteins is not so high. Attention should be paid to drinking water pH, as the solubility and stability of the veterinary medicinal product in drinking water are pH-dependent. The pH of the product is in the high alkali range (Methoxazol 2020). Care should be taken as for proper concentration and willingness to accept water because of the bitter taste in higher concentrations, which can cause insufficient intake and dosing. As the second line antimicrobials aminoglycosides (apramycin, neomycin) with also bactericidal effect are recommended. The possible issue of these molecules is bioavailability in the systemic circulation after oral administration, especially when there are no or minimum lesions of intestinal mucosa. The mean concentration-time profiles of apramycin after a single oral administration to healthy chickens studied with the dose of 25 mg/kg b.w./day, the absorption half-life was 0.11 ± 0.001 h with t_{max} 0.18 ± 0.011 h and c_{max} 0.75 ± 0.03 µg/mL, very low bioavailability of apramycin around 2.50 ± 0.02% was detected, the plasma protein binding was 25.0 ± 2.45%. AUC determined was 0.83 ± 0.02 µg/mL/h. No systemic therapeutic effect could be expected for this drug after oral administration in broilers (Elbadawy and Aboubakr 2017). As for neomycin pharmacokinetic dose of 10–30 mg/kg b.w. is recommended for chickens, turkeys and ducks (EMA 2022), low level of absorption has been described after oral administration (up to 10%) and mostly excretion unmetabolised. Interestingly orally dosed neomycin (in combination with ampicillin, at dose 0.25 (AMP) and 0.5 g/L (NEO) in chicken can lead to a change in immune response to antigens (Murai et al. 2016).

Thanks to the better oral availability of doxycycline as well as the lower rate of resistance of *E. coli* to doxycycline, this molecule is preferred over other molecules from the tetracycline group. Bioavailability of doxycycline in broiler chickens is approximately 41%, after oral administration. In bioequivalence studies of doxycycline in chickens, it has been found that after a dose of 20 mg/kg b.w. administered as a bolus solution, serum concentrations ranged from 0.65 to 0.7 µg/mL, and from 0.25 to 0.35 µg/mL, 12 or 24 h later, respectively (Soliman et al. 2015). Within another study evaluated doxycycline administered at a daily dose of 25 mg/kg body weight for 3 days was shown that doxycycline effectively improved the clinical performance of the chickens and reduced gradually mortality (EMA 2009). For some cases treatment duration is recommended to be prolonged to 5 days. The most recent study of doxycycline pharmacokinetic after oral administration to broiler chickens of Hubbard × Hubbard, both sexes, 35–40 days old, b.w. ranging from 1.5 to 1.7 kg, to which doxycycline solution was administered by oral gavage at dose 20 mg/kg b.w./daily for a total of five doses. Maximum plasma concentration C_{max} 5.65 ± 1.33 µg/mL (first dose) and 5.71 ± 2.11 µg/mL (last dose) was achieved at time T_{max} 3.50 ± 2.06 h (first dose) and 7.25 ± 4.48 h (last dose). Apparent elimination time $t_{1/2}$ was 7.78 ± 1.37 h (first dose) and 19.90 ± 8.46 h (last dose). After first dose, established $AUC_{0\text{-}24h}$ was 80.38 ± 17.57 and after last does 87.29 ± 28.86 µg/h/mL was achieved. Mean residence time was 13.02 ± 2.07 h (first dose) and 21.51 ± 3.39 h (last dose) (Yang et al. 2018).

As mentioned above quinolones, e.g. flumequine or nalidixic acid resistance or reduced susceptibility has been observed in avian isolates of *E. coli* in high extent, what compromise quinolones use.

As last resort antimicrobials are considered aminopenicillins that may promote the selection of organisms that produce "Extended Spectrum Beta-Lactamases" (ESBLs) or AmpC. Aminopenicillins pharmacokinetic properties are also of high importance. Bolus dose of amoxicillin trihydrate in amount of 20 mg/kg b.w. was orally administered to 6 weeks' old, Cobb 500 female chickens, weighing 2.5 kg ± 20 g and lead to maximum serum concentration C_{max} 12.54 ± 0.80 µg/mL at time T_{max} 1.39 ± 0.12 h (Ledesma et al. 2018). Mean residence time established was equal to 2.78 ± 0.04 h and $t_{1/2}$ was 0.96 ± 0.10 h. From these parameters value of AUC was constructed as 44.02 + 0.61 µg/h/mL. Fr = relative bioavailability ($AUC_{interaction}/AUC_E$) * 100 was absolute. Time above or equal to MIC was 6.00 + 0.01 h (MIC considered in the study was for *Pasteurella* spp. at the level of 1 µg/mL). These 6 h for $T \geq$ MIC are of critical importance, as amoxicillin bacterial killing effect is considered as time dependent practically without PAE. Importance is even more pronounced for Gram-negatives, where 80–100% of the interval antimicrobial's concentration should be kept above or equal to MICs of the target pathogen in target tissue in order to obtain optimal clinical results (MacGowan 2011). Also, maximum concentration (C_{max}) as a key parameter is recommended to be 4–5 times the

MIC value (McKellar et al. 2004). Work of Krasucka and Kowalski (2010) reached similar pharmacokinetic parameters after oral dosing of amoxicillin trihydrate 20 mg/kg b.w. to 35 broiler chickens of Hubbart breed, indicating that T_{max} was achieved after 2 h and after 6 h concentration declined below MIC, so $T \geq$ MIC was only 8.33% of 24 h interval. In another work made by Kandeel (2015) amoxicillin of 10 mg/kg was administered orally to healthy chickens and chickens infected with caecal coccidiosis offered a $T \geq$ MIC that ranged from 8.3% to 25%. From the above results is clearly visible, that veterinary medicinal products dosed once, or even twice daily at dose of 20 mg/kg b.w./day cannot pose effective dosing regimen for several pathogens, which MICs exceed 1 μg/mL.

Fluoroquinolones, considered among critically important antimicrobials both by WHO and OIE are listed among last choice, but issue here is also decreased susceptibility or resistance, see also comments above referring to dose and concentration dependence. Data on enrofloxacin pharmacokinetic in broiler chickens are mentioned and commented in the chapter "Selecting Antimicrobial Drugs and Dosing Schedules Optimisation Using PK/PD Characteristics" within this book.

Use of colistin, despite it is still approved for indication of non-invasive *E. coli* in broilers in Europe should be avoided, especially due to the concerns with plasmid transferable resistance encoded by *mcr* genes. If used, it should be considered, that after oral administration it is not absorbed from gastrointestinal tract and therefore systemically unavailable in concentrations necessary for treatment of extraintestinal *E. coli*.

As *E. coli* are common colonising bacteria, for obtaining right sample from clinically diseased animals it is of importance the questions when and how to do sampling to find and confirm APEC presence. Study by Mageiros et al. (2021) worked, among the other with 309 isolates, which were considered as disease-associated APEC and were sampled in extraintestinal site of isolation, and/or symptoms in the bird at autopsy (35 were isolates from the bone marrow, 70 from the liver (perihepatitis), 23 from the heart (pericarditis), 40 from the peritoneum (peritonitis), 26 from blood (septicaemia), 19 from yolk sac infections and 96 from various other infection sites).

Samples for culture should be taken immediately after death or euthanasia of affected organs (If this is not possible, the birds should be frozen). During the full post-mortem analysis, organs suspected of bacterial infection (liver, heart sac, spleen, joints, bone marrow) for septic cases should be sampled. Depending on the clinical form, germs can also reside and be detected in the joints, oviducts, yolk remains, pulmonary alveoli. Sampling of intestine should be avoided as *E. coli* is a normal resident of the intestine and further serotyping as well as virulence factors PCR detection is needed to determine whether it is APEC or not. However, it is extremely difficult to ensure that the *E. coli* culture obtained from intestine sampling, contains really pure APEC. If the pericardium is affected, it is advisable to take a sample from this compartment (to avoid contamination, e.g. from the intestine, compared to the liver, kidneys or spleen). The pericardium should be incised with a sterile scalpel. Each sampling should be carried out as aseptically as possible, with a sterile swab twisting, after penetrating into target site. Sterilisation of the surface with a hot iron spatula/scalpel to limit surface contamination should always be done. In young birds, technique using skin disinfection and careful but quick tearing down the skin, prior sampling of inner organs, can be used. Blood agar primary cultures are advisable, inspection after incubation should examine if E. *coli* pure culture is present. Storage of the swabs directly in a transport medium, preferably at +/− 4 ° C, and sent as quickly as possible to the laboratory.

Study made by Kariyawasam et al. (2004b) performed with artificially collected IgY from eggs obtained from hens under different vaccination regimens demonstrated that induction of the passive maternal protection of day-old chicks could occur and protect the chick, when *E. coli* was used for challenge compared to controls. This study documented that under experimental conditions vaccination of parents may transfer the immunity to the chicks. Within the recent experiments combined vaccination using live attenuated *E. coli* vaccine with autogenous vaccines was tested, resulting with promising outcomes expecting a synergy of protection (Koutsianos et al. 2020).

There exists big pressure on development of alternatives for the future, especially because in infections caused by *E. coli* in poultry there are used antimicrobials classified in category B "restrict". In "B" category are allocated antimicrobials with critical importance in human medicine, which use in animals should be restricted to mitigate the risk to public health (EMA 2019). Namely there is spoken about fluoroquinolones and quinolones and colistin. Also, amoxicillin is recently considered as more risky than in previous times (being in category "D" of low risky antimicrobials what create controversy and signing of divergent position by some member states to AMEG classification (AMEG 2020) being considered as substance able to select for and promote the spread of (or maintain the presence of) Extended-spectrum Cephalosporin resistance. When prevalence of Extended-spectrum Cephalosporin (ESC)-resistance in intestinal *E. coli* in animal productions from several EU countries was considered, there was identified 60–80% flock prevalence of ESC. Especially in poultry flocks it seemed originally as surprising as these last resort antimicrobials have not been licenced in most of EU countries more than two decades and therefore "on label" use cannot be expected. Despite not authorised some use has been reported in hatcheries via spraying eggs or injection *in ovo* years ago. But use of amoxicillin accounts in some countries approximately 25%, therefore contribution of amoxicillin to ESC selection should be taken into consideration and should be subject of confirmation via robust studies.

Among the most promising applications belong those involving the development of antimicrobials selectively targeting a gene, cellular process, or strain of choice, so called "precision antimicrobials" (Paharik et al. 2017). For this targeted action the pathogen sequence should be identified as a well-defined target and nucleic acid-based antibacterials, peptides, bacteriophage therapies, bacteriocins, and anti-virulence compounds could be these appropriate that can exclusively target disease-causing bacteria (Mageiros et al. 2021). For example, CRISPR-Cas technology can be used for sequence-specific bacterial killing through guided nucleases that recognise known DNA sequences (Kiga et al. 2020).

6.5 Infections Caused by Enterococci

Localised or systemic disorders of poultry caused by *Enterococcus* spp. can be associated with different pathologies and target sites affected in the birds' body. Septicaemia caused by enterococci has also been recognised as frequent disease of early life period and causing clinical morbidity and in different scale (0.5–50%) mortality (Borst 2020). After septicaemic phase infection then damage some specific part of blood and circulatory system, digestive tract, musculo-skeletal or nervous system. Based on proper sampling and culturing and isolation, particularly *E. cecorum*, *E. hirae*, *E. faecium*, *E. faecalis* and *E. durans* are found and recognised as ethiological agents (and can be differentiated from other streptococci using high sodium chloride concentrations and bile. One of the exemptions not able to growth on such media is *E. cecorum*, frequently isolated and with increasing causality of poultry disease. In RESAPATH (2021) reporting poultry enterococci isolated in 2019 there were 73% of isolates identified as *E. cecorum*, 21% as *E. faecalis* and 6% as *E. hirae*. *E. faecalis* has been most frequently associated with poultry diseases, especially endocarditis. Other enterococci associated with valvular endocarditis are *E. faecium* and *E. durans*. Arthropathic and amyloidogenic *E. faecalis* was reported as causative pathogens of AA amyloid arthropathy and concomitant systemic amyloidosis in layers (Landman et al. 1994) and in lesser extent, in broiler breeders (Landman et al. 1998; Steentjes et al. 2002). Ascites in hens can also be associated with *E. faecalis* (Huan Shu 1997) and with pulmonary hypertension in broilers (Tankson et al. 2001).

E. durans infection in young chickens has been associated with bacteraemia and encephalomalacia (Cardona et al. 1993). Another enterococcus associated with brain lesions (focal necrosis) in the young chicks is *E. hirae* (Devriese et al. 1991; Randall et al. 1993). *E. hirae* has also been found in cases of endocarditis in 4-week-old broilers (McNamee and King 1996) and diarrhoea in first week layer chicks (Kondo et al. 1997).

Except being an issue in poultry, especially some enterococci are also of importance from "One health perspective". Genetic relatedness was found between human *E. faecalis* isolates and those from poultry, suggesting a possible role of poultry in its spread to humans (Bortolaia and Guardabassi 2015). Also, other species that are either colonising, or causing infectious diseases in poultry were found in human. Within the SENTRY study in EU region *E. faecalis* (62.8%), *E. faecium* (32.8%) and also, occasionally *E. avium*, *E. gallinarum*, *E. durans*, *E. casseliflavus* and *E. cecorum* have been proven to be pathogens causing serious hospital acquired infections (HAI) in human (Pfaller et al. 2019). Case descriptions include endocarditis, septicaemia, peritonitis, and surgical site, skin or urinary tract infections in patients with significant comorbidities. Continuous decrease in the susceptibility to ampicillin and vancomycin was observed in human isolates from HAI (1997–2016, SENTRY study), in some region in the world up to 8% resistance to vancomycin was detected at the same time susceptibility >98% to daptomycin, linezolid, tedizolid and tigecycline—those mostly last resort human antimicrobials was proven. To date, waste majority of reported *E. cecorum* human isolates have been susceptible to glycopeptides as vancomycin and teicoplanin, beta-lactams, quinupristin/dalfopristin, linezolid as well rifampicin (Jung et al. 2018). It has come as an important question to confirm, if there exist any causally proved association among, e.g. broiler strains of enterococci and those isolated in human. Even not directly proven certain proportion of causal link of exact enterococci transfer among human and animals and vice versa, transposons were reported as dissemination tool for resistance genes. Cross link has been proven also to wild birds' population (Stępień-Pyśniak et al. 2018). Mostly spread are genes *tet*M and tetL (R to tetracyclines), *erm*A and *erm*B (R to macrolides), *aph*A-III (R to selected aminoglycosides), *msr* (A/B) with R to macrolides (Hegstad et al. 2010; Stępień-Pyśniak et al. 2018). Interestingly, despite the absence of antimicrobials, pheromone production was reported to induce a high-frequency plasmid transfer in *E. faecalis* (Hirt et al. 2018). Two species *E. faecalis* (namely ST 82 and ST16 lineages) and *E. cecorum* seem to be of importance for the transfer of resistance among poultry, food of poultry origin and human (Torres et al. 2018).

As mentioned above, one of the key importance in pathogenesis and having clinical significance in the last two decades is *Enterococcus cecorum*. Infections by this pathogen in broilers was first described in 2002 (Devriese et al. 2001; Wood et al. 2002). Prior to this time the bacterium was described as being normally resident as a member of the physiological microbiota of the intestinal tract of birds (Devriese et al. 1991) and mammals (Devriese et al. 1992; Devriese et al. 1994) as well as in human. It is unclear why broilers are increasingly suffering from clinical disease caused by this bacterium and why pathological processes raise and *E. cecorum* cross the gut barrier and cause sepsis and in older birds and other morbidities (see below description). Genetic analyses of *E. cecorum* demonstrate that strains with increased pathogenicity are genetically related and share several putative virulence genes (Jung et al. 2018), some hypothesis existed on immunosuppression, as well as coexistence of other pathogens as *E. coli* or *Eimeria* spp. increasing the crossing of the gut barrier rate (Jung et al. 2018). In 2020, selected farms in the Netherlands, necropsies on broilers, *Escherichia coli* was isolated in 89% of the necropsies and *Enterococcus cecorum* in 29%; in 18% of the cases, both *E. coli* and *E. cecorum* were detected (MAH 2020). Genome reduction in the conserved clinical genomes suggested better host adaptability in *E. cecorum* (Sharma et al. 2020).

Approximately 15 years ago *E. cecorum* has been implicated as a possible cause of disease in poultry, most times starting with septic, early phase and later on including spondylitis, vertebral osteoarthritis and femoral osteomyelitis, femoral head necrosis or bacterial chondritis in more and more cases (Jung et al. 2018; Jung et al. 2017). In chicks affected by *E. cecorum* usually mortality increase due to a combination of sepsis early in the growing period and dehydration and starvation of birds suffered from paresis and lameness appearing late in the growing period (Jung et al. 2018; De Herdt et al. 2008). Onset of the signs can be detected within the first 2 weeks, increase of mortality at week 3 of life (in the first 14 days being recovered from yolk sacs or spleens of asymptomatic in flocks that later developed skeletal lesions of *E. cecorum* at approximately 4–5 weeks of age (Wood et al. 2002). The work by De Herdt et al. (2008)) studied ten flocks in Belgium mentioned up to 7% increase in mortality starting from days 21 to 32, with significantly higher rate of the cocks than the hens affected. Other studies mentioned even higher mortality rate (up to 15%) as well as significant morbidity (up to 35%) (Borst et al. 2012; Robbins et al. 2012; Jung et al. 2018).

Isolates were obtained from lesions of osteomyelitis, femoral head lesions as well as, in pure culture, from gross lesions—fibrinous pericarditis and perihepatitis and splenomegaly. Laboratory diagnosis, especially in phase of sepsis is of importance to differentiate *E. coli* and *E. cecorum* caused cases. Typical pathology of later stages are caudal thoracic vertebrae compressing the spinal cord (resulting in the characteristic symmetrical paresis and paralysis typical for spondylitis cause by *E. cecorum*) and/or osteomyelitis of the proximal femur—being referred to as "femoral head necrosis" or "bacterial chondronecrosis and osteomyelitis" (Dolka et al. 2016; Jung et al. 2018). Synovitis and arthritis of the coxofemoral, stifle and hock joints can also occur with yellow-opaque exudate filling and distention of the joint space and in some cases accompanied also by strands and clumps of fibrin (Jung et al. 2018). Mild lesions of the free thoracic vertebra occur with bacterial colonisation of osteochondrosis dissecans lesions, progressive infiltration of heterophils and macrophages into degenerative, necrotic and fragmented articular cartilage and underlying bone. Vertebral bodies are most affected in this pathology and bacterial colonies can be only very rarely observed in the spinal cord itself (Jung et al. 2018).

Borst et al. (2014) proved the chicken embryo lethality assay being able to discriminate between commensal and pathogenic isolates of *E. cecorum*.

Despite *E. cecorum* shows differences comparing to *E. faecium* and *E. faecalis*, it is not an exemption among members of the genus *Enterococcus*, being resistant to multiple classes of antimicrobials. As resistance differs in time and geographically, attention should be paid to susceptibility testing of the samples from lesion and inner organs of clinically diseased animals and performance of antibiogram. Spectrum of antimicrobials to be tested in the cases of infections caused by *E. cecorum* includes aminoglycosides, beta-lactams, macrolides, streptogramins and tetracyclines (Jung et al. 2018). There was published the data 2019 for 333 isolates in French report RESAPATH (2021), which indicates almost all *E. cecorum* as susceptible to amoxicillin (97%) and gentamicin (high level concentration, 98%). The macrolide-lincosamide family and the trimethoprim-sulfonamides combination were on the other hand, less frequently active with 37–59% of susceptible isolates. Only 10% of *E. cecorum* were susceptible to tetracycline.

Prevention of the *E. cecorum* as disease cannot rely on commercially produced vaccine yet. However, inactivated flock-specific vaccines can be administered in broiler breeder flocks, although no studies have investigated whether maternally derived antibodies can protect progeny from enterococcal spondylitis.

Being mentioned already above, also other species of the genus *Enterococcus* can cause infections of poultry known to be sporadic causes of central nervous system disorders, but most frequently in recent years, lameness, arthritis and osteomyelitis have also been observed regularly, starting at the age of 14 days. Metaphylactic use of antimicrobials, especially on farms, where an issue of *Enterococcus* spp. occur is often practiced (data from Belgium, AMCRA 2020). As knowledge on management of this disease by alternative tools, including immunisation is limited, so that antibacterial therapy remains the main control measure for the time being.

Tetracyclines (due to its better oral availability doxycycline, better than older molecules as chlortetracycline or oxytetracycline) are among the substances of choice for the treatment of enterococci, but there is lack of prove of evidence and authorised dosing regimen specifically for this indication in poultry. It is of importance to perform susceptibility testing of enterococci in poultry, because of high prevalence of resistance to tetracyclines, preferably with doxycycline in panel to be tested.

Although a correct clinical effect can be expected (based on in vitro susceptibility testing of aminopenicillins) with amoxicillin, due to its importance to public health, it should be only used when resistance analysis has shown that the first and second choice molecules mentioned above are not effective. The use of aminopenicillins may promote the selection of organisms that produce "extended spectrum beta-lactamases" (ESBLs) or AmpC. Amoxicillin was therefore classified as the third-choice antimicrobial. The issue with proper antimicrobial achieving sufficiently high concentration at the site of infection (especially arthritis or osteomyelitis cases) is also critical.

Of importance is sampling from affected tissues and lesions (from target organs), immediately after death/euthanasia. In the case of sepsis, a sample can be taken with a swab from the liver, kidney, spleen or bone marrow. Attention should be paid to through following of aseptic technique including, e.g. cauterisation of the tissue by heated scalpel, followed careful twisting by the swab through the organ or tissue. By post-mortem examination should be evaluated, which tissue show signs of damage by the disease and therefore will be the best for sampling to gain real causal pathogen: if the pericardium is affected, it is advisable to take a sample (by scalpel incision), this compartment being better protected from possible contaminants, for example from the intestine, compared to the liver, kidneys or spleen. Depending on the clinical form, germs can also reside and be detected in the joints, oviducts, yolk remains, pulmonary alveoli (AMCRA 2020). Sampling of intestine in such cases is not the choice, especially because of concomitant *E. coli* occurrence (even also E. coli can be pathogenic serotypes). The swabs are best to store preferably at +/− 4 °C directly in a transport medium and sent as quickly as possible to the laboratory.

6.6 Salmonellosis

Salmonellosis is percepted especially from the perspective of zoonosis that can be spread by eggs, meat and offal. After analysis, reconsideration of the current target serovars for breeding hens showed that *Salmonella* Enteritidis, *Salmonella* Typhimurium (including monophasic variants) and *Salmonella* Infantis will be kept under the scope of measures, while *Salmonella* Virchow and *Salmonella* Hadar could be replaced by *Salmonella* Kentucky and either *Salmonella* Heidelberg, *Salmonella* Thompson or a variable serovar in national prevalence targets (EFSA 2019). On the other hand speaking about significant pathologies directly in poultry, two key germs should be mentioned *Salmonella* Pullorum and *Salmonella* Gallinarum, as avian host specific and causing the diseases of poultry, with vertical transmission from parent flock to the production one. Due to the highly effective anti-Salmonella programmes in the EU, the prevalence of the diseases in commercial poultry flock has been minimised.

Pullorum disease is characterised as an acute, systemic diseases (chicks and poults affected), often associated with white diarrhoea and high mortality, sometimes signs can be minimal, in other (occasional) cases demonstrated as blindness, laboured breathing and lameness (caused by swelling of the hock joint) (Shivaprasad and Barrow 2013). Unabsorbed yolk sac being detected as mortem sign in newly hatched chicks (Shoaib et al. 2017), sometimes with no gross lesions, but there can also be detected hepatosplenomegaly with white to grey granulomas in the heart, liver and gizzard (Gast and Porter 2020). There can also be found caseous, white to yellow exudate in the caecal lumen and within the joints of legs and wings (Gast and Porter 2020). Misshapen or shrunken ovaries and follicles with fibrous stalks are the post-mortem signs observed in pullorum disease in adult birds (Eswarappa et al. 2009).

The second one, fowl typhoid, can manifest as septicaemic acute as well as chronic disease, mostly in adult poultry. Generalised septicaemia is accompanied and in post-mortem investigation can be characterised as enlarged, dark and friable liver (can be bronze appearance). Splenomegaly and dark brown bone marrow are also observed (Shivaprasad 2000).

For both pullorum and fowl typhoid, the most evident pathological findings are congested and necrotic foci on enlarged liver, splenomegaly, deformed ova, salpingitis and unabsorbed yolk (Hossain and Islam 2004).

From the microbiological perspective *S.* Pullorum and *S.* Gallinarum are typically non-motile, while other Salmonella serotypes (so called paratyphoid salmonellae) are motile and not host specific (among possibly affected belongs practically all animal species (including invertebrates) as well as human beings). Infections caused by paratyphoid salmonellae are rather of concern as food-borne diseases and big attention is paid them in human. In poultry, paratyphoid Salmonella usually affected stressed highly susceptible young birds, but more often with asymptomatic course. Mortality rate, in naturally occurring infection in young chicks reach peak levels at 3–7 days. *Salmonella* Pullorum the infection is generally confined to the first 2–3 weeks; surviving birds may become carriers (bacteria are then localised in reproductive organs as well as splenic macrophages (Wigley et al. 2001). Entrance of infection agent into the body of chick can be either vertically from parent flock, or, as experimentally proven by horizontally via aerosol containing viable bacteria through navel or orally, intratracheally and intraocularly (Cox et al. 1996). After entry, further invasion to internal organs and whole body is more typical for young, weakened chicks, than for older birds. In paratyphoid salmonella, they occur in any stages, but as mentioned already above, without apparent clinical disease. As was demonstrated by several experiments, invasion of the strains of *Salmonella enterica* Enteritidis, *Salmonella* Heidelberg and *Salmonella* Typhimurium, can cause presence of bacteria in liver, spleen, oviduct, ovary, peritoneum, heart blood (Balan et al. 2016), but also in many other parts of the bird's body (kidney, testes, gall bladder, pancreas, synovia, eye) and yolk sac. Numbers of viable bacteria is extremely low after hatching, but eggs contaminated by salmonella, especially when stored in conditions promoting bacterial growth, can finally contain high bacterial load.

Salmonella gain big attention, especially due to persistence and being cause of food-borne zoonosis (through contaminated carcasses and followed by improper storing, handling, kitchen hygiene and insufficiently cooked, or through eggs).

According to the last EFSA report (EFSA and ECDC 2021) Salmonellosis (17.9% of all food-borne outbreaks, 87,923 confirmed human cases/2019/28 EU Member States) was the second most frequent zoonotic disease, following Campylobacteriosis (confirmed 220,682 human cases/2019/28 EU Member States). In total, 926 salmonellosis food-borne outbreaks were reported by 23 EU Member States/2019, 1915 hospitalisations (50.5% of all outbreak-related hospitalisations) and seven deaths. Causative agent of 72.4% outbreaks was *S.* Enteritidis. The four most implicated food vehicles in strong-evidence salmonellosis food-borne outbreaks were "eggs and egg products", followed by "bakery products", "pig meat and products thereof" and "mixed food". Among most frequent serovars "top five" responsible for human infections were distributed as follows from the food–animal sources: *S.* Infantis 29.7%, *S.* Enteritidis 6.9%, monophasic variant of *S.* Typhimurium 4.5%, *S.* Typhimurium 3.9% and S. Derby 3.7% (EFSA and ECDC 2021).

In accordance with original Regulation (EC) No 2160/2003 (in the last amended wording) (European Parliament 2003a) targets have been set for reduction of Salmonella in flocks of breeding hens, laying hens, broilers, breeding turkeys and fattening turkeys by the adoption of several implementing Regulations (Regulation (EU) No 200/2010 related to breeding hens (Com

Table 13 An overview of measures in case poultry populations were still infected with *Salmonella* Enteritidis (EFSA 2021)

Population	Target serotypes	Maximum % remaining postive	Trade restrictions[a]
Adult breeding hens (*Gallus gallus*)	*S.* Enteritidis *S.* Typhimurium *S.* Hadar *S.* Virchow *S.* Infantis	1%	Destruction or safe disposal of (hatching) eggs and birds (Annex II C of Regulation (EC) No 2160/2003
Adult laying hens (*Gallus gallus*)	*S.* Enteritidis *S.* Typhimurium	2%	Destruction or safe disposal of hens birds, marketing of eggs as class B (only for heat treated egg products) Annex II D of Regulation (EC) No 2160/2003
Broilers (*Gallus gallus*)		1%	Absence in 25 g of fresh meat (Point 1.28 of Annex I to Regulation (EC) No 2073/2005[b])
Adult breeding turkeys (*Meleagris gallopavo*)		1%	Destruction or safe disposal of (hatching) eggs and birds (Annex II C of Regulation (EC) No 2160/2003
Fattening turkeys (*Meleagris gallopavo*)		1%	Absence in 25 g of fresh meat (Point 1.28 of Annex I to Regulation (EC) No 2073/2005)

[a]Only in case of detection of *S.* Enteritidis or *S.* Typhimurium
[b]Comission Regulation (EC) No 2073/2005 of 15 Nov 2005 on microbiological criteria for foodstuffs (text with EEA relevance). OJ L 338, 22.12.2005, p. 1–26

Reg 2010), Regulation (EC) No 517/2011 related to laying hens (Com Reg 2011), Regulation (EU) No 200/2012 related to broilers (Com Reg 2012) and Regulation (EU) No 1190/2012 related to turkeys (Com Reg 2012a). The targets are set on two serotypes (*Salmonella* Enteritidis and *Salmonella* Typhimurium, including monophasic *S.* Typhimurium—antigens 1, 4, [5], 12:i:-, except for breeding hens for which the target includes also *S.* Hadar, *S.* Virchow and *S.* Infantis. Salmonella control programmes in these poultry populations are performed in the EU Member States. In addition, a number of trade restrictions have been introduced in case these populations were still infected with S. Enteritidis or S. Typhimurium (EFSA). An overview of measures is provided in Table 13. Based on the EFSA report (EFSA 2021) 18 of the 26 Member States reporting on Salmonella control programmes in poultry populations met all the reduction targets, compared to 14 in 2018.

The EU system for monitoring and collection of information on zoonoses is based on the Zoonoses Directive 2003/99/EC, which obliges EU Member States to collect relevant and, when applicable, comparable data on zoonoses, zoonotic agents, antimicrobial resistance and food-borne outbreaks (European Parliament 2003b). Visualisation (Fig. 1) is provided to ilustrate poultry flocks positve for target *Salmonella* and mostly presented serovars.

6.7 Fowl Cholera

Pasteurella multocida as the causative pathogen of fowl cholera (sometimes also called avian pasteurellosis or avian haemorrhagic septicaemia) is

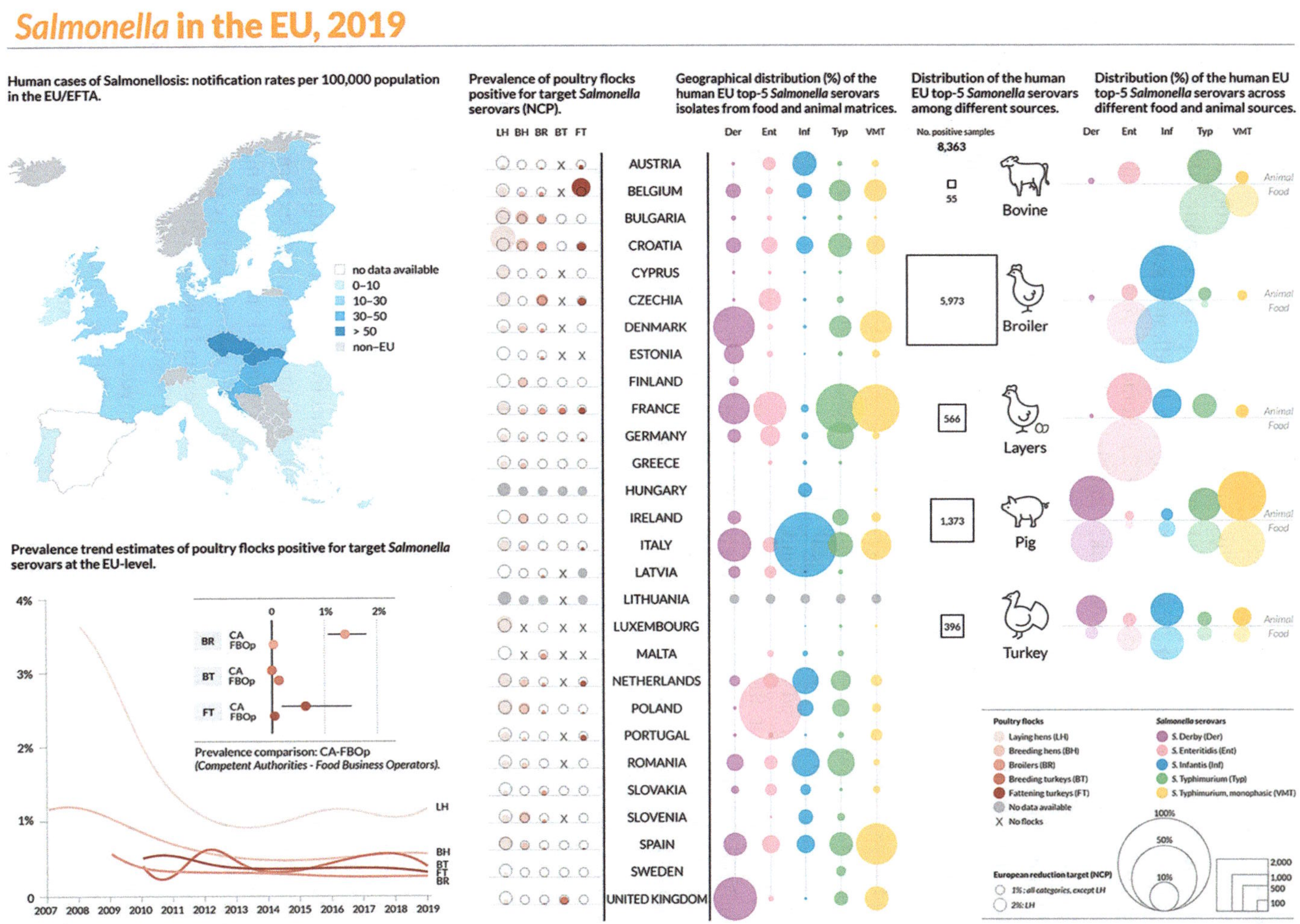

Fig. 1 Summary of the main data reported in the Salmonella in the EU, year 2019: human cases, prevalence of poultry flocks positive for target Salmonella EU-level and in Member States the human EU top-5 *Salmonella* serovars—stratified geographical distribution and distribution according animal sources (EFSA 2019)

contagious disease of poultry as well as wild birds. Sensitivity of poultry in descending order: ducks, geese, turkeys, chickens and guinea fowl. Adult animals are more sensitive than young ones (AMCRA 2020; Sander 2022b). It appears in the nasal and oral cavity of apparently healthy carriers. In the acute phase of the disease, clinical signs may appear shortly before death. Signs that can be frequently observed are anorexia, ruffled feathers, mucous excretion from the mouth/beak, watery and whitish diarrhoea, later mucous and greenish. Mostly also fever is present. High rate of morbidity and mortality can occur. Chronic form of the disease can follow the acute phase or can be caused by less virulent strains. Swollen wattle resulting from localised infection but pathogen and signs of disease can be seen sinuses, leg or wing joints, foot pads and enlarged sternal bursae. Exudative conjunctival and pharyngeal lesions may also be observed (Blackall and Hofacre 2013). Depending on the susceptibility and immunostatus of the host and the virulence of the *Pasteurella multocida* strain, the disease becomes acute or chronic. Especially in chickens, the disease most often becomes chronic, with runny nose, cerebral oedema, inflammation of the wattles and abscess ("Wattle disease"). Carriers who have survived infection play a major role in the transmission of the bacteria. There are also carriers who have never been ill. The infection spreads through the respiratory tract and through skin wounds. Fowl cholera diagnosis combines clinical observations, signs visible via necropsy as well as bacteriological sampling and isolation of *Pasteurella multocida*. A tentative diagnosis of acute form of the disease can be performed via demonstrating bipolar organisms in liver imprints using Wright's staining technique (Blackall and Hofacre 2013). Multiplex PCR (mPCR) that is able to differentiate strains based on the genetic organisation of the lipopolysaccharides (LPS) outer core biosynthesis loci are recommended to be used for the detection of *Pasteurella multocida* (Harper et al. 2015) and can help to detect different serotypes of PM in pure and mixed cultures and clinical samples. This method may help identify carrier animals within flocks. *Pasteurella multocida* strains was originally classified based on serotyping based on, capsular type polysaccharide antigens to five serogroups: A (hyaluronic acid), B (arabinose, mannose and galactose), D (heparin), E (uncharacterised), or F (chondroitin) (Carter 1955), later on came 16 serotypes classification detecting lipopolysaccharide (LPS) antigens using the Heddleston scheme (Heddleston et al. 1972). Some 15 years ago, methodology using multiplex PCR targeting the genes encoding the structure of the LPS outer core biosynthesis loci was elaborated (Harper et al. 2015). This multiplex PCR assay enabled differentiation of 16 LPS serovar strains of *P. multocida* into eight genotypes (L1 to L8) corresponding to Heddleston serovars typing as follows: L1 (1, 14); L2 (1, 2, 5); L3 (3, 4); L4 (6, 7); L5 (9); L6 (10, 11, 12, 15); L7 (8, 13) and L8 (16) (Harper et al. 2015; Hurtado et al. 2020). Several bacterial infections may be confused with fowl cholera based solely on the gross lesions, therefore appropriate sampling (see below) as well as culturing and identification should be performed. *Escherichia coli*, *Salmonella enterica*, *Ornithobacterium rhinotracheale*, Gram-positive cocci, and *Erysipelothrix rhusiopathiae* may all produce lesions indistinguishable from those caused by *Pasteurella multocida* (Sander 2022b).

Good management practices, including a high level of biosecurity, are essential to prevention. Eradication of infection requires depopulation and cleaning and disinfection of buildings and equipment. The premise should then be kept free of poultry for a few weeks. *Pasteurella multocida* can be inactivated by common disinfectants, sunlight, drying or heat, and experiments suggest this pathogen will survive for a maximum of 30 days in the environment (e.g. water or soil) (Christensen and Bisgaard 2000). Phenolic disinfectants, 1% sodium hypochlorite, 70% ethanol, formaldehyde, glutaraldehyde, iodophor, and peracetic acid are effective against *Pasteurella* spp. (Collins and Kennedy 1999). Effort should be paid to avoid entrance and presence of rodents, wild birds, pets, and other animals that may be carriers of *Pasteurella multocida*. Even more is this issue in chickens or layers bread in free yards.

A few bacterin and/or toxin-based vaccines are available commercially (Mostaan et al. 2020). Adjuvant bacterins are widely used and generally effective. Because bacterins are only effective in preventing disease caused by the same serotypes included in the vaccine, proper serotyping is of critical importance with gaining the most prevalent serotypes within an area. For cases where several serotypes are typical for epidemiological/epidemiologically related units an autogenous bacterins are recommended when polyvalent bacterins are found to be ineffective (Sander 2022b). A metagenomics culture-independent approach, as well as a genetic typing scheme for LPS, to ensure an appropriate vaccine strain with a matching predicted LPS structure is used is new very important direction for the selection of strains for use in killed autogenous fowl cholera vaccines (Omaleki et al. 2022).

Attenuated live vaccines are available for administration in drinking water to turkeys and by wing-web inoculation to chickens. These live vaccines can effectively induce immunity against different serotypes of *Pasteurella multocida.* They are recommended for use in healthy flocks only (Sander 2022b). Most of them have homologous protection but a capsular one showed to have heterologous protection partially. Protection range of this kind of formulations is up to 97.5% but route of infection, dose of administrations, quantity of bacteria in one dose, age of animal, type of animal, co-feeding by antibiotics may hinder the protection and its lasting (Mostaan et al. 2020).

A commercial vaccine against fowl cholera (chickens and turkeys), is available and is based on a freeze-dried preparation of a live, avirulent avian isolate of *P. multocida.*

Some studies were also targeted on acellular subunit vaccines that having advantages of no live components inside and therefore low risk of inducing the disease. Three regions of epitope, including the regions of PlpE protein of *P. multocida*, as the appropriate serotype-independent vaccine candidate against pasteurellosis was newly identified through in silico model (Mostaan et al. 2021). This vaccine concept is also considered safer and more stable than live attenuated. Despite advantages also some weaknesses exist as, e.g. isolated proteins, if denatured, may bind to different antibodies than those specific for pathogen protein target structures. Proper combination of antigenic properties producing an effective immune response with the correct pathway is of extreme importance, in some cases several doses must be given for proper life-long immunity (Mostaan et al. 2020). Recombinant vaccines (with some overlap with subunit one) are also another possibility of *P. multocida* vaccination.

It is hardly to find any recent data on the susceptibility of avian isolates of with relevance for European region. Outside Europe, one relatively recent article can be listed, where Nhung et al. (2017) summarising phenotypic resistance in 617 poultry isolates, for which median resistance ordered from lowest up was for: florfenicol (0.6%), ampicillin (2.3%), gentamicin (4.3%), enrofloxacin (4.7%), co-trimoxazole (10.8%), tetracycline (13.8%) and erythromycin (18.0%). Resistance to erythromycin (representing macrolides) and tetracyclines was highest in isolates investigated. Despite that, avian *Pasteurella multocida* susceptibility to antibiotics can be considered as generally good. Due to the fact of lacking regional data is advisable to perform a susceptibility test before starting of the antimicrobial administration. Samples for bacteriological culture should be taken immediately after death or euthanasia. If live sick animals are sent to laboratory, where relevant sample is taken and bacteriological culture can begin swiftly. Ideally, live sick animals will be provided to the laboratory so that sampling can take place and bacteriological culture can begin immediately. Sampling from the affected organs (lungs or pulmonary alveoli) or in acute sepsis, from the liver or spleen as well as heart blood and meninges or bone marrow can be performed. Also sampling of squeeze mucus from nostrils is possible (AMCRA 2020; Blackall and Hofacre 2013). In acute sepsis, it is better to take samples from the liver or spleen. It should be done as aseptically as possible: cauterise the surface with a heated scalpel and rotate the swab in the organ. The swab in the transport medium should be sent as quickly as possible to the laboratory and during the transports should be stored preferably at +/− 4 ° C.

Guidelines (AMCRA 2020; Boereboom et al. 2012) recommend as the first-choice combination of sulphonamide/trimethoprim or doxycycline. Thanks to the better oral availability of doxycycline, this molecule is preferred over chlortetracycline. As low resistance is also reported to florfenicol, this can also be used, but taking into consideration its bacteriostaticity. Despite very low resistance to amino*penicillins as well as enrofloxacin, these are considered as the last, third, choice. This is due to the fact, that the use of aminopenicillins may promote the selection of extended spectrum beta-lactamase (ESBL)/AmpC producing organisms and enrofloxacin, as fluoroquinolone is considered as critically important antimicrobial (WHO 2024) and "B" category (restriction of use) by AMEG (AMEG 2020).

Several taxa linked with similar pathological signs mostly affecting respiratory tract belongs among *Pasteurella*-related microorganisms. Relatively recent reclassification touched these bacterial species mostly based on genome sequencing as well as other molecular methods (Opengard and Boulianne 2020). Strains of avian *Pasteurella haemolytica* were characterised and newly the name is *Gallibacterium anatis* biovar *haemolytica*. *Avibacterium gallinarum* come from original name *Pasteurella gallinarum* and *Avibacterium paragallinarum* from original name *Haemophilus paragallinarum*.

6.8 Erysipelas

Generally, *Erysipelothrix rhusiopathiae* can affect most poultry species with possible acute or peracute course and with high mortality (rates may reach 40–50%, but more frequently with lower mortality up to 15% (Wakenell 2022). The diseases are clinically manifested as septicaemia as well as conjunctival oedema, laying birds are affected with significant egg production losses (Eriksson 2020; Opriessnig et al. 2020). The disease can be fatal in young as well adult turkeys and ducks, with affected birds developing severe haemorrhages in breast and leg muscles (Bickford et al. 1978). Erysipelas in acute septicaemic form mainly affects older turkeys, but disease occurs also in laying hens, serious outbreaks have occurred also in chickens (very rarely in young birds), disease is also reported from ducks, geese, pheasants and quails (Opriessnig et al. 2020). Erysipelas may affect the fertility of males and may contribute to downgrading and processing losses. Birds in contact with the outdoors can be more frequently affected, as *E. rhusiopathiae* belongs among soil-borne germs. Free-range chicken farms outbreaks as well as organic farms outbreaks are recently more reported due to transformation from cage technologies. Especially in rainy years as well as in colder conditions disease outbreaks are more frequent (AMCRA 2020; Eriksson 2020; Wakenell 2022). In the case of an acute infectious course, sudden mortality is observed, with sepsis symptoms and especially a markedly enlarged spleen as well as liver (friable and may be mottled), also petechiae on the muscles and on the heart can be recognised. Clearly visible symptoms that occur just before death include unsteady locomotion and a lack of coordinated movement and an unsteady gait. A chronic form of bacterial infection primarily occurs in turkeys that have been vaccinated against erysipelas, and, in addition to affecting the joints, endocarditis can occur (Bobrek et al. 2013). Once microscopic examination is performed vascular damage and fibrin thrombi are common findings; in smear of the liver or spleen or smear of cardiac blood or bone marrow Gram-positive, slender, pleomorphic rods are detectable (Wakenell 2022). The infection is acquired by mouth or through wounds on the skin, through the mucous membranes such as during artificial insemination (4–5 days after it, death without clinical signs may occur), but also by ingestion of contaminated foodstuffs (particularly cannibalism of infected carcasses), and possibly by mechanical transmission via biting insects (Wakenell 2022). Interestingly the poultry red mite may serve as a mechanical vector of *E. rhusiopathiae* (Chirico et al. 2003). Carriers shedding *E. rhusiopathiae* without showing clinical signs of disease can be among birds as well, e.g. rodents. Faeces, urine, saliva and nasal secretions are considered a vehiculum. Experimental models proposed incubation period among 1–7 days post-infection (Eriksson 2020).

Erysipelothrix isolates have been divided into at least 28 serovars based on specific polysaccharide complexes (double agar-gel precipitation tests used for typing), newer classification is based on consecutive Arabic numbers (1–26) and small letters (Opriessnig et al. 2020). Global distribution of avian strains has been, during the time course reported in individual places/studies in America, e.g. 1b (Crespo et al. 2019), Europe—1a, 1b, 4 (Eriksson et al. 2009), Asia—N, 6, 5 (Nakazawa et al. 1998), Australia—1b, 5, 15 (Eamens et al. 1988). *E. rhusiopathiae* serotypes 1, 2 and 5 have been most frequently isolated from poultry. Diagnosis is based on necropsy, liver or spleen impression smears demonstrating Gram-positive organisms, but isolation and identification of *E. rhusiopathiae* is necessary for confirmation of diagnosis. Identification can be made by immunohistochemistry (detection of antigen, requires availability of antiserum), fluorescent antibody staining, PCR and LAMP (Wakenell 2022, Yamazaki et al. 2014). Quantitative real-time polymerase chain reaction (qRT PCR) assays were developed and validated for tissue samples in combination with enrichment culture for the detection and distinguishing of *Erysipelothrix rhusiopathiae* from other within the genus (To et al. 2009). As chicken blood is rather problematic matrix for PCR because nucleated erythrocytes contribute large amounts of host DNA that inhibit amplification. Therefore, analysis of chicken blood fraction enriched for phagocytic cells is necessary for reliable detection and quantification via real-time and droplet digital polymerase chain reaction (PCR) to detect and quantify *Erysipelothrix rhusiopathiae* DNA in blood samples from infected chickens (Wattrang et al. 2019). One of the methods useful can be LAMP (Loop-mediated isothermal amplification) assay, with advantage of being less expensive, with less instrumentation need for amplification (comparing to RT PCR assays) as isothermal nucleic acid amplification at a constant temperature of 60–65 °C, but again preparing of chicken blood matrix can be an issue. Advantageous of LAMP is naked eye assessment of reaction using fluorescent dyes that intercalate or directly label DNA facilitating its usage both in labs (Gerber et al. 2018).

Prevention of the disease should be based on biosecurity, good cleaning and disinfection. As commented above an issue can raise in outdoor, free range farms to protect the animals from infection. Most routine disinfection methods and disinfectants being effective against *E. rhusiopathiae* used in indoor spaces and are as follows: 0.1% concentration of bichloride of mercury, 0.5% sodium hydroxide solution, 3.5% liquid cresol, 5% solution of phenol, quaternary ammonium, chlorine, or 0.5% formalin as long as it is not in organic matter (Wakenell 2022). Temperatures 55 °C or higher for at least 15 min are proven to be lethal (Stackebrandt et al. 2006).

Management of the disease once it appears needs immediate action, usually combination of treatment via antimicrobial and vaccination, sometimes unfortunately also euthanasia, due to animal welfare reasons. Vaccination can prevent as well as end of an outbreak, but it requires at least 2 weeks before mortality reduction (Eriksson 2020). Thorough checking of health status, euthanasia of animals with fatal signs and removal of dead animals several times per day is highly recommended bearing in mind of zoonotic potential of the disease and therefore wearing all protective equipment (especially gloves, boots and clothes removed after action and disinfect/wash in temperatures exceeding 60 °C) and thoroughly follow hygienic measures. Once the flock is either euthanased, or recovered and finally slaughtered, special care should be paid to all facilities as for mechanical cleaning, disinfection and also technological break. Vaccination is recommended in the following flock. Specific handling with litter/manure is also recommended, due to avoiding of broad spread of the germs leading to spread of the disease to wild animals (storing in closed spaces, not immediate spreading on the fields etc.).

Vaccination belongs among highly important preventive measures. Commercial erysipelas vaccines that are recently available for poultry sector, either attenuated or inactivated, are based on a small number of *E. rhusiopathiae* strains isolated several decades ago. All commercial vaccines available to date for poultry (specifically turkeys) are based on serovar 1a or serovar 2 isolates, where serovar 1a is more commonly used in attenuated vaccines and serovar 2 is more commonly used in bacterins (Opriessnig et al. 2020). Despite these vaccines are generally considered effective in preventing erysipelas, in those cases where residual virulence retains, chronic forms of the disease, which tend to follow the acute form, may develop. As mentioned above, the issue also is the low availability of specific serotypes/target poultry species tailored vaccines, what is probably reason that also off-label use of an attenuated swine vaccine in laying hens (Crespo et al. 2019) or turkeys (Bricker and Saif 1988) has also been reported. Commercially available swine erysipelas vaccine used in layers, two times the recommended dose for swine via injection does not interfere with egg production, but are not approved and no challenge studies are officially available to support this scheme (Wakenell 2022). Due to the lack of commercial vaccine spectra for different serotypes/species specific, autogenous vaccines are used for establishing of active immunity in flocks. Such autogenous vaccines should be produced from pathogenic organisms obtained from birds from the same flock or epidemiologically related flock. Current producers (at least in Europe) exist (https://www.emav.be/members) and moving or already have high-tech production with proper good manufacturing practice processing based on cultures of bacteria following well established and checked purification process. Gaining of pure (non-contaminated) *E. rhusiopathiae* culture as well as appropriate adjuvants selection is critical for final autogenous vaccine produced quality and immunogenic potential. Today, in some parts of the world, vaccine for turkeys, based on freeze-dried technology of production from avirulent culture of *E. rhusiopathiae* is available for oral administration via drinking water which is also less stressful for the birds (Eriksson 2020; Wakenell 2022). As for vaccination schemes, some discussions are in place. There is highly recommended to vaccinate flocks farmed at location of previous disease occurrence. Vaccination scheme as recommended in each individual authorised vaccine should be followed. The use of bacterins in broiler fattening turkeys is useful but labour intensive. For breeders, the bacterin should be given every 4 weeks before onset of egg production (Wakenell 2022). Two doses at intervals of 2–3 weeks are recommended for live vaccines (Wakenell 2022). Treatment of diseased animals should be performed based on susceptibility, which for *E. rhusiopathiae* was reported to penicillin, tetracyclines, quinolones, clindamycin (lincosamides), erythromycin (macrolides), as well as some antimicrobials, which are not authorised for poultry species as cephalosporins or piperacillin (Bobrek et al. 2013). Resistance to tetracyclines has been reported (AMCRA 2020). Rapid-acting penicillin are drug of choice (potassium or sodium penicillin salts). Intramuscular administration at 22,000 U/kg body weight, simultaneously with a full dose of erysipelas bacterin is cited by some sources for certain poultry categories (e.g. Wakenell 2022); it should be noted, that mostly this will be off-label use of the injectable VMP. Administration of penicillin in the drinking water can also be used (at 395,000 IU/L for 4–5 days), but it rather reducing losses (Wakenell 2022). Sulfonamides and oral oxytetracycline/chlortetracycline are not effective, but macrolides (erythromycin) are effective. When aminopenicillins are used, the risk of selection of organisms producing "extended spectrum beta-lactamases" (ESBL) or AmpC is significant, therefore amoxicillin is ranked as third choice by, e.g. AMCRA (2020). There should be taken into account speedy course of the diseases as well as the fact that in feed or water administered antimicrobials intake by birds affected is minimised and therefore only those still eating and drinking normally can be treated. As already mentioned above, antimicrobials can be used concomitantly with vaccination in affected flocks as vaccination with a bacterin helps protect those birds in the flock not yet infected. However, bacterin-derived immunity is not long-lasting and may require two or more injections at intervals of 2–4 weeks. Treatment with antimicrobial or vaccination does not eliminate the carrier state.

Erysipelas can be diagnosed both from clinical signs and autopsy results with help of bacteriological diagnosis, as the lesions are not pathognomic to confirm the diagnosis. Samples should be taken from dead or euthanased animals, directly after death, ideally from affected organs (liver, spleen, kidney, bone marrow (last named is preferred) or heart valves in chronically diseased animals). Aseptical techniques should be very thoroughly followed (cauterisation of the surface with warm scalpel). Swab with sample should be transported in transport medium as quickly as possible to the laboratory (preferable storage/transport temperature +/−4 °C).

6.9 Ornithobacterium Rhinotracheale Infections

Infection of poultry caused by *Ornithobacterium rhinotracheale* (ORT) is contagious disease affecting respiratory tract and causing its distress, mortality of birds and decreased growth. Infections caused by ORT have been yet described in chicken, turkeys, ducks, goose, partridges, pheasants and quails from domestic species, or species which meat or eggs are used for human consumption (Hafez 2002). In commercial poultry flocks was documented that all ages appear to be susceptible (Hafez and Chin 2020). But recently has been described that ORT mainly appears in fattening chickens over 3 weeks of age, and in turkeys from 3 weeks to slaughter age (AMCRA 2020). Study performed by Hauck et al. (2015) documented that ORT incidence was higher in turkeys, with an isolation frequency of 41% compared to 6.9% of broiler chickens. Concomitant respiratory infections were frequently described, mainly with other respiratory bacterial pathogens such as *E. coli*, *Bordetella avium*, *Avibacterium paragallinarum*, *Mycoplasma synoviae*, *Chlamydia psittaci* and *Streptococcus zooepidemicus* (El-Sukhon et al. 2002; Pan et al. 2012; Morales-Erasto et al. 2016; Ellakany et al. 2019; Hafez and Chin 2020). It can also be either concomitant or subsequent in birds primarily affected by Newcastle disease virus, infectious bronchitis, avian influenza and avian metapneumovirus (Travers et al. 1996; Pan et al. 2012a; Hafez and Chin 2020). Especially in chickens and turkeys was proven that presence of other pathogens enhances pathology, lesions and acute course of the disease (Gavrilović et al. 2016). Clinical signs of infection may also differ based on previous or concomitant disease, affected birds show listlessness, decreased food intake and reduced weight gain, transient nasal discharge and sneezing, followed facial oedema. If brain infected sudden death may occur in couple of days up to 20% of affected flock (Hafez and Chin 2020). If laying hens are affected drop of eggs production, smaller eggs with deformities as well as poor eggshell quality are produced by diseased animals. In older turkeys, signs may develop and mortality can reach among 1–15%, but also up to 50% (Roepke et al. 1998), initial symptoms sneezing, coughing, nasal discharge may in advanced and more severe cases progress to dyspnoe, prostration, respiratory distress and sinusitis (Hafez and Chin 2020). The severity of the disease course and mortality rates of ORT-associated infections vary and are influenced also by environmental factors, including high levels of ammonia, inadequate room ventilation, extremes of temperature and humidity, high storage densities, bad litter quality (Chin et al. 2013). Haemagglutinin and neuraminidase are produced by ORT (Kastelic et al. 2013), which both may play an important role in tissue colonisation and inflammation in the host, but despite that haemolysin can play above roles, it has been described that non-haemolytic isolates with longer survival in the poultry host can as well cause pathological course of the disease (Walters et al. 2014). As for laboratory diagnosis, partial sequencing of the *rpo*B gene was used to evaluate different isolates across different avian hosts as well a poultry species (Veiga et al. 2019). The results of 16S rRNA gene sequencing and multilocus sequence typing (MLST) were used to investigate phylogenetic relationship in 65 isolates, identity ranging from 85.1% to 100% was confirmed. This MLST sequencing protocol can identify and differentiate *Ornithobacterium* species (Thieme et al. 2016).

Among preventive measures belongs the control of the possible concomitant and primary pathologies caused by either other bacteria or virus (please see commented above) that can significantly help to minimise clinical severity of the disease. Management measures influencing in complexity proper environmental conditions in housing of birds as well as avoiding of overstocking are therefore essential to prevent ORT. As in other diseases hygiene, disinfection and sufficiently clean facilities belong among essential preventive care. *O. rhinotracheale* strains were proven to be inactivated by 0.5% solution of formic and glyoxylic acid or 0.5% of aldehyde-based solution (containing 20% glutaraldehyde) for 15 min (Hafez and Schulze 2003). Even more strict biosecurity measures should be used in areas with intensive poultry production and in older facilities.

An inactivated vaccine can be used to prevent ORT infections in offspring. Using inactivated vaccine (with formalin-inactivated mineral oil adjuvant bacterin) twice, i.e. first dose at 8 weeks of age and second at 12 weeks of age (a booster dose) produced an early immune response linked with reduced airsac as well as respiratory disease (Murthy et al. 2007). This work was interesting also from the perspective that eight different vaccines (inactivation using formalin and thiomersal) and with or without adjuvant (mineral oil, alum and aluminium hydroxide gel), were produced. Another study as published in Chin et al. (2013) proved for turkeys' vaccination that short duration of presence of antibodies after using of monovalent or trivalent bacterins is still better as for protection (lower mortality and condemnation rates) in vaccinated comparing to non-vaccinated turkeys.

It is authorised and commercially available *Ornithobacterium rhinotracheale* inactivated whole cells, serotype A strain containing vaccine, for which is declared that via active immunisation of female broiler breeders is able to establish (passive) immunisation finally leading to protection against *Ornithobacterium rhinotracheale* serotype A. It was proven that passive immunity transfer during lay for 43 weeks after the last vaccination of broiler breeders, resulting in duration of passive immunity in broilers of at least 14 days after hatching (Nobilis OR inac 2002).

Autogenous vaccines used in young turkeys successfully reduced number of outbreaks (Chin 2013). Orally administered autogenous live vaccine (ORT + spraying with paramyxovirus vaccine) to six-week-old turkeys was proven to result in reduced pathologies (lesions, mortality). French turkey parent flocks specialised companies launched broad flock-specific programme designed to ensure improvement of immunity to ORT infection, which create big pressure and losts. Initiation of the programme was enhanced by occurrence of ORT infections non-treatable via antimicrobials causing also big animal welfare issue. Therefore 47 different vaccine formulations (one to four *Ornithobacterium rhinotracheale* strain/s in each) were used. The process was monitored thoroughly as for measures regulation (of especially manufacturing) autogenous vaccines. Affected prebreeding farms were determined and checked and in total 63 batches of turkeys were vaccinated. No adverse reactions were announced in these farms. Twelve batches from all vaccine were detected for presence of ORT, two of which required antibiotic treatment. This experience confirms that immunisation programme can succeed and can significantly help to minimise the use of antimicrobials. It should be also highlighted that it helps better animal welfare via protection of the animals from suffering by ORT infection and its clinical manifestation (Godener and Le Guennec 2019).

In those cases, where infection by ORT appears repeatedly and more than half of chicken flock introduced on the farm required treatment for the ORT indication the recommendation is the next poultry batch to be imported should be from vaccinated parent-flock mothers. This advice is based on the recent studies carried out in Belgium, which have indicated a lower mortality in chicks born to mothers vaccinated against ORT (AMCRA 2020) as well is supported by above-mentioned French example in turkeys (Godener and Le Guennec 2019).

The rapid and virulent nature of the disease needs the early start of treatment via medication of water. Due to high variability both regional and at strain level, susceptibility to various antibiotics is hardly to be purely empirically predicted. The acute nature of the pathology in contrary with slow growth of pathogen and therefore long-lasting culturing, identification and susceptibility testing cause that the individual therapies are often need to be started before knowledge of the susceptibility/resistance. Due to this reason treatment can be based on the local (or epidemiologically related flocks—e.g. parent flocks) bacteriological culturing and resistance data. The establishment of a resistance profile not older than 1 year is therefore strongly recommended. Except the fact that the degree of acquired resistance of isolates of *O. rhinotracheale* in broilers may be high, situation is even more complicated by frequent presence of concomitant infection of other respiratory pathogens (*E. coli*, *Mycoplasma*, *Avibacterium paragallinarum* etc.). Therefore, treatment with antimicrobials, even in combination may not be successful.

Ornithobacterium rhinotracheale resistance to trimethoprim, sulfonamides and the combination of both is frequently observed. Another difficulty is caused by limited intake of medicated water, especially considering concentrations of trimethoprim/sulphonamides in a dose high enough for therapy of the disease. In the case of tetracyclines (especially doxycycline) and amoxicillin, the resistance to these antibiotics vary greatly and the success of therapy therefore cannot be guaranteed. Thanks to the better oral availability of doxycycline due to its lipophilicity and ease in crossing the biological membranes and also concentration in respiratory tract, this molecule is preferred over chlortetracycline. Similarly, as indicated also in previously mentioned treatment with amoxicillin has been classified as the third choice, especially due to the possibility to promote the selection of organisms that produce "extended spectrum beta-lactamases" (ESBLs) or beta-lactamases AmpC. In review published by Barbosa et al. (2019) susceptibility profile of ORT indicated that tiamulin (Devriese et al. 2001), florfenicol (Watteyn et al. 2013) and gamithromycin (Watteyn et al. 2016) belong among the substances to which this germ is still susceptible.

As indicated also in AMCRA guidelines (2020) the clinical signs and lesions are not pathognomonic. Isolation of the *Ornithobacterium rhinotracheale* from the lesions and its identification or by establishing the presence of antibodies is necessary. This slow growing bacterial species is difficult to isolate, as samples are usually heavily contaminated with other bacteria from the respiratory tract (mainly *E. coli*)—for this reason supplement of gentamicin or polymyxin 10 μg/mL and 5–10% of sheep blood is used to the base of Columbia agar, or soybean casein agar, anaerobic/microaerophilic (7% CO_2) conditions and prolonged culturing up to 48 h. For biochemical identification, some commercial diagnostic kits are also available (e.g. API-20NE and API-ZYM) adapted for ORT identification (Numee et al. 2012; Chin et al. 2013). In susceptibility testing performance of agar dilution or E-test was proven to give quantitative results if correctly performed (Gashe 2017).

Samples taken just at death or euthanasia and from the alveoli of the lungs or trachea are preferred to the infraorbital sinuses or the nose samples (due to concomitant occurrence of fast/more massively growing bacteria, such as *E. coli*, *Proteus* or *Pseudomonas*). Thoroughly aseptic sampling with cauterisation of the surface with a heated scalpel and the swab rotation through the tissue should be performed. In case of inflammation of the joints, sample collection can be obtained after cauterisation, or at least thorough disinfection of the skin surface, followed by sterile scalpel incision the joint and swabbing the joint (AMCRA 2020). The collection of samples for ORT isolation with dry sterile swabs and maintenance at 4 °C for 2 days or −20 °C for 5 days retained the bacterial viability and hindered the growth of other bacteria (Numee et al. 2012).

Blood can also be collected for serological testing, performed by agar gel precipitation and ELISA, to show the presence of humoral antibodies against *Ornithobacterium rhinotracheale*. Genetic characterisation of ORT isolates can also be performed by modified PCR techniques. Enterobacterial repetitive intergenic consensus-PCR (ERIC-PCR), repetitive element palindromic-PCR (REP-PCR), random amplified polymorphic DNA-PCR (RAPD-PCR) and MLST can be used for this purpose (Szabó et al. 2017).

6.10 Mycoplasma Caused Infections

Avian mycoplasmosis belong among the diseases which are distributed worldwide and affect avian as well as domestic species. Their spread via horizontal and vertical (egg) transmission as well as the nature of pathogen makes them difficult to tackle and eradicate. Horizontal transmission can be via direct contact of animals or indirectly birds can be infected via infectious aerosols and through contamination of feed, water, and the environment, and by human activity on fomites (shoes, equipment). The most important mycoplasmas isolated from domestic avian species include *Mycoplasma gallisepticum* (MG) and *Mycoplasma synoviae* (MS), cell penetrating agents, both well known mainly from infections in chickens and turkeys. Another two species of high importance especially in turkey farming are *Mycoplasma meleagridis* (MM) and *Mycoplasma iowae* (MI).

Spectrum of signs and diseases is big, but can be briefly summarised that MG causes chronic respiratory disease of chickens (CRD) and infectious sinusitis (IS) of turkeys. In both poultry species rales (respiratory growling noise of birds heard in close examination), coughing, nasal discharge, sinusitis and the development of severe lesions on the air sacs belong among clinical signs and pathological observations (Stipkovits and Kempf 1996), but uncomplicated cases of MG infections in chickens result in relatively mild catarrhal sinusitis, tracheitis and airsacculitis (El-Gazzar 2020a). Embryo mortality with air sacs lesion can also occur. MG caused diseases are considered to be an important issue in broilers, breeders and commercial layers, causing significant economic losses in the poultry sector. Broilers that are suffered from MG disease have a 10–20% decrease in food conversion efficiency leading to weight gain decrease 20–30% and mortality rate mostly around 5%. In breeders and layers, the disease causes a 10–20% decrease in egg production (Stipkovits and Kempf 1996). Approximately 5–10% of embryo mortality increase has been noted. Very young birds are not usually suffered by the disease (but if yes, experimentally was proven that more severe clinical disease course appeared in young birds). In commercial flocks most of outbreaks appear in birds after 4 weeks. In fattening turkeys, outbreaks were detected among 8–15 weeks of age. Mild signs progress was determined after 2–7 days to cough (80–90% of the flock), followed with sinus swelling and nasal discharge that can occur in minimum of birds, but can also be present in up to 70% of the flock. Vertical transmission via eggs is of importance, but indirect horizontal transfers via dust, droplets, feathers, tools can happen due to the fact of several days survival on contaminated surfaces as proven by Christensen et al. (1994) who investigated survival on feathers MG (2–4 days), MS (2–3 days) and MI (6 days), on other materials in poultry halls approximately 2 days, on human hair (field isolate MI survived for 6 days).

The severity of broiler and broiler breeders' clinical infections caused by MG as well as MS can be enhanced by concomitant appearance and pressure of viruses—infectious bronchitis (IB), avian pneumovirus, avian influenza, Newcastle disease (NWD), but also attenuated vaccines against some of these diseases (IB, NWD). Also, bacteria, *Escherichia coli*, *Pasteurella multocida* (and related bacterial species with impact on respiratory tract) and *Ornithobacterium rhinotracheale* can complicate the disease course (severity, chronicity) and make infection more difficult to manage. *Escherichia coli* coinfections result in severe air sac thickening and turbidity, with exudative accumulations, adhesive pericarditis and fibrinous perihepatitis (El-Gazzar 2020a). Turkeys develop severe mucopurulent sinusitis and varying degrees of tracheitis and airsacculitis.

MS causes infectious synovitis or mild upper respiratory disease, lameness with swollen hock joints (synovitis), growth retardation. In laying birds is reduced egg production as well as deformities in egg shells (Dufour-Gesbert et al. 2006; Kursa et al. 2019). Shell appearance can be typical—shell apex abnormalities or "top coning" due to rough and this shell surface with increase of translucency close to the apex (Dorko 2019), together with deformities also amount of cracked eggs arise. Abnormal eggs occur also when MS and IB are present in laying hens (Rosales 2019). Mortality in chickens is usually low, but can be up to 10%. MS infection also occurs in combination with Newcastle disease (NWD), virus infectious bronchitis attacks or vaccination (Stipkovits and Kempf 1996). MS can spread faster, is more persistent and more difficult to manage than MG and is also considered more virulent.

MM infects only turkeys, causing airsacculitis, skeletal abnormalities and suboptimal production and loss in hatchability around 5% (due to late incubation (D25–28) mortality of embryos) (Naola Ferguson-Noel 2020). Currently MM has been mostly eradicated from all primary breeders and thus not been reported lately from the commercial turkey fattening flocks. Semen used for insemination must be checked and also be free of MM (El-Gazzar 2020a).

MI is associated with late embryo mortality (last 10 days of incubation) causing reduced hatchability in turkey flocks. In young turkeys' leg weakness has also been reported, as well as lameness, stunting and swollen hocks.

In microscopy observations of mucous membranes in the cases of MG infections, they are recognised as thickened, hyperplastic, necrotic and infiltrated with inflammatory cells. The mucosal lamina propria contains focal areas of lymphoid hypoplasia and germinal centre formations (El-Gazzar 2020a). Sampling of MG is performed via tracheal or choanal cleft swab, ideally in Mycoplasma transport broth or directly swabbed to agar medium (see below culturing methods).

Laboratory investigations confirming presence of *Mycoplasma* species mostly use PCR tests, both conventional and real-time PCR. There can also be used quantitative—qPCR, where recently two *M. synoviae*-specific functional genes in the DNA fragment were found, what allows to establish two SYBR green qPCR *M. synoviae* nucleic acid detection methods, which can be promising for rapid molecular identification of *M. synoviae* (Xu et al. 2021). Multiplex PCR covering MG, MS as well as virulence factors of these species has been also described and used in recent studies (Ashraf et al. 2020; Armour 2020). PCR methods are therefore considered as golden standard in Mycoplasma detection and confirmation. Real-time PCR, being sensitive, specific and fast detecting for MG as well as other Mycoplasma, serves also another advantage—possibility of performance directly on clinical swabs taken from site of infection—e.g. choana, sinuses, trachea, airsacs (El-Gazzar 2020a). Targeting and amplifying a specific sequence allows for differentiation between MG isolates and can be particularly useful for epidemiologic investigations and to identify the source of infection (El-Gazzar 2020a; Xu et al. 2021).

Serological examination of host sera by rapid agglutination, haemagglutination inhibition or enzyme-linked immunosorbent assay (ELISA-Ab) tests is also possible. In the confirmation *M. gallisepticum,* haemagglutination-inhibition is used as a confirmatory test, due to non-specific false reactions, which may occur, especially after administration of inactivated oil-emulsion vaccines or infection with *M. synoviae* (El-Gazzar 2020a). Recently an article by Xu et al. (2021) introduced some discussion, if serological tests such as ELISA or haemagglutination inhibition (HI) are fully useful in early stage of the disease caused by *M. synoviae*, commenting specifically on HI test both for antibody detection and pathogen detection that may be is not fully suitable for clinical detection of *M. synoviae* due to unstable haemagglutination activity of this *Mycoplasma* species and fact that *vlh*A encoding gene is mutagenic and horizontally transfers between M. *synoviae* and other mycoplasmas, such as *M. gallisepticum* (Xu et al. 2021).

Culturing of field isolates is extremely difficult and time consuming. It is complicated also by the fact that media selecting for Mycoplasma should be rich as for growth factors (enriched by swine or horse serum, Nicotine adenine dinucleoside (NAD), cystein as well as yeasts derived substrates and glucose) and at the same time inhibiting other more speedily growing microbes. Due to this fact growth medium is supplemented by thallium acetate and penicillin (or ampicillin). For more details can be referred to Ferguson-Noel (2020). In experimental conditions *M. synoviae*, *M. gallisepticum*, *M. iowae* and *M. meleagridis* strains can be cultured on modified Frey's medium at 37 °C with 5% CO_2 as described Dusanić et al. 2009 and used in experiments also by Xu et al. (2021).

As the poultry production is intensified due to the increasing demand for poultry meat, it led to higher density of farms as well as closeness of facilities, where both laying hens and broilers for fattening (chickens/turkeys) are kept.

Due to the vertical transfer via eggs and practically no effective way of eradication, once the pathogen is involved in the flock, main prevention targets are concentrated on repopulation of the primary flocks (Ferguson-Noel 2020). Dipping of eggs in antibiotic (e.g. macrolide—tylosin) was used to reduce the incidence of transmission of MM in infected flocks. Such practices should be avoided due to selection pressure that cause any use of antimicrobials and considering prudent use approach to antimicrobials. In waste majority of breeders were therefore started practices of elimination of the pathogen with the eradication of MM from primary (grandparent) breeder flocks. Programmes containing frequent testing—especially monitoring parent flocks at short intervals (e.g. every 3–4 weeks in turkeys and every 2–3 weeks in chickens) will optimise the detection of the disease and is the most potent prevention tool from both horizontal and vertical transfer prevention (Ferguson-Noel 2020). In the case that in parent-breeders when an outbreak occurs and infection start to spread, one-day-old chicks were injected or first 5–10 days-aged birds were dosed via drinking water, what reduced MM infection and airsacculitis and helps to improve weight gain (El-Gazzar 2020b). Such practices are still used, but in waste majority of the European systems, once Mycoplasma is detected in the flock, the flock is isolated by enhanced biosecurity measures, and depopulated and hatched eggs are destroyed. All facilities in which risk of contamination via Mycoplasma is possible are thoroughly cleaned, disinfected, washed and technological pause is used to minimise/eliminate risk of further occurrence of pathogen and its spread (Landman and van Eck 2017).

At present, two commercial attenuated live vaccines are available to protect against *M. synoviae*. One of the most widely used in the world is the temperature-sensitive MS-H strain, which was developed by the chemical mutagenesis of an Australian field strain 86079/7NS (Morrow et al. 1998). Another vaccine is the NAD-independent MS1 vaccine strain, which was obtained spontaneously during in vitro passages of the type strain WVU 1853 (ATCC 25204, NCTC 10124) (Kreizinger et al. 2017).

Antibiotics (i.e. tetracyclines, macrolides, quinolones and tiamulin) may be used for therapeutic treatment or prophylactic medication. The eradication of Mycoplasma infection can be achieved through improvements in hygiene and management practices, therapeutic treatment of breeder layers and/or of hatching eggs and better monitoring procedures as well as keeping of Mycoplasma-free breeding stock and keeping parent and production flocks free of infection by utilising single-age, all-in all-out farms with high level biosecurity. In many parts of the world such approach has been very successful, and the majority of broiler, turkey and egg production can be kept free of the mycoplasma infection (Kleven 1997).

6.11 Bacteria Associated with Omphalitis

Omphalitis as a non-contagious infection appears in first days of lives of chickens. This infection of the navel and/or yolk sac is mostly associated with poor hygiene and environment as well as poor regulation of incubation temperature and humidity (Sander 2022a). It is also of importance that a hatch is not pulled too early as too many birds will have open and unhealed navels as this point (until approximately 20 days of incubation the yolk sac is outside of the chick's body but at 20 days of incubation the yolk sac should be mostly internalised and the navel should be in the process of closing) (Poultry Health services 2015).

Contamination of hatching eggs, incubators, transportation crates, transporting belts or any other equipment in direct contact with chickens and their unclosed navels is risky for the contamination and burning of infection (Van Meirhaeghe et al. 2019). Chilling or overheating during shipment or early placement, especially to facilities with improper environmental conditions (temperature, humidity, lightening) causing stressful conditions may increase losses.

Exogenous contamination is serving as a source of bacteria, which plays key role in opportunistic bacterial infection (with causative bacteria *Pseudomonas* spp., *S. aureus*, *E. coli*, *Enterococcus* spp., *Salmonella* spp., *Proteus* spp.) either monobacterial, but more frequently mixed infection caused by several bacterial species. Bacteria penetrate before the navel is completely closed and can migrate up the patent yolk stalk and infect the yolk sac.

The most commonly isolated bacterial species is *E. coli*, infection with this bacterial species as well as with others (either individual or more frequently concomitantly affecting navel) can occur when an opportunity arises such as excess contamination of egg shells, cracked hatching eggs, open/unhealed navels at hatch and/or poor sanitation. Frequently reported are also representatives of *Salmonella* spp. or other Enterobacterales representatives. During last years, more and more are also detected navel infections caused by enterococci (usually *E. faecium*, *E. faecalis* or *E. durans*). *Pseudomonas aeruginosa*, bacteria that often contaminate lines that supply drinking water to birds, water utilised in machines used for washing eggs, and water to incubators and hatchers for maintaining humidity is another one associated with omphalitis. As the last one, which should not be omitted is *Staphylococcus* spp., mostly *S. aureus* coming from poultry environments, as part of the normal microflora on the skin of both animals and humans (Poultry Health services 2015).

Clinical signs are characterised as navel inflammation, anorexia as well as growth depression accompanied with reduced weight gain, and increased mortality in the first 2 weeks after hatching (Sander 2022a). Once the navel is inflamed, its closing is usually impossible. Chickens therefore have a wet spot on the abdomen; visible scab may be present. Drooping head birds tend to be close to heat sources, huddled, calm, not approaching to water and feed, what lead to dehydration as well as stagnation or lose weight, further depression and anorexia. Persistent, unabsorbed yolk accumulate in the body cavity (with possible congestion, malodorance and solidified yolk material). Extensive peritonitis may occur. There is increased mortality from hatch to 2 weeks of age. Affected chicks or poults usually appear normal (once not investigated closely) until a few hours before death. Losses of as much as 15% in chickens and 50% in turkeys have been seen (Sander 2022a).

One of the most important tools how to manage disease is prevention. Essential pre-requisite is clean, sanitised hatchery/incubator including hatcher baskets, not allowing to multiply and finally spread in hatched chickens especially opportunistic pathogens as *Escherichia coli* and *Enterococcus* spp., frequently associated with omphalitis. Also following some essential rules (Tables 14 and 15) modified from WOAH (2013) is recommended in laying, hatching and one-day-old chicks handling. Such rules have been summarised in internationally recognised biosecurity procedures in poultry sector being part of the Terrestrial Animal Health Code (WOAH 2013). The incubator and equipment should be cleaned and disinfected thoroughly between hatches. Immediately after emptying, incubators as well as hatcher baskets should be thoroughly cleaned with hot high-pressure water to remove meconium, eggshells, fluff and other particles that can serve as vehiculum/vector for bacteria. After cleaning, sanitisation is recommended. Effective hatchery sanitation is critical factor limiting contamination of naïve neonatal chicks and poults. Formaldehyde fumigation has been traditionally used to reduce the pathogen load in commercial hatch cabinets, but is currently subject of strict regulation and even ban in some territories. Despite that, formaldehyde is considered as the most common and effective disinfectant in the poultry industry, used for hatching eggs in hatcheries disinfection to prevent pathogens transmissions. Formaldehyde kills the most of microbes and pathogens, being low cost, but as well considered as highly hazardous to human health, what made its use quite controversial (Selby et al. 2023). Two methods of formaldehyde use are: the addition of formalin to potassium permanganate ($KMnO_4$) in 2:1 ratio (v/w) recently replaced with formaldehyde volatilisation of the polymer paraformaldehyde under controlled temperature system. Among factors influencing efficacy and maximise germicidal activity belong temperature, humidity, time of exposure, the concentration of formaldehyde, cir-

Table 14 Hygiene measures during the handling of eggs and day-old chicks

Eggs handling	Egg handlers in the hatchery on a poultry farm	Hands wash with soap and water and change to clean outer garments before handling hatching eggs received from the hatching hall
Chicks sexing	Chick sexers and chick handlers	Hands wash and disinfection and change into clean protective clothing and boots before commencing work and between different lots of chicks.
Transfer of chicks	Boxes and equipment for transfer	Delivery/distribution of day-old chicks or other poultry: New boxes or if already used boxes thorough cleaning and disinfection or fumigation is necessary. Important is proper material and design of boxes—easy to clean and disinfect
Chicks delivery	Delivery personnel measures	Delivery directly from the hatchery by personnel wearing clean, disinfected outer clothing. Outer clothing preferably changed between each delivery (and then cleaned and disinfected prior used again)
Trucks	The delivery truck and its equipment	Cleaning and disinfection of truck and its equipment before loading each consignment of chicks, equipment—air conditioning undergoing proper cleaning and disinfection as well

culation of gas, the amount of organic matter on the shell surface and method used of generating formaldehyde vapours (EMTECH 2020). Precautions of using formaldehyde should take in proper practice to avoid any hazard to human health.

Attention should be paid to keep proper temperature and humidity. One of the important predisposing risky factors, not only for omphalitis, but practically for any bacterial disease linked to faecal contamination of eggs/shells is picking eggs hatched on the floor, especially after some time laying there. Only clean, uncraked eggs should be introduced into incubator. If it is not possible to avoid setting of dirty eggs, they should be segregated from clean eggs. Eggs should be treated by sanitisation based on defined programmes, reflecting legal provisions. Time, temperature and frequent changes of water are as critical as the concentration of sanitiser in both wash and rinse water. The rinse should be warmer than the wash water (which should be warmer than the internal temperature of the egg) but should not be >60 °C (Sander 2022b).

Excellent brooding conditions at placement can mean the difference between 1.5% and >2% mortality during the first week. Any added stress that occurs during the first week (improper temperature, poor water quality or temperature, poor ventilation, poor access to feed, etc.) can further set the chick back, making the fight against infection and disease that much more of a challenge (Poultry Health services 2015).

Despite in some farms, preventive use of antimicrobials is performed (in contrary to prudent use policies), treatment using antimicrobials is often ineffective in the case of omphalitis, so all steps of prevention as discussed above are recommended (Sander 2022b). Severely affected sick and/or lame birds, to be identified and it is recommended to cull them from the flock immediately. A concern regarding prevention/early treatment exist that birds, which should be culled are missed (looks temporarily "better". Unfortunately, affected birds are mostly unable to fight effectively infection due to the nature of the initial challenge, short time duration improvement is usually followed by relapse later in the cycle (Poultry Health services 2015).

6.12 Staphylococcus Caused Infection

Overall, 61 staphylococcal species have been described so far, and 56% of them (34/61) have been isolated from clinical cases in poultry (Szafraniec et al. 2022). Despite most known are pathologies caused by *Staphylococcus aureus* as ethological agents of poultry staphylococcosis, other *Staphylococcus* species are mostly present in locomotory system diseases and also cause

Table 15 Recommendations applicable to hatching egg hygiene and transport

From nest through eggs collection-sanitation-transport to final incubation		
Litter and nest boxes	Hygiene and properties of litter and nests	Dry/good quality and condition litter in the laying house clean nest box (if with litter, clean, adequate quantity)
Eggs collection	Automated eggs collection or intervals for eggs collection	Continuous or frequent intervals eggs collection (not less than twice per day) after collection eggs should be placed in clean disinfected containers
Discarding damaged or dirty eggs	Damaged or dirty eggs out of hatching purposes	Dirty, broken, cracked, leaking and dented eggs should be collected in a separate container and should not be used for hatching purposes
Eggs sanitisation	Timeliness of sanitisation after hatching/collection	Sanitisation of clean eggs as soon as possible after collection
Eggs storage	Conditions for storage of eggs prior transport to incubation	Clean, dust-free room used exclusively for the storage of eggs purpose and low temperature less than 20 °C and relative humidity of 70–80%
Eggs transport to incubation	Cleanness and disinfection of all means of transport as routine	New or clean and disinfected/fumigated crates if moved to hatchery by vehicles, those must be cleaned/disinfected (including air conditioning)

further pathologies (which are listed in brackets) examples of species are as follows: *S. agnetis* and *S. simulans* (endocarditis, septic joints and bones lesions), *S. epidermidis* (omphalitis, dermatitis), *S. hyicus* (eye infections and acantholytis dermatitis), and *S. gallinarum*, *S. hemolyticus*, *S. hominis* (septic joint and bones lesion), see also Table 16, modified from Weeks et al. (2000), where are present those frequently associated with lesion affecting locomotory compartments. Examples of such lesions include also these affecting locomotory apparatus of poultry, where Staphylococci readily infect bones and joints and are associated with many pathologies—spondylitis, arthritis, tendinitis, tenosynovitis, osteomyelitis, bumblefoot, dyschondroplasia or chondronecrosis in combination with osteomyelitis and amyloid arthropathy (Szafraniec et al. 2022). Invasivity of Staphylocococci has been proven to the metaphyseal area of joints, leading to arthritis and osteomyelitis. Arthritis, synovitis and osteomyelitis are common forms of staphylococcal infection in poultry. Bacteria can enter to the joints—especially tibiotarsal and stiffle joints, tendons and tendon sheats, or bones through systemic infection or through local injury (Andreasen 2020). Clinical signs include swollen, hot joints with limping and reluctance to walk. Joints contain increased amounts of fluid, tendons and tendon sheaths are inflamed, and bone sections can contain focal areas of necrosis known as osteomyelitis. The femoral head and the proximal head of the tibiotarsus are commonly affected. Vertebrae can be affected as well, leading to spondylitis. For the development of bone and joint diseases fibronectin-binding proteins (FnBPs), collagen-binding protein (Cna) and clumping factor (Clf) play a role.

Among systemic diseases caused by *Staphylococcus* spp. belong septicaemia and gangrenous dermatitis as well as omphalitis (where causative staphylococci are, e.g. *S. aureus*, *S. epidermidis*, *S. haemolyticus*, *S. scuiri*, *S. xylosus* (Szafraniec et al. 2022), as less frequent can be counted infections of sternal bursa, vertebrae, eye lid, testis—orchitis and epididymo-orchitis (Monleon et al. 2008), heart—endocarditis (Stępień-Pyśniak et al. 2017) and granuloma in the liver and lungs (Andreasen 2020). Skin wounds, minor surgical procedures (i.e. beak, toe or comb trimming), vaccine injection, and compromised intestinal mucosa can introduce Staphylococci to local tissue or into the bloodstream. The most common form of staphylococcosis are localised lesions.

Table 16 Infectious diseases affecting locomotory and skeletal system of chickens and laying hens production categories

Infectious diseases of locomotory-skeletal system	Age	Location	Disease/lesions	Causative agent	Usual outcomes
Fattening poultry	Any, usually older	Joints	Arthritis, tenosynovitis	*Staphylococcus aureus*, *Staphylococcus epidermitis* and other coagulase negative staphylococci aviadenoviruses, reoviruses	Lameness
	Any, usually older	Bone	Bacterial chondronecrosis with osteomyelitis	*Staphylococcus aureus*, *Staphylococcus hyicus* and other coagulase negative staphylococci *Enterococcus cecorum* *Escherichia coli*	Lameness, Bones fragility
		Bone	Femoral head necrosis	*Staphylococcus* spp., both coagulase positive and negative Gumboro virus, Ricketsia,	
		Spinal cord Thoracolumbar vertebrae	Spinal osteomyelitis, Spondylitis	*Staphylococcus* spp., both coagulase positive and negative *Enterococcus cecorum*	In spondylitis, due to the impingements on the spinal cord indirectly caused lameness
		Joints	Hock joints	*Staphylococcus* spp. both coagulase positive and negative	
Laying hens including parents/ grandparents' hens	Adults	Bones	Osteoporosis	Retroviridae, avian leucosis/sarcoma virus	
	Adults		Amyloidosis	*Staphylococcus aureus*, *Escherichia coli*, *Salmonella enteritidis*, *Enterococcus faecalis*	
	Adults	Feet	Bumble foot	*Staphylococcus aureus*	Lameness, often chronic

Human health concerns exist in relation to especially *S. aureus* MRSA strains (resistance to methicillin causing an issue of resistance to the whole beta-lactam group as well as containing also several resistance genes, so being multiresistant). Interestingly, based on one genetic study dealing with origin of avian *S. aureus* adapting and widely disseminating from originally single clone (ST5) acquired from humans approximately 30–63 years ago (Lowder et al. 2009) and confirmed by combining phenotype information with evolutionary analyses of staphylococcal genomes, evidence of adaptation, following a human-to-poultry host transition has been given (Murray et al. 2017). In poultry CC5 is considered as common lineage (Monecke et al. 2013), where 44 recombination events in 33 genes along the branch extending to the poultry-specific CC5 cluster were identified and 47 genes being more often in CC5 poultry isolates compared with those from humans were confirmed by team of Murray et al. (2017).

Strategies to minimise occurrence of staphylococcal infections are mainly targeted on management procedures reducing damage of host-defence mechanisms as well as injuries (wounds, food pad ulceration avoidance due to poor litter quality, heat stress) (Andreasen 2020). Hygiene in hatcheries help to reduce infections of navels leading to omphalitis as well as further chronical infections by Staphylococci. Interestingly poultry under mild stress (not the excessive one) may be more resistant to staphylococcosis due to increase of heterophil numbers, which are considered to be the most important defending cells of help in control of bacterial infections including those caused by Staphylococci (Mutalib et al. 1983; Larson et al. 1985; Andreasen Jr et al. 1993). Moreover, the administration of various feed additives influencing gut microbiota has been recognised as possible effective strategy for the prevention and treatment of skeletal disorders in poultry (Jiang et al. 2021).

No effective vaccination exists against *Staphylococcus aureus*, despite that in some experiments—e.g. with vaccines containing staphylococcal cell wall components (peptidoglycan, teichoic acids), capsular polysaccharides and cell wall-anchored proteins, were tested. The only proven concept in turkeys is live vaccine, using avirulent, coagulase-negative isolate of *Staphylococcus epidermidis* strain 115. This approach is based on the principle of bacterial interference and competition, with inhibition of colonisation with *Staphylococcus aureus* as well as killing properties of bacteriocins produced by the avirulent vaccine strain. Above-described vaccine is administered by spraying of aerosol at days 1–10, with repetition at the age of 4–6 weeks (Meyers and Jensen 1987). As described by subsequent research, *Staphylococcus epidermidis* strain 115 harbours plasmid on which is encoded thiopeptide, which belongs among compounds having potent antimicrobial activity against bacteria, fungi and even protozoa (Bennallack et al. 2014). This characteristic of the strain can cause that similar results were confirmed also once this strain was administered to chickens (Nicoll and Jensen 1987).

Among the options for treatment penicillins (both phenoxypenicillin and aminopenicillins—amoxicillin) for strains not producing beta-lactamases as well as for methicillin sensitive strains, erythromycin, lincomycin and spectinomycin (Sato and El-Gazzar 2022). If the infection is in soft tissues and rather at early stages, antibiotic treatment might be effective, but is mostly ineffective for osteomyelitis (difficulties to penetrate in sufficient concentration) and omphalitis (speedy of infection).

6.13 Locomotory Diseases

Leg problems belong among important welfare and production concerns. Walking difficulties (lameness) can be associated with both non-infectious and infectious skeletal disorders. Lameness indicates pain that results in reduced mobility and access to water and feed, what also affect weight gain as well as well-being in complexity. Behavioural observations have shown that lame birds spend less time standing, running, preening and dustbathing and more time sitting (Weeks et al. 2000). Diseases cause also economic loses, being very hardly treatable by antimicrobials, becoming chronic and with limited possibility of immunisation (vaccination).

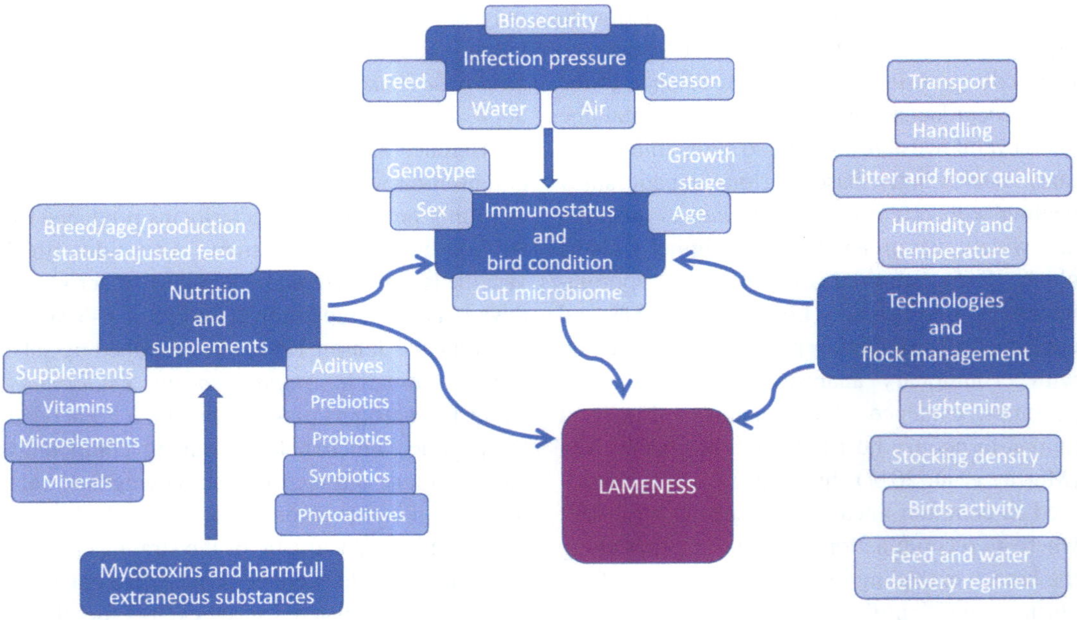

Fig. 2 Lameness and influencing/predisposing factors and interconnections of them

Decreased weight gain, decreased egg production, lameness and birds losses due to osteomyelitis as well as condemnation of carcasses at slaughter have significant impact on economy (Andreasen 2020). Table 16 provides a list of infectious diseases affecting locomotory and skeletal system of chickens and laying hens production categories together with possible causative agents, target site of infection (location) and disease classification. As mentioned in the table, many pathogens can be associated with locomotory diseases and treatment of the infections located in bones, vertebrae, joints, cartilages and other tissue components. For antimicrobials, especially considering oral pharmaceutical forms and properties of the active substances own is very difficult to penetrate and achieve effective concentrations in such locomotory system compartments. So, in many cases despite the pathogen confirmed good susceptibility in vitro, clinical outcomes of the treatment are not in compliance with this laboratory results. This can significantly limit the availability of antimicrobials with adequate pharmacokinetic to the target site of infection and also infections as lameness or chronic infections is hardly to treat. Among antimicrobial able to reach certain concentration in cartilages and bones, especially in growing birds belong enrofloxacin (generally fluoroquinolones), but they belong to the group "B", so should be reserved as the last choice, after susceptibility testing and confirmation that strain concerned is sufficiently susceptible. Many factors in the rearing of broiler chickens therefore influencing the development and function of bone and skeletal system should be checked and the exact issue and possible cause to be identified and targeted by measures to prevent development of the diseases in chickens. These factors are highlighted in Fig. 2. There can also be commented that plenty of these critical factors, can be linked also with other diseases of poultry and therefore be targeted by approaches preventing the animals being ill.

7 Conclusion

Modern production of poultry, especially broiler chickens became complex technology process, rather than standard animals' farming. Pressure on the speedy fattening and high production lead to the production of hybrid clones of breeds with specific characteristics. Enormous pressure on effectiveness of production led to minimising space per animal and for certain time period led also to utilisation of medicinal products or substances enhancing growth. Antimicrobials become one group of growth promoters as no fear exists in this time. More than two decades ago, especially in Nordic countries, with long history of One Health and fight against antimicrobial resistance, started pressure leading finally to ban of antimicrobial growth promoters in the European Union (since 1 January 2006). Despite this important step, pressure on farming intensification continued further, but demands for more sensitive approach to farming and welfare conditions intensified as well, what started an era of defining better conditions of poultry farming, including improvements in welfare and biosecurity. Currently, some countries and farming systems made "step back" from intensity towards more slowly growing chicken' breeds, reared in bigger space per animal and thoroughly monitored conditions. Debate also started on poultry more "friendly" housings. Use of smart technologies allowing daily weight estimates and forecasting, that allows early disease detection, preventing diseases outbreaks and maintaining flock well-being by reducing periods of suboptimal growth patterns has started to be used in advanced farm systems. Consumption of highly balanced feed ratios, free of antimicrobials and free of anticoccidial feed additives together with specific feeding programmes, in which vitamins, minerals, acidifiers, probiotics, synbiotics, herbals and phytoadditives, essential oils, enzymes and other substances of natural origin have started to play bigger role, occur recently in some farming systems. Situation has changed gradually through years. In laying hens, a consumer demand for eggs not being produced in cages, but rather from deep litter or free-ranges farming systems or even organic farms were getting stronger. On the other hand, litter farms were confirmed to have bigger issue with eggs contaminated with, e.g. *Salmonella*, free range farms and organic farms opened to the contacts with the outside environment, including wild birds have started to be problematic considering new threat—occurrence of High Pathogenic Avian Influenza (HPAI), where first confirmed individual outbreaks started at the end 2018 and during 2019. The number of outbreaks had started to grow and overall, the 2021–2022 epidemic of HPAI has been the largest observed in the EU so far. The H5N1 subtype has predominant and other subtypes, like H5N8 and H5N5, have caused only few outbreaks (limited to certain areas in the North of the EU). This epidemic has been detected especially in free range farms (ducks and geese affected mostly). After some time, it was spread also to poultry in halls (still luckily more rarely). HPAI becomes a real challenge endangering production of poultry worldwide. Due to this disease imperative on increasing biosecurity measures caused stepwise improvement of biosecurity in poultry farming. On the other hand, it slows down movement towards (bigger) free range farm systems. Despite vaccination to HPAI exists (till 2023) only one vaccine is authorised in the EU), there is not ensured that vaccinated birds will not be carriers (without clinical signs) and therefore is not commonly accepted over the world also due to the trade reasons. In the EU till 2023 some limited cases of vaccine use were noted and currently if indicated by the epidemiological situation and the particular circumstances in the area where there is a risk of spread of HPAI viruses to poultry, vaccination against HPAI may be used. Such use is considered as a part of the official measures put in place by the competent authorities of the EU Member States for prevention and control of the disease. For this kind of occasions, the specific rules provided by the relevant EU legislation (for HPAI, Delegated Regulation (EU) 2023/361) should be followed. Some new vaccines are under development and testing, with strict supervision. The situation with HPAI brings the poultry farm managers back to the thought of building houses for chickens (and not chicken houses), meaning that more attention is paid to behavioural frameworks and needs of poultry to improve conditions of their life. For these purposes more accurate data about the environment of the poultry halls, including relative humidity, temperature, flow of the air, ammonia concentration, lightening, litter quality and frequency of changing, delivery of clean water and minimising stressors and increasing biosecurity can help farmers to optimise the conditions. New technologies could allow to gain better knowledge on how often chickens eat, drink, rest and play, altogether creating behavioural framework. Good hygiene and biosecurity are extremely important and together with appropriate, system tailored vaccination and welfare programme an essential pre-requisite of the successful farming and healthy flock. But it will cost not only money, but a big effort of all employees on the farm to learn and to understand new technologies, but at the same time to keep sense for animal farming and understanding of animal behaviour. This seems to be the most challenging issue, as finding quality and stable employees is an increasing problem. Veterinarians might be more faced not only with diagnostic followed by decision on antimicrobial therapy and vaccination strategy, but will be more and more involved to interpreting and integrating data from general physiology (measures of stressors, inflammation markers, muscle health and/or blood chemistry) with data from environment and behaviour. Cooperation of veterinarians, farmers and further farm staff will in such way come to the new dimension. Cooperation in poultry sector is might be even more important, as timely start of the treatment is extremely important and therefore speedy diagnostic and rapid susceptibility testing methods development and quick coordinated and complex reaction save flocks of 10–50 thousand birds. Expected is movement towards molecular-biology methods as PCRs of different types (RT, quantitative, multiplex etc.), whole genome sequencing methods will be more used especially for confirmatory analysis and epidemiology, including detection of genes and their locations. As the book and chapter is targeted on antimicrobials, considering threats of antimicrobials resistance, it should be noted as well that veterinarians will probably be more frequently faced with three questions. Should I prescribe antimicrobial and which one and in which dosing schedule to ensure efficacy and minimise resistance selection? What to do next time to avoid the situation of need of administration of antimicrobial? Are there any alternative tools instead of antimicrobials? Are these alternatives safe? Questions, for which finding the correct answers will need huge knowledge, sense for poultry sector, veterinarian and farmer cooperative approach and ability to tailor all of the answers for the exact farm conditions.

Acknowledgement Acknowledgement to Petr Stolář MVD for practical insights to poultry sector and treatment and management as well as Jaroslav Bzdil MVD, PhD for valuable comments to this chapter.

References

Anicon (2021) Diagnostics and custom made vaccines for chickens, Anicon Broschure, p. 3. https://www.anicon.eu/templates/images/documents/263_1_pdf/Anicon%20Broschure%20Vaccines%20CHICKEN.pdf. Accessed 29 May 2021

AMCRA (2020) Vade-mecum pour un usage responsable des produits antibacteriens chez les volailles. https://formularium.amcra.be/a/5 (French language version Accessed 31 May 2023)

Aarestrup FM, Bager F, Jensen NE, Madsen M, Meyling A, Wegener HC (1998) Surveillance of antimicrobial resistance in bacteria isolated from food animals to antimicrobial growth promoters and related therapeutic agents in Denmark. APMIS 106:606–622

Abdelli N, Pérez JF, Vilarrasa E, Luna IC, Melo-Duran D, D'angelo M, Solà-Oriol D (2020) Targeted-release organic acids and essential oils improve performance and digestive function in broilers under a necrotic enteritis challenge. Animals 10:259

AMCRA (2023) AMCRA Formularium Volaille (in French), on line edition https://formularium.amcra.be/a/5 Accessed 16 June 2024

AMEG (2020) Antimicrobial advice ad hoc expert group. Divergent position on the report on Categorisation of antimicrobials in the European Union. https://www.ema.europa.eu/en/documents/other/divergent--position-categorisation-antibiotics-european-union-answer-request-european-commission_en.pdf. Accessed 31 May 2023

Anadón A, Martínez-Larrañaga MR (2013) Veterinary drug residues. Coccidiostats. In: Motarjemi Y, Moy G, Tood E (eds) Encyclopedia of food safety, vol 3. Elsevier, Oxford (U.K.), pp 63–75

Andreasen CB (2020) Staphylococcosis. In: Swayne DE (ed) Diseases of poultry, vol II, 14th edn. Willey Blackwell, Hoboken, NJ, pp 995–1003

Andreasen JR Jr, Andreasen CB, Anwer M, Sonn AE (1993) Heterophil chemotaxis in chickens with natural staphylococcal infections. Avian Dis 37(2):284–289

Apostolakos I, Feudi C, Eichhorn I, Palmieri N, Fasolato L, Schwarz S, Piccirillo A (2020) High-resolution characterisation of ESBL/pAmpC-producing *Escherichia coli* isolated from the broiler production pyramid. Sci Rep 10(1):11123

Argudín MA, Deplano A, Meghraoui A, Dodémont M, Heinrichs A, Denis O, Nonhoff C, Roisin S (2017) Bacteria from animals as a pool of antimicrobial resistance genes. Antibiotics (Basel) 6(2):12

Armour NK (2020) *Mycoplasma gallisepticum* infection. In: Swayne DE (ed) Diseases of poultry, vol II, 14th edn. Willey Blackwell, Hoboken, NJ, p 920

Ashraf AA, Wafaa H, Mohammed El-ordy MS (2020) Detection of virulence factors of *mycoplasma* species isolated from chicken by multiplex PCR. Benha Vet Med J 38

Bailey MA, Macklin KS, Krehling JT (2013) Use of a multiplex PCR for the detection of toxin-encoding genes *net*B and *tpe*L in strains of *Clostridium perfringens*. ISRN Vet Sci 2013:865702

Balan KV, Bigley EC, Gaines DW, Babu US (2016) Tissue colonization and circulating T lymphocytes in laying hens upon oral challenge with *salmonella enterica* serovars. Poult Sci 95(12):2824–2828

Barbosa EV, Cardoso CV, Silva RCF, Cerqueira AMF, Liberal MHT, Castro HC (2019) *Ornithobacterium rhinotracheale*: An update review about an emerging poultry pathogen. Vet Sci 7(1):3

Bennallack PR, Burt SR, Heder MJ, Robison RA, Griffitts JS (2014) Characterization of a novel plasmid-borne thiopeptide gene cluster in *Staphylococcus epidermidis* strain 115. J Bacteriol 196(24):4344–4350

Bennani H, Mateus A, Mays N, Eastmure E, Stärk KDC, Häsler B (2020) Overview of evidence of antimicrobial use and antimicrobial resistance in the food chain. Antibiotics (Basel). 9(2):49

Berg ES, Wester AL, Ahrenfeldt J, Mo SS, Slettemeås JS, Steinbakk M, Samuelsen Ø, Grude N, Simonsen GS, Løhr IH, Jørgensen SB, Tofteland S, Lund O, Dahle UR, Sunde M (2017) Norwegian patients and retail chicken meat share cephalosporin-resistant *Escherichia coli* and IncK/bla_{CMY-2} resistance plasmids. Clin Microbiol Infect 23(6):407

Berrang ME, Northcutt JK (2005) Use of water spray and extended drying time to lower bacterial numbers on soiled flooring from broiler transport coops. Poult Sci 84(11):1797–1801

Bickford AA, Corstvet RE, Rosenwald AS (1978) Pathology of experimental erysipelas in turkeys. Avian Dis 22(3):503–518

Biomin GmBH (2021) Gut performance management: PoultryStar®- Healthy gut, strong chick. https://www.dsm.com/anh/products-and-services/products/eubiotics/poultrystar.html. Accessed 31 May 2023

Blackall PJ, Hofacre CL (2013) Fowl cholera. In: Swayne DE (ed) Diseases of poultry, vol 2020, 13th edn. Wiley-Blackwell, Hoboken, NJ, USA, pp 831–846

Blackall PJ, Soriano EV (2020) Infectious coryza and related bacterial inefctions. In: Swayne DE (ed) Diseases of poultry, vol II, 14th edn. Willey Blackwell, Hoboken, NJ, pp 890–906

Bobrek K, Gawel A, Mazurkiewicz M (2013) Infections with *Erysipelothrix rhusiopathiae* in poultry flocks. Worlds Poult Sci J 69(4):803–812

Borst LB (2020) Streptococcus and Eneterococcus. In: Swayne DE (ed) Diseases of poultry, vol II, 14th edn. Willey Blackwell, Hoboken, NJ, pp 1003–1010

Borst LB, Suyemoto MM, Keelara S, Dunningan SE, Guy JS, Barnes HJ (2014) A chicken embryo lethality assay for pathogenic *enterococcus cecorum*. Avian Dis 58(2):244–248

Borst LB, Suyemoto MM, Robbins KM, Lyman RL, Martin MP, Barnes HJ (2012) Molecular epidemiology of *enterococcus cecorum* isolates recovered from enterococcal spondylitis outbreaks in the southeastern United States. Avian Pathol 41(5):479–485

Bortolaia V, Guardabassi L (2015) Zoonotic transmission of antimicrobial resistant enterococci: a threat to public health or an overemphasised risk? In: Zoonoses–Infections affecting humans and animals. Springer Science+Business Media, pp 407–431

Boudergue C, Burel C, Dragacci S, Favrot MC, Fremy JM, Massimi C, Prigent P, Debongnie P, Pussemier L, Boudra H, Morgavi D, Oswald I, Perez A, Avantaggiato G (2009) Review of mycotoxin-detoxifying agents used as feed additives: mode of action, efficacy and feed/food safety. External Scientific Report Submitted to EFSA. https://www.efsa.europa.eu/en/supporting/pub/en-22. Accessed 31 May 2023

Boereboom MAM, Cornelissen PM, Fabri THF, van Geijlswijk IM, van Geloof HM, Hagenaar M, Verhoeven G (2012) Formularium Pluimvee, Goedgekeurd door WVAB: 21, https://www.knmvd.nl/app/uploads/sites/4/2018/09/formulariumpluimveeversie31juli2012.pdf. Accessed 31 May 2023

Bricker JM, Saif YM (1988) Use of a live oral vaccine to immunize turkeys against erysipelas. Avian Dis 32:668–673

Brugère-Picoux J, Vaillancourt JP, Shivaprasad HL (2015) Manual of poultry diseases, French Association for the Advancement of Science, AFAS

Butaye P, Devriese LA, Haesebrouck F (2000) Incomplete cross resistance against ionophores in *enterococcus faecium* and *enterococcus faecalis* strains from pigs and poultry. Microb Drug Resist 6(1):59–61

Caly DL, D'Inca R, Auclair E, Drider D (2015) Alternatives to antibiotics to prevent necrotic enteritis in broiler chickens: a microbiologist's perspective. Front Microbiol 6:1336

Camarda A, Circella E, Pennelli D, Battista P, Di Paola G, Madio A, Tagliabue S (2008) Occurrence of pathogenic and faecal *Escherichia coli* in layer hens. Ital J Anim Sci 7:385–389

Cardona CJ, Bickford AA, Charlton BR, Cooper GL (1993) *Enterococcus durans* infection in young chickens associated with bacteremia and encephalomalacia. Avian Dis 37(1):234–239

Cardwell M (2019) RWA best practices. Canadian Poultry https://www.canadianpoultrymag.com/rwa-best-practices-31143/. Accessed 31 May 2023

Carter GR (1955) Studies on *Pasteurella multocida*. I. A hemagglutination test for the identification of serological types. Am J Vet Res 16(60):481–484

Castanon JI (2007) History of the use of antibiotic as growth promoters in European poultry feeds. Poult Sci 86(11):2466–2471

Cheng HH, Kaiser P, Lamont SJ (2013) Integrated genomic approaches to enhance genetic resistance in chickens. Annu Rev Anim Biosci 1:239–260

Chin RP, van Empel PCM, Hafez HM (2013) *Ornithobacterium rhinotracheale* infection. In: Swayne DE (ed) Diseases of poultry, 13th edn. Wiley-Blackwell, Hoboken, NJ, USA, pp 807–858

Chirico J, Eriksson H, Fossum O, Jansson D (2003) The poultry red mite, *Dermanyssus gallinae*, a potential vector of *Erysipelothrix rhusiopathiae* causing erysipelas in hens. Med Vet Entomol 17(2):232–234

Christensen H, Bachmeier J, Bisgaard M (2020) New strategies to prevent and control avian pathogenic *Escherichia coli* (APEC). Avian Pathol:1–30

Christensen JP, Bisgaard M (2000) Fowl cholera. Rev Sci Tech Off Int Epiz 19(2):626–637

Christensen H, Bachmeier J, Bisgaard M (2021) New strategies to prevent and control avian pathogenic *Escherichia coli* (APEC). Avian Pathol:1–12

Christensen NH, Christine AY, McBain AJ, Bradbury JM (1994) Investigations into the survival of *mycoplasma gallisepticum, mycoplasma synoviae* and *mycoplasma iowae* on materials found in the poultry house environment. Avian Pathol 23(1):127–143

Christensen H, Blacall, PJ, Bisgard, M (2016). Pasteurellaecae: Pasteurella, Avibacterium, Gallibacterium species. In S.M. Williams, L.D. Jackwood, M.D. Lee, B. Lupiani, W.M. Reed, E. Spackman & P.R. Woolcock (Eds.), A laboratory manual for the isolation, identification, and characterization of avian pathogens 6th edn (pp. 85–102). Athens, GA: American Association of Avian Pathologists, Inc.

CIWF (2020) Dutch slower growing broilers require less antibiotics than fast growing chickens dutch-slower--growing-broilers-require-less-antibiotics-than-fast--growing-chickens-updated-2020.pdf (ciwf.org.uk). Accessed 25 May 2023

Clothier KA, Torain A, Reinl S (2019) Surveillance for *Avibacterium paragallinarum* in autopsy cases of birds from small chicken flocks using a real-time PCR assay. J Vet Diagn Invest 31(3):364–367

Collins CH, Kennedy DA (1999) Decontamination. Laboratory-acquired infections: history, incidence, causes and prevention, 4th edn. Buttersworth, London, UK, pp 160–186

Com Reg (2011) Commission Regulation (EU) No 517/2011 of 25 May 2011 implementing Regulation (EC) No 2160/2003 of the European Parliament and of the Council as regards a Union target for the reduction of the prevalence of certain Salmonella serotypes in laying hens of *Gallus gallus* and amending Regulation (EC) No 2160/2003 and Commission Regulation (EU) No 200/2010 (OJ L 138, 26.5.2011, pp. 45–51)

Com Reg (2010) Commission Regulation (EU) No 200/2010 of 10 March 2010 implementing Regulation (EC) No 2160/2003 of the European Parliament and of the Council as regards a Union target for the reduction of the prevalence of Salmonella serotypes in adult breeding flocks of *Gallus gallus* (OJ L 61, 11.3.2010, pp. 1–9)

Com Reg (2012) Commission Regulation (EU) No 200/2012 of 8 March 2012 concerning a Union target for the reduction of *Salmonella enteritidis* and *Salmonella typhimurium* in flocks of broilers, as provided for in Regulation (EC) No 2160/2003 of the European Parliament and of the Council (OJ L 71, 9.3.2012, pp. 31–36)

Com Reg (2012a) Commission Regulation (EU) No 1190/2012 of 12 December 2012 concerning a Union target for the reduction of *Salmonella Enteritidis* and *Salmonella Typhimurium* in flocks of turkeys, as provided for in Regulation (EC) No 2160/2003 of the European Parliament and of the Council (OJ L 340, 13.12.2012, p. 29–34)

Cooper KK, Songer JG, Uzal FA (2013) Diagnosing clostridial enteric disease in poultry. J Vet Diagn Invest 25(3):314–327

Cox NA, Bailey JS, Berrang ME (1996) Alternative routes for salmonella intestinal tract colonization of chicks. J Appl Poult Res 5(3):282–288

Craven SE, Stern NJ, Cox NA, Bailey JS, Berrang M (1999) Cecal carriage of *Clostridium perfringens* in broiler chickens given mucosal starter culture. Avian Dis 43(3):484–490

Crespo R, Bland M, Opriessnig T (2019) Use of commercial swine live attenuated vaccine to control an *Erysipelothrix rhusiopathiae* outbreak in commercial cage-free layer chickens. Avian Dis 63(3):520–524

Crispo M, Blackall P, Khan A, Shivaprasad HL, Clothier K, Sentíes-Cué CG, Cooper G, Blakey J, Pitesky M, Mountainspring G, Cutler G, Bickford A, Stoute S (2019) Characterization of an outbreak of infectious coryza (*Avibacterium paragallinarum*) in commercial chickens in Central California. Avian Dis 63(3):486–494

Croville G, Foret C, Heuillard P, Senet A, Delpont M, Mouahid M, Ducatez MF, Kichou F, Guerin JL (2018) Disclosing respiratory co-infections: a broad-range panel assay for avian respiratory pathogens on a nanofluidic PCR platform. Avian Pathol Croville 47(3):253–260

CVMP (2018) Questions and answers on use of enrofloxacin-containing veterinary medicines administered via drinking water to chickens and turkeys: Follow-up assessment after the referral under Article 35 of Directive 2001/82/EC (EMEA/V/A/089) EMA/184512/2018. https://www.ema.europa.eu/en/documents/referral/enrofloxacin-article-35-referral-questions-answers-use-enrofloxacin-containing-veterinary-medicines_en.pdf. Accessed 31 May 2023

D'Incau M, Pennelli D, Lavazza A, Tagliabue S (2006) Serotypes of *E. Coli* isolated from avian species in Lombardia and Emilia Romagna (North Italy) Ital. J Anim Sci 5:298–301

Daehre K, Projahn M, Semmler T, Roesler U, Friese A (2018) Extended-spectrum beta-lactamase-/AmpC beta-lactamase-producing *Enterobacteriaceae* in broiler farms: transmission dynamics at farm level. Microb Drug Resist 24(4):511–518

Davies J (2019) Slow-growing birds are fast becoming mainstream. https://www.poultryworld.net/Meat/Articles/2019/7/Slow-growing-birds-are-fast-becoming-mainstream-454287E/. Accessed 31 May 2023

De Baere S, De Mil T, Antonissen G, Devreese M, Croubels S (2018) In vitro model to assess the adsorption of oral veterinary drugs to mycotoxin binders in a feed- and aflatoxin B1-containing buffered matrix. Food Addit Contam A Chem Anal Control Expo Risk Assess 35(9):1728–1738

De Bryine N, Iatridou D, Vanderhaeghen W, Ignate K (2020) Use of antimicrobials in practice (targeted on cattle, pigs, poultry, horses). In: Pokludová L. antimicrobials in livestock 1: regulation, science, practice. Springer Natural, pp 43–79

De Herdt P, Defoort P, Van Steelant J, Swam H, Tanghe L, Van Goethem S, Vanrobaeys M (2008) *Enterococcus cecorum* osteomyelitis and arthritis in broiler chickens. Vlaams Diergeneeskundig Tijdschrift 78(1):44–48

Devreese M, Osselaere A, Goossens J, Vandenbroucke V, De Baere S, De Backer P, Croubels S (2012) Interaction between tylosin and bentonite clay from a pharmacokinetic perspective. Vet J 194(3):437–439

Devriese LA, Hommez J, Wijfels R, Haesebrouck F (1991) Composition of the enterococcal and streptococcal intestinal flora of poultry. J Appl Bacteriol 71(1):46–50

Devriese LA, Laurier L, De Herdt P, Haesebrouck F (1992) Enterococcal and streptococcal species isolated from faeces of calves, young cattle and dairy cows. J Appl Bacteriol 72(1):29–31

Devriese LA, Hommez J, Pot B, Haesebrouck F (1994) Identification and composition of the streptococcal and enterococcal flora of tonsils, intestines and faeces of pigs. J Appl Bacteriol 77(1):31–36

Devriese LA, Herdt P, Haesebrouck F (2001) Antibiotic sensitivity and resistance in *Ornithobacterium rhinotracheale* strains from Belgian broiler chickens. Avian Pathol 30:197–200

Doetkott DM, Nolan LK, Giddings CW, Berryhill DL (1996) Large plasmids of avian *Escherichia coli* isolates. Avian Dis 40(4):927–930

Dolka B, Chrobak-Chmiel D, Makrai L, Szeleszczuk P (2016) Phenotypic and genotypic characterization of *enterococcus cecorum* strains associated with infections in poultry. BMC Vet Res 2016-12:129

Donahue M (2017) Personal communication. AgriStats, Inc., Fort Wayne, IN 46825.25 to Smith JA, Boulianne, Owen RL and Gingerich E.: Disease prevention and control in Antibitic free production. In: Swayne DE et al (eds) Diseases of poultry, vol I, 14th edn. Willey Blackwell, Hoboken, NJ, p 42

Feeks J (2019) More than half of US broilers raised without antibiotics in 2018. Poultry Health Today. https://poultryhealthtoday.com/more-than-half-of-us--broilers-raised-without-antibiotics-in-2018/. Accessed 31 May 2023

Dufour-Gesbert F, Dheilly A, Marois C, Kempf I (2006) Epidemiological study on *mycoplasma synoviae* infection in layers. Vet Microbiol 114:148–154

Dusanić D, Bercic RL, Cizelj I, Salmic S, Narat M, Bencina D (2009) *Mycoplasma synoviae* invades non-phagocytic chicken cells in vitro. Vet Microbiol 138(1–2):114–119

Dwars RM, Matthijs MG, Daemen AJ, van Eck JH, Vervelde L, Landman WJ (2009) Progression of lesions in the respiratory tract of broilers after single infection with *Escherichia coli* compared to superinfection with *E. Coli* after infection with infectious bronchitis virus. Vet Immunol Immunopathol 127(1–2):65–67

Eamens GJ, Turner MJ, Catt RE (1988) Serotypes of *Erysipelothrix rhusiopathiae* in Australian pigs, small ruminants, poultry, and captive wild birds and animals. Aust Vet J 65(8):249–252

EC (2022) EU agricultural outlook for markets, income and environment, 2022-2032. European Commission, DG agriculture and rural development, Brussels. EU agricultural outlook for markets, income and environment 2022-2032–Publications Office of the EU (europa.eu). Accessed 25 May 2023

EFSA (2019) Salmonella control in poultry flocks and its public health impact. EFSA J 17(2):e05596. https://efsa.onlinelibrary.wiley.com/doi/epdf/10.2903/j.efsa.2019.5596. Accessed 25 May 2023

EFSA (2021) EFSA and ECDC (European Food Safety Authority and European Centre for Disease Prevention and Control). The European Union One Health 2019 Zoonoses Report. EFSA Journal 2021;19(2):6406, 286pp https://doi.org/10.2903/j.efsa.2021.6406 Accessed 16 June 2024

EFSA & ECDC (2021) European food safety authority and European Centre for Disease Prevention and Control the European Union one health 2019 Zoonoses report. EFSA J 19(2):6406

El-Sukhon SN, Musa A, Al-Attar M (2002) Studies of bacterial etiology of aersacculitis of broilers in nothern and middle Jordan with special reference to *Escherichia coli*, *Ornithobacterium rhinotracheale* and *Bordetella avium*. Avian Disease 46:605–612

Elbadawy M, Aboubakr M (2017) Pharmacokinetic, bioavailability and tissue residues of apramycin in broiler chickens. Int J Pharam Sci 7(4):1826–1831

El-Gazzar M (2020a) *Mycoplasma gallisepticum* infection in poultry. MSD MANUAL Veterinary Manual. https://www.msdvetmanual.com/poultry/necrotic--enteritis/overview-of-necrotic-enteritis-in-poultry. Accessed 31 May 2023

El-Gazzar M (2020b) *Mycoplasma meleagridis* infection in poultry. MSD MANUAL Veterinary Manual. https://www.msdvetmanual.com/poultry/mycoplasmosis/mycoplasma-meleagridis-infection-in-poultry?query=Mycoplasma%20meleagridis%20Infection%20in%20Poultry. Accessed 31 May 2023

Ellakany HF, Abd-Elhamid HS, Bekhit AA, Elbestawy AR, Abou-Ellif AS, Gado AR (2019) Effect of mixed experimental infection with *Ornithobacterium rhinotracheale* and *mycoplasma gallisepticum* in broiler chickens. Alex J Vet Sci 61:168–178

Elliott C, Connolly L, Kolawole O (2019) Potential adverse effects on animal health and performance caused by the addition of mineral adsorbents to feeds to reduce mycotoxin exposure. Mycotoxin Res 36:115–126

EMA (2009) Referral for the water soluble powders and oral solutions containing doxycycline hyclate indicated for use in poultry and intended for administration via drinking water, Annexes I, II and III. doxycycline-hyclate-article-35-referral-annexes-i-ii-iii_en.pdf (europa.eu). Accessed 31 May 2023

EMA (2013) Summary of the European Public Assessment Report (EPAR) for Poulvac *E. coli*. https://www.ema.europa.eu/en/medicines/veterinary/EPAR/poulvac-e-coli. Accessed 31 May 2023

EMA (2015) Principles on assignment of defined daily dose for animals (DDDvet) and defined course dose for animals (DCDvet), EMA/710019/2014. Principles on assignment of defined daily dose for animals (DDDvet) and defined course dose for animals (DCDvet) (europa.eu). Accessed 31 May 2023

EMA (2016) Defined daily doses for animals (DDDvet) and defined course doses for animals (DCDvet), EMA/224954/2016. Defined daily doses for animals (DDDvet) and defined course doses for animals (DCDvet) (europa.eu). Accessed 31 May 2023

EMA (2019) European Medicine Agency. Categorisation of antibiotics in the European Union EMA/CVMP/CHMP/682198/2017. https://www.ema.europa.eu/en/documents/report/categorisation-antibiotics-european-union-answer-request-european-commission-updating-scientific_en.pdf. Accessed 31 May 2023

EMA (2022) European Medicines Agency, European Surveillance of Veterinary Antimicrobial Consumption, 'Sales of veterinary antimicrobial agents in 31 European countries in 2021. Trends from 2010 to 2021'. (EMA/24309/202). https://www.ema.europa.eu/en/documents/report/sales-veterinary-antimicrobial-agents-31-european-countries-2021-trends-2010-2021-twelfth-esvac_en.pdf. Accessed 30 May 2023

EMA (2021) Veterinary medicine European public assessment reports (EPARs) https://www.ema.europa.eu/en/search/search?search_api_views_fulltext=chicken+vaccine. Accessed 31 May 2023

EMTECH (2020) The use of formaldehyde in fumigation, posted on 14th Aug 2020. https://www.emtech-systems.com/technical-talk/the-use-of-formaldehyde-in--fumigation/. Accessed 25 May 2023

Eriksson H, Jansson DS, Johansson KE, Båverud V, Chirico J, Aspán A (2009) Characterization of *Erysipelothrix rhusiopathiae* isolates from poultry, pigs, emus, the poultry red mite and other animals. Vet Microbiol 137(1–2):98–104

Eriksson H (2020) Erysipelas. In: Swayne DE (ed) Diseases of poultry, vol II, 14th edn. Willey Blackwell, Hoboken, NJ, pp 1010–1018

Eswarappa SM, Janice J, Balasundaram SV, Dixit NM, Dipshikha C (2009) Host-specificity of *salmonella enterica* serovar Gallinarum: insights from comparative genomics. Infect Genet Evol 9(4):468–473

EU-DG Agri Dashboard (2023) Poultry meat: trade, production, word quotations and prices. poultry-meat-dashboard_en_2.pdf (europa.eu). Accessed 30 May 2023

Eurogroup for Animals (2020) Denmark bets on slow growth broiler chicken breeds. https://www.eurogroupforanimals.org/news/denmark-bets-slow--growth-broiler-chicken-breeds. Accessed 31 May 2023

European Parliament (2003a) Regulation (EC) no 2160/2003 of the European Parliament and of the council and regulation of 17 November 2003 on the control of salmonella and other specified food-borne zoonotic agents. OJ L 325:1–15

European Parliament (2003b) Council of the European Union. Directive 2003/99/EC of the European Parliament and ofthe council of 17 November 2003 on the monitoring of Zoonoses and zoonotic agents, amending council decision 90/424/EEC and repealing council directive 92/117/EEC. OJ L 325:31–40

Ewers C, Janssen T, Kiessling S, Philipp HC, Wieler LH (2004) Molecular epidemiology of avian pathogenic *Escherichia coli* (APEC) isolated from colisepticemia in poultry. Vet Microbiol 104(1–2):91–101

Fancher CA, Zhang L, Kiess AS, Adhikari PA, Dinh TTN, Sukumaran AT (2020) Avian pathogenic *Escherichia coli* and *Clostridium perfringens*: challenges in no antibiotics ever broiler production and potential solutions. Microorganisms 8(10):1533

FAO (2023) Gateway to poultry production and products Chickens | Gateway to poultry production and products | Food and Agriculture Organization of the United Nations (fao.org). Accessed 30 May 2023

Ferguson-Noel N (2020) Mycoplasmosis. In: Swayne DE (ed) Poultry diseases, 14th edn. Willey Blackwell, Hoboken, NJ, pp 907–965

Forte C, Acuti G, Manuali E, Casagrande Proietti P, Pavone S, Trabalza-Marinucci M, Moscati L, Onofri A, Lorenzetti C, Franciosini MP (2016) Effects of two different probiotics on microflora, morphology, and morphometry of gut in organic laying hens. Poult Sci 95(11):2528–2535

Frey J (2019) RTX toxins of animal pathogens and their role as antigens in vaccines and diagnostics. Toxins 1(12):719

FVE (2016) FVE position paper on coccidiostats or anticoccidials. https://www.fve.org/cms/wp-content/uploads/FVE-position-paper-on-coccidiostats-or-anticoccidials.pdf. Accessed 31 May 2023

Gashe S (2017) Untersuchungen zum Resistenzverhalten von *Riemerella anatipestifer* und *Ornithobacterium rhinotracheale* des Wirtschaftsgeflügels Inaugural-Dissertation, Freien Universitat Berlin, Journal-Nr.: 3967. Available online at: (in German). https://refubium.fu-berlin.de/bitstream/handle/fub188/1601/Gasche_online.pdf?sequence=1&isAllowed=y. Accessed 31 May 2021

Gast RK, Porter RE (2020) Salmonella infections. In: Swayne DE, poultry diseases 14th edition, in diseases of poultry, vol II, 14th edn. Willey Blackwell, Hoboken, NJ, pp 717–753

Gavrilović P, Gavrilović A, Vidanović D, Parunović J, Jovanović M (2016) Comparative pathomorphological, bacteriological and serological examination of broiler breeders and pheasants experimentally infected with *Ornithobacterium rhinotracheale*. Avian Pathol 45(5):513–519

George BA, Ford AM, Fagerberg DJ, Quarles CL (1982) Influence of salinomycin on antimicrobial resistance of coliforms and streptococci from broiler chickens. Poult Sci 61:1842–1852

Gerber PF, MacLeod A, Opriessnig T (2018) *Erysipelothrix rhusiopathiae* serotype 15 associated with recurring pig erysipelas outbreaks. Vet Rec 182(22):635

Gharib-Naseri K, Kheravii SK, Keerqin C, Morgan N, Swick RA, Choct M, Wu SB (2019) Two different *Clostridium perfringens* strains produce different levels of necrotic enteritis in broiler chickens. Poult Sci 98(12):6422–6432

Ghunaim H, Abu-Madi MA, Kariyawasam S (2014) Advances in vaccination against avian pathogenic *Escherichia coli* respiratory disease: potentials and limitations. Vet Microbiol 172(1–2):13–22

Godener C, Le Guennec J (2019) Field experiences with the use of *Ornithobacterium rhinotracheale* inactivated autogenous vaccination in prebreeding and breeding Turkey flocks in France. Vet Rec Case Rep 7(4):e000864

Gomis S, Babiuk L, Allan B, Willson P, Waters E, Hecker R, Potter A (2007) Protection of chickens against a lethal challenge of *Escherichia coli* by a vaccine containing CpG oligodeoxynucleotides as an adjuvant. Avian Dis 51(1):78–83

Goossens E, Dierick E, Ducatelle R, Van Immerseel F (2020) Spotlight on avian pathology: untangling contradictory disease descriptions of necrotic enteritis and necro-haemorrhagic enteritis in broilers. Avian Pathol 49(5):423–427

Goossens E, Van Erum J, De Gussem M, Van Limbergen T, Callens C, De Zutter L, Ducatelle R, Van Immerseel F (2020a) Incidence and associated risk factors of necrotic enteritis in Belgian layer pullet flocks. Avian Pathol 49(5):476–485

Gray SJ, Ward TL, Southern LL, Ingram DR (1998) Interactive effects of sodium bentonite and coccidiosis with monensin or salinomycin in chicks. Poult Sci 77(4):600–604

Gray P, Jenner R, Norris J, Page S, Browning G (2021) The Australian veterinary association ltd and animal medicines Australia: antimicrobial prescribing guidelines for poultry; Antimicrobial prescribing guidelines for poultry (nih.gov). Accessed 31 May 2023

Gregersen RH, Christensen H, Ewers C, Bisgaard M (2010) Impact of *Escherichia coli* vaccine on parent stock mortality, first week mortality of broilers and population diversity of *E. Coli* in vaccinated flocks. Avian Pathol 39(4):287–295

Guo JR, Dong XF, Liu S, Tong JM (2017) Effects of long-term *Bacillus subtilis* CGMCC 1.921 supplementation on performance, egg quality, and fecal and cecal microbiota of laying hens. Poult Sci 96(5):1280–1289

Hafez H, Schulze D (2003) Examinations on the efficacy of chemical disinfectants on *Ornithobacterium rhinotracheale* in vitro. Archiv Fur Geflugelkunde 67:153–156

Hafez HM, Chin RP (2020) *Ornithobacterium rhinotracheale* infection. In: Swayne DE (ed) Diseases of poultry, vol II, 14th edn. Willey Blackwell, Hoboken, NJ, p 853

Hafez MH (2002) Diagnosis of *Ornithobacterium rhinotracheale*. Int J Poult Sci 1:114–118

Hargis, BM (2014) Overview of necrotic enteritis in poultry. MSD MANUAL veterinary manual. https://www.msdvetmanual.com/poultry/necrotic-enteritis/overview-of-necrotic-enteritis-in-poultry. Accessed 31 May

Harper M, John M, Turni C, Edmunds M, St Michael F, Adler B, Blackall PJ, Cox AD, Boyce JD (2015) Development of a rapid multiplex PCR assay to genotype *Pasteurella multocida* strains by use of the lipopolysaccharide outer core biosynthesis locus. J Clin Microbiol 53:477–485

Hartady T, Syamsunarno MRAA, Priosoeryanto BP, Jasni S, Balia RL (2021) Review of herbal medicine works in the avian species. Vet World 14(11):2889–2906

Hauck R, Chin RP, Shivaprasad HL (2015) Retrospective study on the isolation of *Ornithobacterium rhinotracheale* from chickens and turkeys in Central California: 294 cases (2000-12). Avian Dis 59(1):130–137

Heddleston KL, Gallagher JE, Rebers PA (1972) Fowl cholera: gel diffusion precipitin test for serotyping *Pasteruella multocida* from avian species. Avian Dis 16(4):925–936

Hegstad K, Mikalsen T, Coque TM, Werner G, Sundsfjord A (2010) Mobile genetic elements and their contribution to the emergence of antimicrobial resistant *Enterococcus*. Clin Microbiol Infect 16(6):541–554

Heidarpanah S, Thibodeau A, Parreira VR, Quessy S, Segura M, Meniaï I, Gottschalk M, Gaudreau A, Juette T, Gaucher ML (2023) Immunization of broiler chickens with five newly identified surface-exposed proteins unique to Clostridium perfringens causing necrotic enteritis. Sci Rep 13:5254

Heller ED, Leitner H, Drabkin N, Melamed D (1990) Passive immunisation of chicks against *Escherichia coli*. Avian Pathol 19(2):345–354

Hera A, Bureš J (2004) Veterinary autogenous vaccines. Develop Biol (Basel) 117:19–25

Huevelink A, Wiegel J, Kehrenberg C, Dijkman R, Soriano-Vargas E, Feberwee A (2018) Antimicrobial susceptibility of *Avibacterium paragallinarum* isolates from outbreaks of infectious coryza in Dutch commercial poultry flocks, 2008-2017. Vet Microbiol 217:135–143

Himathongham S, Riemann H, Bahari S, Nuanualsuwan S, Kass P, Cliver DO (2000) Survival of *salmonella typhimurium* and *Escherichia coli* O157:H7 in poultry manure and manure slurry at sublethal temperatures. Avian Dis 44(4):853–860

Hirt H, Greenwood-Quaintance KE, Karau MJ, Till LM, Kashyap PC, Patel R, Dunny GM (2018) *Enterococcus faecalis* sex pheromone cCF10 enhances conjugative plasmid transfer in vivo. MBio 9(1):e00037–e00018

Hossain MA, Islam MA (2004) Sero prevalence and mortality in chickens caused by pullorum disease and fowl typhoid in the government poultry farms in Bangladesh. Bangl J Vet Med 2(2):103–106

Hossain MM, Begum M, Kim IH (2015) Effect of *Bacillus subtilis, clostridium butyricum* and *lactobacillus acidophilus* endospores on growth performance, nutrient digestibility, meat quality, relative organ weight, microbial shedding and excreta noxious gas emission in broilers. Vet Med-Czech 60:77–86

Hu J, Zuo J, Chen Z, Fu L, Lv X, Hu S, Shi X, Jing Y, Wang Y, Wang Z, Mi R, Huang Y, Liu D, Qi K, Han X (2019) Use of a modified bacterial ghost lysis system for the construction of an inactivated avian pathogenic *Escherichia coli* vaccine candidate. Vet Microbiol 229:48–58

Huan Shu (1997) Study on ascites in hens caused by *Streptococcus faecalis* infection. Chinese J Vet Med 23(5):9–10

Hurtado R, Maturrano L, Azevedo V, Aburjaile F (2020) Pathogenomics insights for understanding *Pasteurella multocida* adaptation. Int J Med Microbiol 310(4):151417

Huvepharma (2020) Registration of a new anticoccidial in Europe. https://www.huvepharma.com/news/article/registration-of-a-new-anticoccidial-in-europe/ Accessed 31 May 2023

Jayaraman S, Thangavel G, Kurian H, Mani R, Mukkalil R, Chirakkal H (2013) *Bacillus subtilis* PB6 improves intestinal health of broiler chickens challenged with *Clostridium perfringens*-induced necrotic enteritis. Poult Sci 92(2):370–374

Jeong YW, Kim TE, Kim JH, Kwon HJ (2012) Pathotyping avian pathogenic *Escherichia coli* strains in Korea. J Vet Sci 13(2):145–152

JIACRA III (2021) European Centre for Disease Prevention and Control (ECDC), European Food Safety Authority (EFSA) and European Medicines Agency (EMA). Third joint inter-agency report on integrated analysis of consumption of antimicrobial agents and occurrence of antimicrobial resistance in bacteria from humans and food-producing animals in the EU/EEA, JIACRA III. 2016–2018. Stockholm, Parma, Amsterdam: ECDC, EFSA, EMA

Jiang Y, Mo H, Willingham C, Wang S, Park JY, Kong W, Roland KL, Curtiss R 3rd (2015) Protection against necrotic enteritis in broiler chickens by regulated delayed lysis salmonella vaccines. Avian Dis 59(4):475–485

Jiang S, Yan F-F, Hu J-Y, Mohammed A, Cheng H-W (2021) Bacillus subtilis-based probiotic improves skeletal health and immunity in broiler chickens exposed to heat stress. Animals 11(6):1494

Johnson TJ, Siek KE, Johnson SJ, Nolan LK (2005) DNA sequence and comparative genomics of pAPEC-O2-R, an avian pathogenic *Escherichia coli* transmissible R plasmid. Antimicrob Agents Chemother 49(11):4681–4688

Johnson TJ, Wannemeuhler YM, Scaccianoce JA, Johnson SJ, Nolan LK (2006) Complete DNA sequence, comparative genomics, and prevalence of an IncHI2 plasmid occurring among extraintestinal pathogenic *Escherichia coli* isolates. Antimicrob Agents Chemother 50(11):3929–3933

Johnson TJ, Wannemuehler Y, Doetkott C, Johnson SJ, Rosenberger SC, Nolan LK (2008) Identification of minimal predictors of avian pathogenic *Escherichia coli* virulence for use as a rapid diagnostic tool. J Clin Microbiol 46(12):3987–3996

Jørgensen SL, Stegger M, Kudirkiene E, Lilje B, Poulsen LL, Ronco T, Pires Dos Santos T, Kiil K, Bisgaard M, Pedersen K, Nolan LK, Price LB, Olsen RH, Andersen PS, Christensen H (2019) Diversity and population overlap between avian and human *Escherichia coli* belonging to sequence type 95. mSphere 4(1):e00333–e00318

Jung A, Chen LR, Suyemoto MM, Barnes HJ, Borst LB (2018) A review of *enterococcus cecorum* infection in poultry. Avian Dis 62(3):261–271

Jung A, Metzner M, Ryll M (2017) Comparison of pathogenic and non-pathogenic *enterococcus cecorum* strains from different animal species. BMC Microbiol 17(1):33

Kandeel M (2015) Pharmacokinetics and oral bioavailability of amoxicillin in chicken infected with caecal coccidiosis. J Vet Pharmacol Therap 38:504–507

Karavolias J, Salois MJ, Baker KT, Watkins K (2018) Raised without antibiotics: impact on animal welfare and implications for food policy. Translat Anim Sci 2:337–348

Kariyawasam S, Wilkie BN, Gyles CL (2004a) Construction, characterization, and evaluation of the vaccine potential of three genetically defined mutants of avian pathogenic *Escherichia coli*. Avian Dis 48(2):287–299

Kariyawasam S, Wilkie BN, Gyles CL (2004b) Resistance of broiler chickens to *Escherichia coli* respiratory tract infection induced by passively transferred egg-yolk antibodies. Vet Microbiol 98(3–4):273–284

Kastelic S, Berčič RL, Cizelj I, Benčina M, Makrai L, Zorman-Rojs O, Narat M, Bisgaard M, Christensen H, Benčina D (2013) *Ornithobacterium rhinotracheale* has neuraminidase activity causing desialylation of chicken and Turkey serum and tracheal mucus glycoproteins. Vet Microbiol 162(2–4):707–712

Kaul S (2018) The truth about Coccidiostats and why poultry farmers need to use them? British poultry association. https://britishpoultry.org.UK/the-truth-about-coccidiostats-and-why-poultry-farmers-need-to-use-them/. Accessed 31 May 2023

Kheravii SK, Swick RA, Choct M, Wu SB (2018) Effect of oat hulls as a free choice feeding on broiler performance, short chain fatty acids and microflora under a mild necrotic enteritis challenge. Anim Nutr 4(1):65–72

Kiarie E, Romero LF, Nyachoti CM (2013) The role of added feed enzymes in promoting gut health in swine and poultry. Nutr Res Rev 26(1):71–88

Kiga K, Tan XE, Ibarra-Chávez R, Watanabe S, Aiba Y, Sato'o Y, Li FY, Sasahara T, Cui B, Kawauchi M, Boonsiri T, Thitiananpakorn K, Taki Y, Azam AH, Suzuki M, Penadés JR, Cui L (2020) Development of CRISPR-Cas13a-based antimicrobials capable of sequence-specific killing of target bacteria. Nat Commun 11(1):2934

Kleven SH (1997) Changing expectations in the control of *mycoplasma gallisepticum*. Acta Vet Hung 45:299–305

Knowles TG, Kestin SC, Haslam SM, Brown SN, Green LE, Butterworth A, Pope SJ, Pfeiffer D, Nicol CJ (2008) Leg disorders in broiler chickens: prevalence, risk factors and prevention. PLoS One 3(2):e1545

Kolawole O, Graham A, Donaldson C, Owens B, Abia WA, Meneely J, Alcorn MJ, Connolly L, Elliott CT (2020) Low doses of mycotoxin mixtures below EU regulatory limits can negatively affect the performance of broiler chickens: a longitudinal study. Toxins 12(7):433

Kolawole O, Meneely J, Greer B, Chevallier O, Jones DS, Connolly L, Elliott C (2019) Comparative in vitro assessment of a range of commercial feed additives with multiple mycotoxin binding claims. Toxins (Basel) 11(11):659

Kondo H, Abe N, Tsukuda K, Wada Y (1997) Adherence of *enterococcus hirae* to the duodenal epithelium of chicks with diarrhoea. Avian Pathol 26(1):189–194

Koutsianos D, Athanasiou L, Mossialos D, Koutoulis K (2017) Study on prevalent serogroups and antimicrobial sensitivity profile of *Escherichia coli* strains isolated from commercial layer and layer breeder flocks in Greek territory. In: XXth International Congress of the World Veterinary Poultry Association, Edinburgh, Scotland, p 356

Koutsianos D, Gantelet H, Franzo G, Lecoupeur M, Thibault E, Cecchinato M, Koutoulis KC (2020) An assessment of the level of protection against colibacillosis conferred by several autogenous and/or commercial vaccination programs in conventional pullets upon experimental challenge. Vet Sci. 7(3):80

Krasucka D, Kowalski CJ (2010) Pharmacokinetic parameters of amoxicillin in pigs and poultry. Acta Pol Pharm 67(6):729–732

Kreizinger Z, Sulyok KM, Grózner D, Bekő K, Dán Á, Szabó Z, Gyuranecz M (2017) Development of mismatch amplification mutation assays for the differentiation of MS1 vaccine strain from wild-type *mycoplasma synoviae* and MS-H vaccine strains. PLoS One 12(4):e0175969

Küng E, Frey J (2013) AvxA, a composite serine-protease-RTX toxin of *Avibacterium paragallinarum*. Vet Microbiol 163(3–4):290–298

Kursa O, Pakula A, Tomczyk G, Pasko S, Sawicka A (2019) Eggshell apex abnormalities caused by two different *mycoplasma synoviae* genotypes and evaluation of eggshell anomalies by full-field optical coherence tomography. BMC Vet Res 15:1

La Ragione RM, Woodward MJ, Kumar M, Rodenberg J, Fan H, Wales AD, Karaca K (2013) Efficacy of a live attenuated *Escherichia coli* O78:K80 vaccine in chickens and turkeys. Avian Dis 57(2):273–279

Lambrecht E, Van Coillie E, Van Meervenne E, Boon N, Heyndrickx M, Van de Wiele T (2019) Commensal *E. Coli* rapidly transfer antibiotic resistance genes to human intestinal microbiota in the mucosal simulator of the human intestinal microbial ecosystem (M-SHIME). Int J Food Microbiol 311:108357

Landman WJ, Gruys E, Dwars RM (1994) A syndrome associated with growth depression and amyloid arthropathy in layers: a preliminary report. Avian Pathol 23(3):461–470

Landman WJ, Gruys E, Gielkens AL (1998) Avian amyloidosis. Avian Pathol 27(5):437–449

Landman WJM, van Eck JHH (2017) The efficacy of inactivated *Escherichia coli* autogenous vaccines against the *E. Coli* peritonitis syndrome in layers. Avian Pathol 46(6):658–665

Langemann T, Koller VJ, Muhammad A, Kudela P, Mayr UB, Lubitz W (2010) The bacterial ghost platform system: production and applications. Bioeng Bugs 1(5):326–336

Larson CT, Gross WB, Davis JW (1985) Social stress and resistance of chicken and swine to *Staphylococcus aureus* challenge infections. Can J Comp Med 49(2):208–210

Ledesma C, Rosario C, Gracia-Mora J, Tapia G, Gutiérrez L, Sumano H (2018) Antibacterial activity of amoxicillin in vitro and its oral bioavailability in broiler chickens under the influence of 3 water sanitizers. Poult Sci 97(7):2391–2399

Lee KW, Lillehoj HS, Jeong W, Jeoung HY, An DJ (2011) Avian necrotic enteritis: experimental models, host immunity, pathogenesis, risk factors, and vaccine development. Poult Sci 90(7):1381–1390

Lee Y, Kim WH, Lee SJ, Lillehoj HS (2018) Detection of chicken interleukin-10 production in intestinal epithelial cells and necrotic enteritis induced by *Clostridium perfringens* using capture ELISA. Vet Immunol Immunopathol 204:52–58

Li Z, Wang W, Liu D, Guo Y (2018) Effects of *lactobacillus acidophilus* on the growth performance and intestinal health of broilers challenged with *Clostridium perfringens*. J Anim Sci Biotechnol. 9:25

Liu G, Bogaj K, Bortolaia V, Olsen JE, Thomsen LE (2019) Antibiotic-induced, increased conjugative transfer is common to diverse naturally occurring ESBL plasmids in *Escherichia coli*. Front Microbiol 10:2119

Löhren U (2008) Guidelines for antimicrobial use in poultry. In: Guardabassi L (ed) Guide to antimicrobial use in animals. Blackwell Publishing Ltd., Oxford, UK, pp 126–142

Lowder BV, Guinane CM, Ben Zakour NL, Weinert LA, Conway-Morris A, Cartwright RA, Simpson AJ, Rambaut A, Nübel U, Fitzgerald JR (2009) Recent human-to-poultry host jump, adaptation, and pandemic spread of *Staphylococcus aureus*. Proc Natl Acad Sci USA 106(46):19545–19550

MacGowan A (2011) Revisiting beta-lactams—PK/PD improves dosing of old antibiotics. Curr Opin Pharmacol 11:470–476

Mageiros L, Méric G, Bayliss SC, Pensar J, Pascoe B, Mourkas E, Calland JK, Yahara K, Murray S, Wilkinson TS, Williams LK, Hitchings MD, Porter J, Kemmett K, Feil EJ, Jolley KA, Williams NJ, Corander J, Sheppard SK (2021) Genome evolution and the emergence of pathogenicity in avian *Escherichia coli*. Nat Commun 12(1):765

Manges RA (2016) *Escherichia coli* and urinary tract infections: the role of poultry-meat. Clin Microbiol Infect 22:122–129

Markowiak P, Śliżewska K (2018) The role of probiotics, prebiotics and synbiotics in animal nutrition. Gut Pathog 6(10):21

Mathers JJ, Clark SR, Hausmann D, Tillman P, Benning VR, Gordon SK (2004) Inhibition of resistance plasmid transfer in *Escherichia coli* by ionophores, chlortetracycline, bacitracin, and ionophore/ antimicrobial combinations. Avian Dis 48:317–323

Matthijs MG, Ariaans MP, Dwars RM, van Eck JH, Bouma A, Stegeman A, Vervelde L (2009) Course of infection and immune responses in the respiratory tract of IBV infected broilers after superinfection with *E. Coli*. Vet Immunol Immunopathol 127(1–2):77–84

Matthijs MG, van Eck JH, de Wit JJ, Bouma A, Stegeman JA (2005) Effect of IBV-H120 vaccination in broilers on colibacillosis susceptibility after infection with a virulent Massachusetts-type IBV strain. Avian Dis 49(4):540–545

McKellar QA, Sanchez Bruni SF, Jones DG (2004) Pharmacokinetic/pharmacodynamic relationships of antimicrobial drugs used in veterinary medicine. J Vet Pharmacol Ther. 2004 Dec;27(6):503-514. https://doi.org/10.1111/j.1365-2885.2004.00603.x.

McNamee PT, King DC (1996) Endocarditis in broiler breeder rearers due to *enterococcus hirae*. Vet Rec 138(10):240

McPeake SJ, Smyth JA, Ball HJ (2005) Characterisation of avian pathogenic *Escherichia coli* (APEC) associated with colisepticaemia compared to faecal isolates from healthy birds. Vet Microbiol 110(3–4):245–253

Mehdi Y, Létourneau-Montminy MP, Gaucher ML, Chorfi Y, Suresh G, Rouissi T, Brar SK, Côté C, Ramirez AA, Godbout S (2018) Use of antibiotics in broiler production: global impacts and alternatives. Anim Nutr. 4(2):170–178

Mellata M (2013) Human and avian extraintestinal pathogenic *Escherichia coli*: infections, zoonotic risks, and antibiotic resistance trends. Foodborne Pathog Dis 10(11):916–932

Mellata M, Maddux JT, Nam T, Thomson N, Hauser H, Stevens MP, Mukhopadhyay S, Sarker S, Crabbé A, Nickerson CA, Santander J, Curtiss R 3rd (2012) New insights into the bacterial fitness-associated mechanisms revealed by the characterization of large plasmids of an avian pathogenic *E. coli*. PLoS One 7(1):e29481

Melo EF, Clímaco WLS, Triginelli MV, Vaz DP, de Souza MR, Baião NC, Pompeu MA, Lara LJC (2019) An evaluation of alternative methods for sanitizing hatching eggs. Poult Sci 98(6):2466–2473

Methoxazol (2020), Methoxazol 20/100 mg/ml trimethoprim/sulfamethoxazole: the new standard in oral medication, Eurovet AH, 1–8. http://www.baltivet.com/files/9813/9324/8855/Brochure_Methoxasol.pdf. Accessed 31 May 2023

Meyers CM, Jensen MM (1987) Staphylococcosis of turkeys. 3. Bacterial interference as a possible means of control. Avian Dis 31(1):74–79

Mezhoud H, Chantziaras I, Iguer-Ouada M, Moula N, Garmyn A, Martel A, Touati A, Smet A, Haesebrouck F, Boyen F (2016) Presence of antimicrobial resistance in coliform bacteria from hatching broiler eggs with emphasis on ESBL/AmpC-producing bacteria. Avian Pathol 45(4):493–500

Mishra N, Smyth JA (2017) Oral vaccination of broiler chickens against necrotic enteritis using a non-virulent NetB positive strain of *Clostridium perfringens* type A. Vaccine 35(49 Pt B):6858–6865

Moazzami M, Fernström LL, Hansson I (2021) Reducing *campylobacter jejuni, Enterobacteriaceae* and total aerobic bacteria on transport crates for chickens by irradiation with 265-nm ultraviolet light (UV-C LED). Food Control 119:107424

Møller TSB, Liu G, Boysen A, Thomsen LE, Lüthje FL, Mortensen S, Møller-Jensen J, Olsen JE (2017) Treatment with cefotaxime affects expression of conjugation associated proteins and conjugation transfer frequency of an IncI1 plasmid in *Escherichia coli*. Front Microbiol 8:2365

Mombarg M, Bouzoubaa K, Andrews S, Vanimisetti HB, Rodenberg J, Karaca K (2014) Safety and efficacy of an aroA-deleted live vaccine against avian colibacillosis in a multicentre field trial in broilers in Morocco. Avian Pathol 43(3):276–281

Monecke S, Ruppelt A, Wendlandt S, Schwarz S, Slickers P, Ehricht R, Jäckel SC (2013) Genotyping of *Staphylococcus aureus* isolates from diseased poultry. Vet Microbiol 162(2–4):806–812

MAH (2020) Monitoring animal health: locomotion problems in broilers, Poultry, GD2246/07 2020:1–4.: Highlights report poultry july 2020.pdf. Accessed 31 May 2023

Monleon R, Martin MP, Barnes HJ (2008) Bacterial orchitis and epididymo-orchitis in broiler breeders. Avian Pathol 37(6):613–617
Monson MS, Lamont SJ (2021) Genetic resistance to avian pathogenic *Escherichia coli* (APEC): current status and opportunities. Avian Pathol 25:1–10
Montoro-Dasi L, Villagra A, Sevilla-Navarro S, Pérez-Gracia MT, Vega S, Marin C (2020) The dynamic of antibiotic resistance in commensal *Escherichia coli* throughout the growing period in broiler chickens: fast-growing vs. slow-growing breeds. Poult Sci 99(3):1591–1597
Moore RJ (2016) Necrotic enteritis predisposing factors in broiler chickens. Avian Pathol 45:275–281
Mora ZV, Macías-Rodríguez ME, Arratia-Quijada J, Gonzalez-Torres YS, Nuño K, Villarruel-López A (2020) *Clostridium perfringens* as foodborne pathogen in broiler production: pathophysiology and potential strategies for controlling necrotic enteritis. Animals (Basel) 10(9):1718
Morales-Erasto V, Falconi-Agapito F, Luna-Galaz GA, Saravia LE, Montalvan-Avalos A, Soriano-Vargas EE, Fernández-Díaz M (2016) Coinfection of *Avibacterium paragallinarum* and *Ornithobacterium rhinotracheale* in chickens from Peru. Avian Dis 60(1):75–78
Morrow CJ, Markham JF, Whithear KG (1998) Production of temperature-sensitive clones of *mycoplasma synoviae* for evaluation as live vaccines. Avian Dis 42(4):667–670
Mortensen S, Johansen AE, Thøfner I, Christensen JP, Pors SE, Fresno AH, Møller-Jensen J, Olsen JE (2019) Infectious potential of human derived uropathogenic *Escherichia coli* UTI89 in the reproductive tract of laying hens. Vet Microbiol 239:108445
Mostaan S, Ghasemzadeh A, Ehsani P, Sardari S, Shokrgozar MA, Abolhassani M, Nikbakht Brujeni G (2021) In silico analysis of *Pasteurella multocida* PlpE protein epitopes as novel subunit vaccine candidates. Iran Biomed J 25(1):41–46
Mostaan S, Ghasemzadeh A, Sardari S, Shokrgozar MA, Nikbakht Brujeni G, Abolhassani M, Ehsani P, Asadi Karam MR (2020) *Pasteurella multocida* vaccine candidates: A systematic review. Avicenna J Med Biotechnol 12(3):140–147
Moulin-Schouleur M, Répérant M, Laurent S, Brée A, Mignon-Grasteau S, Germon P, Rasschaert D, Schouler C (2007) Extraintestinal pathogenic *Escherichia coli* strains of avian and human origin: link between phylogenetic relationships and common virulence patterns. J Clin Microbiol 45:3366–3376
Murai A, Kitahara K, Okumura S, Kobayashi M, Horio F (2016) Oral antibiotics enhance antibody responses to keyhole limpet hemocyanin in orally but not muscularly immunized chickens. Anim Sci J 87(2):257–265
Murray S, Pascoe B, Méric G, Mageiros L, Yahara K, Hitchings MD, Friedmann Y, Wilkinson TS, Gormley FJ, Mack D, Bray JE, Lamble S, Bowden R, Jolley KA, Maiden MCJ, Wendlandt S, Schwarz S, Corander J, Fitzgerald JR, Sheppard SK (2017) Recombination-mediated host adaptation by avian *Staphylococcus aureus*. Genome Biol Evol 9(4):830–842
Murthy TR, Dorairajan N, Balasubramaniam GA, Dinakaran AM, Kalaimathi R (2007) The effect of vaccination of pullets against *Ornithobacterium rhinotracheale* infection. Avian Pathol 36(6):481–485
Mutalib A, Riddell C, Osborne AD (1983) Studies on the pathogenesis of staphylococcal osteomyelitis in chickens. I. Effect of stress on experimentally induced osteomyelitis. Avian Dis 27(1):141–156
Naemi A-O, Hymonti D, Nosheen K, Torbergsen Sandvik S, Schau Slettemeås J, Nesse LL, Simm R (2020) NarAB is an ABC-type transporter that confers resistance to the polyether Ionophores Narasin, Salinomycin, and Maduramicin, but not Monensin. Front Microbiol 11:104
Nakazawa H, Hayashidani H, Higashi J, Kaneko K, Takahashi T, Ogawa M (1998) Occurrence of *Erysipelothrix* spp. in broiler chickens at an abattoir. J Food Prot 61(7):907–909
Nedbalcová K, Zouharová M, Matiašková K, Bzdil J (2020) Monitoring of resistance of avian pathogenic isolates *Escherichia coli* (APEC) to antimicrobials in poultry farms in The Czech Republic in 2019. Veterinářství 70(10):613–617. (avalable in Czech language only)
Nhung NT, Chansiripornchai N, Carrique-Mas JJ (2017) Antimicrobial resistance in bacterial poultry pathogens: a review. Front Vet Sci 4:126
Nicoll TR, Jensen MM (1987) Preliminary studies on bacterial interference of staphylococcosis of chickens. Avian Dis 31(1):140–144
Nobilis OR inac (2002) European public assessment report: Nobilis OR inac emulsion for injection for chickens. Avaialable online at: https://www.ema.europa.eu/en/documents/product-information/nobilis--inac-epar-product-information_en.pdf. Accessed 31 May 2023
Nolan LK (2020) Colibacillosis. In: Swayne DE, Disease of poultry, 14th Wiley Blackwell II: p. 770–830
Norm/Norm-Vet (2001-2014). NORM/NORM-VET 2000-2014, in Usage of antimicrobial agents and occurrence of antimicrobial resistance in Norway, (Tromsø)
Norm/Norm-Vet (2019) Usage of antimicrobial agents and occurrence of antimicrobial resistance in Norway. NORM/NORM-VET 2018, Tromsø
VKM (2015) Norwegian Scientific Committee for Food Safety: The risk of development of antimicrobial resistance with the use of coccidiostats in poultry diets. Available online at https://vkm.no/download/18.2994e95b15cc5450716152d3/1498142579152/0025301628.pdf. Accessed 31 May 2023
Numee S, Hauck R, Hafez HM (2012) Influence of different storage media, temperatures and time duration on susceptibility of *Ornithobacterium rhinotracheale*. Pak Vet J 32:438–442
OIE (2020) Report of the meeting of the oie working group on antimicrobial resistance Paris–Annex III: technical

reference document listing antimicrobial agents of veterinary importance for poultry. https://www.oie.int/fileadmin/Home/eng/Internationa_Standard_Setting/docs/pdf/WGAMR/A_WG_AMR_Oct2020.pdf. Accessed 31 May 2023

Okorafor ON, Anyanwu MU, Nwafor EO, Anosa GN, Udegbunam RI (2019) Multidrug-resistant enterobacteria colonize commercial day-old broiler chicks in Nigeria. Vet World. 12(3):418–423

Olnood CG, Beski SSM, Iji PA, Choct M (2015) Delivery routes for probiotics: effects on broiler performance, intestinal morphology and gut microflora. Anim Nutr. 1(3):192–202

Omaleki L, Blackall PJ, Cuddihy T, White RT, Courtice JM, Turni C, Forde BM, Beatson SA (2022) Phase variation in the glycosyltransferase genes of Pasteurella multocida associated with outbreaks of fowl cholera on free-range layer farms. Microb Genom 8(3):000772

Opengard K, Boulianne M (2020) Necrotic Entereitis. In: Swayne DE (ed) Diseases of poultry, vol II, 14th edn. Willey Blackwell, Hoboken, NJ, pp 972–973

Opriessnig T, Forde T, Shimoji Y (2020) *Erysipelothrix* spp.: past, present, and future directions in vaccine research. Front Vet Sci. 15(7):174

Osman KM, Kappell AD, Elhadidy M, ElMougy F, El-Ghany WAA, Orabi A, Mubarak AS, Dawoud TM, Hemeg HA, Moussa IMI, Hessain AM, Yousef HMY (2018) Poultry hatcheries as potential reservoirs for antimicrobial-resistant *Escherichia coli*: a risk to public health and food safety. Sci Rep 8(1):5859

Osselaere A, Devreese M, Watteyn A, Vandenbroucke V, Goossens J, Hautekiet V, Eeckhout M, De Saeger S, De Baere S, De Backer P, Croubels S (2012) Efficacy and safety testing of mycotoxin-detoxifying agents in broilers following the European food safety authority guidelines. Poult Sci 91(8):2046–2054

Ovi F, Zhang L, Nabors H, Jia L, Adhikari P (2023) A compilation of virulence-associated genes that are frequently reported in avian pathogenic *Escherichia coli* (APEC) compared to other E. coli. J Appl Microbiol 134(3):lxad014

Oviedo-Rondón EO (2020) Available online at: Dysbacteriosis, its causes and its impact, Prestage Department of Poultry Science, College of Agriculture and Life Sciences. North Carolina State University, Raleigh, NC 27695–7608. https://www.dsm.com/content/dam/dsm/anh/en_US/documents/Dysbacteriosis,%20its%20causes%20and%20its%20impact.pdf. Accessed 31 May 2023

Paharik AE, Schreiber HL 4th, Spaulding CN, Dodson KW, Hultgren SJ (2017) Narrowing the spectrum: the new frontier of precision antimicrobials. Genome Med 9(1):110

Pan Q, Liu A, He C (2012) Co-infection of *Ornithobacterium rhinotracheale* with *streptococcus zooepidemicus* in chickens. Avian Dis 56(4):680–684

Pan Q, Liu A, Zhang F, Ling Y, Ou C, Hou N, He C (2012a) Co-infection of broilers with *Ornithobacterium rhinotracheale* and H9N2 avian influenza virus. BMC Vet Res 8:104

Panel on additives and products or substances used in animal feed (EFSA FEEDAP Panel) (2011a) Scientific opinion on the safety and efficacy of bentonite (dioctahedral montmorillonite) as feed additive for all species. https://efsa.onlinelibrary.wiley.com/doi/full/10.2903/j.efsa.2016.4623. Accessed 31 May 2023

Panel on additives and products or substances used in animal feed (EFSA FEEDAP Panel) (2011b) Scientific opinion on the efficacy of bentonite (dioctahedral montmorillonite) for all species. https://efsa.onlinelibrary.wiley.com/doi/full/10.2903/j.efsa.2017.5096. Accessed 31 May 2023

Papouskova A, Masarikova M, Valcek A, Senk D, Cejkova D, Jahodarova E, Cizek A (2020) Genomic analysis of *Escherichia coli* strains isolated from diseased chicken in The Czech Republic. BMC Vet Res 16(1):189

Pfaller MA, Cormican M, Flamm RK, Mendes RE, Jones RN (2019) Temporal and geographic variation in antimicrobial susceptibility and resistance patterns of enterococci: results from the SENTRY antimicrobial surveillance program, 1997-2016. Open Forum Infect Dis 6(Suppl 1):54–62

Pham VH, Kan L, Huang J, Geng Y, Zhen W, Guo Y, Abbas W, Wang Z (2020) Dietary encapsulated essential oils and organic acids mixture improves gut health in broiler chickens challenged with necrotic enteritis. J Anim Sci Biotechnol 11:18

Pikkemaat MG, Rapallini MLBA, Stassen JHM, Alewijn M, Wullings BA (2022) Ionophore resistance and potential risk of Ionophore driven co-selection of clinically relevant antimicrobial resistance in poultry, Wageningen Food Safety Research, WFSR report 2022. 005, pp. 30

Poultry Health Services (2015) Early mortality (Omphalitis), https://www.poultryhealth.ca/early--mortality-omphalitis/. Accessed 31 May 2023

Poultry world (2020) Russian poultry farmers want more leeway on coccidiostats. https://www.poultryworld.net/Health/Articles/2020/9/Russian-poultry-farmers--want-more-leeway-on-coccidiostats-644896E/. Accessed 31 May 2021

Prapapanpong J, Udomkusonsri P, Mahavorasirikul W, Choochuay S, Tansakul N (2019) In vitro studies on gastrointestinal monogastric and avian models to evaluate the binding efficacy of mycotoxin adsorbents by liquid chromatography-tandem mass spectrometry. J Adv Vet Anim Res 6(1):125–132

Prátová H (2020) Enhancement of the quality of the use of antimicrobials in veterinary medicine (Zvýšení kvality používání antimikrobik ve veterinární medicíně), Brno, Nov 2020, Ministry of Agriculture FÚ 8/2020; p: 72–79 (available in Czech language only)

Prescott JF, Parreira VR, Mehdizadeh Gohari I, Lepp D, Gong J (2016) The pathogenesis of necrotic enteritis in chickens: what we know and what we need to know: a review. Avian Pathol 45(3):288–294

Projahn M, Daehre K, Roesler U, Friese A (2016) Extended-spectrum-beta-lactamase- and plasmid-encoded cephamycinase-producing Enterobacteria in the broiler hatchery as a potential mode of pseudo-vertical transmission. Appl Environ Microbiol 83(1):e02364–e02316

Projahn M, Daehre K, Semmler T, Guenther S, Roesler U, Friese A (2018) Environmental adaptation and vertical dissemination of ESBL-/pAmpC-producing *Escherichia coli* in an integrated broiler production chain in the absence of an antibiotic treatment. Microb Biotechnol 11(6):1017–1026

Randall CJ, Wood AM, MacKenzie G (1993) Encephalomalacia in first-week chicks. Vet Rec 132(16):418

Rautenschlein S, Cheng HH, Lamont SJ (2020) Host factors for disease resistance: genetics of disease resistance. In: Swayne DE (ed) Disease of poultry, vol I, 14th edn. Wiley, Blackwell, pp 90–108

Razmyar J, Peighambari SM, Zamani AH (2017) Detection of a newly described bacteriocin, perfrin, among *Clostridium perfringens* isolates from healthy and diseased ostriches and broiler chickens in Iran. Avian Dis 61(3):387–390

Redgrave LS, Sutton SB, Webber MA, Piddock LJ (2014) Fluoroquinolone resistance: mechanisms, impact on bacteria, and role in evolutionary success. Trends Microbiol 22(8):438–445

Redweik GAJ, Jochum J, Mellata M (2020) Live bacterial prophylactics in modern poultry. Front Vet Sci. 7:592312

RESAPATH (2021) Anses: Resapath - French surveillance network for antimicrobial resistance in bacteria from diseased animals, 2019 Annual report, Lyon et Ploufragan-Plouzané-Niort, France, April 2021 https://resapath.anses.fr/resapath_uploadfiles/files/Documents/Rapport%20annuel/2019_Resapath_Annual%20report_GB.pdf Accessed 16 June 2024

Ricke SC, Lee SI, Kim SA, Park SH, Shi Z (2020) Prebiotics and the poultry gastrointestinal tract microbiome. Poult Sci 99(2):670–677

Ritter GD, Acuff GR, Bergeron G, Bourassa MW, Chapman BJ, Dickson JS (2019) Antimicrobial-resistant bacterial infections from foods of animal origin: understanding and effectively communicating to consumers. Ann N Y Acad Sci 1441(1):40–49

Rizzi L, Simioli M, Roncada P, Zaghini A (2003) Aflatoxin B1 and clinoptilolite in feed for laying hens: effects on egg quality, mycotoxin residues in livers, and hepatic mixed-function oxygenase activities. J Food Prot 66(5):860–865

Robbins KM, Suyemoto MM, Lyman RL, Martin MP, Barnes HJ, Borst LB (2012) An outbreak and source investigation of enterococcal spondylitis in broilers caused by *enterococcus cecorum*. Avian Dis 56(4):768–773

Roepke DC, Back A, Shaw DP, Nagaraja KV, Sprenger SJ, Halvorson DA (1998) Isolation and identification of *Ornithobacterium rhinotracheale* from commercial Turkey flocks in the upper Midwest. Avian Dis 42(1):219–221

Rood JI, Adams V, Lacey J, Lyras D, McClane BA, Melville SB, Moore RJ, Popoff MR, Sarker MR, Songer JG, Uzal FA, Van Immerseel F (2018) Expansion of the *Clostridium perfringens* toxin-based typing scheme. Anaerobe 53:5–10

Rood JI, Keyburn AL, Moore RJ (2016) NetB and necrotic enteritis: the hole movable story. Avian Pathol 45(3):295–301

Rosales AG (2019) Mycoplasmosis prevention and control in broiler breeders and broilers. Ross note-Mycoplasmosis Prev Control. https://eu.aviagen.com/assets/Tech_Center/Ross_Tech_Articles/RossNote--Mycoplasmosis-EN-19.pdf. Accessed 31 May 2023

RSPCA Approved Farming Scheme Standard–Meat Chickens (v. 1.1) (2020). https://rspcaapproved.org.au/wp-content/uploads/2020/08/2020_08_MEATCHICKENS_Standard_v1.1.pdf. Accessed 31 May 2023

Russell JB, Houlihan AJ (2003) Ionophore resistance of ruminal bacteria and its potential impact on human health. FEMS Microbiol Rev 27(1):65–74

Sadeyen JR, Kaiser P, Stevens MP, Dziva F (2015) A cyclophosphamide-sensitive cell compartment is essential for homologous protection conferred by licensed vaccines for the control of avian pathogenic *Escherichia coli* in chickens. Vaccine 33(31):3624–3627

Sander JE (2022a) Omphalitis in poultry. MSD manual veterinary manual. https://www.msdvetmanual.com/poultry/omphalitis/omphalitis-in-poultry?query=Omphalitis%20in%20Poultry. Accessed 31 May 2023

Sander JE (2022b) Fowl cholera. MSD manual veterinary manual. https://www.msdvetmanual.com/poultry/fowl-cholera/fowl-cholera Accessed 31 May 2023

Sargeant JM, Bergevin MD, Churchill K, Dawkins K, Deb B, Dunn J, Logue CM, Novy A, O'Connor AM, Reist M, Winder CB (2019) The efficacy of antibiotics to control colibacillosis in broiler poultry: a systematic review. Anim Health Res Rev 20(2):263–273

Sato Y, El-Gazzar M (2022) Staphylococcosis in poultry. MSD manual veterinary manual. https://www.msdvetmanual.com/poultry/staphylococcosis/staphylococcosis-in-poultry?query=el-gazzar. Accessed 31 May 2023

Sawata A, Kume K, Nakai T (1984) Relatiship between anticapsular antibody and protective activity of a capsular antigen of Haemphilus paragallinarum. Jap J Vet Sci 46:475–486

Selby CM, Beer LC, Forga AJ, Coles ME, Graham LE, Teague KD, Tellez-Isaias G, Hargis BM, Vuong CN, Graham BD (2023) Evaluation of the impact of formaldehyde fumigation during the hatching phase on contamination in the hatch cabinet and early performance in broiler chickens. Poult Sci 102(5):102584

Sgariglia E, Aconiti Mandolini N, Napoleoni M, Medici L, Fraticelli R, Conquista M, Gianfelici P, Staffolani M, Fisichella S, Capuccella M, Sargenti M, Perugini G (2019) Antibiotic resistance pattern and virulence genesin avian pathogenic *Escherichia coli* (APEC) from different breeding systems. Vet Ital 55(1):26–33

Sharma P, Gupta SK, Barrett JB, Hiott LM, Woodley TA, Kariyawasam S, Frye JG, Jackson CR (2020) Comparison of antimicrobial resistance and pangenome of clinical and non-clinical *enterococcus cecorum* from poultry using whole-genome sequencing. Food Secur 9(6):686

Shivaprasad HL (2000) Fowl typhoid and pullorum disease. Rev Sci Tech 19(2):405–424

Shivaprasad HL, Barrow PA (2013) Salmonella infections: pullorum disease and fowl typhoid. In: Diseases of poultry, vol 3, 13th edn. John Wiley & Sons, Inc, The United States, pp 678–693

Shoaib M, Dasti JI, Shah MAA, Zafa MA, Hasan MU, Riaz A, Rehman SU (2017) Salmonellosis in poultry, new prospects of an old disease: a review. Pakistan J Sci 69(4):361–368

Schouler C, Schaeffer B, Brée A, Mora A, Dahbi G, Biet F, Oswald E, Mainil J, Blanco J, Moulin-Schouleur M (2012) Diagnostic strategy for identifying avian pathogenic *Escherichia coli* based on four patterns of virulence genes. J Clin Microbiol 50(5):1673–1678

Simjee S, Heffron AL, Pridmore A, Shryock TR (2012) Reversible monensin adaptation in *enterococcus faecium*, *enterococcus faecalis* and *Clostridium perfringens* of cattle origin: potential impact on human food safety. J Antimicrob Chemother 67:2388–2395

Simm R, Slettemeås JS, Norstrom M, Dean KR, Kaldhusdal M, Urdahl AM (2019) Significant reduction of vancomycin resistant *E. Faecium* in the Norwegian broiler population coincided with measures taken by the broiler industry to reduce antimicrobial resistant bacteria. PLoS One 14(12):e0226101

Singer RS, Johnson TJ, Hofacre CL (2020) Antimicrobial therapy (including resistence). In: Swayne DE (ed) Disease of poultry, vol 1, 14th edn. Wiley, Blackwell, p 52

Singer RS, Porter LJ, Thomson DU, Gage M, Beaudoin A, Wishnie JK (2019) Raising animals without antibiotics: U.S. producer and veterinarian experiences and opinions. Front Vet Sci. 6:452

Singh M, Lim AJ, Muir WI, Groves PJ (2021) Comparison of performance and carcass composition of a novel slow-growing crossbred broiler with fast-growing broiler for chicken meat in Australia. Poult Sci 100(3):100966

Skyberg JA, Horne SM, Giddings CW, Wooley RE, Gibbs PS, Nolan LK (2003) Characterizing avian *Escherichia coli* isolates with multiplex polymerase chain reaction. Avian Dis 47(4):1441–1447

Smith JA (2019) Broiler production without antibiotics: United States field perspectives. Anim Feed Sci Technol 250:93–98

Smith JA, Boulianne ORL, Gingerich E (2020) Disease prevention and control in Antibitic free production. In: Swayne DE et al (eds) Diseases of poultry, vol I, 14th edn. Willey Blackwell, Hoboken, NJ, pp 40–57

Soliman AM, Aboubakr M, El-Hewaity M (2015) Bioequivalence study of two oral doxycycline formulations (Doxysol® and Doxymed®) in healthy broiler chickens. Pharmacol Phar 6:1–8

Soriano-Vargas E (2022) Infectious Coryza in chickens, MSD Manual, Veterinary manual. https://wwwmsd-vetmanualcom/poultry/infectious-coryza/infectious--coryza-in-chickens Assessed 30 Jun 2023

Stackebrandt E, Reboli AC, Farrar WE (2006) The genus Erysipelothrix. In: Dworkin M, Falkow S, Rosenberg E, Schleifer KH, Stackebrandt E (eds) The prokaryotes, vol 4, 3rd edn. Springer, New York, pp 492–510

Steentjes A, Veldman KT, Mevius DJ, Landman WJ (2002) Molecular epidemiology of unilateral amyloid arthropathy in broiler breeders associated with *enterococcus faecalis*. Avian Pathol 31(1):31–39

Stępień-Pyśniak D, Hauschild T, Nowaczek A, Marek A, Dec M (2018) Wild birds as a potential source of known and novel multilocus sequence types of antibiotic-resistant *enterococcus faecalis*. J Wildl Dis 54(2):219–228

Stępień-Pyśniak D, Wilczyński J, Marek A, Śmiech A, Kosikowska U, Hauschild T (2017) *Staphylococcus simulans* associated with endocarditis in broiler chickens. Avian Pathol 46(1):44–51

Stipkovits L, Kempf I (1996) Mycoplasmoses in poultry. Rev Sci Tech 15(4):1495–1525

Swormink BK (2019) Coccidiostats: antibiotic or feed additive? Poultry world, nutrition background. Avaialable online at: https://www.poultryworld.net/Nutrition/Articles/2019/3/Coccidiostats-Antibiotic-or-feed-additive-401585E/. Accessed 31 May 2021

Szabó R, Wehmann E, Makrai L, Nemes C, Gyuris É, Thuma Á, Magyar T (2017) Characterization of *Ornithobacterium rhinotracheale* field isolates from Hungary. Avian Pathol 46(5):506–514

Szafraniec GM, Szeleszczuk P, Dolka B (2022 Dec) Review on skeletal disorders caused by *staphylococcus* spp. in poultry. Vet Q 42(1):21–40

Tankson JD, Thaxton JP, Vizzier-Thaxton Y (2001) Pulmonary hypertension syndrome in broilers caused by *enterococcus faecalis*. Infect Immun 69(10):6318–6322

Tebrün W, Motola G, Hafez MH, Bachmeier J, Schmidt V, Renfert K, Reichelt C, Brüggemann-Schwarze S, Pees M (2020) Preliminary study: health and performance assessment in broiler chicks following application of six different hatching egg disinfection protocols. PLoS One 15(5):e0232825

Temmerman R, Garmyn A, Antonissen G, Vanantwerpen G, Vanrobaeys M, Haesebrouck F, Devreese M (2020) Evaluation of fluoroquinolone resistance in clinical avian pathogenic *Escherichia coli* isolates from Flanders (Belgium). Antibiotics (Basel) 9(11):800

Thieme S, Mühldorfer K, Lüschow D, Hafez HM (2016) Molecular characterization of the recently emerged poultry pathogen *Ornithobacterium rhinotracheale* by multilocus sequence typing. PLoS One 11(2):e0148158

Timbermont L, De Smet L, Van Nieuwerburgh F, Parreira VR, Van Driessche G, Haesebrouck F, Ducatelle R, Prescott J, Deforce D, Devreese B, Van Immerseel

F (2014) Perfrin, a novel bacteriocin associated with *net*B positive *Clostridium perfringens* strains from broilers with necrotic enteritis. Vet Res 45(1):40

Timbermont L, Haesebrouck F, Ducatelle R, Van Immerseel F (2011) Necrotic enteritis in broilers: An updated review on the pathogenesis. Avian Pathol 40:341–347

Tivendale KA, Logue CM, Kariyawasam S, Jordan D, Hussein A, Li G, Wannemuehler Y, Nolan LK (2010) Avian-pathogenic *Escherichia coli* strains are similar to neonatal meningitis *E. Coli* strains and are able to cause meningitis in the rat model of human disease. Infect Immun 78(8):3412–3419

Tofteland S, Lund O, Dahle UR, Sunde M (2017) Norwegian patients and retail chicken meat share cephalosporin-resistant *Escherichia coli* and IncK/ blaCMY-2 resistance plasmids. Clin Microbiol Infect 23(6):407

To H, Koyama T, Nagai S, Tuchiya K, Nunoya T (2009) Development of quantitative real-time polymerase chain reaction for detection of and discrimination between *Erysipelothrix rhusiopathiae* and other Erysipelothrix species. J Vet Diagn Invest 21(5):701–706

Torres C, Alonso CA, Ruiz-Ripa L, León-Sampedro R, Del Campo R, Coque TM (2018) Antimicrobial resistance in *enterococcus* spp. of animal origin. Microbiol Spectr 6(4)

Travers AF, Coetzee L, Gummow B (1996) Pathogenicity differences between south African isolates of *Ornithobacterium rhinotracheale*. Ondersspoort J Vet Res 63:197–207

USDA (2016) FSIS food safety and inspection service labeling guideline on documentation needed to substantiate animal raising claims for label submissions. https://www.fsis.usda.gov/wps/wcm/connect/6fe3cd56-6809-4239-b7a2-bccb82a30588/RaisingClaims.pdf?MOD=AJPERES. Accessed 31 May 2023

U.S. Food and Drug Administration (2020a) Animal drugs @ FDA. https://animaldrugsatfda.fda.gov/adafda/views/#/search. Accessed 31 May 2023

U.S. Food and Drug Administration (2020b) Veterinary feed directive. https://www.fda.gov/animal-veterinary/development-approval-process/veterinary-feed-directive-vfd. Accessed 31 May 2023

USKVBL (2021) List of authorised immunologicals for *Gallus gallus* in the Czech Republic. http://www.uskvbl.cz/en/authorisation-a-approval/marketing--authorisation-of-vmps/list-of-vmps/currently--authorised-vmp/vmp-search. Accessed 31 May 2023

Van den Bogaard AE, London N, Driessen C, Stobberingh EE (2001) Antibiotic resistance of faecal *Escherichia coli* in poultry, poultry farmers and poultry slaughterers. J Antimicrob Chemother 47(6):763–771

Van Goor A, Stromberg ZR, Mellata M (2017) A recombinant multi-antigen vaccine with broad protection potential against avian pathogenic *Escherichia coli*. PLoS One 12(8):e0183929

Van Hoek AH, Stalenhoef JE, van Duijkeren E, Franz E (2016) Comparative virulotyping of extended-spectrum cephalosporin-resistant *E. Coli* isolated from broilers, humans on broiler farms and in the general population and UTI patients. Vet Microbiol 194:55–61

Van Immerseel F, Rood JI, Moore RJ, Titball RW (2009) Rethinking our understanding of the pathogenesis of necrotic enteritis in chickens. Trends Microbiol 1:32–36

Van Meirhaeghe H, Schwarz A, Dewulf J, Van Immerseel F, Vanbeselaere B, De Gussem M (2019) Transmission of poultry dIseases and biosecurity in poultry production. In Biosecurity in animal production and veterinary medicine, Vetworks, 9880 Poeke, Belgium, pp. 330–356

Vanni M, Meucci V, Tognetti R, Cagnardi P, Montesissa C, Piccirillo A, Rossi AM, Di Bello D, Intorre L (2014) Fluoroquinolone resistance and molecular characterization of gyrA and parC quinolone resistance-determining regions in *Escherichia coli* isolated from poultry. Poult Sci 93(4):856–863

Veiga IMB, Lüschow D, Gutzer S, Hafez HM, Mühldorfer K (2019) Phylogenetic relationship of *Ornithobacterium rhinotracheale* isolated from poultry and diverse avian hosts based on 16S rRNA and rpoB gene analyses. BMC Microbiol 19(1):31

Vila-Donat P, Marín S, Sanchis V, Ramos AJ (2018) A review of the mycotoxin adsorbing agents, with an emphasis on their multi-binding capacity, for animal feed decontamination. Food Chem Toxicol 114:246–259

VMD-DEFRA (2020) Nobilis *E.coli* inac, summary of product characteristic. https://www.vmd.defra.gov.uk/ProductInformationDatabase/files/SPC_Documents/SPC_123124.PDF. Accessed 31 May 2023

Wade B, Keyburn AL, Haring V, Ford M, Rood JI, Moore RJ (2016) The adherent abilities of *Clostridium perfringens* strains are critical for the pathogenesis of avian necrotic enteritis. Vet Microbiol 25(197):53–61

Wade B, Keyburn AL, Haring V, Ford M, Rood JI, Moore RJ (2020) Two putative zinc metalloproteases contribute to the virulence of *Clostridium perfringens* strains that cause avian necrotic enteritis. J Vet Diagn Invest 32:259–267

Wakenell PS (2022) Erysipelas in poultry, MSD Manual veterinary manual. https://www.msdvetmanual.com/poultry/erysipelas/erysipelas-in-poultry. Accessed 31 May 2023

Walsh C, Fanning S (2008) Antimicrobial resistance in foodborne pathogens–a cause for concern? Curr Drug Targets 9(9):808–815

Walters J, Evans R, LeRoith T, Sriranganathan N, McElroy A, Pierson FW (2014) Experimental comparison of hemolytic and nonhemolytic *Ornithobacterium rhinotracheale* field isolates in vivo. Avian Dis 58(1):78–82

Wattrang E, Jäderblom V, Jinnerot T, Eriksson H, Bagge E, Persson M, Sørensen Dalgaard T, Söderlund R (2019) Detection and quantification of *Erysipelothrix rhusiopathiae* in blood from infected chickens—addressing challenges with detection of DNA from infectious agents in host species with nucleated red blood cells. J Med Microbiol 68:1003–1011

Watteyn A, Russo E, Garmyn A, De Baere S, Pasmans F, Martel A, Haesebrouck F, Montesissa C, De Backer P, Croubels S (2013) Clinical efficacy of florfenicol administered in the drinking water against *Ornithobacterium rhinotracheale* in Turkey housed in different environmental conditions: a pharmacokinetic/pharmacodynamic approach. Avian Pathol 42:474–481

Watteyn A, Devreese M, Plessers E, Wyns H, Garmyn A, Reddy VR, Croubels S (2016) Efficacy of gamithromycin against *Ornithobacterium rhinotracheale* in Turkey poults pre-infected with avian metapneumovirus. Avian Pathol 45:545–551

Weeks CA, Danbury TD, Davies HC, Hunt P, Kestin SC (2000) The behaviour of broiler chickens and its modification by lameness. Appl Anim Behav Sci 67(1–2):111–125

WHO (2024) WHO List of Medically Important Antimicrobials: a risk management tool for Licence: CC BY-NC-SA 3.0 IGO. https://www.who.int/news/item/08-02-2024-who-medically-important-antimicrobial-list-2024 Accessed 16June 2024

Wigley P, Berchieri A Jr, Page KL, Smith AL, Barrow PA (2001) Salmonella enterica serovar Pullorum persists in splenic macrophages and in the reproductive tract during persistent, disease-free carriage in chickens. Infect Immun 69(12):7873–7879

Wilde S, Jiang Y, Tafoya AM, Horsman J, Yousif M, Vazquez LA, Roland KL (2019) Salmonella-vectored vaccine delivering three *Clostridium perfringens* antigens protects poultry against necrotic enteritis. PLoS One 14(2):e0197721

Wilson KM, Chasser KM, Duff AF, Briggs WN, Latorre JD, Barta JR, Bielke LR (2018) Comparison of multiple methods for induction of necrotic enteritis in broilers. I. J Appl Poultry Res 27:577–589

WOAH (2013) Biosecurity procedures in poultry production in terrestrial animal Helath code, as amendded by update adopted in 2013 https://www.woah.org/fileadmin/Home/eng/Health_standards/tahc/2018/en_chapitre_biosecu_poul_production.htm. Accessed 25 May 2023

Wong A (2019) Unknown risk on the farm: does agricultural use of ionophores contribute to the burden of antimicrobial resistance? mSphere 4:e00433–e00419

Wood AM, MacKenzie G, McGiliveray NC, Brown L, Devriese LA, Baele M (2002) Isolation of *enterococcus cecorum* from bone lesions in broiler chickens. Vet Rec 150:27

Xu B, Chen X, Lu F, Sun Y, Sun H, Zhang J, Shen L, Pan Q, Liu C, Zhang X (2021) Comparative genomics of *mycoplasma synoviae* and new targets for molecular diagnostics. Front Vet Sci. 8:640067

Yamazaki Y, Oba E, Kashiwagi N, Sugita K, Shiiba K, Baba Y, Shimoji Y, Yamazaki W (2014) Development of a loop-mediated isothermal amplification assay for rapid and simple detection of *Erysipelothrix rhusiopathiae*. Lett Appl Microbiol 58(4):362–369

Yang F, Yang F, Wang G, Kong T (2018) Pharmacokinetics of doxycycline after oral administration of single and multiple dose in broiler chickens. J Vet Pharmacol Ther 41(6):919–923

Yang QE, Sun J, Li L, Deng H, Liu BT, Fang LX, Liao XP, Liu YH (2015) IncF plasmid diversity in multi-drug resistant *Escherichia coli* strains from animals in China. Front Microbiol 6:964

Yin D, Du E, Yuan J, Gao J, Wang Y, Aggrey SE, Guo Y (2017) Supplemental thymol and carvacrol increases ileum lactobacillus population and reduces effect of necrotic enteritis caused by *Clostridium perfringens* in chickens. Sci Rep 7(1):7334

Yoshida H, Bogaki M, Nakamura M, Nakamura S (1990) Quinolone resistance-determining region in the DNA gyrase gyrA gene of *Escherichia coli*. Antimicrob Agents Chemother 34(6):1271–1272

Zanella A, Alborali GL, Bardotti M, Candotti P, Guadagnini PF, Martino PA, Stonfer M (2000) Severe *Escherichia coli* O111 septicaemia and polyserositis in hens at the start of lay. Avian Pathol 29(4):311–317

Zhu Ge X, Jiang J, Pan Z, Hu L, Wang S, Wang H, Leung FC, Dai J, Fan H (2014) Comparative genomic analysis shows that avian pathogenic *Escherichia coli* isolate IMT5155 (O2:K1:H5; ST complex 95, ST140) shares close relationship with ST95 APEC O1:K1 and human ExPEC O18:K1 strains. PLoS One 9(11):e112048

Zhu Y, Hassan Y, Watts C, Zhou T (2016) Innovative technologies for the mitigation of mycotoxins in animal feed and ingredients–a review of recent patents. Anim Feed Sci Technol 216:19–29

Cattle

Lucie Pokludová

Abstract

Dairy and beef as well as veal calves create three main subsectors of cattle and have specificities not only in farming targeted on milk or meat production but also for therapeutical or metaphylactic use of antimicrobials. Group oral treatment (especially in pre-ruminating calves) has recently started to be replaced by long-acting injectables. Specific pharmaceutical forms for intramammary and intrauterine therapy are used in cows, where high-yield dairy cows create significant proportion of animals treated with antimicrobials. Considerations linked to pharmacokinetics after different routes of administration are therefore more pronounced in the cattle medicine and use of antimicrobials, especially highlighting the need to reach effective concentration of antimicrobials in udder tissues in mastitis treatment. In both beef and dairy cattle, old breeds as well as breeds with high genetic and top yields bring benefits as well as risk of suffering by the diseases, from which not only in Europe but also worldwide, most frequently treated by antimicrobials is mastitis in dairy cows and respiratory diseases in beef and veal calves' sectors. In young animals, infections caused by *Escherichia coli* and protozoa are main indications for use of antibiotics and anticoccidials, respectively. Considering these major diseases, but not omitting those less frequent, this chapter tries to highlight also preventive tools, vaccination, improved hygiene and alternatives to antimicrobials. More responsible and justified use, preference of the first-choice antimicrobials, selective drying off and supporting targeted antibiotic treatment with tools as on-farm culturing of mastitis pathogens, followed by antimicrobial susceptibility testing accompanied with strengthening the role of veterinarians working with farmers together, are promoted. Finally, an attempt to introduce alternatives to antimicrobials and future directions with this respect in the bovine sector are briefly summarised.

Keywords

Antimicrobials in cattle medicine · Antimicrobials in mastitis treatment · Susceptibility of cattle pathogens · Cattle pharmacokinetics and pharmacodynamics · Alternatives to antimicrobials in cattle

L. Pokludová (✉)
Institute for State Control of Veterinary Biologicals and Medicines, Brno, Czech Republic
e-mail: pokludova@uskvbl.cz

L. Pokludová (ed.), *Antimicrobials in Livestock 2: Choices of Prevention and Treatment*,
https://doi.org/10.1007/978-3-031-60807-0_4

1 Characterisation of the Cattle Sector

The European Union was ranked according to cattle population in 2021 in the TOP 5 regions with highest production, following India, Brazil, China and the USA that belong among the biggest producers of beef, veal and milk. India was the country with the highest number of dairy cows in the word in 2022 (approx. 59.5 million heads), followed by the EU (over 20 million heads). Interestingly, the production of milk in 27 member states in the EU in total is nearly twice the amount of milk in India. The top producer of beef and veal is the USA, according to 2022 data (Shahbandeh 2023). Substantial population of livestock was kept in the 27 EU member states as well, with 76 million bovine animals in 2021. The majority of bovine livestock are kept in just a few member states—three quarters of the EU's 2021 total bovine population was kept in France (22.9%), Germany (14.59%), Ireland (8.8%), Spain (8.7%), Poland (8.4%) and Italy (8.3%). Half of the EU's beef was produced in three member states (2021), i.e. France (21.1%), Germany (17.5%) and Italy (11.3%), and about 70% of the EU's veal meat was produced in three member states: the Netherlands (26.1%), Spain (23.7%) and France (19.7%).

1.1 Beef Cattle

It should be noted that certain cattle breeds are reared specifically for their beef, although beef can also come from dairy cattle. Beef is meat from the slaughtered bovine of at least 1 year old, and veal is meat from bovine younger than 1 year (usually male calves (dairy) and young cattle and calves from "beef breeds"). The EU produced nearly 6.8 million tonnes of bovine meat (beef and veal carcasses) in 2021, which signals a continuing decrease of production.

Import to the EU increased (83.9 million tonnes in 2021 to 103.6 million tonnes in 2022). The main importers (2022) to the European Union were the UK (approx. 30%), Brazil (nearly 25%), Argentina (approx. 24%), Uruguay (11%) and other countries in total (approx. 10%). As a note can be considered, in the Americas, based on the data from 6th report (WOAH 2022), 17 countries still allowed (2020 data) using of certain antimicrobials as growth promoters (AGP), including 8 classes in the WHO category of highest-priority critically important antimicrobials. But it should be also mentioned that during the last 5 years the situation has started to be changed in some of them, e.g. Uruguay (Azevedo 2019) is mentioned as declaring ban of AGP in cattle, sheep and goats, and in Brazil since 2016, colistin, tylosin, lincomycin and tiamulin were withdrawn, and according to the data available (Cardoso 2017), the following antimicrobials were allowed as growth promoters in cattle in Brazil: bacitracin, flavomycin, lasalocid, monensin, narasin, salinomycin and virginiamycin. Within the European Union antimicrobial growth promotion (AGP) was banned, and since January 2006 no antimicrobials are allowed to be used as AGP. In addition, the regulation (EC) 2019/6, applicable since January 2022 involves, together with a set of delegated and implemented acts, more strict rules for use of antimicrobials (Regulation (EU) 2019/6).

Four main specialised beef cattle production systems exist in the EU: suckler cow farms (breeder farm, BF) with production of calves to be sold to other farms for fattening; breeder and fattener (B&F) farms having both breeders and calves and fatteners; fattening-only farms (FF) that purchase weanlings to be fattened; and finally meat producers from dairy herds (MDH). Only around 50% of male cattle sold at between 1 and 2 years of age are raised on specialised beef farms with specialised fatteners (35%), predominating over breeder and fattener farms (17%), and minority of beef is gained as male calves from dairy herds. Traditional systems co-exist alongside more advanced, modern technology systems of farming. A considerable variation among farms exists within countries as well among them (Savoia et al. 2019).

1.2 Dairy Cattle

Around 23.5 million dairy cows are kept by EU farmers on 1.7 million farms. In 2021, an average milk yield per dairy cow was 7.57 tonnes/head/year, with highest production exceeding 10 tonnes/head/year (EUROSTAT 2023). The number of dairy cows has decreased steadily during the last 30 years in the EU. But using new and improved genetic selection, new technologies and advanced feeding concepts growing the most productive dairy herds as well as contracting the less productive ones—all in combination—cause the milk production per cow to still increase annually. Between member states variation exists on whether they increase or decrease cow numbers in recent years. As stated in DG Health and Food Safety Overview Report (DG Health and Food Safety 2017), the European dairy farms differ as for largeness, average milk-specialised farms keep an average of 54 dairy cows (some "west 10 old" EU countries, including west Germany), while in East Germany as well as in the Czech Republic, larger farms are typical (average herd size in CZ, 2017, was 314 cows per herd, when 70% of the Czech dairy cows are kept in barns with a capacity of 255 cows and above in 2017) (EDF Congress 2017). In contrast, farms in Bulgaria, Romania and Poland keep rather individual small family farming systems.

As for the diseases of the dairy cows, in the USA clinical mastitis affects at least 25% of cows during each lactation period (Goncalves et al. 2022), while in the recent study (Böker et al. 2023) performed in different regions of Germany, it indicates that annual median incidence of mild clinical mastitis on the farms was 14.8% (north), 16.2% (eastern) and 11.8% (South) and that of severe clinical mastitis was 4.0% (north), 2.0% (eastern) and 2.6% (south). Some examples from other EU countries on the incidence of mild mastitis signal an average of 14.4% in Austria (Firth et al. 2019) and 19% in Finland (Hiitiö et al. 2017). Although the exact incidence is, among the other factors, influenced by husbandry systems, also a methodology of reporting should be considered. Other bacterial diseases being treated with antimicrobials were, in the study by Goncalves et al. (2022) summarising US data, metritis (7–11% of cows), retained placenta (5–8% of cows) and respiratory disease (3% of cows).

2 Antimicrobial Stewardship in Cattle

Antimicrobial stewardship as a proactive approach is taken to emphasise the prevention of bacterial diseases in cattle (but generally in any animal species), to minimise the need for use of antimicrobials and finally to preserve the efficacy of antimicrobials, which are used within the framework of evidence-based treatment protocols. Such approach lowers the selection pressure on resistance spread. Recently antimicrobial stewardship principles, which were firstly introduced in human medicine, have been developed and started to be detailed and accommodated for different specific sectors of animal production. This is as well case for the cattle sector, where based on common principles of antimicrobial stewardship in bovine medicine (AABP 2022) antimicrobial stewardships for beef producers (BQA 2020) and dairy producers (Ruegg 2022) have been defined. Figure 1 provides some examples and flow chart within the main steps of antimicrobial stewardship in dairy cattle herds.

Despite not directly touching antimicrobial stewardship, an interesting outcome can be noted from the study performed via a survey carried out under leadership of the Federation of Veterinarians of Europe (FVE), where most effective alternative measures to prevent and avoid the need for metaphylactic treatment were identified (2329 respondents, majority were from mixed practices, followed by cattle, pigs, poultry, goat/sheep and rabbit farming sectors). Majority of respondents identified three key areas to prevent the use of antimicrobials via metaphylaxis: vaccination, followed by biosecurity, including hygiene measures and improved health management of importance, followed by two further factors, i.e. improved feeding and improved housing (Jerab et al. 2022).

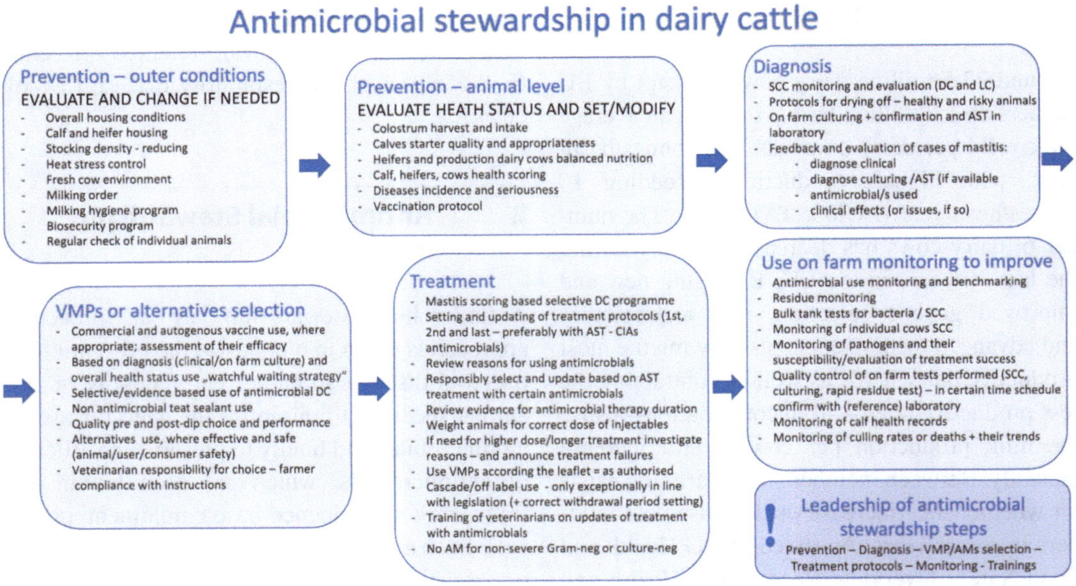

Fig. 1 Key elements and examples of steps in antimicrobial stewardship in dairy cattle (Ruegg 2022, modified and amended)

All the principles mentioned above highlight the fact that nowadays veterinary medicine is broadened towards preventive medicine, continuing with precise diagnosis and rational and responsible prescription and use of antimicrobials once needed. For good performance of veterinary medicine in practice, good partnership of the farmers, farm employees and attending veterinarian, with huge long-term knowledge of livestock's health and conditions on the farm, is vital. Regular veterinarian visits can end up being less expensive than emergency visits, as regular ones allow farmers to reduce extra veterinarian visit and costs of urgent care as well as choose the right moment for calling him/her. Trust created among veterinarian and farmer allows to understand better the herd health plan as well as rules for treatment with antimicrobials and follow the correct administration of VMPs, including off-label use of them. These exemptional cases need more close cooperation, including correct establishment of the sufficient withdrawal period. Farm-tailored prevention (vaccination protocols, biosecurity and hygienic measures, alternative tools to use of antimicrobials) should be shared and thoroughly discussed. Farm-tailored treatment protocols elaborated by attending veterinarian for the most common conditions as well as through recording of the VMPs used (including antimicrobials) should be golden standard, and both partners' sides should cooperate on following the valid legal conditions.

3 Groups of Antimicrobials and Their Purposes for Use in Cattle, Including Considerations on Routes of Administration of Antimicrobials

Antimicrobials, in the broad sense of meaning (i.e. according to the regulation (EU) 2019/6) involve antibiotics, antimycotics, antivirotics and antiprotozoal substances. In fact, within the portfolio used, most of the active substances belong to the group of antibiotics and much less among antiprotozoal. Veterinary medicinal products containing antimycotic and antiviral active substances for use in cattle are not available or very rarely available (cascade use) to practitioners.

A spectrum of antibiotics contained in VMPs for cattle is quite broad, but some of them (e.g. florfenicol, tildipirosin, tulathromycin) are not intended for dairy cows producing milk for human consumption (more information available in Volume I, p.86). Antibiotics are authorised in different pharmaceutical forms administered via different routes of administration, oral for non-ruminating calves only administered in milk, milk replacer or drinking water or specific pharmaceutical forms of bolus, injection, intramammary, intrauterine and topical.

A spectrum of authorised antiprotozoal substances mainly targeting coccidia (*Eimeria* spp., *Eimeria bovis*, *Eimeria zuernii*, *Eimeria alabamensis*) or on *Cryptosporidium parvum*, used in calves in authorised VMPs, is quite limited. Substances authorised for use in cattle in veterinary medicinal products are toltrazuril and diclazuril (targeting coccidia) and halofuginone (prevention or reduction of diarrhoea caused by *Cryptosporidium parvum* in newborn calves). From antimicrobials used for indications both on bacterial as well as *Cryptosporidium* spp. in non-ruminating calves, paromomycin sulphate and antimicrobials from sulphonamide group (against bacteria and coccidia) are the actives in authorised VMPs. The last substance to be mentioned targeting also *Cryptosporidium* spp. infections (but having as well antiviral activity) is nitazoxanide (which is banned for use in animals, based on the Commission Implementing Regulation (EU) 2022/1255 designating antimicrobials or groups of antimicrobials reserved for treatment of certain infections in humans, being in action since February 2023).

In all groups of antimicrobials, resistance is an issue, but considering the current level of knowledge, zoonotic potential of certain bacterial pathogens, frequency of use of antibiotics as well as portfolio of antibiotics authorised in cattle, further part is mainly targeted on antibiotics (being aware that some substances may have potential to inhibit agents from other groups of microorganisms and viruses).

3.1 Oral Water/Milk/Milk Replacer

It is a well-known fact that antimicrobials cannot be used orally in ruminating animals due to harmful effect on rumen microbiota determined using orally administered antimicrobials (or rather more specifically antibiotics) in pre-ruminating animals/calves only. Unfortunately, even in the absence of known disease, antibiotics were used extensively in calves for both therapeutic and prophylactic purposes, and although selling milk replacer containing antibiotics has been prohibited in the European Union for almost 30 years, it was used, in some extent in practice, in many countries (Eibl et al. 2021). Milk replacers with antibiotics—mostly tetracyclines, amoxicillin, tylosin but also aminoglycosides and even critically important antimicrobial colistin (De Briyne et al. 2014)—were mostly prescribed and used in diarrhoea in calves. Based on a study by Bokma et al. (2019), based on the defined daily dose for animals (DDDvet)/y calculations, 76.2% of orally administered antimicrobials and 23.8% of parenteral administrations were recorded in the veal calves' sector. From those orally administered amoxicillin, doxycycline, macrolides, oxytetracycline, aminosides, colistin and sulphonamide-trimethoprim were recorded (78 veal farms, 2014–2016). Prophylactic use of antimicrobials has been significantly restricted by Regulation (EU) 2019/6, but still a lot of veterinary medicinal products containing antimicrobials continue to be authorised for indication of calf diarrhoea (some of them with prophylaxis as well). Despite that fact, e.g. AMCRA (2023) guidelines do not advise the use of antibiotics as the first-line treatment of neonatal calf diarrhoea. Concerning prophylaxis for digestive infections or dysbacteriosis and neonatal dairy calf diarrhoea, studies suggest that calves receiving prophylactic antibiotics in their milk replacer (or even drinking discarded milk from treated dairy cows) during the first 2 weeks of life have a 28% greater risk for diarrhoea compared to calves free of any antibiotic intake. Fluid therapy and correct colostrum administration are recommended as a part of treatment and prophylaxis of neonatal calf diarrhoea (Meganck et al. 2014). For other types of infections, use of orally administered antimicrobials for treatment and metaphylaxis is quite limited, considering either pharmacokinetic (very small or near to zero proportion absorbed from gastrointestinal tract after oral administration namely in colistin and aminoglycosides) or limited intake of orally administered drugs in animals with serious course of infection.

Even if considered all limitations mentioned above, some factors are of importance, when oral treatment is needed. If not administered individually, it should be considered that bioavailability can differ as significantly as variation occurs in the absorption rate, depending also on individual and group animals' behaviour (hierarchisation). The consequence is large inter-individual differences in pharmacokinetic profiles, which affect markedly the systemic exposure to antimicrobial and hence therapeutic outcome within the group (Toutain et al. 2017). Insufficiently dosed and finally not bacteriologically cured animals can be a source of the spread of infection. In addition, spreading of resistant bacteria can be an issue, especially in some antimicrobials with single-step resistance selection. If such microbes are exposed to subinhibitory concentrations, moreover for shorter or longer time than appropriate for treatment, selection pressure on resistance can increase. Another recommendation for sufficient medicated liquids intake, important for receiving sufficient dose, is to prepare the solution in the milk/milk replacer of approximately 37–40°C, shortly before administration, because calves prefer to drink milk, which is warmed. Milk replacer needs to be dissolved in hot water. This led to preparing of the medication at farm level with hot water sometimes exceeding 70°C, which could be an issue especially in thermolabile antimicrobials. In calves, medicated milk replacer/milk may be fed using nipple bottles, individual nipple buckets, multi-feeder buckets (with nipples), milk lines with nipple bars or automated feeders. A lack of proper cleaning of all these devices as lines, reservoirs, nipples and surfaces both around and below nipples could lead to the persistence of residues of veterinary medicinal products and the formation of biofilms, slimes and moulds (EMA, 2020). Such improper hygienic and technology condition can be an issue both from hygiene and spread of disease view but can also create hot spots for selection of resistance. Yet no specific rules are provided by the legislation, although the feed hygiene legislation (Regulation (EC) 183/2005) makes the regular cleaning of watering systems mandatory, but this provision does not mean specifically cleaning before/after the use of VMPs in milk or milk replacer. In conclusion, in the case when oral administration is reasonable for proven medicinal conditions, all effort should be paid to provide administration of antibiotics in correct dose, frequency and total duration.

3.2 Injectable (Including Infusion and Fluid Therapy)

Nowadays injection administration is reasonably expected being the most frequently used route of administration in cattle generally (involving all production categories: beef cattle, dairy cows and veal calves) and considering also switch from oral administration to LA injectables in the veal calves' sector (Bokma et al. 2019). Injections are administered via intramuscular (IM), subcutaneous (SC) and less frequently intravenous (IV) route. A decision of which route will be more suitable is based on the exact VMP used (including considerations on the antimicrobial active, its bacterial killing effect, but also excipients and local tolerance). Also, from pharmacokinetic point of view of more speedy systemic availability, an IV route can be used as the first shot, followed by a SC route (acute mastitis, fluoroquinolone marbofloxacin) or IM only (cattle respiratory infections).

In the case of correct administration of antimicrobials via injection, aspect of sufficient dose, proper site of injection and correct administration as well as volume injected to one injection site, both from perspective of local tolerance and also efficacy (sufficient extent and speed of distribution from the injection site) and consumer safety (injection site residues), should be considered. Once thinking about the choice of antimicrobial administered via injection, the physicochemical properties of the substance itself as well as the formulation of exact VMP and also the slow release (long-acting VMPs) or depot formulations properties are of importance. Important for pharmacokinetics is if the product is administered subcutaneously or into the muscle itself or into connective tissue between muscles. Following administration, tissue reactions such as fibrosis, encapsulation, abscesses, cysts and necrosis contribute to retarded release of substances from injection sites. Local intolerance can have an impact on the pharmacokinetics of antimicrobials as well as food hygiene (Pyörälä et al. 1994). Adverse reactions, even transient, can also influence timeliness of sufficient concentration of antimicrobial bioavailability after IM or SC administration. Especially those with longer persistence can also have an impact on the quality of carcasses or residues of antimicrobials at injection sites and surroundings. Injection site lesions were detected in 4.1% of carcases (UK abattoirs inspections), consisting of cysts, discolouration, nodules, abscesses or scarring (Cresswell et al. 2017). As for dose, it was already pointed out (above under antimicrobial stewardship comment) that weighting of the animal should belong to standards of good veterinary/farm practice, especially ensuring not to underdose (risk of lack of efficacy) or overdose (risk of residues) the animal treated with antimicrobial VMPs.

As many of the VMPs used in cattle are designed to be injected into the muscle with the most common intramuscular injection site used being the gluteal region (the rump), instructions appear due to previously identified risks (Kirkwood et al. 2018). Both text and illustrations, as well as on-line videos, as for the best practice for injection (Barry 2019), injection techniques for beef cattle and broader, but detailed SOP for SC, IM and IV injections in cattle (UBC 2020) are available. As it has been shown that injection site lesions tend to occur more frequently and severely in gluteal region compared with the neck region (Cresswell et al. 2017), some authors recommend rather using this area (whenever appropriate also from the vet safety and suitable facility equipment). Injection to the neck area (intramuscular and subcutaneous neck area) as well as areas to be avoided (nuchal ligament, shoulder, spinal column, jugular furrow) are illustrated by Barry (2019). Certain authorised VMPs (based on patent of this route of administration) are used specifically at the base of the ear location (described for the VMP containing ceftiofur crystalline (EU, NAXCEL), where it is of importance to administer it properly in order to comply with the meat withdrawal period.

In severe mastitis, mainly those caused by coliforms, supportive therapy using NSAIDs via injection and fluid therapy via using solutions for infusions are used. Hypertonic saline 7.2% in concentration 5 mL/kg body weight in rapid infusion can be administered as supportive treatment.

On the other hand, slow infusion is recommended for calcium gluconate 23%, where amounts among 250–500 mL are given as adjunctive therapy.

3.3 Intramammary

This route of administration is authorised almost for cattle (*Bos taurus*) or buffaloes. Specific applicators = intramammary injectors (syringes), with cannulas of different length and with dosing of content of one injector for administration to individual quarter are used. Within intramammary administration of VMP with antimicrobial active substance, high local concentrations and exposures may be achieved at or near the site of infection (based on different ability of the substance to penetrate, e.g. intracellularly), with a relatively low dose of antimicrobials. Systemic side effects are therefore minimised as well as distribution, which in systemically available VMPs is a concern from tissue residues' perspective. From the perspective of evaluation, the efficacy against target pathogen concentration in milk is relevant (and not in plasma). However, the udder cannot be simply considered as tissue with fully homogenous distribution of the antimicrobial. Different concentrations are achieved in milk within milk cisterna and also in other udder tissue compartments and several concepts of excipients and active substance/s differs across different forms (suspensions, oitments, foams) and PK and PD issues should be therefore considered. A situation is even more complicated once considering possible differences for healthy udder or subclinical mastitis udder or udder with acute clinical mastitis. What is of importance as well is localisation of the causative pathogen. The distribution through the udder is dependent not only on the active substance physicochemical properties but on the exact formulation of the product, including excipients, viscosity, particle size and other individual VMP characteristics.

It should be also noted that for the effective treatment, especially in the cases of serious mastitis, both parenteral and intramammary treatment can be used concomitantly. Systemic treatment may be indicated when the parenchyma is intensely swollen, since the milk duct system is then either compressed or blocked by inflammatory secretions and cell debris. Such conditions inhibit the distribution throughout the udder and make the access of antimicrobial drug to the site of infection limited or fully blocked (Gruet et al. 2001).

This approach should be considered with care as for the establishment of the safe withdrawal period, especially for milk, but in some substances (with ability to accumulate), also in the inner organs, especially kidneys. Attention should be paid to consideration of the health status of the animal that can affect speed of the residue depletion.

It should be also noted that intramammary treatment schedules are designed and VMPs authorised (based on both efficacy and withdrawal period studies) for twice per day milking routines. Some milking routines or robotic milking systems are established for more than two times per day of milking (mostly three times per day), but also flexible milking systems exists. In such cases, dosing schedule (especially frequency of administration) is questioned as well as withdrawal period to be established after different dosing schedules. This case is typical example, where it is hard to follow the Article 106 (1) of the Regulation (EU) 2019/6, which stipulates that "Veterinary medicinal products shall be used in accordance with the terms of the marketing authorization".

The above commented information is targeted on cows in lactation. Intramammary treatment is used also in dry cow therapy, where mostly antibiotics from the same groups are used but either in different forms of active substances (e.g. benzathine salts) or in modified formulations (long-acting oil bases), which allows stability and more long-acting effect.

3.4 Intrauterine

Specific, topical route of administration is intrauterine, where several pharmaceutical forms are authorised for use in cattle (intrauterine foaming tablet, intrauterine capsule or bolus, intrauterine foam or spray, intrauterine gel, emulsions, suspensions or solutions), some of them administered via specific devices. Within the EU, surprisingly a lot of antimicrobial substances are authorised for intrauterine use in mostly monocomponent intrauterine VMPs (amoxicillin, ampicillin, cephapirin benzathine, chlortetracycline, formosulfathiazole, gentamicin, kanamycin, oxytetracycline, rifaximin, tetracycline and thiamphenicol) but also VMPs containing two actives in fixed combination (amoxicillin/clavulanic acid, ampicillin/cloxacillin, ampicillin/oxacillin, ampicillin/neomycin, oxytetracycline/neomycin, oxytetracycline/sulfamonomethoxine/trimethoprim/sulfadiazine). Increasing the severity of metritis, accompanied by a compromised barrier of the endometrium and a delay in involution, can lead to a higher absorption of some antimicrobials (e.g. oxytetracycline from the uterus), which can result in higher milk concentrations of the drug (Gorden et al. 2016). The pharmacokinetics of antimicrobials in the uterus and uterine secretion and compartments was mostly studied in healthy cows after intrauterine administrations; therefore it does not reflect the changes in inflamed tissue (Mileva et al. 2022) Attention should be paid to the changes as increased blood delivery to the uterus, levels of inflammatory proteins, epithelial barrier breakdown and uterine secretion changes, as these factors influence the disposition of antimicrobials in effective concentration to target causal pathogen (Gorden et al. 2016).

4 Pharmacological Considerations with Respect to Use of Antimicrobials in Cattle

Different pharmaceutical forms are used in cattle, and age is one of the factors that can influence the pharmacokinetics of the active substances, both in positive (increased penetration to the site of infection) and in negative way (higher toxicity, especially in neonates). Below some examples are given for different pharmaceutical forms, routes of administration and composition models (e.g. long-acting formulations of injectable and intramammary VMPs). To give an overview of antimicrobials used in the cattle sector in the EU, but also bringing the information, of which active substances are available for peroral or parenteral treatment, Table 1 lists DDDvet$_{catte}$ for parenteral and peroral treatment, which were collected and set based on the "Principles on assignment of defined daily dose for animals (DDDvet) and defined course dose for animals (DCDvet)" (EMA/710019/2014; EMA 2015) for the cattle. Technical DDDvet units reflect the exact authorised doses in SPCs and therefore can be many times considered very close/similar to doses which were authorised/used in the treatment in this period, and with high probability most of them stay valid, as investment to variation of marketing authorisations of antimicrobials to change the dose (and consequently withrawal period as well) is hardly expected. Several antimicrobials have, however, authorised doses, which significantly differ from those in DDDvet, these are (where relevant) highlighted in italic-bold and placed in brackets in Table 1. It is therefore necessary to search and, based on Article 106/1 of Regulation (EU) 2019/6 to follow, the doses as indicated in SPCs of authorised veterinary medicinal products.

4.1 Calves

As already commented above, pre-ruminating calves' stage is the only administered with antimicrobials via oral route in the cattle sector. Because of reticular groove closure when a calf suckles, most drugs (including antimicrobial) administered in milk/milk replacer via suckling completely bypass the immature rumen. This fact substantially reduces the time it takes for a drug to reach the small intestine. Care should be taken to interaction with milk/milk replacer that may affect certain antimicrobials' effect (e.g. calcium divalent cations interaction with oxytetracycline). In the neonates, the first weeks of the animals' life are characterised with increased antimicrobial absorption from the gastrointestinal tract (e.g. aminoglycoside apramycin absorption in 1-day-old calf is sevenfold higher than in 21-day-old calf (Buck et al. 1990). In 1-week-old calves fed 4 mg/kg Na-penicillin G together with milk, the bioavailability was 11.7% and at the age of 5 weeks it was only 6.6%, with C_{max} 102 ng/mL and 56 ng/mL, respectively and AUC decrease from 870 to 362 ng/mL h (Musser and Anderson 2001).

Another route of administration that can be used both in calves and adult cattle is injection. Parenteral absorption varies in neonates due to the altered blood flow to intramuscular and subcutaneous injection sites, with the decreased muscle mass/body weight in neonates compared to adults (Brown 1992). In calves, especially in the neonatal period, increased bioavailability and altered disposition (wider distribution and slower elimination) can be expected. The altered pharmacokinetic behaviour of antimicrobials in neonates affects the plasma drug concentration and finally also the concentration attained at the target site of infection. In calves it concerns mainly the period since birth to approximately 8 weeks of age. As indicated by different authors (Baggot 2001; Brown 1992; Buck et al. 1990), some differences in the key PK characteristic features, which are more pronounced at the earliest period after the birth (first 24 h) and going more towards adult pharmacokinetics in later stages, can be described. For IV administration, the area under the concentration-time curve (AUC), 7456 and 5508 ng/mL h, and systemic clearance, 0.54 and 0.73 L/kg h, were significantly different ($P < 0.05$) at 1- and 5-week-old calves, respectively (Musser and Anderson 2001).

As for injection administration, despite the route of administration, in neonates lower binding to plasma proteins (especially albumin)

Table 1 The defined daily dose for animals (DDDvet) for antimicrobial agents for use in cattle administered orally or parenterally (EMA/710019/2014; EMA 2015)

Antimicrobial	Oral (PO) DDDvet in mg/kg/day	Parenteral (INJ) DDDvet in mg/kg/day	Antimicrobial	Oral (PO) DDDvet in mg/kg/day	Parenteral (INJ) DDDvet in mg/kg/day
Amoxicillin	20	8.3 (***15***)	Penethamate hydriodide	Not used	12.0 (***9, 15***)
Ampicillin	29	11.0	Phthalylsulfathiazole in combi (trimethoprim)	25.0	Not used
Apramycin	30	20.0	Procaine benzylpenicillin	Not used	13.0
Benzylpenicillin	Not used	14.0	Spectinomycin (sole)	20.0	28.0
Benzathine benzylpenicillin in combi	Not used	2.9	Spectinomycin + lincomycin[b]	Not used	18.0
Cefalexin	Not used	7.0	Spiramycin	35.0	21.0
Cefquinome	Not used	1.5	Streptomycin	70.0	10.0
Ceftiofur	Not used	1.0	Sulfachlorpyridazine_trimethoprim[c]	20.0	Not used
Chlortetracycline	22.0	Not used	Sulfadiazine + trimethoprim [c]	24.0	13.0
Colistin	4.8	2.6	Sulfadimethoxine	40.0	30.0
Danofloxacin	Not used	1.9	Sulfadimethoxine + trimethoprim[c]	23.0	19.0
Difloxacin	Not used	3.8	Sulfadimidine	105.0	101.0
Dihydrostreptomycin	48.0	25.0	Sulfadimidine + trimethoprim[c]	30.0	16.0
Doxycycline	10.0	10.0	Sulfadoxine + trimethoprim[c]	25.0	14.0
Enrofloxacin	4.7	4.2 (***5***)	Sulfaguanidine	77.0	Not used
Erythromycin	20.0	12.0	Sulfamethoxazole + trimethoprim[c]	Not used	15.0
Florfenicol	Not used	13.0 (***20, 30, 40***)	Sulfamethoxypyridazine in combi not with trimethoprim	35.0	Not used
Flumequine	14.0	9.8	Sulfamethoxypyridazine + trimethoprim[c]	23.0	Not used
Framycetin	Not used	5.0	Sulfamethoxypyridazine	Not used	51.0
Kanamycin	Not used	15.0	Tetracycline	28.0	2.5
Lincomycin + spectinomycin[a]	Not used	13.0 (***18***)	Thiamphenicol	45.0	Not used
Marbofloxacin	1.0	3.6 (***2, 8, 10***)	Tilmicosin	21.0	4.0 (***10, 25***)
Neomycin	21.0	Not used	Trimethoprim + sulfa	4.8	2.8
Neomycin in combi	Not used	7.4	Tylosin	41.0	13.0 (***10, 20, 40***)
Oxolinic acid	17.0	Not used	Gamithromycin_LA (single-dose active in mg/kg bw listed)	Not used	6.0
Oxytetracycline	20.0	6.5 (***20, 25***)	Tildipirosin_LA (single-dose active in mg/kg bw listed)	Not used	4.0
Paromomycin	38.0	12.0	Tulathromycin_LA (single-dose active in mg/kg bw listed)	Not used	2.5
			Ceftiofur_LA (crystalline-free acid, micronised, adjuv) (single-dose active in mg/kg bw listed)	Not used	5.0

"in combi" (after the antimicrobal´s name) means the dose of the substance, once administered as fixed combination with other antimicrobial/s mostly is substantially lower; numbers in ***italic-bold*** in brackets mean the doses which are currently (2024) authorised for certain indications in injectable VMPs for cattle.
LA for gamithromycin, tildipirosin and tulathromycin: Single dose as indicated in the product SPC in mg of active/kg body weight is listed
LA for ceftiofur: Single dose as indicated in the product SPC in mg of active as base "ceftiofur"/kg body weight is listed
[a]When in combination with spectinomycin, because the dose of lincomycin is substantially lower in single-substance products than in combination with spectinomycin
[b]When in combination with lincomycin, because the dose of spectinomycin is substantially higher in single-substance products than in combination with lincomycin
[c]When in combination with trimethoprim, due to synergistic effect

occurs. Protein binding in organic bases (lincosamides, erythromycin and other selected macrolides and trimethoprim) decreases as calves mature.

In lipophilic molecules (macrolides, lincosamides fluoroquinolones (e.g. enrofloxacin), trimethoprim) of antimicrobials, lower ratio of body fat to fluids has an effect, which causes such antimicrobials to remain in the vascular system to a larger degree in neonates. In hydrophilic molecules (e.g. penicillins, cephalosporins, aminoglycosides, oxytetracycline, sulfadiazine), larger extracellular fluid volume of neonates leads to the distribution of more antimicrobials out of the bloodstream compared with adults. Therefore, a larger volume of extracellular fluid causes dilution, which finally results in a lower extracellular concentration of hydrophilic antimicrobial (Brown 1992).

There has been as well-documented increased permeability of the blood-brain barrier (enhanced by inflammatory processes), as this barrier is not fully morphologically developed at the birth and some hours after, therefore moderating lipid-soluble antimicrobials could as well as entering the cerebrospinal fluid. Slower elimination of the drugs was described in neonates (especially in very early stages (2–3 days) or if the calf is born premature) due to immature glomerular filtration, and in drugs demanding active renal tubular secretion can be up to 20–30 days of life (Baggot 2001; Brown 1992). Also, hepatic drug metabolism plays a role and differs in calves, e.g. in trimethoprim and sulfadiazine, fixed combination (ratio of 1:5) is authorised and therefore administered. However, the metabolism of trimethoprim and sulfadiazine differs and, in neonatal and maturing calves, does not maintain the same ratio in the bloodstream. Question can be therefore raised both on the ratio and on bactericidity (which is functional due to the combination of both, and mostly in seriously ill calves, bactericidal antibiotics are preferred to be used).

4.2 Cattle and Long-Acting Injectables

Injectable veterinary medicinal products are frequently used in treatment of many diseases in cattle. There are medicinal products, which are designed, by prolonged absorption, to provide an extended dosage interval. Historically first of such products were oxytetracyclines, later amoxicillin and ampicillin, then followed by florfenicol and ceftiofur crystalline and finally macrolides or related classes (gamithromycin, tulathromycin). Some examples of PK parameters of the selected LA injectable products are given in Table 2. For amoxicillin, one example of two long-acting VMPs is visible difference, where one of the factors can be different excipient composition. In gamithromycin, only one VMP is authorised in the EU, and PK data shows the difference among parameters achieved in plasma and PELF, where data by Giguere et al. (2011) clearly show higher Cmax, MRT and AUC are achieved, and therefore the product is indicated against respiratory tract pathogens.

4.3 Dairy

In dairy cattle injectable, intramammary, intrauterine and topical administration of antimicrobials is performed.

Considering the information presented by Gruet and accounting as well the portfolio of antimicrobials sold in 2021 as intramammary VMPs for lactating cows in Europe and antimicrobials authorised for treatment of mastitis administered via injection route, Table 3 categorises antimicrobials (parenterally and intramammary administered) based on their potential distribution throughout the udder.

The ideal candidate drug for parenteral treatment of mastitis (once target pathogen is suscep-

Table 2 Examples for PK parameters of selected LA antibiotics

Active substance	Excipients	Dosing schedule	Compartment	C_{max} [μg/mL]	T_{max} [h]	$T_{1/2}$ [h]	MRT [h]	Cl [mL/h]	AUC [h . μg/mL]
Amoxicillin I[a]	Butylated hydroxytoluene Butylated hydroxyanisole Aluminium stearate Propylene glycol dicaprylocaprate	15 mg/kg bw/single inj/ volume divided into 2 inj sites/calf	Plasma	2.91 ± 0.93	1.30 ± 0.35	12.9 ± 10.9	17.65	428.4 ± 92.3	36.3
Amoxicillin II[b]	Silica, colloidal anhydrous Sorbitan oleate Propylene glycol dicaprylocaprate	15 mg/kg bw/single inj	Plasma	5.02	2	7.8			
Gamithromycin[c]	Monothioglycerol Succinic acid Glycerol formal	6 mg/kg bw/single inj/SC	Plasma	0.265	1	50.8 ± 3.8	41.3	654.0	9.25 ± 1.10
		6 mg/kg bw/single inj/SC	Plasma	0.433	1	62	43.1	830.0	7.95
		6 mg/kg bw/single inj/SC	PELF	4.61	24	50.6	71.1		348.0
		6 mg/kg bw/single inj/SC	Plasma	0.13 ± 0.003			52.8 ± 0.18	1140 ± 27	
		6 mg/kg bw/single inj/SC	PELF	3.04 + 0.09			79.0 ± 0.29		
Tulathromycin[d]	Monothioglycerol Propylene glycol Citric acid Hydrochloric acid Sodium hydroxide Water for injection	2.5 mg/kg bw/single inj/SC	Plasma	0.5	0.5	90			
		2.5 mg/kg bw/single inj/SC	Plasma	0.718 ± 0.405	0.25–1.0		171.5 + 17.53	139 + 27.8	
		2.5 mg/kg bw/single inj/SC/ calf	Plasma	0.75	2.2	84	91.4		47.9
Florfenicol[e]	Methylpyrrolidon Propylene glycol Macrogol 300	20 mg/kg bw/single inj/IM/ calf (Lobell et al. 1994)	Plasma	3.07	3.33	18.3			42

[a]Betamox LA, Lees et al. (2015), composition according SPC; single inj = single injection; 2 inj sites = two injection sites
[b]Vetaxam LA, composition according SPC; single inj = single injection
[c]Zactran, Huang et al. (2010), De Donder et al. (2016), Giguere et al. (2011), composition according SPC; single inj = single injection
[d]Draxxin, Rivera et al. (2019), Toutain (2017), composition according SPC; single inj = single injection
[e]Nuflor, Lobell et al. (1994), calves, composition according SPC; single inj = single injection

Table 3 Categorisation of parenterally and intramammary administered antimicrobials based on potential distribution throughout the udder

	AMEG	Antimicrobial group	Parenteral	Intramammary
Good	D	Penicillins	Penethamate	Penethamate
		Aminopenicillins[a]		Ampicillin[a]
				Amoxicillin[a]
				Hetacillin
		Diaveridines	Trimethoprim	Trimethoprim
		Sulfonamides (selected)	Sulfanilamides	
	C	Ansamycins[b]		Rifaximin[b]
		Cephalosporins 1st gen		Cefalexin
		Macrolides	Tylosin	Tylosin
			Erythromycin	Erythromycin
			Oleandomycin	Oleandomycin
			Spiramycin	Spiramycin
		Lincosamides	Lincomycin	Lincomycin
				Pirlimycin
		Pleuromutillins	Tiamulin	
	B	(Fluro)quinolones	Enrofloxacin	Enrofloxacin
			Marbofloxacin	Marbofloxacin
				Cefoperazone
		Cephalosporins 3rd gen		Ceftiofur[c]
		Cephalosporins 4th gen		Cefquinom
	Unclassified			Novobiocin
Limited	D	Penicillins	Penicillin G	Penicillin G
		Isoxazolyl penicillins	Cloxacillin	Cloxacillin
				Oxacillin
		Aminopenicillins[a]	Ampicillin[a]	
			Amoxicillin[a]	
		Cephalosporins 1st gen	Cefalexin	Cephalonium
				Cephapirin
				Cephacetril
		Tetracyclines	Oxytetracycline	Tetracycline
		Steroid antibacterials	Fusidic acid	
	C	Ansamyins[b]	Rifamycin[b]	
	Unclassified		Novobiocin	
Poor distribution	D	Cyclic polypeptides		Bacitracine
	C	Aminoglycosides	Dihydrostreptomycin	Dihydrostreptomycin
			Gentamicin	Gentamicin
			Kanamycin	Kanamycin
			Neomycin	Neomycin
		Aminocyclitols	Spectinomycin	
	B	Polymyxins	Polymyxin	Polymyxin
	Unclassified			Tyrothricin

[a]Some authors recommended it as a last choice (ESBL selection pressure)
[b]Some authors recommended it as a last choice (single-step resistance development, importance as antituberculotic)
[c]Not authorised in intramammary VMPs in the EU yet

tible) should have good bioavailability after IM administration, be a weak base (or otherwise non-ionised in serum) with good lipid solubility, low degree of plasma proteins binding and long half-life in the body and should retain activity in inflammatory secretions and not accumulating in organs (i.e. have clearance from body organs and tissues similar to the clearance of the drug from the blood (Ziv 1980b; Erskine et al. 2003). The issue of crossing the blood-milk barrier can be linked with pH (plasma approx. 7.4) vs. normal milk (mildly acidic, 6.4–7.0) vs. mastitis (up to 7.5, according mastitis severity), and the ionisation of the drug is dependent on the severity of infection (Gruet et al. 2001; Bansal et al. 2005).

Intramammary antimicrobials released from the pharmaceutical formulation distribute into the aqueous and lipid fractions of the milk by passive diffusion. In hydrophilic sector formed by milk and more or less modified mammary secretions filling udder cisterns, antibiotics with hydrophilic molecules will be distributed, while lipid-soluble drug will concentrate in the lipophilic structures of the udder, mainly the membranes. The drug's solubility in lipids is of critical importance for the effect of the antibiotic, since it determines how readily it will move through the lipid-rich membranes that separate small ducts from the alveoli, the milk secreting tissue and small milk-collecting tubules (Gruet et al. 2001).

From some studies (McDougall 1998; Serieys et al. 2005) targeted on clinical mastitis treatment and comparing the efficacy of intramammary vs. parenteral administration, it can be suggested that after both routes, similar outcomes can be reached for certain antimicrobials in certain infections. But robust data for the whole portfolio of antimicrobials used in treatment of mastitis are limited, and generally speaking, in mastitis, mostly higher concentration of antimicrobial is achieved in mammary tissue using intramammary infusion in cases where intensive swelling is not present. For example, the concentration of penicillin G in milk after intramammary administration is 100–1000 times as high as the concentration after systemic (parenteral) administration (Pyörälä 2009). Less total amount of antimicrobial is administered in teat canal via intramammary route, and only certain part from this amount is distributed systemically. Therefore, load for the gastrointestinal tract (GIT) is smaller after intramammary than after parenteral administration of antimicrobials. As GIT is considered as "hot spot" for resistance selection and transfer of resistance within gut microbiome, intramammary administration seems less risky from the resistance selection and transfer perspective.

Despite this, parenteral (systemic) therapy may have clinical advantages in certain cases as, e.g. when the parenchyma is intensely swollen, since the milk duct system is then either compressed or fully blocked by inflammatory secretions and cell debris. The site of infection in udder compartments is therefore unreachable for the antibiotic administered intramammary (Gruet et al. 2001); if *Mycoplasma bovis* outbreak occurs within the herd (to decrease pathogen spread at the time of treatment); when multiple quarters of a cow are infected concurrently or when large numbers of cows need treatment into all four quarters (AVA-AMA 2022); when administration of intramammary infusion is not possible due to pain; from practical point of view, injectable product administration once daily (or once per 48 h) vs. administration via intramammary infusion twice daily and safety of administering person (handling with hardly manageable animals).

Penicillin antibiotics (benzylpenicillins or penethamate hydroiodide) are frequently used as first-line treatment in mastitis caused by susceptible Gram-positive bacteria via both parenteral and intramammary routes. When three different doses of procaine benzylpenicillin intramammary are administered to cows, the penicillin concentration in the udder tissue exceeded the MIC of the susceptible *Staphylococcus aureus* and *Streptococcus uberis* for at least 24 h: after administration of 200,000 IU the drug concentration was 20 times > MIC, after 300,000 IU it was 22 times, and after 600,000 IU it was 70 times > MIC (Abbeloos et al. 2018). In the case of penethamate, due to its weakly basic properties and high lipophilicity, its absorption is eight times faster than penicillin G after intramammary administration (Ziv 1980a). Penethamate-lipophilic antibiotic, with pKa 8.5, is present in the blood in an undissociated form, and unlike penicillin G, it can cross natural barriers such as the blood-brain barrier or the blood-milk barrier (Ziv 1980b; Kroker 2003). After parenteral administration, high concentrations are therefore reached in milk and in the lungs and central nervous system as well. In the udder tissue, penethamate is partially cleaved into active penicillin G and diethylaminoethanol, which spread diffusionally. The remaining undissociated penethamate enters the parenchyma and cells, where rapid dissociation to penicillin G prevents its back diffusion into the interstitial fluid. Increased concentration of penicillin G in the udder tissue and cells occurs (Ziv 1980a; Rasmussen 1959). In milk (slight acidic, compared to serum), penicillin G is released from penethamate (Kroker 2003). In comparison to penethamate and the corresponding dose of penicillin G administered, the concentration of penicillin G in milk is up to 40 times higher (Demuth and Müntener 2006). In cattle to which 15 mg/kg penethamate had been administered intramuscularly, the maximum concentration in milk was 5.91 μg/mL after 0.811 h, and mean AUC of 8.558 μg h/mL was about 2.1 times higher than the AUC in plasma. Based on regimen with a first dose of 15 mg/kg, followed by twice (24 and 48 h after) by a dose of 7.5 mg/kg at 4.65 h after the third injection, a maximum concentration of 0.437 μg/mL was measured (Friton et al. 2003). In another study, where penethamate hydroiodide at the dose equivalent to 5 million IU penicillin G was administered intramuscularly for 4 consecutive days, an average concentration of 1.20 ± 0.36 μg/mL was measured in milk on day 1 and 1.12 ± 0.04 μg/mL on day 4 (Ziv and Storper 1985). Undissociated penethamate rapidly and easily penetrates the udder whether inflamed or not, and its liposoluble nature gives it advantage, compared with other beta-lactams (e.g. amoxicillin) and aminoglycosides to diffuse through the parenchyma of the udder, pass into the milk and penetrate the lactogenic cells. The mechanism of "ion trapping" explains the different penetration of penethamate compared with penicillin G (Friton et al. 2003).

Staphylococcus aureus survives in acidic media, including phagolysosomes. In a study by Moncada Cárdenas et al. (2009), the MIC of *S. aureus* field strains and reference strain *S. aureus* ATCC 25923 were tested at different pH: 7.4, 6.5 and 5.0. Such acidity simulates conditions in subcellular structures which are commonly associated with *S. aureus* intracellular persistence. For penicillin G, detected MIC_{90} at pH 7.4 was consistent with those reported by CLSI 2007 (0.5 μg/mL), but at pH 5.0 (phagolysosomes) the activity of penicillin G increased markedly and almost linearly (~tenfold decrease in MIC of 0.06 μg/mL).

Mestorino an Errecalde (2012) also performed tests with cloxacillin in lactating cows with lower and higher yields. They detected slower half-life and prolonged elimination in cows with low milk production, which seems quite logical.

An interesting molecule that can be administered intramammary is thiamphenicol, where high concentrations in the nonmedicated quarters and in serum were determined. Complete transfer from the treated quarter to the homolateral quarter (rear left (RL)) was achieved by simple diffusion (short time 0.25 h), while transfer to the heterolateral quarters was achieved through circulation (longer time with 0.58 h to rear right (RR) and 1.92 h to front right (FR)). In the homolateral quarter, the main C_{max} and the highest bioavailability were recorded 75.30 (RL), 15.06 (RR), 19.03 (FR) and 15.79 μg mL (in serum). The authors suggested transfer to the contralateral quarters is via the blood circulation due to that the longitudinal medial intermammary septum is a highly impermeable (Mestorino et al. 1993).

As for Gram-negative bacteria, there can be some first-line treatment performed also by injectable VMPs. As for combination of sulphonamide and trimethoprim, it was recently noted that after extravascular injection sites, poor absorption leads to failure in treatment due to not reaching effective concentrations in the udder (AVA-AMA 2022). On the other hand, an improved clinical outcome for cows with coliform mastitis was achieved by once intramuscular loading dose of 8 mg trimethoprim/kg and 40 mg sulfadiazine/kg, followed by administration of three daily doses of 4 mg trimethoprim/kg and 20 mg sulfadiazine/kg (Shpigel et al. 1998). It should be noted that first increased dose exceeds doses as currently approved, but there are authorised VMPs with a dose of 25 mg/kg bw in total (combination of trimethoprim and sulfadoxine in a ratio of 1:5). It should be as well noted that IV administration is possible. After administration of 40 mg/kg sulfadiazine and 8 mg/kg trimethoprim, a therapeutic concentration of 2 µg/mL sulfadiazine is reached in the udder after 0.5–4 h (IV), 2–12 h (IM) and 8–12 h (SC) (Kaartinen et al. 1999).

One more aspect should be considered. Practically all intramammary VMPs are authorised for dosing regimen that is based on administration twice daily. With the increasing frequency of dairy farms milking three times per day and with the popularity of automated milking systems, increasing interest on how milking frequency affects the distribution and elimination of intramammary antibiotics has been noted. Recently, in other systems and production stage in milking phase regimen, ten milkings in 7 days can also occur. Therefore, some discussion was held on how to cope with administration ensuring efficacy and adequate withdrawal period setting for another regimen different from those authorised. In some systems, cows can be placed to quarantine, with treatment and milking scheme changed to twice daily, which demand extra spaces, extra staff care and extra costs. In dairy cows with high milk production, it is even more difficult to switch to another regimen, especially in animal suffering from a disease.

There are only few studies discussing three times daily milking regimen and administration of antibiotic. In one study (Lindquist et al. 2015) with hetacillin, which is rapidly converted back to ampicillin and acetone in the mammary gland after administration, treatment schedule once per day (dose of 62.5 mg potassium hetacillin) administered exceed MIC_{90} for at least 65% of the 24-h dosing interval (8- and 16-h concentrations above MICs of Gram-positive pathogens). In another study using intramammary cephapirin, cows dosed at a 16-h interval maintained antibiotic concentrations for a longer period of time than cows dosed at an 8-h interval in Holsteins that were milked three times per day (Stockler et al. 2009). The frequency of intramammary treatment in that study had no effect on milk withdrawal period.

A situation with pharmacokinetics differs in VMPs administered at drying off. The process of release of the antibiotic active substance is slow and influences in large scale the initial shape of the concentration versus time curve. The excipients used, although very variable, may include vegetable or mineral oils. The mineral oil, such as paraffin or glycerine oils, causes delay in the release of the active ingredient and prolongs time for which the antibiotic stays inside the gland. Other agents that increase the retention, such as aluminium monostearate in an oily base and hydroxystearin, can also be used (Ziv 1980b). After the DC intramammary administration at drying off, the elimination of antibiotic from the mammary gland can be expressed by a monoexponential or biexponential curve. The elimination rate is affected by the different parameters—the dose, the nature of the excipient and physicochemical properties, the particle sizes, the antibiotic characteristics and the extent of binding of the antibacterial agent to the gland content and to the mammary tissue (Mestorino and Errecalde 2012; Mattie et al. 1997). Overview of sales of DC intramammary VMPs in Europe in 2021 has shown that cloxacillin benzathine is the most used active at drying off (personal communication of the author, ESVAC data). A study by Mestorino et al. (2018) shows that 2 days after treatment, concentrations of cloxacillin of 19.96 ± 12.14 µg/mL were detected, which decreased slowly towards day 29 at levels close to 0.12 µg/mL, after the intramammary infusion of formulation containing 500 mg of cloxacillin benzathine. The level of binding of cloxacillin to dry udder secretion detected was high 86 ± 6.5%, acting as a reservoir and increasing the persistence of the ATM inside the udder (Mestorino and Errecalde 2012).

5 Main Protozoal Pathogens in Calves and Beef Cattle and Their Management and Treatment with Antimicrobials with Antiprotozoal Activity

Among very common agents causing diarrheal disease in calves belong to *Eimeria* spp., a single host, intracellular protozoal parasite damaging the lining of the intestine (small or large intestine). The most pathogenic and widespread species of coccidia affecting calves include *Eimeria bovis*, *Eimeria zuernii* and *Eimeria alabamensis* (Staněk 2023; Gilleard 2021). Calves between 2 and 5 months of age mostly suffered from coccidiosis, but disease can occur in calves younger than 2 months or calves older than 6 months. Coccidiosis is usually encountered 2–4 weeks after calves are moved into new housings and grouped after weaning. Farms with continuous operation of rearing calves, where there is no proper sanitation of the breeding environment, are faced with coccidiosis frequently. Calves are infected exclusively via the oral route: sporulated (mature) oocyst ingestion through food, water or direct contact with contaminated litter and faeces or licking of other animals.

Coccidiosis typically presents as a diarrhoeal disease that can vary from acute and severe to low grade and chronic. Subclinical forms can cause reduced growth rates and food conversion, possible future fertility impairment and a predisposition to intercurrent disease.

The most important factors that influence whether the calf will develop the disease and what the course of the disease will be, including prevention and treatment options, are listed in Table 4.

Following groups and individual substances were or still are present in authorised VMPs targeted on coccidia: quinolones (decoquinate; calves use), triazine derivatives (toltrazuril and diclazuril; both calves use) and sulphonamides (e.g. sulfadimidine; calves use). Other substances as ionophores (monensin and lasalocid) are approved as feed additives. Toltrazuril is effective against all the different internal stages of *Eimeria* and is used strategically as a single dose (15 mg toltrazuril/kg bw), during the pre-patent period. The aim is to break the life cycle before disease and larger levels of environmental contamination occur. Interestingly such strategy allows the infection to progress sufficiently, so it allows immunity to develop. Once the same breed/age of calves of different weights are treated, the dose is recommended to be calculated according to the heaviest from the group intended to be treated. One recent case report of interest is the development of toltrazuril resistance in *Eimeria* spp. in sheep in Norway (Odden et al. 2018).

Another protozoal gastrointestinal disease, which occur and cause pathologies in young ruminants, partly neonatal calves, is cryptosporidiosis. The causal agent *Cryptosporidium parvum* is a protozoal parasite with life cycle containing stages (oocysts), which allows either direct transmission via faecal-oral route or indirect colonisation through contaminated environment and being ingested via water, food, or licking the housing equipment contaminated as a few oocysts only can cause infection. Infected animals may suffer from profuse watery diarrhoea, inappetence, lethargy and dehydration, and in severe cases death can occur (Thomson et al. 2017). Diarrhoea usually starts around 3–4 days after ingestion of infective oocysts, and the duration is 1–2 weeks. Oocyst shedding occurs approximately from 4 to 12 days post-infection (depending on the initial challenge dose (Zambriski et al. 2013). Shedding is not only in diarrhoeic animals. Except suffering of the calves and causing economic lost, the issue with cryptosporidiosis is its zoonotic potential, with young children being particularly vulnerable (mainly *Cryptosporidium hominis*, but *Cryptosporidium parvum* can cause human infections as well) (Hadfeld et al. 2011). As quite limited prevention is available (no vaccines and extremely difficult elimination of oocysts from the environment), the only control measure is to try to reduce the environmental contamination. Measures are very similar to those listed above for coccidia, which are mainly high standards in cleaning, housing hygiene, removal of faeces and contaminated bedding from calving areas and calf houses, combined with steam cleaning and disinfection with a suitable disinfectant (e.g. hydrogen peroxide disinfectants). As already described in *Eimeria* spp. prevention (parts of Table 4), strategies based on thorough cleaning (Harp and Goff 1998) with very hot water followed by drying can also be effective as oocysts are susceptible to extremes of temperature (down to −20 °C and up to 60 °C) and desiccation (Robertson et al. 1992). As for diagnostics, *Cryptosporidium* spp. can be detected via faecal flotation and microscopy, antigen-capturing enzyme-linked immunosorbent assay (ELISA) or polymerase chain reaction (PCR) (Cho and Yoon 2014; Quinn et al. 2011). As for the treatment and

Table 4 Main measures to do to minimise the risk of coccidiosis development, including prevention, diagnosis and proper treatment (Staněk 2023; Gilleard 2021, amended)

Main measures to do	Recommendations/explanations
Ballance nutrition to positively influence health condition	Poor nutrition status with influence on health condition. Poor health condition makes the animals more prone to different diseases
Try to improve immunostatus of the calves	Presence of specific antibodies influencing how prone they will be for infestation; of importance is adequate amount of quality colostrum intake
Improve hygiene to lower infection pressure	Level of hygiene (feeding and water) and hygiene of both animals and environment (level of contamination of the environment with coccidia (influencing rate of infectious pressure)
Ensure key steps in disease prevention	Selected elementary steps to avoid coccidiosis in calves: – Calves of same/similar ages (all in-all out system) – Removal of bedding with thorough removal of debris – Subsequent mechanical cleaning of housing and technology, including removal of feed residues and cleaning of feeding devices – Cleaning, disinfecting (product effective against coccidia oocysts), drying, steam "wap" systems and total drying of all spaces and equipment – New bedding (alkalisation substance application may help)
Improve clime in housings	Dry, clean, proper temperature, with appropriate/comfortable airflow; regular changing the litter prevents the disease (in opposite vet and dirty conditions and wrong litter quality promote the disease spread)
Consider pathogen properties and ecology	Level of pathogenicity of coccidia, stages present (repeated cycle causing re-infestation of calves) and conditions in housing enabling to develop and spread developmental stages (matured oocysts) infecting calves
Watch the animals for symptoms of coccidiosis	The following signs can be present in calves with the coccidiosis: – Despondency and apathy – Bristly coat – Very heavily soiled rump and root of the tail – Diarrhoea (up to watery appearance), which may contain both mucus (usually at the beginning of the disease) and blood (in the later stages of the disease) – Frequent straining to defecation (with risk of rectal prolapse) – Loss of weight/condition of calves and severe dehydration
Collect and investigate samples properly as this is an important part of diagnosis	Repeated coprological examination (a risk of high inaccuracy with a single sample taken from the calves at the beginning of the infection; very small number of oocysts may be detected) or if only one mixed sample of faeces is sent for diagnosis. Of importance is specification of pathogenic species of coccidia and quantification of oocytes (flotation of the number per 1 g of faeces; faecal oocyst counts > 5000 oocysts/gram is considered significant, in acute infection peak can be 100,000 oocyst/gram, with rapid falling down after the peak) (AVA-AMA 2022)
Perform differential diagnostics	For differential diagnostics distinguishing from bacterial causes of diarrhoea in calves, bacteriology sampling and culturing should be performed
Start treatment timely	Watching and sampling the animals, especially around 3–6 days after weaning in new housing, if risk of coccidiosis is known. Metaphylactic administration of sulfadimidine is possible, and treatment of animals with proven presence of coccidia, but prior clinical signs, is recommended in toltrazuril. Be careful with diclazuril, which may need repeated administration. SPC instructions of individual VMPs should be followed
Use supportive therapy in clinically ill animals	Clinically ill animals, whenever possible, should be separated from the rest of calf heard and given supportive oral and parenteral fluid therapy
Rotate anticoccidials and follow resistance development	Similarly, as in bacteria, also coccidia can develop resistance; therefore care should be taken to use anticoccidial still effective to pathogen present in the herd
Consult regularly all steps with veterinarian	Veterinarian in cooperation with farmer and staff should explain the importance of preventive measures and establish herd health plan including management and treatment of coccidiosis

prevention via veterinary medicinal products, halofuginone lactate and paromomycin containing VMPs are authorised. As already mentioned, nitazoxanide is not possible to use any more in animals, due to being reserved for human cryptosporidiosis and viral diseases only. Halofuginone should be given prophylactically within the first 24–48 h of age. Therapeutically, for improving diarrhoea course, it must be given within 24 h of the onset of symptoms. It must be administered for 7 consecutive days, which can be difficult to manage, particularly for beef calves that are kept with the dam (Thomson et al. 2017). Halofuginone also limits the spread of the disease by preventing the formation of oocysts which are a stage in the life cycle of the parasite that are passed in the faeces. The bioavailability of halofuginone, following single oral administration, is about 80%, T_{max} is 11 h and C_{max} is 4 ng/mL, V_d is 10 L/kg, and $t_{1/2}$ is 30.84 h. The plasmatic concentrations of halofuginone after repeated oral administrations are comparable to the PK after single oral treatment (Summary of Product Characteristics of VMPs containing halofuginone). Paromomycin containing VMPs for pre-ruminant calves are authorised as oral solution containing e.g. 140,000 IU/mL, with indication reduction of the severity and the duration of diarrhoea associated with *Cryptosporidium parvum* and reduction of faecal oocyst shedding. The bioavailability of paromomycin following a single oral dose of 35,000 IU paromomycin/kg bw (calves 2–6-week-old) declared for this VMP is 2.75%, as the main part of paromomycin stays in intestine. Paromomycin has age-dependent pharmacokinetics, with the greatest systemic exposure occurring in newborn animals. With respect to the absorbed fraction, PK parameters—mean T_{max} is 4.5 h, C_{max} was 1.48 mg/L, and $t_{1/2}$ was 11.2 h—were determined (Summary of Product Characteristics of VMP containing paromomycin). Recently, vaccine authorised with claim of the active immunisation of pregnant heifers and cows to increase antibodies in their colostrum and targetting calves via passive immunisation to reduce clinical signs (diarrhoea) caused by *Cryptosporidium parvum* was approved.

6 Main Bacterial Pathogens and Diseases in Calves and Beef Cattle, Where Antimicrobials Are Used

Infectious diseases caused or associated with bacteria belong to the most frequent indications, where pharmaceutical veterinary medicinal products are used either for the treatment, prophylaxis and metaphylaxis claims or where immunological products (mainly vaccines) are used for the prevention claims. Bacteria being considered as having pathogenicity in cattle and associated with cattle diseases are listed by the groups (VOL I, page 334), with specific attention to those responsible for mastitis (VOL I, p. 335). As for indications, where antimicrobials are used in cattle of different age and production categories, Table 5 gives a brief overview linking the disease with the most frequent species of the infectious agents (where bacterial pathogens prevail). Speaking about indications, where antibiotics are mostly used for the treatment in Europe, there were two big surveys organised by FVE in Europe in 2012 and 2016, in which major diseases and pathogens were identified as mentioned below (De Briyne et al. 2014; FVE 2016). In neonatal and veal calves, respiratory diseases (*Mannheimia haemolytica, Pasteurella multocida* and *Histophilus somni*) and diarrhoea (*E. coli* and *Salmonella* spp.) are the most frequently treated with antimicrobials. In dairy cattle, therapy of mastitis (including dry cow therapy), whereas the major causal agent was identified *Staphylococcus aureus* and second one was *Streptococcus agalactiae,* which has been and is nowadays broadly replaced by *Streptococcus uberis* (Kabelitz et al. 2021), was and still is acknowledged as the most frequent cause of use of antibiotics. The second one was lameness (anaerobes identified as mostly causative), and third one was metritis (*E. coli* and *Trueperella pyogenes*). In beef cattle, diseases mostly treated using antimicrobials include respiratory diseases (*Mannheimia haemolytica, Pasteurella multocida* and *Histophilus somni*), lameness and arthritis. Despite those named, specified parts of this chapter will also target some other diseases

Table 5 Main causative or disease-associated pathogens in various diseases of cattle, where antimicrobials are used

Disease	Microorganisms (causative or disease associated)
Omphalitis (umbilicus and/or surrounding tissue infections)	Gram-positive bacteria such as *Trueperella pyogenes*, *Staphylococcus aureus*, *Streptococcus pyogenes*, *Enterococcus* spp. and gram-negative bacteria such as *Escherichia coli*, *Klebsiella pneumoniae* and *Proteus* spp. (Painter et al. 2023). If maternal infection with chorioamnionitis is suspected, anaerobic bacteria such as *Bacteroides fragilis*, *Clostridium perfringens, and Clostridium tetani* can also contribute to infection (Brook 2002)
Calf diarrhoea (bacterial)	Enterotoxigenic *Escherichia coli* (ETEC)—the first 3–4 days of life attaching and effacing *E. coli* (AEEC)—typically in age 4–28 days
Calf diarrhoea (protozoal)	*Eimeria bovis*, *Eimeria zuernii* and *Eimeria alabamensis* and *Cryptosporidium parvum*
Calf septic arthritis	*E. coli*, *Salmonella* spp., *Mycoplasma* spp., *Streptococcus* spp. and *Trueperella pyogenes. Trueperella pyogenes* is more frequently isolated from the joints of older calves
Neonatal sepsis	Approx 80% caused by Gram-negative bacteria, mostly *Enterobacterales*, predominating is *E.coli, followed by Salmonella* spp., *Klebsiella* spp., *Pseudomonas aeruginosa* and *Mannheimia haemolytica*, and from Gram-positive bacteria such as *Staphylococcus* spp., *Streptococcus* spp. and *Trueperella pyogenes*
Calf pneumonia	*Mannheimia haemolytica*, *Pasteurella multocida*, *Histophilus somni*, *sporadically Trueperella pyogenes*, *Mycoplasma bovis* and *Mycoplasma dispar*
Calf meningitis	Gram-negative enteric bacteria, such as *E. coli*, *Enterobacter* spp. and *Salmonella* spp.
Necrotic laryngitis (calf diphtheria)	*Fusobacterium necrophorum*
Respiratory disease	*Mannheimia haemolytica*, *Pasteurella multocida*, *Histophilus somni* and *Mycoplasma bovis*
Lung abscesses	*Trueperella pyogenes* and *Fusobacterium necrophorum*
Mastitis (bacterial)	*Predominating species: Streptococcus uberis*, *Staphylococcus aureus*, non-aureus *Staphylococci*, *Streptococcus dysgalactiae*, *Streptococcus agalactiae*, *Enterococcus* spp., *E. coli*, *Klebsiella* spp., *Raoultella* spp., *Pseudomonas aeruginosa*, *Corynebacterium bovis*, *Trueperella pyogenes* and *Mycoplasma bovis*
Mastitis (non-bacterial)	Yeasts, *Prototheca* spp. and *Aspergillus* spp.
Metritis (acute puerperal)	*E. coli*, *Trueperella pyogenes*, *Fusobacterium necrophorum*, *Prevotella* spp. and *Bacteroides* spp.
Endometritis	*Trueperella pyogenes*
Urinary tract infections (mostly pyelonephritis and cystitis)	Pyelonephritis : *Corynebacterium renale* Pyelonephritis and cystitis: *Escherichia coli*, *Trueperella pyogenes*, *Corynebacterium cystitidis*, *Corynebacterium pilosum*, *Staphylococcus* spp., *Streptococcus* spp., *Enterococcus* spp., *Klebsiella* spp. *and Pseudomonas* spp.
Interdigital necrobacillosis (foot rot)	*Fusobacterium necrophorum*, *Prevotella melaninogenica* and *Porphyromonas levii*
Endocarditis	*Trueperella pyogenes* (most common), *Streptococcus* spp., *Staphylococcus* spp. and aerobic Gram-negative bacteria
Actinomycosis (lumpy jaw)	*Actinomyces bovis*
Actinobacillosis (wooden tongue)	*Actinobacillus lignieresii*
Infectious bovine keratoconjunctivitis	*Moraxella bovis and Moraxella bovoculi*, rarely *Mycoplasma bovis*, *Listeria monocytogenes*, single-outbreak *Chlamydophila* spp., *Pasteurella multocida*
Otitis	*Mycoplasma bovis* (young animals in herds); other cases' main respiratory pathogens: *Pasteurella multocida*, *Mannheimia haemolytica*, *Histophilus somni*, *Trueperella pyogenes* and rarely *Streptococcus* spp.
Skin (wounds and surgical sites infections)	*Trueperella pyogenes*, *Enterobacterales* and anaerobes such as *Bacteroides*, *Fusobacterium* and *Prevotella* spp.

associated with certain categories and being reason for use of antimicrobials as omphalitis and neonatal septicaemia, which are many times causally linked with previously mentioned infections (respiratory or gastrointestinal) and calves' necrotic laryngitis. In adult cattle main attention will be paid to respiratory disease, infections of the foot, mastitis (including dry cow therapy) and metritis, and diseases of skin and sensory organs are also covered.

7 Consideration for Diseases Mostly Treated with Antimicrobials in Calves and Adult Cattle

Diseases caused or associated with bacteria are subject to treatment with antibiotics. Nowadays, being aware not only of benefits but risks associated with use of antimicrobials, including resistance to antimicrobials, means huge effort is paid to prevent these diseases without the use of antimicrobials, but the occurrence of infection also necessitates proper diagnosis and finally responsible use of the most appropriate antibiotic. Below subsections pay attention to the diseases of calves and adult cattle creating majority of use of antibiotics in the cattle sector. Some more sporadic diseases were not involved.

7.1 Calves

Historically, based as well on information from bibliography, various antimicrobials (including ampicillin, chlortetracycline, oxytetracycline, streptomycin, neomycin and furazolidone) were routinely used to prevent diarrhoea and septicaemia in calves (Constable 2003). Although the results of these studies varied, their results mostly do not support positive outcomes of prophylaxis with antimicrobials. As noted by Smith (2015) within the overview, only one study in which chlortetracycline was orally dosed, shows a decrease in mortality rate from diarrhoea; a few studies only show shorter duration of diarrhoea, while increased rates of diarrhoea after neomycin use and in antibiotic-treated calves were noted; or no change of the incidence of diarrhoea in calves compared with untreated controls was detected. Currently, especially for uncomplicated neonatal diarrhoea, antimicrobials are not the first treatment choice, as they can cause more microbiome disbalance and increase diarrhoea, suppress growth and reduce feed intake (AMCRA 2023). Broad and routine use of antimicrobials in calves is no more acceptable especially due to the issue of resistance to antimicrobials. Although it is many times linked especially with human medicine or more specifically there is a real lack of potent antimicrobials in life-threatening infections caused by Gram-negative bacteria in hospitals, during the last years a similar situation has started to appear on the veterinary side. This can be exampled on the calf septicaemia, where even with the timely administration of the last-choice antimicrobial, we are faced with failure and are therefore not able to save the life of the animal. So even if it is stipulated that "Any calf, which appears to be ill or injured, must be treated appropriately without delay, and veterinary advice must be obtained as soon as possible for any calf that is not responding to the stock keeper's care", veterinarians sometimes stay with "weak weapons" to fight the infection. This above-mentioned information supports the strong need for the targeted treatment to diseased animals or targeted metaphylaxis to those with high risk of infection. Moreover, it supports also all appropriate, safe and effective preventive tools (colostrum intake, vaccination of mothers and calves as well). The rules for metaphylaxis are given by Regulation (EU) 2019/6, Article 107/4 and could be further detailed by the national law of individual member states. This above information also supports the need for more (rapid, less costly and available) diagnostic methods, including susceptibility testing becoming standard also in veterinary medicine. Finally, it should be noted that currently, the era of oral broad medication of calves is over, treatment and limited metaphylaxis should be punctually targeted, and the preferred route of administration is injection of the antimicrobials with known susceptibility, once necessary. Moreover, the word "prevention" has nowadays gained new meaning, and the meaning of preventive (= prophylactic) use of antibiotic should be in fact mostly forgotten and replaced by the broad meaning of preventive tools and measures, based on which mostly healthy animals will be kept, which need minimum antibiotics.

Key Prevention Step Start in Newborn Calf
It should be noted that the first 2 months of life are the most key for the (dairy) heifer health and performance as well. Preventive herd health programme with specific part paying attention to newborn calves is of paramount importance altogether with good and working cooperation among attending veterinarian (with detailed knowledge of the herd in the care) and farmer (owner investing into such programme) and staff taking care of animals from their birth (nowadays the most critical part to ensure high-quality staff). Except proper standards of biosecurity and hygiene, ensuring newborn calf is not immediately after the birth faced with high infectious pressure from hugely contaminated environment, early, timely and sufficient colostrum intake is vital. Without sufficient and proper colostrum quality intake (making calves to have adequate passive immunity), the risk of different infectious diseases can be 6.4 times higher, and the risk of death is 5.4 higher (within first 28 days of life) (Burt 2008).

Colostrum intake and impact can be indirectly measured and checked by the IgG level detection. High standard hygiene of quality colostrum with at least 50 g/L of IgG (test of this level needed) is of importance, and feeding an amount of approximately 3–4 L (or 10–12% of bw) within first 2 h after birth is recommended (Godden et al. 2019; AHDB 2019). This amount should be ensured (if not from nipple bottles then from oesophageal tube feeder, which needs trained staff for administration). Calves should be tested for IgG (>12 g/L is needed for good protectivity). Colostrum quality decreases shortly (>6 h). Contamination should be avoided as it influences the quality. If colostrum is frozen, care should be taken to warm it carefully (ideally up to 45 °C not to have detrimental effects on IgG/antibodies). This goes to a controversy with recommendation to heat colostrum to a maximum of 60 °C for 60 min, for cases of herds with *Mycoplasma bovis*, *Mycoplasma avium* subsp. *Paratuberculosis* and *Salmonella* spp., to minimise the risk of intake of these germs by calves with colostrum and not harm IgG concentration (Godden et al. 2019; Donahue et al. 2012). In a study by Donahue et al. (2012), although some loss of IgG was determined and in higher-quality batches a higher magnitude of loss of IgG as a result of heat treatment was noted compared with lower- or intermediate-quality batches of colostrum, the colostrum IgG concentrations of high-quality batches remained high overall and still within acceptable limits for feeding and building of immunity of calves.

7.1.1 Neonatal Septicaemia

Septicaemia (neonatal septicaemia and bacteriemia) is a life-threatening condition associated with invasion of bacteria to the bloodstream, where they persist and may produce toxins. Septicaemia may result in multiple organ dysfunction syndrome, septic shock and death. Sepsis is a combination of an infection (or suspected infection) and a systemic inflammatory response (Fecteau 2023). The presence of toxins worsens the course of the disease and increases the mortality rate due to seriously harmful effect (Fecteau 2023). Due to the presence of the cell wall lipopolysaccharide of Gram-negative bacteria, septic shock can quickly occur. Septic shock is an exacerbation that results in hypotension and/or hypoperfusion.

Usually newborn or very young calves suffer from septicaemia, as the lack of passive immunity in this age is of high importance in the disease outbreak. From a total of 230 critically ill calves involved in the study, 87% were less than 1 month of age, 34.3% developed sepsis, and 61.3% died (Pas et al. 2023). A less acute form of septicaemia occurs in calves 7–28 days old, where rather localised infection occurs (arthritis, growth plate infection, hypopyon, meningitis, pneumonia), more frequently associated with mild clinical signs and more chronic evolution (Fecteau 2023). Although the importance of early detection of sepsis and accurate estimation of the causative agent is life-saving, in common veterinary practice it is a big challenge, especially as clinical signs alone are not sufficient for diagnosing septicaemia. Fecteau (2023) proposes the early straightforward diagnostic approach, which uses four essential criteria to diagnose septicaemia: abnormal core temperature, heart rate abnormality (tachycardia or bradycardia), tachypnoea and leukocyte count abnormality (leukopenia or leucocytosis). More broad risk models were also defined utilising scoring of clinical factors. Pas et al. (2023) used the following in their study: temperature [°C], heart rate [beats/min], respiratory rate [breaths/min], mucosa colour [pink, pink pale, congested], capillary filling time [seconds], skin pinch test [seconds], enophthalmos [mm], abnormal behaviour, congestion of scleral vessels and (co) morbidities such as arthritis, phlebitis, omphalitis, diarrhoea, enteritis and pneumonia. From these factors finally hypothermia, pale mucous membranes and abnormal behaviour were selected as the most promising indicators (due to significance in logistic regression, survival rate and decision clinical tree). Clinically abnormal behaviour in calves was frequently associated also with acidosis (d-lactatemia), hypoglycaemia, and hypercapnia—these are all identified as significantly linked with mortality (Pas et al. 2023).

To predict this invasive disease, mostly associations between septicaemia and diarrhoea/enteritis and pneumonia are noted in the bibliography (Lofstedt et al. 1999; Fecteau 2023). In a recent study (Pas et al. 2023), there was a determined 39% association with sepsis and total pneumonia cases (including further comorbidities) and diarrhoea/enteritis (d/e) with sepsis in 29.4% (total d/e cases including further comorbidities), while omphalitis and neonatal respiratory distress syndrome were observed in approximately 22%. *Escherichia coli* is generally considered as the most frequent causal agent (Fecteau et al. 2009), and in a study by Pas et al. (2023), it was isolated from blood cultures in 30.6%, of which 90.9% were multidrug-resistant isolates. Nevertheless, very different bacteria (*Staphylococcus* spp., *Salmonella* spp., other *Enterobacteriaceae*, *Streptococcus* spp., *Trueperella pyogenes*, *Pseudomonas aeruginosa*, *Mannheimia haemolytica*) were isolated from cases of septicaemia in association with neonatal diarrhoea as well (AMCRA 2023; Pas et al. 2023). In addition to this, more individual forms of septicaemia, certain bacteria, such as *Histophilus somni* and *Salmonella* spp., and certain strains of *Pasteurella multocida* have virulence factors that promote polyserositis. Virulence factors that drive the invasiveness and seriousness of infection are adhesion capacity, serum resistance, aerobactin production and toxin synthesis (Fecteau 2023). Septicaemia can also develop iatrogenically from non-sterile medication administrations or from contamination of stay catheters.

Although timely administration of antimicrobial is of critical importance in septicaemia, blood sample should be taken before administering of antibiotics to avoid false-negative results. Blood culture collection from live animals is recommended. Recently an approach recommending a large volume of blood to be collected (at least 35 mL), due low sensitivity of small volumes, was published (AMCRA 2023). Blood sampling from the jugular vein, under thoroughly aseptic collection (scrubbing with, e.g. chlorhexidine, and disinfection with, e.g. ethanol and sterile gloves), and bacteriology sampling using special blood cultures vials with within 2 h delivery to lab (room temperature) are recommended, followed by culturing in automatic detection systems (Pas et al. 2023). Useful can be cultures of other body fluids (joint, cerebrospinal, peritoneal or pleural) and abscesses (Fecteau 2023) or from the suspected source of infection and considering cases of infections mostly associated with sepsis (lung, navel or bladder (AMCRA 2023), which, despite being no help in immediate treatment decision, could help in treatment corrections and in de-escalation towards a narrower-spectrum targeted antimicrobial treatment. Intestine or throat sampling is not recommended. Especially in cases where more animals are endangered by septicaemia and mortality occurs, the liver is preferably sampled aseptically and examined histologically to show sepsis. In addition, samples of other affected tissues (spleen and kidney) can also be collected. These results can give the basis for the treatment of animals in the group that can be endangered by septicaemia later (as infected calves excrete a large number of bacteria in their urine and also in nasal secretions and saliva as well). A common source of virulent isolates could contribute to the incidence of the disease on a particular farm or in a particular area of a farm. An aerobic culture is mostly recommended (due to expected pathogens); if signals of possible anaerobic bacteria and yeasts/fungi are present, specific medium/bottle/culturing conditions of automated system are necessary.

The most rational way to use antimicrobials for sepsis in calves is considering the model used in the human medicine in septic patients. To maximise survival, de-escalation model can be recommended, where treatment is started with intravenous administration of a broad-spectrum bactericidal antimicrobial (Gram-positives and Gram-negatives targeting). As soon as antimicrobial susceptibility test results are available, a narrow-spectrum antimicrobial can be selected, limiting selection pressure on commensals (Rhodes et al. 2017).

Especially once the recent situation in susceptibility on the farm is favourable, sulphonamide with trimethoprim can be used (but not in anaerobes as causal agents and not in exudative inflammation cases). Further possibility is fixed combination veterinary medicinal product (rather aminopenicillin with aminoglycoside), which is preferred to monotherapy due to expected broader coverage and better efficacy against Gram-negative bacteria. Gram-negative bacteria are mostly intrinsically resistant to benzylpenicillin, and fixed combination of penicillin with dihydrostreptomycin can broaden action to them; unfortunately common high proportions of resistance to this aminoglycoside disqualify the combination from the treatment options. Single, more narrow-spectrum substances can be used in good clinical response only (especially in group of animals with expected or confirmed same ethiology and good susceptibility). Whenever possible, critically important antibiotics (third- and fourth-generation cephalosporins and fluoroquinolones) should be kept as the last choice. Especially in aminoglycoside nephrotoxicity, hypovolaemia, dehydration or pre-existing kidney damage can complicate treatment outcomes (the highest risk in neonates) (Fecteau 2023). Extrapolation of AST results in *E. coli* samples from adult animals (even within the same herd) cannot be recommended, as resistance in *E. coli* isolates in both commensal and clinically diseased animals is generally more common in young veal calves (AMCRA 2023).

Primary preventive measures, which are, unfortunately, many times abandoned and cause finally septicaemia, are hygiene and thorough care about neonates. Newborn calves with the high risk of umbilical infection, and in a heavily contaminated environment being colonised by infectious pathogens earlier than normal flora can be established, are more endangered by sepsis. Further key neonatal septicaemia effective prevention is an optimal colostrum management (healthy cow and clean udder to reduce the probability of bacterial high load even in colostrum). Conditions minimising the risk of pneumonia and diarrhoea (see below in this chapter) and adequate umbilical disinfection (e.g. iodine) will help to minimise the occurrence of sepsis as well. Hygienic measures both with regard to the housing of the animals and the administration of medication are necessary. Mothers can be vaccinated, but in calves, vaccination is not performed.

As in another serious infections and to help better manage animals suffering as well, NSAIDs can be recommended, being described as helpful for reducing mortality. Ketoprofen was shown to reduce all symptoms (endotoxin release cases), while corticosteroids were not proven to be helpful (Plessers et al. 2016). As adverse effects may occur with NSAID intensive treatment (abomasal ulcers and nephrotoxicity that can be potentiated by concomitant use of aminoglycosides), a short course necessary for survival is highly preferred. As noted by Fecteau (2023), supportive care and nutritional and respiratory perfusion should be provided as needed. This includes administration of IV fluids (dextrose (2.5–5%) combined with physiological saline (0.9%) and, for acidosis, buffering hypertonic bicarbonate (5.4%), further plasma transfusion, oral or parenteral nutrition and oxygen administration may be useful. Moreover, calves should be under the care and monitoring of milk intake, where at least 10% or more ideally 15–20% of body weight every 24 h is advised, with feeding 3–5 times per day (Fecteau 2023). Enteral feeding by suckling or tube feeding is of importance to keep gastrointestinal system function. Noting rather of zootechnic care nature, the attention and care should be paid to environment in stable (ambient temperature, no draft, proper ventilation, avoidance of too wet conditions), bedding conditions (straw thick layer) and regular cleaning of the boxes.

7.1.2 Neonatal and Later Age Calf Diarrhoea

Enteric disease, with economic significance, suffering both beef and dairy cattle, occurs in the first 3 weeks (neonatal diarrhoea) or up to 30 days of the calf's life. There should be distinguished nature of the disease: nutritional vs. infectious. Nutritional background occurs after improperly prepared milk replacer (including being too cold) or in the case of too high fat levels in natural cow's milk. In infectious diarrhoea in the first 3–4 days of life, enterotoxigenic *Escherichia coli* (ETEC) is the only infectious agent. Further agents may appear as well and belong to attaching and effacing *E. coli* (AEEC)—typically in age 4–28 days. For infectious diarrhoea that develops later, but still within the first week of life, bovine coronavirus, rotavirus and *Cryptosporidium parvum* are the most common agents, and bovine torovirus, norovirus and calcivirus are the less common agents. Coccidia (*Eimeria* spp.) can occur as well but in calves >3 weeks old. *Salmonella* spp. (*S. typhimurium* or *S. Dublin*) appear in later age of 2–6 weeks and are to be detected exceptionally (rare detection can be also caused by the intermittent shedding together with difficulties in selective culturing, especially in subclinical cases caused by *Salmonella Dublin*). Also further enteropathogenic *E. coli* strains can cause diarrhoea at these young animals. The most common type of enterotoxemia (toxic intestinal infection) in young calves is caused by *Clostridium perfringens*.

The lesions at the level of the intestine are not pathognomonic, with isolation of enterotoxigenic *E. coli* (with virulence factors) from samples of faeces/swab sampling from the rectum or from the small intestine post-mortem. Exposure to sunlight and dehydration of samples (influencing growth ability of the germ) should be avoided, and transporting/storing of intestine in ±4° is of importance.

Detection of the adhesion factor F5 can be enough (enterotoxin STa and/or STb enterotoxin presence well corelating with it) (AMCRA 2023). Differential diagnostics' methods can be used—rapid agglutination, ELISA tests or PCR, to indicate bacterial, viral, protozoal or parasitic causality. Both toxins and adherence factors can be confirmed via real-time PCR (rtPCR). Multiplex methods were currently tested and have started to be used recently. Enterit4Calves multiplex qPCR test for the detection of *Clostridium perfringens*, *Clostridium perfringens* types B and C, *Salmonella* Dublin, *Escherichia coli* F5, bovine rotavirus, bovine coronavirus, *Cryptosporidium parvum* and *Eimeria* spp. is commercially available (DNA diagnostics, 2023) and was tested by Panesri et al. (2022) determining fair to good detection abilities and quantification of pathogens of calf diarrhoea.

The best prevention is immune system building with early and sufficient quality colostrum intake, which can be further enhanced/prolonged by vaccination of mothers (vaccination once or twice during the drying-off period, anti-rota/coronaviruses and *E. coli*) or by smaller portion of the colostrum being supplemented during the period of risk (at least 10–12 days in vaccinated mothers). Ideally bigger farms have some quality colostrum portions frozen, to be gently warmed and used in calves, whose mothers have suffered from mastitis or died during parturition. Good care of animals, including monitoring of health status and faeces consistence and hygiene, is essential to keep the animals in good health conditions and minimise infection pressure. Biocides effective against *Cryptosporidium* spp. can help avoid diarrhoea or pathologies caused by this germ.

From the treatment options, rehydration therapy is of critical importance (dose/frequency of electrolyte solutions depends on the severity of diarrhoea; Table 6). Specific rehydrants and products for drinking solutions can contain alkaline-buffering ingredients (propionate, citrate, bicarbonate), probiotic strains and selected natural compound combinations (e.g. rice flour, algae, fructooligosaccharides, yeasts or their extracts, psyllium), vitamins (B vitamin, fat-soluble vitamins) L-carnitine and other additives with positive effect on gut health and minimising diarrhoea. Clinical symptoms and suffering from acute diarrhoea can be reduced by NSAIDs of the "oxicam group" (calves > 1 week).

As indicated by authors of the recent study among farmers (N = 873) in Austria, Belgium, Portugal and Scotland, there is an urgent need for more scientific evidence to define best practices for the treatment of neonatal calf diarrhoea, as there still exists an overuse of antimicrobials, which is without any evidence serving firm background for doing it (Eibl et al. 2021). The authors of this study summarised that fluoroquinolones, sulphonamides and penicillins as the most frequently specified classes for the antibiotic treatment of NCD based on farmer responses. High-level warning signals were released by publication of Prieto et al. (2022) that in Spanish ETEC strains (N= 41) investigated for susceptibilty/resistance in 14 antimicrobials, from 4 pharmacological classes (beta-lactams, aminoglycosides, fluoroquinolones, tetracyclines) and sulfa/trim, all isolates were multidrug-resistant (MDR), and with the exception of gentamicin and cephalosporins of third and fourth generation, less than 40% susceptibility was detected for the rest of antimicrobials tested. Another study of 8713 isolates (2015–2019) from calves with severe or fatal diarrhoea in Germany indicated 84.2% MDR and 15.8% XDR. The first-, second- and third-line antimicrobials for the treatment (amoxicillin/clavulanate, enrofloxacin and trimethoprim-sulfamethoxazole) exhibited mean non-susceptibility rates of 28.1%, 37.9% and 50.0%/5-year period (Feuerstein et al. 2022). National monitoring of the target pathogens in the Czech Republic indicated (2022; N = 87; calf diarrhoea and inner organ isolates) resistance in amoxicillin/clavulanate of 13.8%, in enrofloxacin 12.6% and in trimethoprim-sulfamethoxazole 27.6 %; highest resistance was detected in tetracycline, slightly lower for doxycycline (54% and 43.7%, respectively (CZ NMTP 2022).

Infections caused by invasive *E. coli*, where systemic disease endangers the calf, are indicated for parenterally administered antimicrobials. Similarly, in neonates with fever and another

Table 6 Categorisation of diarrheic calves into treatment groups

Categorisation of diarrheic calves	Therapy
Calves less than 7% dehydrated: – Skin elasticity test: less than 5-s reaction – Extent of enophthalmos up to 3 mm Slightly dry mucosa, warm mouth, good suck reflex, good muscle tone, standing	Oral fluid + electrolyte therapy, supervision and help to ensure regular intake of sufficient amount of solutions
Calves with 7–9% dehydration: – Skin elasticity test: 5 to max 6-s reaction – Extent of enophthalmos 4 to max 5 mm Slight acidosis, moderately dry mucosa, warm mouth, weak suck reflex, good muscle tone but starting to stand insecurely	Hypertonic saline solution (7.5%) IV at 3–4 mL/kg bw in 5 min Fluid/electrolytes Bad status calves Stomach tube administration (40–60 mL/kg bw) Better status: Voluntary fluid/electrolyte intake under supervision, amount to be checked
Calves more than 9% dehydration: – Skin elasticity more than 6-s reaction – Extent of enophthalmos more than 5 mm Dry and cool oral mucous membranes, recumbent, no suck reflex, depression, need of dextrose	Necessary IV fluid and electrolytes, alkalising agents involved 6–8 h, max 24–36 h

Modified from Constable et al. (2017b)

concomitant bacterial disease (omphalitis or respiratory disease), antimicrobials are indicated. Also calves with blood or mucosal shreds in their faeces, as it marks a breakdown of the blood-gut barrier, are candidates for injectable antimicrobial treatment (Constable 2009). Antibacterial agents that can be used (with knowledge of patterns of resistance on farm or at least regionally) are generally those with excretion in bile and reaching an effective level in the gastrointestinal tract (Constable 2004).

Gram-negatives (*E. coli* and *Salmonella* spp.) are generally susceptible to sulphonamide/trimethoprim, aminopenicillins (amoxicillin, ampicillin, amoxicillin/clavulanic acid), aminoglycosides (neomycin, gentamicin). In cases of concomitant infections and good causal agent's susceptibility, procaine benzylpenicillin and aminoglycoside (dihydrostreptomycin or neomycin) can be used as well. In *E. coli* aminopenicillin resistance occurs frequently and the fact that aminopenicillin group can select for extended-spectrum beta-lactamases (ESBL)/AmpC production in bacteria should be considered in the choice of antimicrobial. The last-choice antimicrobials are (fluoro)quinolones, where it should be noted that resistance is growing, and if resistance is detected for enrofloxacin (ciprofloxacin in vitro test), cross-resistance can be expected within the (fluoro)quinolone group.

In *Clostridium perfringens* infections, quick diagnosis and timely start of treatment are necessary to avoid status, which cannot be reversed, as once the toxins penetrate through the damaged gut wall into the bloodstream, the calf goes into shock and dies within a few hours. Clostridial gut infection has a sudden onset, starting without diarrhoea but rather with distended belly or acute gut pain, getting up and down and staggering around. It can be noted that prevention with vaccination in *Clostridium perfringens* (specific toxoids' types) can be used; as for antimicrobials, IM injectable benzylpenicillin, as-second choice amoxicillin and cefalexin, can be used.

7.1.3 Pneumonia in Calves

Pneumonia in calves is an infectious disease caused by both viruses and bacteria—species from *Pasteurellaceae*, namely, *Mannheimia haemolytica* (MH) and *Pasteurella multocida (PM)*, as well as *Histophilus somni* (HS), are common pathogens. More sporadically can occur also *Trueperella pyogenes (TP)*. From bacteria with more difficult susceptibility testing, mycoplasmas (*Mycoplasma bovis* and *M. dispar (MY)*) can be causative agents. Calves that suffered from respiratory diseases are either immunocompromised (e.g. by insufficient colostrum intake or by stress) or under the pressure of risk factors, which are frequently present concomitantly. Risk factors are environmental conditions: ammonia, dust, drafts and housing type (high humidity in cold stables, frequently associated with wet and cold floor, where animals rest). Another high-risk factor is merging of calves. Calves of different ages, prone to hierarchy fights or different sources associated with mixing of different microbiomes, including pathogens as well as overstocking in transporting lorries or even in the stable, are important health-compromising factors. Merged animals have different immune status and fight with microbial pressure of the new space. All such conditions lead to mass prophylactic administration of antimicrobials, either injectable or oral. It should be highlighted that all such conditions are involved under the terms "poor hygiene", "inadequate animal husbandry" or "poor farm management" described in the new regulation on VMPs (EU)2019/6 under the Article 107 (1), where it is stipulated that such conditions should not be compensated by routine administration of antimicrobials. Even in avoiding such risks in maximum scale possible, the animals can get sick, and, in such cases, either treatment or metaphylaxis is applied. Not only due to economy but also to guarantee optimal animal welfare, early treatment in pneumonia is necessary, which explains the observed success of metaphylaxis.

Best results are obtained with bacteriological sampling via proper technique and performed in animal/s with acute disease, prior treatment and not after treatment failure, where culturing/interpretation is limited. Nasopharyngeal swab (deep nasal swabs caudal part, only predictive value, risk of contamination), or transtracheal wash

(TTW) or broncho-alveolar lavage (BAL—highly recommended, cleaning the nose mirror, opening external nasal using cotton wool, disinfecting it with alcohol, endoscopically, or BAL catheter only) are options for sampling. In post-mortem sampling, the whole animal or the complete lung package or samples of the lungs (sterile techniques, at the level of lesions' sampling, and alcohol disinfection) are to be sent to the lab. Sampling several animals (based on the group size), culturing of each sample and real-time PCR (especially of microbes with difficult culturing, *HS* or *MY*) on a pool of samples from one group should be performed. If both bacteriology and PCR are negative, histology is recommended. To get adequate results, necessary steps are both sampling and proper transport of samples within rinsing fluid (24 h/4–8 °C) or preferably "flocked" swab with bronchial rinse or lung tissue, preferably in sterile transport medium/charcoal for bacteriology and with clear media for PCR (preferably up to 48 h), is necessary. Culturing for bacteriology and non-selective media (for *Pasteurellaceae*) but also PPLO agar for *Mycoplasma* spp. should be requested (results available within 2–5 days). If *Mycoplasma bovis* (as a known pathogen) is cultured, result is expected within 7–14 days, and due to technical difficulties, routine performance of antibiogram is not possible. *Mycoplasma bovirhinis* is commensal, and the pathogenic role of *Mycoplasma arginini* and *M. dispar* is currently unknown.

According to the pathogen and its susceptibility, antimicrobials should be selected, but generally as the first choice florfenicol is considered. Some authors include also *M. bovis* under the spectrum of florfenicol, which may be correct provided that a sufficient dosing regimen ensures the efficacy. There should be noted information from the divergent opinion that treatment with some single-shot dosing of florfenicol formulation may not be adequate to reach sufficiently high and long-lasting levels for the treatment of BRD associated with *M. bovis* (EMA 2014). Among the first choices are benzylpenicillin (selected susceptible pathogens as *Pasteurella multocida,* but not appropriate for treatment of *M. bovis*) and trimethoprim/sulphonamide (not *M. bovis*). The second choices include aminopenicillins (amoxicillin and ampicillin, but not in *M. bovis* infections), tetracyclines (oxytetracycline and doxycycline; including *Mycoplasma* caused infections), penicillin + aminoglycoside (e.g. penicillin + neomycin) and macrolides (gamithromycin and tildipirosin and including *Mycoplasma* caused infections also tulathromycin, tylosin and tilmicosin). If previous choices cannot be used and considering susceptibility testing results, the third choice includes fluoroquinolones (enrofloxacin, marbofloxacin, danofloxacin, difloxacin), flumequine or cephalosporins ceftiofur (third generation cephalosporin, not effective in *M. bovis* caused infectons) and cefquinome (fourth generation cephalosporin). Fluoroquinolones are effective also for *Mycoplasma* (AMCRA 2023; Smola et al. 2022). Currently (status at 2023) it should be noted that MRL for milk has not been established yet and florfenicol, tilmicosin and tulathromycin are therefore not authorised for use in lactating dairy cattle producing milk for human consumption (in tulathromycin also 2 months prior to the expected parturition in animals expected to produce milk for human consumption).

7.1.4 Necrotic Laryngitis of Calves

Diphtheroid necrobacillosis is an inflammatory disease of the larynx and oropharynx, affecting mucosa as well as cartilages, with the causal agent *Fusobacterium necrophorum*, which enters through microscopic traumas via oral route. Oral infection, exacerbated as stomatitis in the young up to 3 moths of age animals and as laryngeal form (begins with mechanical mucosal lesions in the larynx) in the young up to 18 months (exceptionally up 1 year) of age, was described, more frequently in the autumn and winter. The prevalence is approximately 1–2% (Campbell 2023). Laryngitis with inspiratory dyspnoea and stridor, moist, painful cough and purulent nasal discharge, excessive salivation and painful swallowing motions, together with necrotic lesions and arytenoid cartilages accompanied with systemic clinical signs as fever (41 °C), anorexia and depression, occurs (Constable et al. 2017a;

Campbell 2023). On-the-spot diagnosis based on clinical signs (inspiratory to mixed stridor), together with inspection of the larynx via speculum or laryngoscope, is essential. Where feasible, final confirmation by bacteriology of *Fusobacterium necrophorum* (*Trueperella pyogenes* or other (anaerobic) bacteria) is recommended. Early and long-lasting antibacterial treatment is necessary; due to the frequent failure of it, surgical interventions, as, e.g. debridement of necrotic lesions, need to be performed and in serious cases tracheostomy to allow breathing. Cortisone and NSAIDs can also be used concomitantly to antimicrobials. Hyperthermia can occur, and therefore wet spraying of animals could be an important tool to lower the body temperature. There is no vaccine available. Lesion sampling, using deep swabbing, with immediate removal to anaerobic transport medium avoiding contact with oxygen and followed by as quickly as possible transport to the laboratory, under ±4 °C to prevent overgrowing by facultative anaerobic concomitant bacteria, is recommended.

To increase the chance of success and minimise the suffering of the animal, antibacterial treatment together with NSAIDs or corticosteroid, laryngostomy and external cooling is used. Procaine benzylpenicillin and sulphonamide/trimethoprim combinations are the first-line choices. Campbell (2023) listed also oxytetracycline either in 12-h scheme (11 mg/kg bw) or long acting with 72 h (20 mg/kg bw). In procaine benzylpenicillin treatment, a frequency of 12 h for benzylpenicillin 22,000 IU/kg and a duration 7 days seem to work for uncomplicated cases, while prolonged treatment might be necessary in severe, complicated cases (Campbell (2023) indicated 2–3-weeks´ duration). Among the second choices, florfenicol and aminopenicillins are recommended by several bibliographical sources (AMCRA, 2023; Smola et al. 2022), and macrolide administration can be of choice as well (tilmicosin, tylosin). NSAIDs (flunixin or ketoprofen injectable or oral administration of acetylsalicylic acid) to decrease fever and laryngeal inflammation and oedema can be recommended. In animals with severe respiratory distress, due to the laryngeal oedema, corticoids (e.g. single-dose dexamethasone) can be used. Intravenous fluids may be required in dehydrated animals. Once severe inspiratory dyspnoea occurs with other serious clinical symptoms, tracheostomy can be performed. Of special importance is good nursing highly individualised care.

7.2 Diseases in Adult Cattle

7.2.1 Respiratory Diseases

Multifactorial respiratory disease resulting from the complex interaction of respiratory pathogens with the immune system, caused by different ethiological agents, is involved in the complex called bovine respiratory disease (BRD). An epidemic or endemic occurrence is described in cattle of all ages but predominantly in calves. Calves, without clinical signs, can be carriers of potentially pathogenic bacteria, such as the *Pasteurellaceae*, in the upper respiratory tract. Predisposing conditions, including viral infection, promote bacterial colonisation of the deeper respiratory tract leading to pneumonia. Interaction with defence factors of the lungs especially with virulence factors (e.g. adhesion proteins and fimbriae, lipopolysaccharides, outer membrane proteins or lipoproteins) or toxins (mainly leukotoxin from *M. haemolytica*, produced in lag growing phase, but also glycoproteases hydrolysing IgG and neuraminidase) is involved in pathology processes and tissue damages as necrosis, thrombosis and exudation, culminating as pneumonia (Constable 2017a; Campbell 2023; AMCRA 2023). Different pathologies are associated with the opportunistic pathogen *P. multocida* (frequently already damaged lungs with lesions) or *H. somni* (except affection of the bronchi and lung is able disseminate to brain, myocardium, synovium and pleural and pericardial tissues, with late deaths (up to 2 months after arrival)) (Campbell 2023; Miller and Gal 2017). BRD is a typical group-related problem and therefore requires a group-based approach. A major practical problem is the late detection of the initial symptoms of pneumonia, finally leading to too late start of the treatment with insufficient results. Both in order to achieve good economic results and to guarantee optimal animal welfare, early treatment in pneumonia seems necessary, while prophylaxis and metaphylaxis in BRD problems have been brought under the scrutiny, not only due to excessive overuse of antimicrobials and current legal provisions (Regulation (EU) 2019/6 on VMPs) but also due to scientific evidence as well (Baptiste and Kyvsgaard 2017; Abell et al. 2017). Recently, "Whisper On Arrival Technology" (WOAT) has been investigated and in the trial in field shown (Nickell 2022) maintenance of the benefits of a conventional BRD control programme with reduction of BRD control antibiotic use by 11–43%. This WOAT machine learning algorithm (working with heart sounds, lung sounds, rectal temperature, and BW) generates a probability estimate of developing BRD, which then drives the decision to administer antimicrobials (as metaphylaxis). Together with various management measures that are initially taken, this can limit the BRD outbreaks, together with minimising of the risk factors for BRD, which are mainly poor practices in weaning, transportation with stress and mixing animals from different sources, unfavourable weather, unfavourable conditions in the housing of the animals (drafts, dust, ammonia) and further stressful events such as dehorning and castration. Biosecurity principles should be followed accompanied with prevention via vaccination (including autogenous vaccines). Proper diagnostics, with recently used methods, involve, except standard bacteriology (culturing and AST) also PCR tests, *Mannheimia haemolytica* (serotypes A1, A2/A5, A6), *P. multocida* (capsule typing), *H. somni* and *Mycoplasma bovis* identification, as well as recently identified acute BRD causal agent *Bibersteinia trehalosi*. Different authorised commercial vaccines for the major viruses (BRSV, PI-3, BHV-1 and BVDV) and bacteria (mostly *M. haemolytica* and *H. somni*) are available and used. To reflect epidemiological situation customised autogenous vaccines covering the spectrum of expected bacterial pathogens (e.g. *M. haemolytica*, *H. somni*, *P. multocida*, *T. pyogenes)* are used in practice as well (CZ autogenous vaccines use 2021–2023, unpublished data). Use of *M. haemolytica* vaccines in cattle and measurement of IgG and IgM antibody levels in a herd might be used to determine the effectiveness of vaccination programmes (Poonsuk et al. 2023). Even though no commercially produced authorised vaccine has been used yet for *M. bovis* in Europe, some promising results signal future possibilities. One of them could be novel attenuated vaccine strains of *Mannheimia haemolytica* (St1 and St6), expressing and secreting inactive immunogenic leukotoxin (leukotoxoid) fused to antigenic domains of *Mycoplasma bovis* elongation factor Tu and heat shock protein 70, which can control both pathogens, which cause respiratory diseases (Briggs et al. 2021). As a supportive therapy, the use of (N)SAIDs may be indicated. Under risky conditions, infection caused by viruses is mostly followed by bacterial pneumonia. In such cases, once clinical disease is manifested, the severity of infection is characterised by depression, fever (40–41 °C, no signs attributable to other systemic/organ diseases) and toxaemia; moreover, it is frequently accompanied with mucopurulent nasal discharge and moist cough, together with increased bronchial sounds, crackles and wheezes. Pleurisy development, characterised by an irregular breathing pattern and grunting on expiration, has been described in severe cases. Pulmonary abscesses further complicate the BRD cases, especially once pneumonia becomes chronic. *Trueperella pyogenes* is frequently cultured from these abscesses.

Diagnosis using sampling from death or live (acutely ill) animals that are previously untreated should be performed (as in treated animals results cannot be fully representative and should be interpreted with caution). In case of live animals Broncho-alveolar lavage (BAL), starting with cleaning/alcohol disinfection the nose mirror and external nasal opening or transtracheal wash (TTW), is recommended to obtain a representative, deep respiratory tract sample as nasopharyngeal swab (even deep nasal swab) is not fully reliable due to possible contamination. In case of dead animals, a whole animal or lung sample can be obtained using sterile equipment, with the surface disinfected prior the excision. Several animals to be sampled in groups affected, bacteriology/AST per individual and PCR of pooled samples can be performed. Histology can give an indication of a viral or bacterial infection, if PCR gives a negative result. Transporting of samples with expected *P. multocida* and *M. haemolytica* from pulmonary rinsing fluid (8 h/room temperature or 24 h/4–8 °C), or swab soaked in the bronchial rinse, or swab taken from the lung tissue (in both swabs, sterile transport medium, little or no nutrients, preferably by 48 h to be cultured) is recommended in the accompanying protocol, investigation of *Pasteurellaceae* as well as *Mycoplasma* spp. (demanding specific media and conditions) should be requested (results available within 2–5 days, or for *M. bovis* 7–14 days). The role of *M. dispar* in BRD has not been clearly elucidated, but some authors consider that *M. dispar* has a role in pathology due to the virulence factors and ability of biofilm production (Bottinelli et al. 2020; Chen et al. 2019). Due to the fact that results for *M. bovis* are not easily accessible in the routine practice, results of recent studies are of importance to give at least essential overview. In Belgium (2016–2019) were collected 141 isolates (29 dairy, 69 beef, 12 mixed, 31 veal farms), which were tested for MICs in 9 antimicrobials and acquired resistance. The macrolides were identified with frequently obtained acquired resistance, while no acquired resistance to oxytetracycline and doxycycline, minimal acquired resistance to florfenicol and tiamulin, and a limited acquired resistance to enrofloxacin were detected (Bokma et al. 2020).

As for the treatment possibilities, some authors (AMCRA 2023) suggest florfenicol as the first-choice antibiotic, with the recommended dose of 40 mg/kg bw (single shot) against common BRD pathogens (*P. multocida, M. haemolytica* and *H. somni*). Although indication and dosing regimen remain authorised as well for *M. bovis*, there was expressed the opinion signed by several experts (EMA 2014) that consider florfenicol as a time-dependent, bacteriostatic antibiotic, and based on several arguments thoroughly listed in the published opinion, they indicated that treatment with single dose may not be adequate to reach sufficiently high and long-lasting levels for the treatment of BRD associated with *M. bovis*. The second-choice antimicrobial (*M. haemolytica, M. bovis*)—but could be considered first line as well in favourable susceptibility patterns and low resistance—is oxytetracycline (injectable form, adequate doing regimen). For *P. multocida* and *H. somni* the first line can be also benzylpenicillin and amoxicillin. As for macrolides, despite being listed among second choice (AMCRA 2023), they should be used with caution as resistance can emerge relatively quickly, which can be especially observed in herds with prophylactic or metaphylactic use of members of macrolide or related groups (tulathromycin, gamithromycin, tildipirosin, tylosin, tilmicosin).

7.2.2 Infections of the Foot

Interdigital phlegmon (syn: interdigital necrobacillosis, foot rot, necrotic pododermatitis) and digital dermatitis (papillomatous digital dermatitis), leading to the lameness, are important cattle diseases with prevalence across the worldwide dairy industry ranging from 8% in pasture-based systems to 15–30% in confinement, with 80–90% of cases due to a few common infectious and non-infectious hoof lesions. Diseases of the foot can be distinguished to different noseological units (see Fig. 2). Diseases associated with limping and causing significant discomfort of the animals also affect the milk production in dairy herds, due to impact on feeding and total condition of the animals.

The primary bacterial causes of interdigital necrobacillosis and diseases of the foot in cattle are *Fusobacterium necrophorum*, *Prevotella melaninogenica*, *Dichelobacter nodosus* and *Porphyromonas asaccharolytica.* In current days new treatment protocols appeared, but still for the invasive cases, antimicrobial therapy is indicated (Van Amstel and Shearer 2006). For invasive digital dermatitis leading to lameness in dairy cattle, topical antimicrobial treatment, together with bandage and proper care and hygiene, can help to resolve this disease mostly caused by spirochaetal bacteria (*Treponema* spp.; see Fig. 2 for individual species associated with the disease) (Bruijnis 2012).

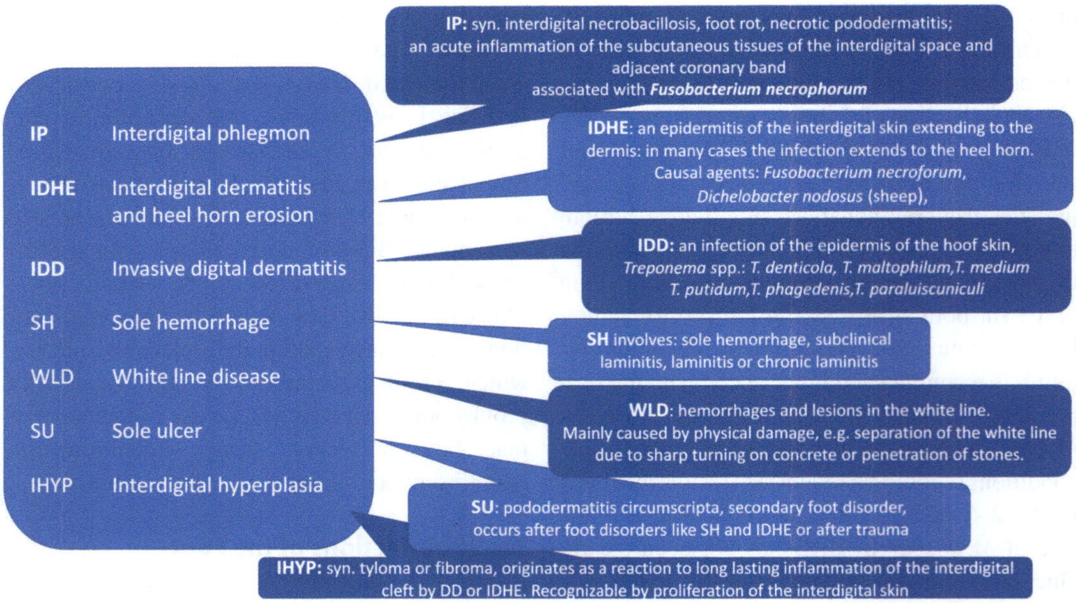

Fig. 2 Noseological units of diseases of the foot in cattle (based on summary by Bruijnis 2012)

Panaritium (interdigital phlegmon or necrobacillosis), associated mostly with intensive farming and dirty (uncleaned manure), wet and rough unfavourable surfaces, is mostly associated with *Fusobacterium necrophorum* as an etiological agent but could be associated with several other species (listed above), which may also be present and their exact contribution is not exactly known yet. Panaritium is usually diagnosed when the animal suddenly becomes limping or can be predicted using smart technologies (as pedometers) in very early stages. Swelling of the intermediate claw slit and crown edge occur as well together with interclaw fissure with cellulite and liquid necrosis presenting with a foul-smelling odour. Fever may be present. Except essential prevention (cleaning floors, good floor surface, regular care on foot), foot baths with disinfectant or alternatively other substances that decrease infection pressure and prevent the spread of the infectious agents can be done. It should not be forgotten to clean the claws prior to entry to baths as well as clean/use of intensive disinfectant in the bathtub and change the bath disinfectant solution regularly (instructions of disinfectant leaflet should be followed—change of the solution must be carried out after each 100–200 animals). The bathtub was recommended to be 3.5 m long, 60 cm wide and 25 steps in height to allow proper disinfection. Baths with automated management (disinfectant change, stop the cow in and further regulatory processes) have been introduced on the market. Regular trimming of the claws together with inspection of its status and early detection of the diseases help to the effective disease management.

Clinical signs are the bases for the diagnosis as isolation of *Fusobacterium necrophorum* and other (strictly) anaerobic germs from the affected tissues is difficult due to sensitivity to oxygen, which complicate sampling, transporting and culturing. Routine susceptibility to these bacteria is not performed as well. At the level of the affected tissues of the lower foot (claw slit and crown edges), the material can be collected by aspiration of tissue fluid using a hypodermic needle or by using a swab. Swabs are preferably kept refrigerated in an anaerobic transport medium (±4 °C) and have them examined as soon as possible (AMCRA 2023).

Antibacterial therapy at an early stage is recommended, with the first-line treatment using cephalexin, procaine benzylpenicillin and second-line treatment with aminopenicillins or amoxicillin in combination with clavulanic acid, oxytetracycline, penicillin/aminoglycoside combination, tilmicosin or tylosin. Most of the causal agents are susceptible to the first- or the second-line treatment options. Last-choice antimicrobial can be ceftiofur or cefquinome. It should be emphasised that despite zero withdrawal period for milk in ceftiofur containing veterinary medicinal product, ceftiofur should be kept as the last choice and only for the cases where other antimicrobials with lower risk for antimicrobial resistance do not work. The reason is that ceftiofur selects for extended-spectrum beta-lactamases (ESBL) across microbiota of the animal treated (including commensal *Enterobacteria*) and its use is therefore considered of high risk (AMCRA 2023).

7.2.3 Metritis

Post-partum acute (septic) metritis, as an inflammation of the uterus, occurs in cows and as the predisposing risk factors can be encountered conditions linked with partus: abnormalities, or complications in calving, failure to get rid of the placenta afterbirth in time, metabolic imbalances around the time of the partus and lastly mentioned, but manageable by good hygiene and farming practice, bad sanitary conditions at the partus. Acute status can be associated with foul smelling uterine discharge, can be with or without retention of placenta and due the time course associated with fever. Forms of (endo)metritis and basic timing of occurrence and signs are summarised in Table 7. Predominant first days' causative agent is *E. coli*, followed later especially by *Trueperella pyogenes*, but several Gram-negative anaerobes such as *Bacteroides* spp. and *Fusobacterium necrophorum* and *Prevotella* spp. can act as causal or associated pathogens as well. Metritis often involves a polybacterial condition (Constable et al. 2008). The causative pathogen(s), i.e. those considered responsible for inflammatory reaction, cannot always be identified.

Clinical endometritis may be associated with trauma of the genital tract and the subsequent disruption of physical barriers rather than with bacterial contamination. Subclinical forms of endometritis are rather detected as fertility disorders and based on cytology investigation. Pyometra is acute, immediate post-partum serious life-threatening disease that need urgent treatment.

Some studies show that there is often a spontaneous recovery from metritis, especially in cows with basically good condition and feeding status, accompanied with functioning immune defence mechanisms. Most bacteria invading the uterus at the partus are eliminated in 2–4 weeks after calving.

The severity of the inflammatory reaction and the accompanying deterioration of the health and condition status of the cow can vary significantly. The causal conditions can be premature birth (sometimes with a dead calf), twin birth, immature fetal membrane due to induced parturition, hard parturition (big calf, weak contraction) and cow condition-derived factors (ketose, lipomobilising syndrome, weak contraction due to subclinical or clinical post-partum paresis). The situation can be worsened by insufficiency of vitamins E and A, selenium or other elements ensuring energetic, vitamin and microelements/minerals balance.

Different opinions and continuing discussion are linked to the treatment possibilities to be used for different (endo)metritis forms, as listed above. Controversy occurs on the successful treatment definition (clinical and bacteriological cure as well as return to full reproduction/production status).

Acute metritis is often treated through intrauterine administration of antibacterial agents (e.g. oxytetracycline, cephapirin). Intrauterine VMPs differ as for pharmaceutical forms, there are available intrauterine tablets, suspensions, sticks, foams or powders for intrauterine lavages. In more mild cases, iodine-containing veterinary medicinal products can be used. In the cases of severe, generalised symptoms a parenteral, systemic treatment with injectable antibacterials can be used. In acute metritis, NSAIDs and possibly fluid administration can be administered concomitantly to antimicrobials as a supportive treatment. In the cases caused by anaerobes and *Trueperella pyogenes* with expected susceptibil-

Table 7 Forms of (endo)metritis and basic timing and signs

Endometritis form	Timing post-partum	Signs
Acute metritis (lochiometra; puerperal or post-partum or toxic or septic metritis)	Within 10 days after calving	– Foul smelling – Watery and reddish-brown to purulent excretion, in chronic cases become purulent – Elevated body temperature >39.5 °C – Large uterus with weak wall through rectal examination – Apathy, inappetence, increased heart rate, decreased milk yield
Clinical endometritis		– No signs of systemic disease or fever – Associated with trauma of the genital tract
	21 or more days post-partum	– Purulent uterine discharge, detectable in the vagina
	26 or more days post-partum	– Mucopurulent excretion into the vagina
Subclinical endometritis	5 weeks post-partum cytology diagnosis	– Inflammation of the endometrium – No visible symptoms of clinical endometritis occur – No (muco)purulent outflow observed – Impaired fertility – Cytology: the number of neutrophils increases
Pyometra		– Pus accumulation in the uterus – Corpus luteum present

Table adapted from Constable et al. (2017b), Sheldon et al. (2009) and AMCRA (2023)

ity, penicillins can be used as the first choice, followed by cefalexin and cephapirin and oxytetracycline (last two available in intrauterine pharmaceutical forms). As the second choice, amoxicillin, amoxicillin and clavulanic acid as well as tylosin can be recommended. Ceftiofur can be used as the last choice, being aware that its use should be avoided for uncomplicated cases of the retained placenta. It is indicated for the treatment of the acute post-partum (puerperal) metritis within 10 days after calving associated with *Escherichia coli*, *Trueperella pyogenes* and *Fusobacterium necrophorum*, in cases where treatment with another antimicrobial has failed (AMCRA 2023).

Sampling techniques differ, but preferable is long uterine swab in cannula (e.g. Metricult set, by LabmediaService, patent 305628), where swab is in canula, and by pushing, swab removes the biodegradable cap from the front part of canula in the uterus; therefore the sample is taken safely and free of vaginal contamination. Vial with Amies transport medium, where swab is placed for the transport, is recommended for facultative anaerobes as well as microaerophilic or anaerobic bacteria, and a temperature of 5–22 °C secures most of samples for 24–48 h.

7.2.4 Skin and Sensory Organ Infections

7.2.4.1 Skin

Bacterial skin infections in cattle belong among the frequent diseases as the large surface of the skin is susceptible to various injuries, possibly accompanied by the entrance of infective agents into the body. Depending also on part of the skin and source of infection (even accompanied with specific body areas as skin of the udder and skin in digital hoof and foot area), the following bacteria are frequently found as pathogens of cattle skin: *Actinomyces bovis*, *Prevotella melaninogenica*, *Staphylococcus aureus*, *Staphylococcus hyicus*, *Streptococcus dysgalactiae*, *Fusobacterium necrophorum*, *Moraxella bovis* and *Trueperella pyogenes*. Some bacteria are also associated with chronic wounds—*Staphylococcus* spp., *Streptococcus* spp. and

Corynebacterium spp.—and are able to form biofilms. Superficial staphylococci (mostly *S. aureus* and more rarely non-aureus staphylococci) can cause infection and impetigo and may affect the teats and udder of dairy cows, causing pustules, erosions and crusts (Scott 2007). Furunculosis, folliculitis and rain scald (dermatophilosis) are other superficial skin conditions caused by bacteria, all dependent on the breakage of the protective barrier of the skin that may appear anywhere on the body (Scott 2007). Treatment should be based on knowledge of infectious agent/s present (which is often hard to identify due to skin/fur contamination) and the extent of lesions or infection locus or wounds (including those caused by injury or by surgical techniques). Topical antimicrobials, but as well supportive, e.g. local disinfectants (povidone-iodine or chlorhexidine), herbal or enzyme-containing products, can be administered to help healing and ideally eliminate infection. Wound infections can be prevented by several steps prior-, at and post-surgeries, such as preparation of animals for surgery, using aseptic surgical techniques with adequately sterilised tools and material and via post-operative care, where also housing in a hygienic environment is necessary. Injectable VMPs can also be used, and their selection should be based on the expected infectious agents (e.g. cephalexin or procaine benzylpenicillin + aminoglycoside (dihydrostreptomycin or neomycin), sulphonamide + trimethoprim as the first choices and amoxicillin (alone or with clavulanic acid) or fixed combination of lincomycin/spectinomycin (anaerobic infections can be targeted by lincomycin) as the second choices.

Generally skin microbiota can contaminate the sample; therefore the surrounding of wound and abscess surface should be disinfected, and wound fluid/pus should by collected with swab at the level of lesions/directly from wound (it is not necessary to perform massive swabbing, which rather catch contaminants). In the case of deep wounds, fluid can be collected via a sterile needle and syringe. In abscesses, aspirate the contents in a syringe or collect pus on a swab (sampling from the wall of the abscess—not in the center of the abscess—gives better results). Negative culturing can be due to sterile abscesses or long-lasting abscesses with limited viability of germs (caution is necessary for strictly anaerobic bacteria that are very sensitive to contact with oxygen and can easily die). Immediate refrigeration is recommended for samples in transport medium (±4 °C) to prevent residential bacteria from overgrowing any pathogens. Swabs should be sent to the laboratory as soon as possible. If an anaerobe is suspected as a causal agent, preferably use an anaerobic swab/transport medium set and contact lab for performance culturing under strictly anaerobic conditions.

7.2.4.2 Ears

Otitis media (ear infection) affects mostly young animals (1 week to 18 months of age), but can also be suffered by adult cows (Yeruham and Elad 2004). Inclement weather has been associated with increased incidence of otitis (Bertone et al. 2015). Although beef calves are affected more often, cases were reported among dairy calves as well (Van Biervliet et al. 2004). Several forms of the disease manifest as the inflammation of the outer ear (otitis externa), middle ear (otitis media) and inner ear (otitis interna). Some cases can be classified as otitis media-interna (OMI) and occur in calves 1–10 weeks of age and typically accompanied with head tilting and cautious walking to maintain the balance. Otitis media inflammation due to different causal agents can occur once upper respiratory tract infection further spreads through the Eustachian tube. A severe otitis externa can also develop to otitis media. Scratching of the ears at the beginning of disease stage and depression, anorexia, drooping ears and head tilting at the later stages are clinical signs of otitis media. In some animals, fever can be present. In the young calves with suppurative otitis (accompanied as well with pneumonia or lung lesions), *Mycoplasma bovis* is the main cause. Pathogens such as *Pasteurella multocida*, *Mannheimia haemolytica*, *Histophilus somni* and *Trueperella pyogenes*, but sometimes also species of *Streptococcus* spp., are either causal or associated pathogens, mostly affecting also the respiratory tract. Clinical diagnosis is based on physical and neurological evaluation and can be confirmed by otoscopic examination as well as via diagnostic imaging techniques and necropsy (Morin 2004). Infected milk may cause an infection, both if calves are bottle-fed and in nursing on a cow with mastitis, as infection may ascend to their Eustachian tube. In up to 75% of ear infections, the calf will also have a concurrent pneumonia.

As a preventive tool, of key importance is management of properly stored and non-contaminated, ideally fresh colostrum intake by calves from healthy cows. Proper vaccination programme for respiratory diseases (BRD complex) is reported to help as both diseases are to be associated. Identifying and culling cows with ethological agent presence (mainly *Mycoplasma bovis*) help to control infection in calves. Feeding of contaminated milk should be avoided.

Treatment starting early in the course of the clinical disease or at subclinical phase contributes to the success of cure. Concomitantly NSAIDs (e.g. carprofen, flunixin or ketoprofen) can be administered. Nebulisation therapy to loosen mucus can be provided as well (AMCRA 2023). If eardrum rupture occurs, the tympanic cavity should be carefully cleaned (flushes with heated saline solution and low vacuum suction), as often a droopy ear with smelly discharge from the affected ear can occur. Eardrum perforations of small scales usually heal in 2–3 weeks. Once the disease becomes chronic, the probability of the success of non-invasive treatment decreases dramatically, so surgeries (myringotomy or tympanic bulla osteotomy) can be finally the only possibility. Since pneumonia is very often present, the severity of pneumonia should be monitored, and treatment should cover both diseases.

Florfenicol can be the first-choice antibiotic (in sufficient dose also for *M. bovis* infections). Among second-choice antibiotics are oxytetracycline, doxycycline (oral) and macrolides (tilmicosin, tylosin, tulathromycin). The last choice can be fluoroquinolones (e.g. marbofloxacin).

Sampling at any time is performed by inserting a sterile swab (if possible, in the canula, so the swab can touch just directly the sampled area to avoid external ear contamination) into the external ear canal. Swab (in transport medium) is recommended to be examined within 48 h. Postmortem the bulla tympanica can be opened. Based on anamnesis, request for culturing on both media for non-fastidious microorganisms and a selective media for *Mycoplasma* spp. (or PCR detection of *M. bovis*). If culturing was performed, results can be available within 2–5 days for non-fastidious microorganisms, including AST, and 7–14 days for *M. bovis*, where antibiograms are not performed.

7.2.4.3 Eyes

Conjunctivitis can have infectious as well as non-infectious (irritating gases, e.g. ammonia) causality. Infections with various viruses such as BVDV, OVH-2, BHV-1 and various primary respiratory viruses (BRSV, PI-3) can occur. Keratitis can also develop as a secondary infection or concomitantly, characterised by the blurring of the cornea, possibly with ulceration of the cornea. Infectious bovine keratoconjunctivitis (IBK), commonly known as pinkeye, can occur, where causal agents are mostly *Moraxella bovis* and *Moraxella bovoculi* (Rogers et al. 1987; Angelos 2010; Gould et al. 2013). Infections are frequent mainly in summer and autumn with flies being vectors for pathogen transfer. As a clinical syndrome, IBK has a typical progression from a watery eye (epiphora) with sensitivity to light and redness (conjunctivitis) to an ulcerated cornea with cloudiness due to oedema and inflammation. Infiltration of blood vessels along the outer edge of the cornea causes this "pink appearance". In worst cases, the resulting ulcer can extend through the cornea, allowing the iris to protrude through the ulcer, or the eye may rupture completely, resulting in permanent blindness (Angelos 2015). Moreover, direct sunlight, dust and fly invasions can cause more serious course of disease and injuries. Rarely, *Mycoplasma bovis, Listeria monocytogenes* and in single outbreaks also *Chlamydophila* spp. and *Pasteurella multocida* were confirmed to be associated with keratoconjunctivitis. Differential diagnosis is necessary to distinguish with mycotic keratoconjunctivitis. The diagnosis of infectious bacterial keratoconjunctivitis can be confirmed by culturing following proper sampling on a conjunctival swab. Attention should be paid also to the possibility of traumatic causality, as keratoconjunctivitis may be due to the presence of foreign object or physical/chemical injury, which needs a different approach in treatment.

As there are no commercial vaccines available, autogenous vaccines can be the choice, and if not accessible, prevention needs to rely on repellent-containing products (sprays or ear tags impregnations) to combat flies, together with avoidance of housing in a dusty/irrigating gases environment. Frequent inspection of animals on pasture is important in order to quickly detect the symptoms and thus be able to treat them in a timely manner. As IBK is a group-related condition, in one case all animals on the pasture should always be examined and possibly treated. Animals that have had a clinical or subclinical infection can remain carriers for months. These carriers can introduce the germ into an originally unaffected farm, resulting in the rapid spread within the farm. In strains, which are not producing beta-lactamases (as AMR gene encoding resistance to beta-lactam affects penicillins as well as aminopenicillins), procaine benzylpenicillin and cephalexin can be used, while in beta-lactamase producers sulphonamide/trimethoprim combination is the first choice, followed by tulathromycin, tylosin, florfenicol and oxytetracycline as the second choice (AMCRA 2023; Smola et al. 2022). The use of florfenicol may facilitate selection and spread of resistance to oxazolidinones (linezolid) in Gram-positive bacteria. Its use should therefore be restricted. Animals with clinical signs (epiphora) can be sampled by taking a sterile eye swab. It is recommended to pull the eyelid down/up and sample the conjunctival sac well. The swab/in transport medium must be stored at ±4 °C for a maximum of 48 h. Susceptibility testing (including beta-lactamase test) is recommended.

7.3 Mastitis

Mastitis is a disease of high impact on animal welfare as well as on economy. It adversely affects the gains of animal producers/farmers and leads to a large production loss in the dairy sector worldwide (Izquierdo et al. 2017; Aghamohammadi et al. 2018; Sharun et al. 2021). Mastitis has been considered to create almost two-thirds of the economic loss in an average herd (Halasa et al. 2007), with approximately 70–80% of the financial losses by its subclinical form (Seegers et al. 2003; Halasa et al. 2007) and predominantly through bacterial infection (e.g. Schwarz et al. 2010). In the European conditions, depending also on the country as well as on the severity of mastitis and extent and costs of the intervention, the range of the total costs of mastitis per cow and year may vary. Direct annual economic losses per cow in the event of clinical mastitis range from 115.4 to 193 € as calculated for Holstein dairy cows in Czechia, which included the increased need for veterinary treatment and medicines, breeder's time, discarded milk and increased cow culling (Krupová et al. 2019). Authors of another study bring dataset and describe in detail individual cost items and estimate average total costs of mastitis at 240 €/lactating cow per year (van Soest et al. 2016), while another study, which involves additionally the costs of chronic mastitis, modelled the median total costs of mastitis to be 230 € (Bonestroo et al. 2023). The sum of 240 € consists of failure costs contributed by 120 €/lactating cow per year and prevention costs with another 120 €/lactating cow per year. Milk production losses, discarded milk and culling as the main contributors to failure costs have been detected at 32, 20 and 20 €/lactating cow per year, respectively. The main contributor to prevention costs is labour, next to consumables and investments, at 82, 34 and 4 €/lactating cow per year, respectively. In more details can be described all influencing factors contributing to the economic losses due to mastitis. Direct costs include payment of diagnostic testing, veterinary care and prices of veterinary medicinal products administered, discarded milk and labour of farmers taking care. Indirect costs include future milk production loss, decreased milk quality, reduced reproduction and increased death or premature culling and replacement of mastitic cows (Hogeveen et al. 2011; Rollin et al.

2015; Heikkilä et al. 2018, van Soest et al. 2016). The costs of preventive measures should also be considered in the total costs of mastitis (van Soest et al. 2016). Mastitis, however, impacts milk quality as well. Residues of veterinary medicinal products (including antimicrobials) can pose risks for human health, especially if cow is treated and milk is not discarded for a sufficiently long period. Residues, among other properties, can also have an impact on the microbiome, especially once exceeding MRLs. Moreover, implications for the public health are not only due to residues of antimicrobial/s in milk but also thinking within One Health perspective that mastitis is perceived as a threat both for human and animal health due to the zoonotic potential of several causative agents as well as from the perspective of resistance spread (Piras et al. 2020; Sora et al. 2022). Two exact examples of antimicrobial resistance that can persist in the bacterial community are *Streptococcus agalactiae* and *Staphylococcus aureus* being an issue also in human medicine, especially if the strains are multiresistant or genes of specific resistance (e.g. to methicillin in staphylococci) are present (Sharma et al. 2017).

Cows with mastitis infection can show a wide range of symptoms: localised for udder (swelling, heat and pain) and systemic (increased body temperature, lethargy and anorexia) as well as milk with abnormal appearance (Kibebew 2017). In principle, mastitis, as an infectious process, is an immune response to pathogens affecting the udder (Gurjar et al. 2012). The udder can be regarded as the milking cow's shock organ as it is more burdened than most other bodily functions. Any negative and stressful load on the cow, metabolic disorders or infection, results in an increased incidence of both clinical and subclinical mastitis (Landin et al. 2020). A healthy udder is characterised by immune homeostasis due to the presence of complex community of diverse bacteria creating balanced intramammary microbiota (Rainard 2017; Derakhshani et al. 2018; Andrews et al. 2019). As the phyla creating the core structural microbial flora of healthy cow milk were identified as *Firmicutes*, *Proteobacteria*, *Bacteroidetes* and *Actinobacteria*, the same phyla were detected in milk from cows with mastitis with addition of *Tenericutes* (Luo et al. 2023), at the family level *Ruminococcaceae*, *Lachnospiraceae* and in more detailed genus level *Propionibacterium*, *Stenotrophomonas*, *Corynebacterium*, *Pseudomonas*, *Streptococcus*, *Comamonas*, *Bacteroides*, *Enterococcus*, *Lactobacillus and Fusobacterium* are involved (Rainard 2017). Similarly, also congenital anomaly of the udder or teat, such as teat fistula, teat spider, leaky teat and retention of milk, predisposes the animals to more ease of development of this disease. Moreover, udder wounds (including micro-wounds) once exposed to the attack of bacteria, especially those with host specificity and virulence factors, speed up the inflammatory process and mastitis development (Sharun et al. 2021). The effect of severe teat-end hyperkeratosis on both contagious (especially *Staphylococcus aureus*) and environmental clinical, as well as subclinical mastitis, emphasises the importance of teat health (Pantoja et al. 2020) and proper setting of all parameters of (semi)automated milking processes for mastitis control. Landin et al. (2020) commented, based on Swedish experience, that a healthy udder is considered with a constant SCC level below 100,000 cells/mL of milk. Based on long-term experience as well as evaluation of the production and SCC parameters, it was proven that cows with SCC levels below 50,000 are the ones that have the highest yield and almost never suffer from mastitis (Landin et al. 2020). As the most common SCC level of cows in Sweden, regardless of breed and age, the authors indicated 20,000–50,000 cells/mL of milk. On the other hand, average bulk tank milk SCC (BTSCC) in Sweden was well above 200,000 cells/mL, which suggested that many cows with increased SCC were present in the herds that indicated the need for continuous work on improvement of udder health (Landin et al. 2020). This led the board of Swedish experts to elaborate the "udder health pyramids" (please refer to Fig. 7, later in this chapter).

7.3.1 Classification of Mastitis and Microbes Playing Role as Mastitis Causal Agents

Essential traditional classification to environmental or contagious mastitis according to epidemiology and mode of transmission from the source of infection can be used. According to the degree of inflammation in the mammary gland, mastitis is classified as clinical, with visible abnormalities in the udder or milk, or subclinical, where no clinical signs but elevated somatic cell count (SCC) is detected; based on the disease course and clinical symptoms, per-acute, acute, sub-acute and chronic mastitis can be distinguished. As for the European continent from scientometric metanalysis of prevalence studies of 1979–2019 data analysed by Krishnamoorthy et al. (2021), major bacterial bovine mastitis pathogens are as follows: *Staphylococcus* spp. (23%), *Streptococcus* spp. (12%) and *Escherichia coli* (10%). Further possible (major and less frequent) ethiological agents are summarised in Table 8 and briefly commented below in the text of this subsection.

Contagious mastitis is caused by bacteria such as *Staphylococcus aureus (S. aureus)*, *Streptococcus agalactiae (S. agalactiae)*, *Streptococcus dysgalactiae (S. dysgalactiae)*, *Mycoplasma* spp. and *Corynebacterium bovis*. Transmission routes, due to which spread from infected cow to a healthy cow occurs, can be influenced by thorough cleaning, disinfection ordering of animals for milking. Contagious pathogens usually spread through the hands of the staff during milking, towels and/or the milking machine serving as the in-contact reservoir of pathogens.

Environmental mastitis was traditionally considered as caused by bacteria that spread primarily outside of the milking parlour and bacteria being mostly ubiquitous and coming from the housing environment as a bedding material and from soil, manure, faeces and contaminated water. However some more recent bibliographies (e.g. Biggs 2018) consider this classification as too simplistic, especially for certain ethiological agents. The ground for the new thoughts is evidence of the existence of host-adapted strains of certain udder pathogens such as *Streptococcus uberis* and *Klebsiella* spp.; also for *S. dysgalactiae* new classification considers it rather as an intermediate pathogen (Biggs 2018; Wente and Krömker 2020). Highly mammary gland-adapted strains own adhesins or other factors of virulence and have many properties that are the same or closely similar to contagious pathogens (see description below for *S. uberis* and *S. dysgalactiae*). Notwithstanding these new considerations, we could enumerate a set of bacterial species, especially bacteria from Enterobacterales as, e.g. *Escherichia coli*, *Klebsiella* spp., *Raoultella* spp., *Enterobacter* spp. or genus *Streptococcus* or *Enterococcus*—some of them highly adapted for invasion to the udder compartments. A specific group can be considered coagulase-negative *Staphylococci* or, more broadly, recently rather listed as non-aureus *Staphylococcus* spp., which are also considered as ethiological agents of environmental mastitis. One of the very critical pathogens is also *Pseudomonas aeruginosa*, causing fatal mastitis cases. Among elementary steps to control environmental mastitis generally are limiting water use in parlour (to avoid increasing humidity and spreading of microbes via aerosol), avoidance of overmilking keeping teat tops healthy, removal of hair from the udder and keeping udder and teats clean, using specific bedding materials lowering environmental pathogens and scraping alleys frequently and in certain scale also vaccination (commercial/autogenous vaccines).

Despite not belonging among those most frequent, also other microorganisms can cause mastitis with fatal ending as those belonging among *Prototheca* and yeast pathogens, where antibiotic treatment is clinically not effective. *Prototheca zopfii*, where specifically the lineage known as genotype 2, recently raised to species status as *Prototheca bovis* (Jagielski et al. 2019a) and the current species *Prototheca wickerhamii*, which are both the main causative agents for cattle and human infections respectively (Jagielski et al. 2019b; Kano 2020; Libisch et al. 2022). Recently in some countries, *Prototheca* became not that negligible counts. Based on the data available, examples can be listed in Poland, with

Table 8 Overview of mastitis pathogens, highlighting the major pathogens being most frequently isolated

	Gram-positives	Gram-negatives
Major	**G+ cocci**	**G-*Enterobacterales***
	Streptococcus uberis	*Escherichia coli*
	Staphylococcus aureus	*Klebsiella* spp.
	Streptococcus dysgalactiae	*Klebsiella pneumoniae*
	Streptococcus agalactiae	*Klebsiella oxytoca*
Minor/other	**Non-aureus staphylococci**	**G-*Enterobacterales***
	Staphylococcus chromogenes	*Raoultella ornithinolytica*
	Staphylococcus simulans	*Raoultella planticola*
	Staphylococcus haemolyticus	*Enterobacter* spp.
	Staphylococcus epidermidis	*Serratia* spp.
	Staphylococcus xylosus	*Citrobacter* spp.
	Staphylococcus sciuri	**Other bacteria**
	Staphylococcus cohnii	*Pseudomonas* spp.
	Staphylococcus capitis	*Mycobacterium* spp.
	Staphylococcus warneri	*Chlamydia* spp.
	Staphylococcus equorum	*Chlamydophila* spp.
	Other G+ cocci	**Yeasts and fungi**
	Enterococcus spp.	*Candida krusei*
	Aerococcus viridans	*Candida rugosa*
	G+ rods	*Aspergillus* spp.
	Trueperella pyogenes	**Algae**
	Corynebacterium bovis	*Prototheca* spp.

8.3% samples (n = 400/milk), where *P. bovis* was detected on eight farms out of 16 examined and in Italy with a prevalence of 11.2% in Italian herds tested —Jagielski et al. 2019b; Zecconi et al. 2020). Yet in vitro data only has been published, and it is awaited in vivo studies and clinical trials (Shave et al. 2021), where in vitro *Prototheca bovis* strains were observed to be susceptible to colistin sulphate, gentamicin and netilmicin (100% of susceptible strains). Inhibiting 31% of the strains by nisin shows another possibility, and nisin inhibition exceeds the inhibition of lysozyme or potassium sorbate. RAPD-PCR was used as a rapid, inexpensive and highly discriminating tool for *Prototheca* strains (Morandi et al. 2016). When protothecal infection is detected in a cow, the elimination of the animal from the herd is usually necessary, hygienic conditions with respect to milking equipment should be improved, and the proper management of animal feeding and drinking water is advised.

Unfortunately, it has to be noted that treatment success in mastitis is close to zero in clinical mastitis caused by *Pseudomonas* spp., *Mycoplasma bovis*, *Mycobacterium* spp., *Prototheca* spp., yeasts and fungi.

7.3.2 Diagnostic Methods

All diagnostic methods should be based on reasonable sample taking. By naked eye investigation can be detected variations in colour and consistency and presence of clots and blood. Physicochemical changes influencing the increase of conductivity, pH and changes in water content can occur in milk as well (Haxhiaj et al. 2022). The California Mastitis Test (CMT) is one of the oldest and easiest tests to use as a cow-side test. It evaluates the milk's alkalinity using a detergent (alkyl aryl sulfonate) combined with a pH indicator, bromocresol purple. If the test is positive and indicating mastitis, a purple viscous mass due to nucleic acids and other constituents is released from the lysis of somatic cells.

Portable devices, such as the DeLaval cell counter, Porta SCC and Fossomatic, are useful for on-farm SCC evaluation as well. Even though these portable systems are easy to use and fast, their lower sensitivity at low SCC was noted (Haxhiaj et al. 2022). Despite being a well-recognised indicator of the processes indicating mastitic changes and very frequently used as quantitative descriptor, SCC can be influenced by different factors such as lactation number, milk yield, stress, season and breed that could finally influence SCC's interpretation. Other methods, mostly those detecting via different ways potential causative pathogen/s, are therefore used.

For other tests based on culturing and molecular biology methods, proper, strictly aseptic milk sampling techniques minimising contamination and appropriate handling with sample (either on-farm, with direct inoculation to plates, or transfer in sterile tubes to laboratory for further investigation) are vital. For basic rules to be followed, refer to Fig. 3. Aseptic collection is the key factor as well as storage of samples (4–8 °C); sometimes a preserving agent is used in the test tubes (Ly20, containing boric acid) GVA (2009).

In most practices two systems are currently used for bacterial culturing. One of them is direct on-farm culturing systems (point-of-care tests), with use of different media—specifically designed (e.g. chromogenic, with or without selective compounds) and specific protocols of interpreting the results. Comprehensive overview of "point-of-care tests" and their characteristics, including sources of information, is available as a supplementary file in the publication by Malcata et al. (2020). Approaches may differ also in targeting such testing—individual quarter/cow level or bulk tank test—to investigate the situation in the herd, including monitoring of the results of preventive measures. The second system is laboratory conditions´ culturing, once milk is transported to the laboratory, where not only bacterial culturing, identification with MALDI-TOF (matrix-assisted laser desorption/ionization-time of flight) mass spectrometry system and using expertise of microbiologists but also antimicrobial susceptibility testing is performed.

From data by Duse et al. (2021), it can be confirmed that with thorough, validated performance of all steps in the whole "chain" of sampling-

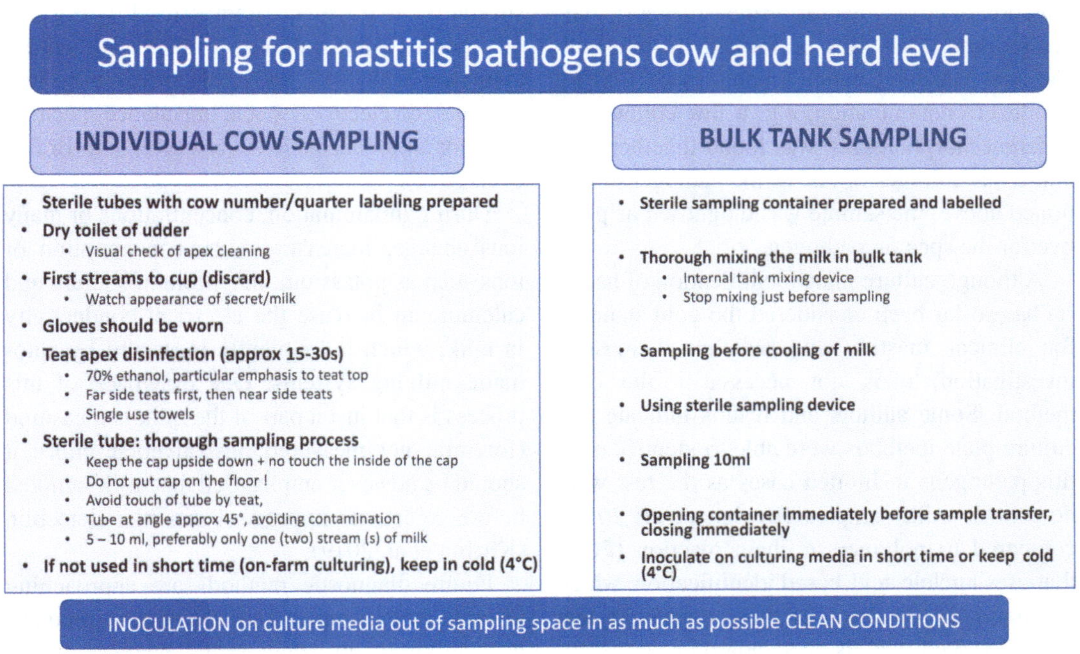

Fig. 3 Summary of elementary rules of sampling for culturing of mastitis pathogens

transporting-culturing-expert first reading, confirmation of identification via MALDI-TOF and susceptibility testing is still the gold standard, routinely feasible and cost-acceptable method. Within the monitoring performed in Sweden, finally assessed total of 775 sampled quarters of 734 cows suffered from mastitis, from which 675 samples (89%) were positive (i.e. contained one or more microbial species). Some principal moments of the study can be noted, as that clinical mastitis, acute cases, the most affected quarters were aseptically sampled with proper transport ensurance and culturing. In brief, culturing can be summarised: 10 µL milk streaked onto blood agar (5% blood) with esculin, incubated at 37 °C, and evaluated after 18–24 but as well after prolonged culturing 36–48 h, what is of importance for bacteria with slower growth and can be the cause of negative results if not followed. Then finally thorough reading was performed (1 CFU interpreted as positive sample specifically for *S. aureus* and non-aureus Staphylococci, *S. uberis, S. agalactiae, S. dysgalactiae, S. parauberis, T. pyogenes, E.coli, Klebsiella* spp., which could be identified by experienced microbiologist), attention to more than one colonies type was paid and thorough selection of colony/ies for further testing (identification and AST) was performed. In line with National Mastitis Council guidelines (2017), if a low-level contamination, e.g. a few colonies of different morphologies, was found together with numerous colonies of one of the pathogens mentioned above, the sample was diagnosed as positive for the specific pathogen.

Although culture plate identification of bacteria has so far been considered the gold standard for clinical mastitis and subclinical mastitis investigation, it is not necessarily the only method. Some authors and results indicate that culture plate methods were able to identify mastitis pathogens in limited cases as the rest were no-growth milk samples (Bexiga et al. 2011), compared to polymerase chain reaction (PCR) that uses nucleic acid-based identification, which facilitates the detection of those pathogens that cannot be identified using standard bacterial culturing, however it shoudl be considered that detection of bacterial DNA does not imply the presence of viable bacteria. Although some PCR tests are commercially available (e.g., PathoProof PCR assays from Thermo Fisher Scientific), they are not intended for on-farm use, which limits broadening of use. Investments in the availability of rapid, preferably on-farm, PCR tests will increase options to adopt selective clinical mastitis treatment protocols (de Jong et al. 2023).

Electrospray ionisation mass spectrometry (ESI-MS) is another diagnostic method that can detect other microorganisms such as parasites, yeasts and viruses. But both ESI-MS and PCR are more expensive methods than bacterial culturing, where the cheapest is on-farm culturing. Other molecular techniques, rather used as confirmatory methods or in scientific studies, include real-time quantitative PCR (qPCR), loop-mediated isothermal amplification (LAMP) and next-generation sequencing (NGS) methods (Sharun et al. 2021). Also further innovative, rapid methods have been developed as paper-based multiplex LAMP assays, which may provide a fast and low-cost option, microbiologically safe (paper can be burned after use) and environmentally sustainable (Reboud et al. 2019) rapid 3-h assay with a set of specific antibodies that recognise tags attached to specific PCR fragments immobilised by printing on porous nitrocellulose membrane slides (Mujawar et al. 2013) and an electrochemical sensor based on electrochemical impedance spectroscopy for *Staphylococcus aureus* detection (Braiek et al. 2012).

During inflammation, concentrations of many ions change. Increases in the concentration of ions such as potassium, magnesium, sodium and calcium can increase the electrical conductivity in milk, which is frequently measured by automatic milking systems. One drawback of this process is that just a part of the milk is measured (foremilk not measured, just alveolar milk); it should be however emphasised that milk sampled before ejection improves mastitis detection (Khatun et al. 2019).

Future diagnostic methods are approaching towards more complex and advanced techniques, not relying on the milk analysis only. Metabolomics by analysis of other fluids (blood, serum, urine) and metabolic biomarkers can identify animals that are susceptible to or that suffered from mastitis (Haxhiaj et al. 2022), but there have not been developed sufficiently robust, achievable and cheap tests that can be used routinely; therefore further large-scale validation and production of predictive biomarkers for routine diagnostic are foreseen. One of possible directions can be innovative rapid immunodiagnostic assays for bovine mastitis based on proteins with confirmed immunoreactivity. Very recently such proteins were described by Dobrut et al. (2023a) for *E. coli* (three proteins), *S. uberis* (four proteins) and *S. agalactiae* (six proteins). Similarly in other work by Dobrut et al. (2023b), *S. aureus* fibronectin-binding protein was proposed as a possible immunoreactive protein, for which immunoreactivity to antibodies present in serum from cows with diagnosed mastitis was demonstrated.

7.3.3 Clinical Mastitis

7.3.3.1 Clinical Mastitis Characterisation and Risk Factors Contributing to It

Clinical mastitis can be identified as visible symptoms signalising udder inflammation, where redness in the affected part or complete udder, swelling, warmth, pain upon touch, milk clots, discolouration and change in consistency of milk are typical. Systemic symptoms are pyrexia, when 39.5 °C is exceeded, and loss of appetite.

Clinical mastitis treatment should be carried out as quickly and as targeted as possible, on the basis of the results of the clinical examination and animal health status consideration, symptoms, milk appearance and considerations of the recent bacteriological examination results of previous cases of clinical mastitis in the respective herd. Despite the performance of in-house testing (with rapid screening identification of the pathogens concerned), especially once performed accurately and well trained and experienced person is of important help, it should be emphasised that (at least in reasonable time intervals) confirmation of identification (MALDI-TOF) and performance of susceptibility testing in laboratory conditions to gain knowledge of resistance profiles and their trends within the herd during the time course is vital for proper treatment and mastitis control. It is also vital to consult the veterinarian, who regularly attends the herd and knows the history, anamnestic data and trends in susceptibility and disease prevalence and can discuss with the zootechnician further aspects related to mastitis cases in individual animals, groups of animals and the whole herd. Treatment should be held according to the guidelines recommended in the operating treatment plan established by the veterinarian and tailored for the herd/holding management practices, including frequency of milking (two, three times/day or robotic). Treatment with antimicrobials (sole intramammary, or combined intramammary and injection administration) is performed, and some bibliographies also present results of concomitant, supportive treatment, such as antioxidants (vitamins C and E and β-carotene), lysozyme dimer or NSAID that can be useful in treatment of mastitis and improving fertility of cows that suffered from mastitis compared to treatment with antibiotic only (Smulski et al. 2020).

As mentioned above, rapid (in-house) diagnostic culturing or newly also flow dipstick scan tests may also be used to perform screening for the presence of mastitis causative agents in the infected area and in case of culturing systems on chromogenic media to identify them. Recently (June 2023) Vetscan® Mastigram+, a rapid on-farm mastitis diagnostic test, with flow dipstick test to detect Gram-positive mastitis in 8 h, enabling results before the next milking has been introduced. The milk sample is collected aseptically in a test tube and incubated for 7 h, then the lateral flow stick is inserted and the test run for 10 min, and results are readable from the test strip. Pending its result, considering recent herd anamnesis as well as successfulness of treatment of previous similar cases, treatment can start in cows with mild mastitis (only milk abnormalities) or moderate (milk and udder abnormalities) with a non-steroid anti-inflammatory veterinary medicinal product, and in Gram-positives the farm veterinarian can decide on the basis of rapid scan test or culture test results and considering in complexity also other factors of importance, whether and which antibiotics are indicated in these cows. The decision can be made after 8 h in the rapid scan test and/or after confirming by culturing test mostly after 24 h or after 48 h if needed prolonged culturing and considering udder/cows status or after >72 h in specific, non-bacterial pathogens. In experienced, well-cooperating farmers (with help of up-to-date algorithm set by the veterinarian), a similar approach can be used in cases where an attending veterinarian is not able to visit the farm in an everyday manner. In the case of mild and moderate mastitis caused by Gram-positive bacteria, antibiotic treatment is recommended, unless they are non-aureus *Staphylococci*. Antibiotic treatment is not indicated, once culturing is negative, however attention should be paid to quality control of sampling, handling with sample, culturing conditions as well as prolongation of culturing which might be necessary for some pathogens on certain type of culturing media (e.g. *S. aureus, P. aeruginosa*) to minimise cases of false-negative results. After 18–24 h it is mostly not possible to evaluate growth of the following pathogens: *Corynebacterium bovis*, *Trueperella pyogenes*, *Campylobacter* spp. and *Psychrobacter* spp., but also some other microbes need more than 72-h culturing: yeasts (e.g. *Candida* spp.), moulds (e.g. *Aspergillus* spp., 168 h) and algae (e.g. *Prototheca* spp.). Except short culturing and inadequate conditions (e.g. missing some specific supplements in "universal" culturing media) for some bacteria with specific growth demands (e.g. *Mycoplasma bovis*), there are several reasons for no growth of microbes from clinical mastitis milk. One example can be *E. coli* strains from mastitis infections that are weakened by the cow's immune system (Shinozuka et al. 2018). Further cases reported show that also freezing the milk samples prior to culturing decreases the survival rate and growth of certain bacteria such as *E. coli* but not *S. aureus* (Hubackova and Rysánek 2007). The likelihood of a negative culture also depends on the stage of the infection and inflammation, and sampling of acute phase of clinical infection within the stage when microbes actively multiplicate increases the chance of positive culturing of the pathogen causing such infection. On the other hand, periodic shedding increases the probability of no growth result.

Severe mastitis (abnormalities of the milk and udder accompanied by systemic symptoms of disease in cows) should be immediately treated with antibiotics. Sampling should be performed at the moment prior to the first dose of the treatment, as later, once the antibiotic starts to act, microbes can be inhibited and are not able to grow in culturing media. In serious cases, which respond poorly on already started treatment, an antibiogram can help to switch to another antibiotic, once necessary.

The treatment of these cows can be adjusted in the light of the results of the rapid test or laboratory test due the time course. Simplified decision algorithm for clinical mastitis is proposed in Fig. 4, which considers elementary parameters as seriousness of clinical mastitis based on clinical examination of cow and udder, milk appearance and start and choice of the treatment based on culturing (both options on-farm and laboratory investigation including antimicrobial susceptibility testing (AST)); it should also be noted that milk from any mastitis cases should be discarded under normal appearance and if cows are treated, until the end of the withdrawal period relevant for veterinary medicinal products used to ensure consumer safety.

Clinical mastitis can become recurrent; on the other hand recent data (Wente et al. 2020) shows that while recurrence rate can be around 22% (445 from 2043 diagnosed mastitis cases), recurrence with identical strain can be quite limited (11%; 49/445). Interestingly dynamic shifts of microbiome composition and genomic features with different pathological conditions of mastitis were proven, and based on the knowledge this can contribute to developing microbiome-based diagnostics and therapeutics for bovine mastitis.

Jamali et al. (2018) summarised in the review of 57 manuscripts risk factors for clinical mastitis (CM) and its recurrence, which can be stratified as follows (below risk factors were amended by other bibliography as indicated in the text):

1. *Parity/age*: Increasing parity increases prevalence of intramammary infections

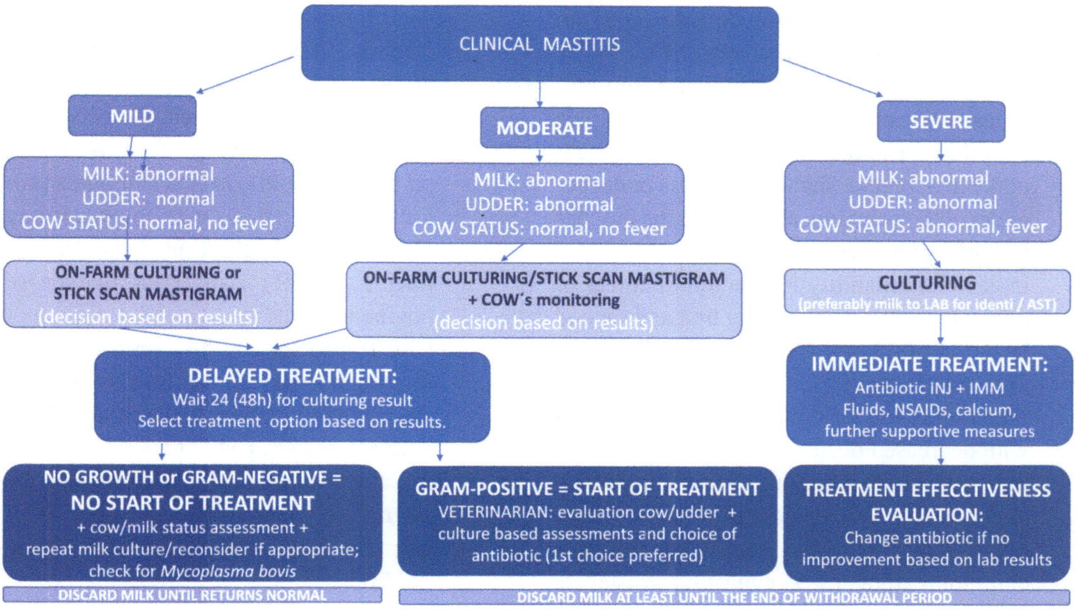

Fig. 4 Simplified clinical mastitis' decision algorithm for treatment start and evaluation based on seriousness, culturing and clinical examination (amended/updated from Zigo et al. 2021)

(Fukushima et al. 2020) regardless of causative pathogen (Taponen et al. 2017), and increase of clinical mastitis is generally more reported in older cows and cows ≥3 lactations with 15-fold higher risk of recurrent CM (Pinzón-Sánchez and Ruegg 2011); percentages of cows reported according to parity experienced with two and more clinical mastitis in a given lactation according to Wolf et al. (2010) was established as follows: 19% (first), 27%(second), 36% (third) and 39% (fourth). Possible reasons for the increase in clinical mastitis incidence with parity could be changes in the udder and the contractility of the teat sphincter finally influencing invasion of pathogens into the teat canal and udder (Boujenane et al. 2015). Streptococcal mastitis frequency for old cows (>10 years) is approximately 24% higher than for young cows (3–6 years) due to wider or permanently partially open teat canal as a result of frequent milking (Cheng and Han 2020). Cows with a high calving number (>6) developed a streptococcal mastitis 30% more often than cows with few calvings (1–3) (Kabelitz et al. 2021).

2. *Early lactation period* (Jamali et al. 2018; Cheng and Han 2020): A period of 0–3 months, where changes in immune functions as well as nonspecific host defence mechanisms, increased oxidative stress and low antioxidant defence could be considered as a cause of the high incidence rate of clinical mastitis in early lactation. As for recurrent cases, a study by Wente et al. (2020) shows that from 2043 mastitis cases recorded on a quarter level, 1598 were first cases and 445 were recurrent, from which 145 (32.6%) cases were confirmed to harbour the same pathogenic species as previous infections. RAPD PCR confirmed the same species strain in 49 (11%) of the recurrent infections, so only 11% was total recurrence rate with the same strain confirmed.
3. *First 30 days of milking* are considered of higher risk (4.6 and 2.0 times higher) than during the remaining lactation in multiparous and primiparous cows, respectively (van den Borne et al. 2010; Hammer et al. 2012).
4. *Genetics:* Genetically correlated indicator traits highlighted in a symposium review targeted on the novel strategies to genetically

improved mastitis resistance (Martin et al. 2018) were somatic cell counts (SCC), udder depth and fore udder attachment. From previous work made by Nash et al. (2000), a selection for longer productive life, stronger udder attachment, shallower udders or deeper udder cleft seem may also reduce the number of CM events per lactation. This study also indicated heritability for the total number of clinical mastitis cases during first and second lactation of 42 and 15%, respectively. Also, Taponen et al. (2017) indicated that better breeding index for milk SCC was associated with decreased occurrence of intramammary infections, indicating breeding for improved udder health as a successful tool for reduction of mastitis incidence caused by the most common pathogens. For mastitis resistance, it is difficult to identify candidate genes that determine traits due to the polygenic nature of the disease. However, a database of candidate bovine genes and genetic markers for milk production and mastitis has been compiled containing 943 genes and aiming for an integrated research tool inclusion (Ogorevc et al. 2009). A total of 117 candidate single-nucleotide polymorphisms (SNPs) and 27 quantitative trait loci (QTLs) associated with mastitis resistance within a population of phenotypically well-characterised dairy cattle were also identified (Kurz et al. 2019). An important point is also the prevention of mastitis, and some genes were recommended to be selected among the most promising resistance genes for mastitis and involved in the genetic programmes (Zemanova et al. 2022), which could be a starting point to minimise recurrence as well:

(a) CD14: product of this gene is an important part of host innate immunity, which provides sensitivity to lipopolysaccharides to individual cell types, such as neutrophils, monocytes and macrophages (Wu et al. 2019). This product mediates host defences against Gram-negatives, provides immunity against viral infections and determines the effect of changes in protein expression on the surface of monocytes and neutrophils in healthy dairy cows (Wu et al. 2019).
(b) CXCR1: interleukin 8 (IL-8) α receptor, encoded by the *CXCR1* gene, is present on the neutrophil surface and binds pro-inflammatory IL-8 with high affinity (Teijeira et al. 2020). Hence, the gene for bovine CXCR1 can be a marker of bovine mastitis.
(c) LACTOFERRIN: this represents an important part of the host's first-line defence in the infected mammary gland, where it makes up about 5% of milk during acute inflammation and ensures the migration, maturation and function of immune cells (Shimazaki and Kawai 2017).
(d) LACTOGLOBULIN: this is a beta-lactoglobulin encoded by *BLG* gene, which binds iron through siderophores; thus, it is fundamentally involved in the immune response in the fight against pathogens in milk (Alim et al. 2015).

5. *Production level:* Results from the meta-analysis made by Jamali et al. (2018) within scientific literature sources suggest that high-producing dairy cattle are more prone to have clinical mastitis recurrence. Despite that, Taponen et al. (2017) concluded that increasing prevalence of automatic milking is not a reason for special concern of occurrence of intramammary infections.
6. *Housing and pasture seasons*: In some pathogens seasonality can occur, as documented, e.g. by Duse et al. 2021 in a study performed in Sweden, where *S. aureus* was more commonly detected in mastitis cases occurring during the late housing season (January to April) compared with the early housing season (September to December), in contrary to *S. uberis* and *S. dysgalactiae* (more in early housing season), as documented also by a US study (Makovec and Ruegg 2003). In the cases of clinical mastitis caused by non-aureus staphylococci, higher probability of isolation was determined during the late housing and pasture

(May to August) seasons (similarly also by studies in other Nordic countries—Finland (Koivula et al. 2007) and Norway (Østerås et al. 2006)), and *Klebsiella* spp. as causative agent of clinical mastitis was identified rather during the pasture season and more rarely in early housing season in Sweden (Duse et al. 2021). However, it should be noted that seasonal effects in cross links differ between different studies and countries, being influenced probably by other confounding factors, possibly masked by season (e.g. climate with sharper/milder difference among seasons, indoor/outdoor housing and farming systems and technologies).

7. *Successful bacteriological cure:* One can expect that cows with clinical mastitis cured clinically but not fully successfully bacteriologically can be followed by recurrent CM cases. This was confirmed by one study (Pinzón-Sánchez and Ruegg 2011), according to which cows with successful bacteriological cures were eight times less likely to have clinical mastitis recurrence than those without success in bacteriological cures. Certain pathogens pose a risk of recurrent CM especially due to the insufficient bacteriological cure and persistence in the udder (*S. uberis*, intracellularly located *S. aureus* as well as *E. coli*) (Zadoks et al. 2003).
8. *Treatment strategy*: Many contradictory results exist on treatment strategy with respect to risk of CM recurrence. Unfortunately studies thoroughly comparing different designs of treatment (systemic + intramammary, intramammary alone), considering also bacterial killing effect dependency on time (T) or concentration (C) or co-dependent (CT) for different antimicrobials as well as level of absorption and ability to target, e.g. intracellularly located pathogen, are not available. It seems that treatment strategy is an important determinant of clinical mastitis recurrence. In some pathogens extended treatment has been described as successful and useful for avoiding recurrent episodes of clinical mastitis (e.g. *S. uberis*, Biggs 2018) and in other pathogens less successful (*S. aureus*, depending also, if early started prior biofilm formation or if *S. aureus* with resistance to methicillin and other antimicrobials occurs in the herd). McDougall et al. (2019) have reported that the combination of amoxicillin + clavulanic acid + prednisolone administered intramammary over short (3 times, q12 h) and long periods (5 times, q 12 h) significantly reduced rate of clinical failures, but cure prolongation had no effect on cure proportion, SCC or new infection rate.
9. *Clinical mastitis in current/previous lactation*: According to the results of recent study from Sweden (Duse et al. 2021) *Klebsiella* spp. was 3.6 times more commonly present in clinical mastitis in cows that had a previous case of clinical mastitis in the current lactation. History of clinical mastitis in current lactation was also determined for *Trueperella pyogenes*-caused cases. Interestingly clinically manifested mastitis associated with *T. pyogenes* may in some cases require a concomitant infection with another pathogen (e.g. *S. dysgalactiae*), which was confirmed by findings in a study by Duse (2021), in which this pathogen was isolated almost 18 times more likely in concomitancy with another pathogen.
10. *Factors related to SCC:* SCC indicates the total number of cells per millilitre in milk. Increased numbers are associated with inflammation, mostly associated with infection. Milk is considered abnormal when SCCs are higher than 200,000 cells/mL (thresholds differ; 150,000 or even 120,000 cells/mL is considered in some systems). Normally, milk should not have more than 150,000 cells/mL, and if the SCC is greater than 200,000 cells/mL, it shows some level of abnormality and that the immune system is activated (Vissio et al. 2014). Under normal health conditions 66–68% of detected cells are macrophages. Other somatic cells may include neutrophils (increasing in intramammary infection progress), mononuclear and epithelial cells. The presence of high SCCs with leukocytes and specifically high

numbers of neutrophils is seen in almost 90% of mastitis cases (Pilla et al. 2012). There is a chance that high SCCs sometimes can be caused by factors other than an infection. High levels of animal stress, for example, can produce high SCCs, thus giving false-positive results (Schukken et al. 2003; Ruegg and Pantoja (2013). In a study (Blagitz et al. 2015) increasing SCCs, determined with an automated, fluorescent, microscopic somatic cell counter, were present in milk samples containing *S. dysgalactiae*. Indirect fluorescence labelling targeting neutrophils showed that there was a particular increase in neutrophil recruitment as well as notable increase in *Il-1beta* and *TNF-alpha* expression (RT PCR), both being potent inducers of the acute immune response.

11. *Factors-related causative agent*: Although studies (mostly historically, without genetic confirmation the identical strains) have spoken about recurrence in some pathogens, e.g. in works of Lam et al. (1996) and Swinkels et al. (2013), where *Staphylococcus aureus* and *Streptococcus uberis* recurrent clinical mastitis was noted more frequently than *E. coli* cases, some other studies do not confirm this (*E. coli* genotype was identified in 86% of *E. coli* clinical mastitis by Döpfer et al. (1999)). Therefore Jamali et al. (2018) summarised that it cannot be concluded that the pathogen initially involved in the first clinical mastitis case is an important determinant of the subsequent clinical mastitis recurrence risk. This summary can be recently more explored not only thanks to NGS and "omics" techniques confirming the identity of the same strain but also by recognising the role of microbiome and not only individual pathogen. It was documented that microbiomes in recurrent clinical mastitis cases had overexpression of genes encoding biofilms adhesins (Gomes et al. 2016; Fleitas Martínez et al. 2019), glutathione non-redox reactions (Kelly et al. 2019), regulation of oxidative stress (Bar-Or et al. 2015) and transposable elements (Saleh et al. 2019), which play an important role in many chronic-recurrent bacterial infections, including recurrent clinical mastitis (Hoque et al. 2020). Other above-mentioned factors play a role in creating conditions supporting certain pathogens as well as whole microbiome ecology and by these means also recurrence of clinical mastitis.
12. *Teat characteristics and teat affections that need attention*: This includes teat-end callosity and tramped teats (Elbers et al. 1998; Zadoks et al. 2001). The teat length, shape and closure time (both after each milking and during drying off) vary among cows and are further important genetic traits influencing mastitis susceptibility (Kibebew 2017; Dego 2020). Prior to drug therapy, the main cause of udder infection needs to be clarified. The affections of the teat or udder, like teat fistula, leaky teat, teat spider and udder wounds, need immediate attention. Since such affections break the effective barrier and tend to expose the teat canal or udder to the external microbes, the earliest rectification is needed (Sharun et al. 2021).

7.3.3.2 Major Pathogens Acting as Causative Agents of Clinical Mastitis and Considerations on the Treatment Strategy

Among the bacterial species, *S. aureus* (25%), coagulase-negative (non-aureus, NAS) *Staphylococcus* species (20%), *Escherichia coli* (11%), *S. agalactiae* (9%) and *S. uberis* (9%) are the most important mastitis pathogens present in the world in the period evaluated via scientometric meta-analysis of the dataset in 1979–2019 (Krishnamoorthy et al. 2021). Very recent data (Astrup et al. 2022) reported among 147 isolates 20.4% of *S. aureus*, 18.4 of *S. uberis* and 17.7% of *E. coli* and 14.3% of NAS isolates, which were detected in pure cultures gained from culturing of milk from clinical mastitis cases. Among the clinical mastitis pathogens that occur in most studies performed in Europe, notwithstanding management practices are *S. aureus*, *E. coli*, *S. dysgalactiae*, *S. uberis* and non-aureus staphylococci with a dominance of *S. epidermidis*, *S. chromogenes*, *S. simulans* and *S. haemolyticus*, while the prevalence of previously more frequent *S. agalactiae* dropped down (Nonnemann et al. 2019; Duse et al. 2021). Above pathogens spread via different routes, and some species, or even strains, are considered and classified as contagious and some rather environmental. Transmission by common cattle fly *Hydrotaea irritans* being a vector for establishment and maintenance of some mastitis infections has been described for infections caused by *Staphylococcus aureus*, *Streptococcus dysgalactiae* and *Trueperella pyogenes* (Chirico et al. 1997) as well, especially for certain environmental conditions.

As for the treatment strategy, a plethora of factors can be considered (influence of pathogen/s involved, health and production status, SCC number status, acute or chronic infection, clinical or subclinical infection, treatment in lactation or at drying off). Relatively uniform clinical mastitis' decision algorithm for treatment start and evaluation based on seriousness, culturing and clinical examination exists and is described in Fig. 3. On the other hand, controversy exists on the success of treatment with local (i.e. intramammary antibiotic administration) vs. combined (partly or fully concomitant) treatment with locally and systemically administered antibiotics. One of the very recent studies performed in Denmark (Svennesen et al. 2023) was carried out as a non-inferiority trial with a non-inferiority margin set to a relative reduction in bacteriological cure of 15% between these two treatment groups. Only clinical mastitis with growth (24 h) on Gram-positive bacteria on culturing media were enrolled to the study. Bacteriological cure was assessed based on the bacterial species identified in the milk sample from the clinical mastitis case prior to the start of treatment and in two follow-up samples collected approximately 2 and 3 weeks after the end of treatment. Intramammary (IMM) administered procaine benzylpenicillin alone (dosing schedule 0.6 mg procaine benzylpenicillin /day; 3–5 days) and in the second arm the same intramammary VMP/same dosing schedule in combination with injectable (INJ) penethamate hydroiodide (9 g penethamate hydroiodide/day/IM/600 kg Holstein cow/3–5 days) were administered to diseased, culturing-positive animals. For *S. uberis*, isolated with highest prevalence of 40% (138/345), 72 cases were treated locally (IMM) and 66 with combined IMM + INJ antibiotic administration. The results of treatment showed that 81.9% ($n = 113$) were successfully bacteriologically cured and 18.1% ($n = 25$) were not cured. Local (IMM) treatment cure rate was 76.8 and 83.1% was cure rate for IMM + INJ treatment schedule. Consdering full dataset (all mastitis cases, all pathogens) average 79.95% cases were cured in total by both schedules. Based on these results and using 15% non-inferiority margin, conclusion was drawn that bacteriological cure of local penicillin treatment for mild and moderate clinical mastitis cases was non-inferior to the combination of local and systemic treatment (Svennesen et al. 2023). A simple comparison shows 6.3% more cured cows in combined treatment. The work cited above is a significant piece of information that can be involved in the whole mosaic of the consideration of the treatment possibilities, especially for Gram-positive mild and moderate mastitis treatment. Despite the results of MICs detected in pathogens involved, comparing 3-day vs. 5-day duration (as both antibiotics used are time-dependent), PK data signalising concentration achieved in udder tissue (in cows that suffered from mild/moderate mastitis) and performance of a complex in-depth analysis would be very useful to better understand the whole treatment possibilities and outcomes (being aware that gaining all data for such a collection of pathogens/mastitis cases is nearly impossible). Other studies bringing data on oxytetracycline administered via injection route, where both LA injectable veterinary medicinal products (administered intramuscularly) and those with composition that allow also intravenous administration are available. After IM administration, especially in low-dose regimen (4 mg/kg bw), the concentration achieved in milk is insufficient to cover mastitis target pathogens. However, following IV, oxytetracycline was detected to be distributed into milk at similar concentrations to serum (AVA-AMA, 2022). Parenteral treatment used concomitantly with intramammary therapy can be recommended for susceptible target pathogens. Concomitant administration of oxytetracycline (IV) and cephapirin (IMM) with further supportive therapy (flunixin, fluid therapy) in field trial was proven being effective for the treatment of severe coliform mastitis (Morin et al. 1998). In a clinical trial. where *S. uberis* was predominant, similar results were achieved treating cows with tylosin base (5 g/dose/animal/IM, 3 times at 24-h intervals) compared to treating cows with penethamate hydroiodide (5 g/dose/animal/IM, 3 times at 24-h intervals) as noted by publication of AVA-AMA (2022).

7.3.3.3 *Streptococcus uberis*

Streptococcus uberis occurs primarily in the cow's environment. Prevalence of mastitis caused by *S. uberis* is increasing as can be seen from recent data showing a situation in Europe, where approximately 38% of mastitis is caused by the streptococci. *Streptococcus uberis* contribute by 64% to all streptococci clinical mastitis (Kabelitz et al. 2021). Due to this high proportion *S. uberis* has been introduced among the main pathogens also to the scopes of mastitis pathogen or cattle pathogen monitorings (El Garch et al. 2020; Mader et al. 2022). The results of susceptibility testing are reflected in Table 9. Grazed pastures have a bigger load of *S. uberis* than ungrazed pasture, and a similarly high load is confirmed in bedding contaminated by faeces in loose yards or cubicle beds in housed cattle with particularly organic cow bedding such as straw, which was determined to be one order of magnitude higher than sand or sawdust. The presence of pathogen in high density in the environment and lying and bedding material results in the increase of probability of teats and teat ends contamination and subsequent intramammary infection as well as increased infection prevalence within the herds affected (Biggs 2018; Kabelitz et al. 2021). Very closely related species *S. parauberis* appears with a very low abundance (0..5% mentioned in Kabelitz et al. 2021), which is phenotypically indistinguishable but can be differentiated as *S. uberis* produces β-D-glucuronidase (Krömker et al. 2014; Pitkäla et al. 2008). *Streptococcus uberis* is typically alpha-haemolytic (but partial haemolytic as well as non-haemolytic strains appear). Christine-Atkinson-Munch-Peterson (CAMP) test phenotype combined with aesculin, sodium hippurate and inulin degradation tests allows identification, which can be also relatively easily performed on specifically designed chromogenic media. Lancefield classification of *S. uberis* to group E, G, P or U can be confirmed. Of great importance are virulence factors. A recent study by Vezina et al. (2021) determined from 24 to 35 virulence factors per isolate (every isolate has 22 putative virulence factors including biofilm putative glycosyltransferase, biofilm putative glycosyltransferase 2 and relevant genes *cps*B, *cps*C, *cps*D/*cps*4D, *cyl*A, *fab*G 1/*cyl*G, *fbp*S, *gta*B/*has*C, *has*C homologue *gps*A, *lmb*, *mga*, *mtu*A, *opp*F, *pau*A, putative surface-anchored protein, *rqc*H/*fbp*54, *sca*R, *scp*A, *sua*, *tag*U 3/*cps*4A and *srt*A). Another publication lists three most prevalent factors as those encoded by genes *has*C, *sua* and *gap*C (Kabelitz et al. 2021) with different ways of action influencing invasiveness and host adaptation. Data from at least three references (Kabelitz et al. 2021; Vezina et al. 2021; Zouharova et al. 2022) can be considered once speaking about essential virulence genes—some of them can be considered as good targets for vaccine development due to frequent occurrence and their highly conserved nature:

- *has*C gene encoding protein HasC (hyaluronic acid capsule), capable of masking the surface antigens of *S. uberis*; hyaluronic acid capsule protects from opsonisation and finally therefore protects from phagocytosis by bovine neutrophils. Strains with this gene expressed can resist clearance within neutrophil extracellular traps. A similar function is for genes *has*A and *has*B (present in 94% of isolates) and being part of profile also containing genes *sua*, *skc*, *pauA*, *gapC and opF*, which were detected in 127 out of 135 farms sampled in the Czech Republic in 2019–2021 (Zouharova et al. 2022).
- *cps*BCD and *cps*4a encoding products involved in capsular polysaccharide formation

Table 9 Overview of MIC_{50} and MIC_{90} distribution for *Streptococcus uberis* isolates from acute mastitis in dairy cows

MIC_{50} and MIC_{90} distribution for *Streptococcus uberis* (*n* = 208) from acute mastitis in dairy cows						
Belgium (32), the Czech Republic (15), France (29), Germany (30), Italy (3), the Netherlands (31), Switzerland (28), the UK (40)						
Antimicrobials class	Antimicrobial	AMEG	MIC_{50}	MIC_{90}	S[%]	R[%]
Penicillins beta-lactamase sensitive		**D**	**0.06–0.25**	**0.25–0.5**		
	Penicillin G		0.06	0.25	87.1	
Penicillin in combination	Penicillin/framycetin		0.12	0.25		
Penicillin in combination	Penicillin/streptomycin		0.25	0.5		
Isoxazolyl penicillins		**D**	**2**	**4**		
	Cloxacillin		2	4		
Aminopenicillins		**D**[a]	**0.25**	**0.25–0.5**		
	Ampicillin		0.25	0.25	94.7	
	Amoxicillin		0.25	0.5		
	Amoxicillin/clavulanic acid (2:1)		0.25	0.25	Concentration of amoxicillin	
Cephalosporins 1st gen		**C**	**0.06–0.5**	**0.06–0.5**		
	Cefazolin		0.5	0.5		
	Cephalonium		0.06	0.06		
	Cephapirin		0.25	0.25		
	Cephalexin		0.25	0.5		
Cephalosporins 4th gen		**B**[a]	**0.12**	**0.25**		
	Cefquinome		0.12	0.25		
Fluoroquinolones		**B**[a]	**0.5 to 1**	**1–2**		
	Danofloxacin		1	2		
	Enrofloxacin		0.5	1		
	Marbofloxacin		1	2		
Macrolides		**C**	**0.06–1**	**>64**		
	Erythromycin		0.06	> 64	76.4	23.6
	Tylosin		1	>64		
Aminoglycosides		**C**	**2 to > 64**	**0.5–2**		
	Kanamycin	C[a]	64	>128		
	Neomycin		>64	>64		
Aminoglycosides in combination	Kanamycin/cephalexin (10:1)		2	4	Concentraton of kanamycin	
Lincosamides		**C**	**0.12–2**	**8–128**		
	Lincomycin		2	128		
	Pirlimycin		0.12	8	84.1	15.9
Ansamycines		**C**[a]	**0.12**	**0.12–0.5**		
	Rifaximin		0.12	0.5		
Ansamycines in combination	Rifaximin/cefacetrile (1:2)		0.12	0.12	Concentraton of rifaximin	

S/R expressed where CLSI interpretive criteria available
Based on data presented in El Garch et al. (2020)
[a]In some countries, more strict conditions of use are stipulated for these antimicrobials

and two putative glycosyltransferases involved with biofilm formation.

- *cyl*AG contributing to haemolytic activity, corresponding to an ABC transporter and an enoyl reductase involved in fatty acid biosynthesis.
- *sua* gene encoding *S. uberis* adhesion molecule (SUAM protein) responsible for lactoferrin binding creating a less hostile environment for the pathogen; this protein is probably also important for cell adhesion and invasion of the pathogen into mammary epithelial cells. Also *vru* gene has an influence on pathogenicity and ability to invade (Almeida et al. 2015).
- *opp*F, an oligopeptide permease with importance for auxotrophic amino acid acquisition from bovine, described also as a solvent active transfer.
- *sca*R, another virulence regulator upregulating virulence-associated genes in response to

mastitis infection during low levels of Mn^{2+} in cow's milk.

- *neu*ABCD locus is involved in sialic acid synthesis, and their presence indicates it may produce a capsule to assist with immune system evasion.
- *rqc*H/*fbp*54, a fibronectin-binding protein and putative fibronectin-binding protein *fbp*S, which are of importance for adhesion.
- *lmb* encoding a laminin binding protein important for bacterial colonisation.
- *Mga* as a virulence regulator with importance for a number of phenotypes including biofilm formation, growing in whole human blood and soft tissues and phagocytosis resistance.
- *gap*C gene responsible for plasmin binding as well as immunomodulation, known as encoding surface dehydrogenase protein, playing role in colonisation.
- *skc* and *pau*A, encoding PauA and SkC proteins, confer the activation of plasminogen to plasmin leading to the degradation of the extracellular matrix, finally facilitating bovine mammary epithelial cell invasion by *S. uberis* (Kabelitz et al. 2021). These genes were the most prevalent virulence-associated genes, with 98% (*skc*) or 97% (*pau*A) of isolates from clinical and subclinical mastitis (Zouharova et al. 2022).

Among other characteristics is the ability of casein lysis that provides extra nutrients and thereby favours biofilm formation. *Streptococcus uberis* genes important for biofilm formation showed that certain genes were more prevalent than others; *luxS* with responsibility for quorum sensing (with 42.8% prevalence) and *comEA* corresponding to bacterial competence (prevalence of 21.4%) were found in a study by Reinoso (2017).

Prevention of *S. uberis* infections is quite complex, so here only a brief summary is given. Strict udder health management (hygiene on farm, milking hygiene (pre- and post-milking) and proper care, drying off management) is highly recommended. Measures tailored for the exact farm conditions involve on-farm diagnostics, including at least, e.g. bulk tank test on chromogenic media (if properly performed) that can show dynamics in occurrence of pathogens, including *S. uberis*, on the farm and effectivity of the measures taken and where appropriate individual cows on farm culturing are accompanied with sending the samples for correct identification (MALDI-TOF) and susceptibility testing. Specific data (e.g. NGS, MALDI-TOF; RAPD PCR), machine learning processing of them (Maciel-Guerra et al. 2021) and creating probabilistic models are expected to help to set certain scenarios that could help to set management, prevention and treatment strategies:

- Udder/cow-adapted strains, with persistence and contagiosity patterns with resistance to certain antibiotic portfolio
- Udder/cow-adapted strains, with persistence and contagiosity patterns but still good susceptibility to available antibiotic portfolio
- Environmental strains with expected short-lived infections

For example, if cow-adapted strains predominated, then contagious management control protocols could be narrowed to key elements, e.g. wearing gloves during milking and optimising post-milking disinfection, whereas if environmental strains are the most common in the herd, then hygiene of housing (or pasture management) and premilking disinfection could be emphasised (Biggs 2018); see also Fig. 5.

Vaccination with both commercial and autogenous vaccines is possible. Vaccinations for mastitis mostly reduce the severity of the condition and cannot prevent reoccurring infections, and vaccines should be used in conjunction with herd udder health programme and mastitis prevention programme.

Treatment options of *S. uberis* infections (clinical and subclinical mastitis) in chronic cases should be considered, including culling of the cow. Treatment protocols for lactating cows and dry cows (prevention teat sealant only, or antibiotic, depending also on the pressure of *S. uberis* and the characteristics of strains—contagious vs. environmental) can be set. For lactating cows, in most cases first-line treatment is sufficient, and whenever possible monocomponent VMPs should be preferred. Once justified, especially for cow-adapted strains, where can be also expected

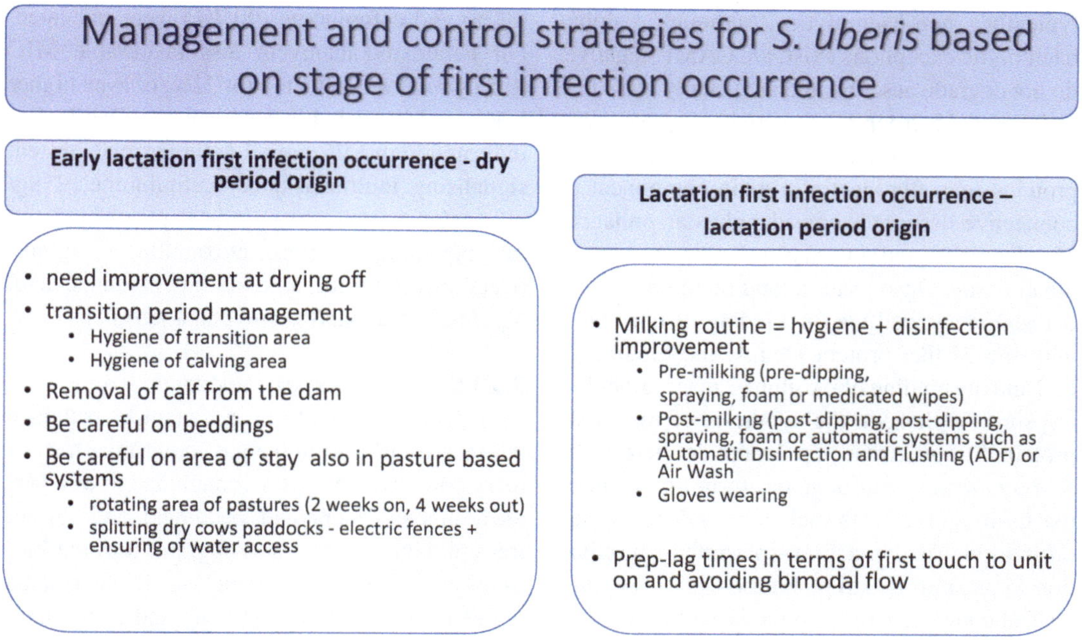

Fig. 5 Management and control strategies for *S. uberis* based on the stage of first infection occurrence (graphics based on Biggs 2018)

internalisation of pathogen, more broad treatment can be used. It is either combination of actives (with different spectrum and considering also PK characteristics) or in some treatment protocols also combination of intramammary (IMM) + injection (INJ) administration. Treatment and prevention udder health protocol should be designed, so the first-line treatment together with preventive measures should cover most cases of *S. uberis* on the farm. First-, second- and last-choice antimicrobials are listed below, but national antimicrobial policies should be followed, and therefore shift of some substances/combinations to the last choice is possible (AMCRA, 2023; Biggs 2018; Smola et al. 2022). It should be also noted that recent data (Duse et al. 2021) indicated various degrees of acquired resistance in some isolates, where detected distributions of MICs were trimodal (erythromycin, penicillin and tetracycline) or bimodal (cefotaxime, clindamycin, gentamicin, kanamycin and oxacillin); also 16 isolates (from total 89 isolates) had deviating high MICs (0.12–0.25 μg/mL) for penicillin (i.e. exceeding 0.06 μg/mL) and 1 isolate had MIC = 4 mg/L. Such data indicate the need for monitorings to recognise trends in resistance development.

First-choice antimicrobials: penethamate (IMM and INJ are available), cephalosporins first generation (cefacetrile, cefalexin (also INJ), cefalonium, cephapirin and cefazolin IMM are authorised and used in the EU), procaine benzylpenicillin (IMM and INJ), cloxacillin (IMM) and sulphonamide/trimethoprim (INJ)

Second-choice antimicrobials: amoxicillin/clavulanic acid, lincomycin/neomycin, pirlimycin, procaine benzylpenicillin/neomycin, penethamate/procaine benzylpenicillin/neomycin (all IMM), cephalexin/kanamycin (in some countries kanamycin as third line antimicrobial), tylosin (INJ) and lincomycin/spectinomycin (INJ)

Third-choice antimicrobials: cefoperazone (IMM), cefquinome (IMM) and rifaximin IMM + macrolide (INJ)

7.3.3.4 *Streptococcus dysgalactiae*

Streptococcus dysgalactiae, sometimes also considered as an intermediate strain—not easily allocated as environmental or contagious. Strains are typically non-haemolytic, although alpha-haemolytic exceptions exist; are CAMP negative, do not degrade aesculin and are usually classified as Lancefield group C. *Streptococcus dysgalactiae* capability of binding extracellular matrix proteins (e.g. fibronectin, a main component of connective tissues) is considered as an enhancer of adherence and internalisation in mammary gland tissue. Opsonisation (and therefore phagocytosis) is avoided via the binding of fibrinogen with the M-like protein DemA, the binding of IgG and the binding of the alpha 2 macroglobulin-trypsin complex via the synthesis of the MAG protein. Streptokinase, a polypeptide secreted by *S. dysgalactiae*, can degrade fibrin and connective tissue and cause in such manner deeper tissue infiltration. Another action—plasminogen activation to plasmin by streptokinase—causes hydrolysis of connective tissue proteins and subsequent tissue penetration (Calvinho et al. 1998). Hyaluronic acid degradation by hyaluronidase and damage of polysaccharide present in connective tissues contribute to the tissue-invasive properties of streptococci. *Streptococcus dysgalactiae* is often found in herds with high levels of teat-end lesions and where teat skin condition is poor. Spread is observed between cows indirectly via contamination of hands or milking cloths. As a high prevalence of this bacteria is often present in herds with poor post-milking teat disinfection routines, post-milking routine should be improved (also using the dips with emollient). Cows should be protected from cold conditions, if possible, as the very cold conditions cause teats to become chapped and sore. Treatment options (first, second and last choice) can be the same as mentioned above for *S. uberis*, considering clinical status and knowledge of *S. dysgalactiae* recent regional susceptibility profiles, ideally those detected in strain/s being present in the herd. Some susceptibility data results can be extracted from recent studies. Data from Bolte et al. (2020), coming from Germany, with 54 isolates of *S. dysgalactiae* show that still most strains tested are susceptible to penicillin. Within this dataset, susceptibility and MICs ≤0.06 μg/mL for penicillin (94.4%), cephapirin (87%), oxacillin (74%), amoxicillin/clavulanic acid (87%) and cefquinome (90.7%) were indicated. For cefalexin/kanamycin less favourable MIC (0.5 μg/mL) was detected for 72%, or even higher MICs were detected in nine isolates. One isolate indicated high MICs in all antimicrobials tested, signalising multiresistance (cefquinome, 2 μg/mL; cefoperazone, 16 μg/mL; cephapirin, 2 μg/mL; penicillin, 8 μg/mL; cloxacillin, 32 μg/mL; oxacillin, 32 μg/mL; amoxicillin/clavulanic acid, 4 μg/mL; cefalexin/kanamycin, 4 μg/mL).

7.3.3.5 *Streptococcus agalactiae*

Streptococcus agalactiae is a contagious pathogen but may occur also in the gastrointestinal tract of dairy cows, which causes contamination of cows' environment (floors and beddings) and further transmission. Longer surviving and forming biofilms predispose it for being heavily associated with subclinical mastitis (Cheng and Han 2020). This pathogen is Lancefield group B classified and generally considered being beta-haemolytic (some non-haemolytic) with CAMP positivity. Ten distinctive serotypes of *S. agalactiae* labelled Ia, Ib, II, III, IV, V, VI, VII, VIII and IX have been described yet, linked also with pathogenicity rate in cattle, but some (most frequently serotype V but also serotypes Ia, II and III) are associated with bacterial meningitis of neonatal children (Kabelitz et al. 2021). Of importance is that most human STs are quite distinct from bovine STs, with some exemptions (Kabelitz et al. 2021). Examples of genetical relationship/developmental link for ST-23 bovine/ST-23 human and ST-61/ST-67 bovine to ST-17 human strain (serotypes Ia and III) as well as links to occurrence in fish were intensively discussed (Sørensen et al. 2010; Li et al. 2018) but also serotype IV (Lyhs et al. 2016). Controlling of the spread within the herd is based on thorough milk hygiene (changing towels, teat dipping, dry cow handling and treatment if appropriate). Lactation treatment is more successful than in *S. uberis*. Treatment options (first, second and last choice) can be the same as mentioned above for *S. uberis*, considering clinical status and knowledge of *S. agalactiae* recent regional susceptibility profiles, ideally those detected in strain/s being present in the herd. With respect to *S. agalactiae* (44 isolates, 2015–2016, Europe, El

Garch et al. 2020) MICs profiles for the portfolio of antimicrobials relevant for Gram-positives were tested. For the majority of the β-lactam antibiotics MICs were ≤0.25 µg/mL (with exemption of cephalexin, where MIC_{90} was 4 µg/mL and cloxacillin with MIC_{90} 2 µg/mL); penicillin MICs were 0.015–0.06 µg/mL, so penicillin as the first choice can be used. High $MICs_{90}$ were reported for aminoglycosides and lincosamides (neomycin and pirlimycin > 64 µg/mL; kanamycin and lincomycin > 128 µg/mL). Fluoroquinolone MIC $_{90}$ was among 1–2 µg/mL and rifaximin 0.5 µg/mL. Specification of the resistance rate was only possible for erythromycin and pirlimycin (both at 27.3%).

7.3.3.6 *Staphylococcus aureus*

Staphylococcus aureus is a causal agent of subclinical, clinical as well as chronic mastitis. Clinically infected quarters often show moderate swelling or hardness and visible clots (chunks) in the milk, especially in forestrippings. Acute infections generally develop late in the lactation. However, the clinical symptoms do not show up until calving or early in the next lactation. *Staphylococcus aureus* is a toxin producer. Toxins destroy cell membranes and can directly damage milk-producing tissue. Damage of the tissues lining the teat and gland cisterns within the quarter and possible formation of scar tissue, followed by moving up into the duct system and establishing of deep-seated pockets of infection in the milk-secreting cells (alveoli), are frequently present. Finally, the formation of abscesses can further complicate the success of the treatment by affecting the penetration of an antibiotic to the focus of infection. *Staphylococcus aureus* is also able to escape the killing effects of some antibiotics by hiding, e.g. within neutrophils and other host cells. While leukocytes attempt to remove bacteria, many organisms survive and become dormant within the neutrophils. Up to 2 days, when immune cells die, the bacteria are released to resume the infection process.

Within a farm, multiple epidemiological types of *S. aureus* might be present (Leuenberger et al. 2019; Rainard and Foucras 2018). On farm with *S. aureus* occurrence and troubles (bulk tank on farm culturing and monitoring, as an economical and feasible way), special attention should be paid then to individual cows (culturing of all high SCC cows) but also to handling of colostrum (not to be contaminated with *S. aureus*, e.g. treatment via careful heating as described in this chapter above). From culturing in individual cows, isolates can be confirmed in laboratory including antimicrobial susceptibility testing and distinguishing of strains present in the herd.

Staphylococci do not persist on healthy teat skin but readily colonise damaged skin and teat lesions, where they multiply, which increases the risk of infection of further udder tissues and teat canals, internalisation into cells and spread to colonise/infect also muzzles and nostrils. Spread to uninfected quarters can be by teat cup liners, milkers' hands, washcloths and flies. Care should be therefore taken with hygiene of milking and all equipment, personnel and towels used. Preventive pre- or post-milking teat disinfectants can contain various compounds, such as lactic acid, chlorhexidine or iodine. Several *S. aureus* mastitis control plans exist, but some elementary steps can be summarised as follows:

- Focusing on the teat tops and their appearance (damage and apparent signs of infection) and monitoring milk filters (simple visual monitoring signalises troubles).
- Milking hygiene (wearing gloves, dry toilet prior to thorough pre-dipping for 15–30 s, and using clean application devices (key for mostly environmental pathogens: *S. uberis*, *E. coli* and *Klebsiella* spp.) and post-dipping as soon as possible after removal of milking machine (key for mostly contagious pathogens: *S. aureus*, *S. agalactiae*, *Mycoplasma bovis* and *Corynebacterium bovis*)). The National Mastitis Council publishes guideline on the effectiveness of teat dips (2-year interval). Attention should be also paid to rotation of the dips, watching any undesirable reactions (irritation including hypersensitisation, chapping, lesions, teat skin drying and caustic reactions). Attention should be paid once some germs (*Pseudomonas* spp., *Prototheca* spp.) occur on farm to choose effective dips.
- Minimising the risk via culling of carriers or at least milking infected cows separately; if not feasible, the last option is milking of

infected animals from the milked group at the end of the milking cycle (after that follows very thorough cleaning of devices and hygiene of parlour).

- Optimisation of nutrition, including vitamins and microelements ideally tailored on the groups of cows with the same status level.
- Drying-off process and individual care of animals (considering also time point of drying off, as (very) high-yield cows are problematic) including very thorough administration of teat sealant or "alternative" to antimicrobials or antimicrobial/s containing medicinal product/s.
- Monitoring new cases of *S. aureus* infections, particularly in early lactation and in younger cows, collecting the samples, performing the test (on-farm culturing/other rapid test) and immediately after SA confirming, starting the treatment to catch the initial phase of infection, when treatment can be successful and can lower the probability of infection becoming chronic and of the animal being a carrier.

Improvement of the prevention measures includes a ten-point schedule:

1. Bulk tank testing and monitoring of SCC and evaluation of results.
2. If *S. aureus* is positive from bulk and cows with high SCC, perform targeted monitoring of individual cows. Teat ends should be examined for evidence of chapping, cracks or lesions, which may harbour *S. aureus*.
3. Sorting of cows being positive (marking with colour strip on the leg) to be milked separately or at least at the end of herd milking. Clean and disinfect very thoroughly (potent disinfection, time for its action, drying) all equipment after milking of these cows.
4. Wearing gloves and changing them frequently, especially when dirty or after stripping an infected animal.
5. Forestripping five squirts of milk from each quarter and checking for abnormal milk or flakes.
6. Dirt brushing off teats with a dry, single-use towel. Scrub teats five times or for 20 s.
7. Using of water as part of any milking procedures (even if a sanitising solution is added) should not be performed (sanitisers do not maintain activity throughout a milking; also if contaminated water—pathogens introduction causing serious, difficult to treat infections).
8. Appropriate, effective pre-dip administration via a dipper or cup and giving at least 30 s of contact time (spraying less appropriate—not covering all teat surfaces, hardly accessible to teats farthest from personnel treating). Foamers can be used once generate foam with proper concentration of active substance to sufficiently disinfect.
9. Milking units should be attached 90 s after first tactile stimulation (stripping or wiping, whichever comes first).
10. Appropriate, effective post-dip administration after milking use, ensuring that the entire barrel of each teat is covered. At the end of each milking, any teat dip left in the dip cup should be discarded, and cups should be rinsed with water and allowed to dry.

Commercial and autogenous vaccines were expected to be of help in prevention of *S. aureus* (SA) mastitis, but results reported by studies publicly available for the only EU-authorised commercial vaccine differ, as noted in a recent EFSA (2022) document summary:

- Schukken et al. (2014): two large dairy herds observed for 21 months, with moderate reduction of new infections and reduction of the "basic reproduction ratio" of 45% for SA
- Bradley et al. (2015): seven farms (but only 2.5% clinical SA mastitis), with no reduction in the incidence and prevalence of clinical and subclinical mastitis but significant reduction of the severity of clinical cases and milk losses
- Landin et al. (2015): herds with SA prevalence with no beneficial effect on udder health, milk production or culling rate
- Freick et al. (2016): herds with SA as a predominant pathogen, with the final statement indicating that the vaccine was not an appropriate tool to manage *S. aureus* issue

Another complex meta-analysis (Mata et al. 2023) working with the efficacy of vaccination studies in dairy cattle (working with vaccines authorised through the years worldwide) has shown dubious results as well (see below in the part of this chapter dedicated to vaccines).

One of the major protective antigens of the above commented vaccine is slime-associated antigenic complex (SAAC) (primarily consisting of the surface polysaccharide poly-*N*-acetyl β-1,6 glucosamine, the product of *ica* operon). Not all SA lineages encode this operon (e.g. CC71 has a large genomic rearrangement near the origin of replication and therefore does not encode it). Therefore, this vaccine may not provide a protection for the strains from this lineage (Budd et al. 2015; Keane 2019).

Intracellular live *S. aureus* have been observed in alveolar cells in milk samples from chronic bovine mastitis (Hebert et al. 2000). Interestingly, strains of the most prevalent and persistent *S. aureus* genotype displayed both high biofilm production and high cellular invasiveness (Grunert et al. 2018). Some studies dealing with biofilm and invasiveness factors are targeted on the development of new vaccination technologies or drug delivery systems. Recently studies are working with nanoparticles, in specific delivery formulations. Some promising results can already be found, e.g. for macrolide tilmicosin (TIL) formulation, indicating the cure rate of half dosage and normal dosage of TIL-nanogel was 58.3% and 75.0%, which was higher than that of commercial injection (50.0%) at normal dosage (Zhou et al. 2019). Studies with tilmicosin further continued, and the last published results have shown enhanced antibacterial activity of another macrolide—tylosin against *S. aureus* small colony variants by specific composite nanogels (with chitosan and sodium carboxymethyl cellulose). Further studies work with the florfenicol composite nanogel, with concentration-dependent effect against small colony variant of *S. aureus* (Liu et al. 2022) and enrofloxacin in the form of gelatine sodium alginate composite nanogels (Luo et al. 2022).

Other preventive measures involve hygienic, biosecurity and health management practices. Those should be tailored to the specific farm setting. In dairy cattle, *S. aureus* is considered a major contagious pathogen, for which mastitis control programmes have been developed.

Certain measures, the same as against environmental pathogens, can be also involved (see above) that are of importance especially for the protection against the spread of *S. aureus* within the herd.

Antibiotic treatment will not control this disease within the herd; although clinical cure seems to appear, bacteriological cure is an issue, as surviving bacteria can evolve in the new infection or be further spread. Cows with a history of chronic clinical or subclinical mastitis should not be treated, as chronic cases are not possible to effectively cure and as it is known that not antimicrobial resistance but more typically factors like immune evasion, cell internalisation, biofilm formation and the forming of small colony variants are barriers to success in the treatment (Grunert et al. 2018; Rainard and Foucras 2018). Culling is advised for chronic infections, especially in multiparous cows (Rainard and Foucras 2018). Cure rate decreases with increasing age of the cow, increasing somatic cell count, increasing duration of infection, increasing bacterial colony counts in milk before treatment and increasing number of quarters infected. Mastitis caused by *S. aureus* in hind quarters has a low cure rate compared with front quarters. In older cows and in high SCC score, the probability of cure, even in new infections, is low. Timely start of treatment is essential especially in first clinical mastitis of first lactation cows, as it increases the probability of success in treatment and minimises tissue damage. Some authors recommend extended duration of treatment, which can increase *S. aureus* infection cure rates, but it is not firmly justified and especially in chronic infections. Moreover, considering also economy (milk discarding during the treatment and during prolonged withdrawal period) as well as low probability of being back to production and risk of unsuccessful bacteriological cure/recurrence of infection and spread of pathogen to further animals makes this prolonged treatment extremely problematic.

EFSA (2022) summarises that for antimicrobial therapy of dairy cow with *S. aureus* mastitis, various antimicrobial compounds are available in authorised VMPs (both for lactating (LC) and dry cows (DC)) within the EU. Monocomponent beta-lactam antimicrobials are among the most used (EU dataset, ESVAC, unpublished data

2021): penicillins (benzylpenicillin, amoxicillin, oxacillin, cloxacillin), first-generation cephalosporins (cefacetrile, cefalexin, cefalonium, cephapirin, cefazolin) as well as last-choice third-generation (cefoperazone) and fourth-generation antimicrobials (cefquinome) were used in practice in LC therapy. In DC mostly used are VMPs containing cloxacillin benzathine, followed by first-generation cephalosporins (cefalonium, cephapirin, cefalexin, cefazolin), but also the last choice cefquinome (VMPs with *S. aureus* in indication) was used. The last beta-lactam (in fixed combination) to be mentioned is amoxicillin with clavulanic acid. Beta-lactams were reported to have especially high cure rates when administered at drying off (Schnitt and Tenhagen 2020). But it should also be noted that beta-lactams are not appropriate for the treatment of methicillin-resistant *Staphylococcus aureus* (MRSA). Based on ESVAC (unpublished data 2021), aminoglycosides (neomycin, kanamycin, gentamicin, framycetin), lincosamides (pirlimycin, lincomycin), macrolides (spiramycin, tylosin, erythromycin), rifaximin, bacitracin and thiamphenicol belong to the portfolio used in different extent and in fixed combination (e.g. cefalexin/kanamycin; lincomycin/neomycin; cefalexin/rifaximin) with VMPs indicated for use in mastitis caused by Gram-positives, including *S. aureus* infections. Except intramammary use, which predominate, some substances can also be administered systemically as injections (with portfolio broadened by e.g. sulphonamide/trimethoprim combination; also fluoroquinolones to be administered systemically for treatment of *S. aureus* mastitis as indicated by Alfonseca-Silva et al. 2021). As for combination treatment (IMM + INJ), there are available results (Mestorino and Errecalde 2012) summarising the results of a study by Lucas et al. (2009), in which the experimental animals received 3 IMM infusions of amoxicillin/clavulanic acid (200–50 mg) in each quarter in combination with 3.5 mg kg^{-1} of 15% amoxicillin/clavulanic injection administered intramuscularly every 12 h. Individual quarter milk samples were collected until 96 h after the third administration. The incidence of the health status of the quarters (mastitic quarters vs. healthy ones) and the level of milk production (quarters of high-producing cows vs. quarters of low-producing cows) in the PKs of the drugs was evaluated. Concentrations in mastitis quarters were significantly higher both for amoxicillin and clavulanic acid compared to healthy quarters ($P = 0.0384$). The percentage T>MIC_{90} of the period between 0 and 12 h post-administration, when the concentration was ≥ the MIC_{90}, varies among quarters assessed and was among 62.64 ± 8.87 (mastitic) and 78.25 ± 23.16 (healthy). Perner et al. (2002) found the combination treatment to be superior than parenteral or intramammary treatment only, with bacteriological cure rate for all causing agents and mastitis types (acute, subclinical and chronic) of 75.3%. Lucas et al. (2009) confirm bacteriological cure of 100% for combination therapy. Interesting was also the comparison of susceptibility results in different pH conditions (*S. aureus* survives in acidic media). Penicillin was tested in strains of *S. aureus* (field strains isolated from mastitic quarters) and *S. aureus* ATCC 25923 at pH 7.4, 6.5 and 5.0, in order to simulate the conditions of acidity of subcellular structures which are commonly associated with *S. aureus* intracellular persistence (Moncada Cárdenas et al. 2009). MIC_{90} of penicillin at pH 7.4 was consistent with those reported by CLSI (2007) (0.5 μg/mL); at pH 5.0 (phagolysosomes) the activity of penicillin increased markedly and almost linearly (~tenfold decrease in MIC 0.06 μg/mL). Another beta-lactam used is cloxacillin (isoxazolyl penicillin), where sodium salt is used in LC IMM and benzathine is the most frequently used in DC IMM in the EU. After IMM administration, sodium cloxacillin binds scarcely to the mammary tissue (25%), with a moderate passage of milk: plasma (Mestorino and Errecalde 2012) and PK parameters were determined after intramammary administration (250 mg sodium cloxacillin, 3 times, q12 h, into each quarter) in healthy lactating cows of low production. Milk elimination half-life was rather prolonged ($T_{½β} = 4.37 \pm 0.458$ h), with levels in milk at 60 h post-last administration ≥0.35 μg/mL. Cows with staphylococcal subclinical mastitis at drying off were enrolled in a study (Lucas 2009) determining the PK of cloxacillin benzathine in the dry udder secretion after a single IMM infusion. The permanence of the ATM in the mammary compartment or glandular tissue for a long time and binding level to dry udder secretion was very high (86 ± 6.5%) corelating with another study, where approximately 80% was calculated (Ødegaard and Sviland 2001). In the experimental animals, concentrations above the limit of quantitation by microbiological methods (0.03 μg/mL) during the first 24–31 days were maintained in a study by Lucas (2009). It should be noted that variations after intramammary administration of antimicrobials at drying off exist especially due to the composition of VMP (including particle size, viscosity, nature of active substance and excipients), udder size and milk production, body condition of cows, physiological changes in the body condition at the start of the dry period and other factors affecting both the mammary gland and body. An interesting substance is thiamphenicol (also authorised and used in LC intramammary VMP in the EU), which is more stable in solution than chloramphenicol, is not appreciably protein bound in the body (16%), does not undergo significant biotransformation and diffuses into intracellular spaces. After intramammary administration in an individual quarter, high concentrations in the nonmedicated quarter and in serum were determined. In animals milked every 12 h, levels exceeding MICs for the majority of pathogens were maintained during 36 h in the treated quarter and between 8 h and 12 h in the nontreated quarters and during less than 4 h in the case of blood (Mestorino et al. 1995). Mastitis caused by *S. aureus* cannot be successfully treated with oxytetracycline, because the concentrations reached in milk are insufficient. Moreover, there is an irreversible binding between tetracycline and large molecules of milk, which might be due to a hydrophobic interaction (Mestorino and Errecalde 2012). Reduction of tetracycline activity in milk was mainly attributable to the milk protein casein as well as to other components of milk (Kuang et al. 2009).

For non-MRSA strains, similar treatment options (first, second and last choice) can be used as mentioned above for *S. uberis*, with knowledge of susceptibility patterns and trends of strains being present in the region and ideally susceptibility profiles with the herd of concern. Table 10 demonstrates the susceptibility patterns of 247 clinical mastitis *Staphylococcus aureus* (SA) isolates from 8 European countries in the year 2015 and 2016 (El Garch et al. 2020). Results presented in another datasets show MICs in Sweden (2013–2018 monitoring data; 227 SA isolates) (Duse et al. 2021), Northern Germany (2013–2018; 85 SA isolates from 58 farms) (Bolte et al. 2020) and the Czech Republic (national monitoring 2020–2021; 217 SA isolates) (CZ NMTP 2020, 2021). Penicillin G susceptibility data indicates European, German and Czech isolates with MIC_{50} 0.06 μg/mL, while Swedish isolates even lower: MIC_{50} 0.03 μg/mL. Values of MIC_{90} differ: European 4 μg/mL, Czech 1 μg/mL, German 0.5 μg/mL and Swedish 0.06 μg/mL. Considering especially PK of penethamate hydroiodide after intramuscular administration, this antimicrobial can be used in timely treatment of infections caused by *S. aureus* susceptible to penicillin, especially in subclinical mastitis cases and treatment regimen of six doses/q24 h (Steele and McDougall 2014). Favourable PK of penethamate, being hydrolysed to penicillin G, which is finally ionised in the udder, limiting its return to the circulation and keeping it "trapped" in the udder in increasing concentrations for sufficient time (Mestorino and Errecalde 2012), helps to keep concentrations above MICs for a sufficient time span. In vitro susceptibility tests are poor predictors of clinical efficacy (especially in chronic cases), and timely start of the treatment until internalisation of the pathogen and biofilm formation is of importance. It needs to be noted as well that if resistance is detected in vitro by standardised, well performed testing, it will with high probability mean that treatment with such antimicrobial will not be successful, and it brings the warning of shift from susceptibility to resistance (e.g. acquired resistance, mutation after selection pressure). For MRSA strains, it is not possible to use beta-lactams and authorised cephalosporins and even if susceptibility testing is performed, sometimes is hardly to find an effective treatment (as occurrence of mul-

Table 10 Overview of MIC_{50} and MIC_{90} distribution for *Staphylococcus aureus* isolates from acute mastitis in dairy cows

MIC_{50} and MIC_{90} distribution for *Staphylococcus aureus* ($n = 247$) from acute mastitis in dairy cows						
Isolates from Belgium (29), the Czech Republic (15), France (28), Germany (32), Italy (51), the Netherlands (31), Switzerland (31) and the UK (30); 2015–2016						
Antimicrobial class	Antimicrobial	AMEG	MIC_{50}	MIC_{90}	S[%]	R[%]
Penicillins beta-lactamase sensitive		**D**	**0.06**	**4**		
	Penicillin G		0.06	4	74.5	25.5
Penicillin in combination	Penicillin/framycetin		0.06	0.5		
Penicillin in combination	Penicillin/streptomycin		0.06	1		
Isoxazolyl penicillins		**D**	**0.25**	**0.5**		
	Oxacillin		0.25	0.5	99.2	0.8 (MRSA)
	Cloxacillin		0.25	0.5		
Aminopenicillins		**D**[a]	**0.12–0.25**	**2**		
	Ampicillin		0.12	2		
	Amoxicillin		0.25	2		
	Amoxicillin/clavulanic acid (2:1)		0.25	2	Concentration of amoxicillin	
Cephalosporins 1st gen		**C**	**0.12–4**	**0.25–8**		
	Cefazolin		0.25	0.5		
	Cephalonium		0.12	0.25		
	Cephapirin		0.25	0.25		
	Cephalexin		4	8		
Cephalosporins 4th gen		**B**[a]	**0.5**	**1**		
	Cefquinome		0.5	1		
Fluoroquinolones		**B**[a]	**0.12–0.25**	**0.25–0.5**		
	Danofloxacin		0.12	0.25		
	Enrofloxacin		0.12	0.25		
	Marbofloxacin		0.25	0.5		
Macrolides		**C**	**0.25–1**	**0.25–2**		
	Erythromycin		0.25	0.25		
	Tylosin		1	2		

Aminoglycosides		**C**	**0.5–2**	**0.5–2**		
	Kanamycin	C[a]	2	2		
	Neomycin		0.5	0.5		
Aminoglycosides in combination	Kanamycin/cephalexin (10:1)		2	2	Concentration of kanamycin	
Lincosamides		**C**	**0.25–1**	**0.5–2**		
	Lincomycin		1	2		
	Pirlimycin		0.25	0.5	96.8	3.2
Ansamycines		**C[a]**	**0.25**	**0.5**		
	Rifaximin		0.25	0.5		
Ansamycines in combination	Rifaximin/cefacetrile (1:2)		0.12	0.25	Concentration of rifaximin	

S/R expressed where CLSI interpretive criteria available
Based on data presented in El Garch et al. (2020)
[a]In some countries, more strict conditions of use are stipulated for these antimicrobials

tiresistance to other antimicrobials from different pharmacological classes appears). Such MRSA strain is very risky for the whole herd as well as for the farmers and veterinarians; therefore culling of infected cow/s is recommended.

In addition to antimicrobials, various other (non-conventional) compounds are available for mastitis treatment or supportive treatment: prednisolone, bovine lactoferrin, nisin (bacteriocins), ginseng, cytokines, essential oils and other herbal and homoeopathic remedies (Barkema et al. 2006; Rainard and Foucras 2018).

7.3.3.7 Non-aureus Staphylococci

The newly accepted name non-aureus staphylococci (NAS), previously rather called coagulase-negative staphylococci, represents the group of staphylococci other than *S. aureus*, where also some coagulase-positive and coagulase-variable mastitis pathogens are included (e.g. *Staphylococcus hyicus* and *Staphylococcus agnetis*). Interestingly also some coagulase-negative variants of *S. aureus* are known to exist, some of which can have similar pathogenicity to their coagulase-positive variants (Fox et al. 1996). On the other hand, also some NAS have the gene *vWbp* (encoding von Willebrand factor-binding protein, exhibiting coagulating activity), which was detected in *S. agnetis*, *S. hyicus* and *S. chromogenes* (Naushad et al. 2019). NAS have been the most frequently isolated bacteria from udder quarters with subclinical mastitis, and their ability to cause clinical mastitis cannot be omitted, but still exist knowledge gaps or uncertainties that need to be elucidated to fully understand the role of different species of NAS and their role in the pathology of udder or, on the other hand, in balance of microbiome. *Staphylococcus chromogenes* is recently considered as a dominating NAS in subclinical mastitis cases and based on indices collected is considered host adapted (udder and skin occurrence only). *Staphylococcus simulans, S. xylosus, S. haemolyticus and S. epidermis*, followed by *S. sciuri*, *S. warneri*, *S. equorum*, *S. cohnii*, *S. capitis*, *S. arlettae, S. vitulinus* are NAS mostly associated with mastitis. The prevalence and distribution differ across regions being impacted by many environmental and management factors such as geographic region, climate, water sources, access to pasture, barn type, bedding and host factors: parity, quarter location and antibiotic use (De Buck et al. 2021). There have been identified 191 virulence factors by means of whole-genome sequencing (study by Naushad et al. 2019). It can be listed examples of genes responsible for the following:

- Methicillin resistance (*mec*A) and biofilm formation (*eno*)
- Intracellular adhesion (*ica*A, B, C)
- Iron regulation and uptake—genes encoding iron-regulated surface determinant (*isd* A/B/C/I)
- Haemolysis causing—genes encoding haemolysis (*hla*)
- White and red blood cells lysis and linked also to biofilm formation and stimulation of inflammation—genes encoding phenol-soluble modulins (*psm*β 1/2/3/4)

Host adaptation relates to colonisation and persistence of isolates as well as the level of inflammation caused. Species and frequency of isolation differ between teat canal and milk samples. Four udder-adapted NAS species, two of which were considered persistent and some demonstrated characteristics of contagious pathogens (*S. chromogenes* and *S. simulans*), were described by a recent longitudinal study (Jenkins et al. 2019). Even more difficult to understand in full complexity is the interaction with the microbiome, which has commensal interaction dimension as well as pathogenic interaction dimension. As for the second one, the role of NAS in the pathology and influencing other "major pathogens" is still not fully understood. There is also a need for further research elucidating the role of NAS as a reservoir of resistance for major and minor pathogens (de Buck et al. 2021). Resistance profiles vary among NAS; in some species strains multiresistance occurs, and many times resistance to beta-lactams was determined (based either on penicillinase presence or by *mec*A gene encoded methicillin resistance). Based on published data analysed in review by Ruiz-Romero and Vargas-Bello-Pérez (2023), species of non-aureus staphylococci and *Mammaliicoccus* spp. may exhibit resistance to the following antimicrobials: penicillin, cefoxitin, erythromycin, clindamycin, gentamicin, streptomycin, tetracycline, ciprofloxacin, chloramphenicol, fusidic acid, oxacillin, vancomycin and trimethoprim-sulfamethoxazole. Decision to treat with antimicrobials should be closely linked to clinical findings, pathology and considerations of the prevalence of NAS and other pathogens within the herd, together with the dynamics of both groups. The situation within the herd should be monitored, and if in clinical or subclinical mastitis and in bulk tank samples (indicative for herd situation) some NAS species/strains prevail, susceptibility testing will be appropriate and useful for further decision on the antibiotic of choice. Within the herds that have successfully controlled other mastitis pathogens, controlling cases of clinical and subclinical mastitis caused by NAS may be of importance for further lowering of bulk tank milk SCC (de Buck et al. 2021). Antibiofilm and targeted antimicrobial agents use is considered as one of the ways to minimise the negative impact of NAS in the dairy herds (Ruiz-Romero and Vargas-Bello-Pérez 2023).

7.3.3.8 *Trueperella pyogenes*

Trueperella pyogenes is often referred to as "dry cow mastitis", and infection occurred more readily and with more severe clinical signs during the dry period and during the early lactation than during lactation (Duse et al. 2021). Clinical mastitis makes the quarter more susceptible to a new intramammary infection (Derakhshani et al. 2018). This is in line with the findings of Duse et al. (2021) of three times more likely isolation of *T. pyogenes* in cows that had had a previous case of clinical mastitis. This causal agent was frequently detected in mastitic cows in loose housings than in tie stalls (Duse et al. 2021). Concomitant isolation with one or two other udder pathogens is a relatively common finding (Duse et al. 2021). The mammary gland inflammation caused by this bacterium is characterised by severe pyogenic lesions in the mammary tissue and malodorous and purulent milk, especially in case of co-infection with anaerobes (*P. indolicus*, *F. necrophorum* and *Prevotella melaninogenica*). A decrease of the milk yield and a low recovery rate were also documented (Ishiyama et al. 2017). It should be also noted that further infections in cattle can be caused by this pathogen (e.g. metritis). The mostly unimodal MIC distributions of *T. pyogenes* indicate that acquired resistance does usually not occur in this species. With this respect, the only exceptions were in trimethoprim alone and in combination with sulphamethoxazole as well as in tetracycline, where bimodal distributions were seen (Duse et al. 2021). Susceptibility testing of *Trueperella pyogenes* (94 isolates, 2015–2016, Europe, El Garch et al. 2020) confirmed low MIC values for various beta-lactam antimicrobials, with penicillin MIC range detected of $\leq$0.004–0.06 μg/mL and lowest MIC_{90} from cephalosporins being detected for cephapirin, 0.12 μg/mL and relatively low also for cefquinome 0.5 μg/mL, while for cephalexin MIC_{90} was 4 μg/mL. In addition, for tylosin, recommended for therapy of *T. pyogenes* infections, the majority of isolates were characterised by low MICs (with MIC_{90} of 0.12 μg/mL, the only one isolate had a deviating high MIC = 16 μg/mL); similarly also lincosamides have shown low MIC_{90} of 0.25 μg/mL. High MICs ($\geq$32 μg/mL) were detected for tetracycline. Treatment options and choice of antimicrobial for *T. pyogenes* (and in the case of another concomitant udder pathogen) should, except clinical investigation and experience, be based on knowledge of pathogen/s recent regional susceptibility profiles, ideally those detected in strain/s being present in the herd.

7.3.3.9 Enterobacterales with Emphasis of *E. coli* and *Klebsiella* spp.

Generally, germs from *Enterobacterales* (i.e. *Escherichia coli*, *Klebsiella* spp., *Raoultella* spp., *Enterobacter* spp., *Citrobacter* spp., *Serratia* spp. and other members of this taxonomical unit) cause both clinical and subclinical mastitis. The environment is mostly considered as a source, but probability of spreading increases in farms with frequently diseased animals, especially in dirty and wet conditions, spreading via contaminated bedding, milking wet cows, wet, uncleaned and not dipped teats and milking machines not properly cleaned and sanitised and also with air slips. Main preventive and controlling steps should be targeted on the huge improvement of hygiene (mechanic cleaning and removal of dirt first), followed by hot water high-pressure cleaning and sanitation of spaces (without the presence of animals) as well as changing bedding regularly. Milking of clean animals with thoroughly cleaned and dried udder and teats (mainly their tops) and avoidance of air leaks and liner slips are essential prerequisites to decrease the risk of infection. Production loss of approximately 15% was calculated in Holstein cows with clinical mastitis caused by *E. coli* (approximately 200L per cow during 305 days), but exceeding 30% was reported in persistent and chronic mastitis courses (Blum et al. 2014).

E. coli

Escherichia coli infections without systemic clinical signs, with mild or moderate course of infection, are mostly of short duration and can be resolved by immune defence mechanisms with the help of NSAIDs and without administration of antimicrobials; however, serious, systemic infections are very difficult to resolve (Günther et al. 2017). Approximately 10% of *E. coli* mastitis have the potential to cause systemic life-threatening infections, accompanied with fever, anorexia, depression, lethargy, cows' weakness (cows are unable to rise), loss of appetite, a reduction in grooming, dehydration, large clots or serum-like appearance or blood in milk and markedly decreased milk production associated with increased SCC. Acute clinical signs and speedy worsening of the cow's status are due to the toxins produced and released by the bacteria.

Recently so-called mammary pathogenic *E. coli* (MPEC) are characterised with virulence attributes resisting host defence mechanisms. Goulart and Mellata (2022) summarised various virulence-related features being suggested to explain the pathogenicity of *E. coli* in bovine mastitis, among them the activation of the innate immune system by pathogen-associated molecular patterns (PAMPs), resistance to serum and neutrophil killing, adhesion and invasion of mammary epithelial cells and survival and proliferation in milk through the secretion system. But these authors also noted that despite these features benefitting *E. coli* to attack the mammary gland, high genetic abundance and variability of mastitis-causing strains were detected, and the features are also common in Extraintestinal pathogenic *E. coli* (ExPEC) strains. Iron scavenging by the ferric dicitrate system is strongly associated with mastitis-associated *E. coli* (Blum et al. 2015), and for overcoming the low levels of free iron in milk, mastitic strains express a variety of iron transport systems, such as the high-affinity siderophores (e.g. enterobactin, aerobactin and ferrichrome) and iron-regulated outer membrane proteins (IROMP). Blum et al. (2018) proved that the *Fec* system is essential for pathogenicity in udder conditions, as MPEC without this system were unable to induce mastitis. It can be hypothesised that this feature could be target for novel treatment options (Goulart and Mellata 2022). Moreover, persistent mastitis-associated *E. coli* strains have a higher ability to adhere, invade and survive in bovine mammary epithelial cells (importance of *lpf* genes encoding for long polar fimbria is expected). Another interesting observation is also inducing immune response by different *E. coli* groups, where in those causing acute mastitis the immune response was higher than in *E. coli* strains associated with the chronic infections. The severity of mastitis is determined by the speed with which neutrophils act at the site of infection and how effectively they phagocytose the pathogen (Chengolova et al. 2021). Lactoferrin, having bacteriostatic properties and being produced by epithelial cells and leukocytes, belongs among the key components of defence mechanisms/host's innate immunity, together with cytokines including interleukins, colony-stimulating factor, interferon and tumour necrosis factor (TNF). Interestingly, TNFα was identified as the main mediator of endotoxic shock that occurs in cases of super-acute mastitis caused by coliforms (Hogan and Smith 2003). Phagocytosis, in which neutrophils release chemicals that eliminate MPEC, is of importance, but it should be pointed that this process causes mammary epithelial cell damage, resulting in the continuous decrease of milk production. Apoptosis of neutrophils and release of chemokines that attract macrophages, which "remove" apoptotic neutrophils via efferocytosis, minimise the release of granular contents that damage the tissue. Once this process is ineffective, dendritic cells are able to clear apoptotic neutrophils. More rescued mammary tissue and less extent of damage of the mammary gland can lead to a more successful treatment.

Non-serious, bovine *E. coli* mastitis treatment, under veterinary care and supervision, can be built on fluid therapy and NSAIDs (or sometimes also steroids) or alternative supportive treatment, which could help mastitis being resolved via immune defence mechanisms. As a prevention, available commercial or autogenous vaccines could be used and evaluated for incidence rate of coliform mastitis in certain farm conditions.

Vaccination is often timed to correspond with coliform mastitis risky pre- and immediate postpartum period (Rainard et al. 2021). Effectivity of vaccination against *E. coli* mastitis in real clinical practice is still subject of discussion. Recently, results of the study (Stewart et al. 2022) comparing two commercial vaccines, Enviracor™ J-5 and Bovilis® J-5, were released, which were administered at approx. 60 days prior to anticipated calving (day of drying off for cows), approx. 30 days after the primary immunisation and on day 14 of lactation (5 mL subcutaneously, approx. 100 mm anterior of the scapula on the neck). Health and production data for 506 and 479 animals were analysed. Efficacy of vaccines differed in reducing coliform clinical mastitis caused by coliform bacteria during the first 100 days of lactation (0.0444 cases vs. 0.0183 cases per 100 cow-days) and in total clinical mastitis rate (0.1138 cases vs. 0.0690 cases per 100 cow-days). Control group was not present, so it was hard to confirm the situation with and without vaccination in the comparable conditions. As for incidence of coliform and total clinical mastitis, results can be given on the mean incidence rate of total clinical mastitis, which was 0.278, 0.257 and 0.252 cases per 365 cow-days at risk in herds with low (≤150,000), medium (150,000–250,000) and high (250,000–400,000 cells/mL) bulk milk SCC, respectively (Barkema et al. 1998).

Treatment of serious severe infections caused by *E. coli* is based primarily on systemic administration of antimicrobials (Oliveira and Ruegg 2014; Krömker and Leimbach 2017) to mitigate the risk of bacteraemia and subsequently risk of sepsis. Systemic antimicrobial administration for severe *E. coli* caused less severe milk production decreases in cows treated with systemic cefquinome (Shpigel et al. 1997; Suojala et al. 2013) and faster recovery from reduced milk production in cows receiving systemic fluoroquinolone only (compared with no treatment) after *E. coli* challenge (Poutrel et al. 2008). Although third- and fourth-generation cephalosporins and fluoroquinolones are recommended for the treatment of severe coliform mastitis, these belong among critically important antimicrobials, and their use to treat animals must be rigorously assessed and regulated according to bacteriological diagnoses. Differences in legislation among countries affect the portfolio of antimicrobial drugs, which are authorised and finally available. In the USA, no antimicrobials are currently FDA authorised for systemic clinical mastitis treatment. This currently leads to off-label administration (individually possible according to legislation), but considering the perspective approximately 10 years ago, systemic antimicrobials were commonly used in treatment of severe clinical mastitis in the USA (Oliveira and Ruegg 2014). In Europe systemic treatment is authorised and allowed. Also, AVA-AMA (2022) guidelines (Australia) pointed out the possibility of parenteral treatment of mastitis. Guidances (as well as summary of product information) of authorised VMPs recommend in the case of serious mastitis the first dose to be administered intravenously (sulphonamide/trimethoprim or fluoroquinolone-containing injectable VMPs). Generally, it can be concluded that it is hard to assess treatment effectivity and success, as there is a lack of well-controlled studies for serious *E. coli* mastitis to compare the most effective treatment regimens with antimicrobials (IMM vs. systemic or systemic and IMM) and supportive therapy.

Regardless of antimicrobial treatment, administration of NSAID to alleviate pain and discomfort is recommended for all severe clinical mastitis cases and is common practice in many countries (Hewson et al. 2007; Breen 2017). As summarised by de Jong et al. (2023), administration of NSAIDs (e.g. flunixin, carprofen or ketoprofen) with lower rectal temperatures, lower heart rate and improvements in rumen motility and clinical signs was determined in *E. coli* mastitis. Concomitant administration of meloxicam with cefalexin or penethamate improved bacteriological cure (both Gram-positive and Gram-negative) (McDougall et al. 2016) and lowered SCC and culling (McDougall et al. 2009). No differences in bacteriological cure or recurrence rate were reported in Gram-negative cases treated with ketoprofen compared with no antimicrobial treatment (Latosinski et al. 2020). It should be however noted that scarce evidence of severe mastitis cases can be found as the studies are mostly performed with mild to moderate mastitis. Another study (Sintes et al. 2020) also concluded that NSAIDs do not prevent the mammary epithelial barrier opening by lipopolysaccharides from *E. coli*; however, ketoprofen, flunixin meglumine and meloxicam obviously support the re-establishment of the barrier integrity. Long-term treatment with NSAID non-selective for the COX-2 receptor (such as flunixin and ketoprofen) has been associated with side effects such as decreased renal perfusion and gastric ulceration (Orr et al. 2014). In such cases discussion can be held, once concomitantly administered antimicrobial (with elimination mostly via renal route), what will the scale of residue depletion from kidneys. Thus, as pointed out by de Jong et al. (2023), consideration of NSAID for mild and moderate CM cases should be thoroughly discussed between herd veterinarian and producer, considering farm production goals, cow welfare and costs (those influenced also by the length of the withdrawal period).

Other types of supportive treatments (including rehydration fluids (see below under *Klebsiella* spp. part), frequent milk-out, oxytocin, calcium, hypertonic saline and corticosteroids) are frequently considered as part of clinical mastitis treatment protocols (Oliveira and Ruegg 2014; Persson Waller et al. 2016). Evaluation of effectiveness of different approaches in supportive therapy is scarce (Leslie and Petersson-Wolfe 2012; Francoz et al. 2017). Erskine (2022) commented on the possibility to administer glucocorticoids, at the beginning of the course of disease caused by endotoxin-producing coliforms, as these drugs inhibit pro-inflammatory pathways related to the acute-phase response. An example of treatment with dexamethasone (30 mg, IM) immediately after the introduction of *E. coli* into the mammary gland has been reported to reduce mammary gland swelling and improve rumen motility. Despite such treatment being risky for pregnant cows, also severe clinical *E. coli* or *Klebsiella* mastitis itself may cause pregnancy loss in cattle. No added benefit of oxytocin and frequent milking on resolving clinical signs and achieving cure was commented by Francoz et al. (2017).

Klebsiella spp.

Klebsiella spp. (mainly *K. pneumoniae* and *Klebsiella oxytoca* and new species *Raoultella*) belong to other frequently occurring bacteria from the *Enterobacterales* group causing very severe mastitis. Differentiation of *Klebsiella* spp. from other Gram-negative pathogens that do not require treatment is of importance, because for this species antimicrobial treatment is highly indicated (de Jong et al. 2023). Laboratory investigation is of importance for these cases. Mild (abnormal milk), moderate (abnormal milk and swollen udders) and severe cases (systemic signs, including fever, off feeding, steeply decreased milk production, shock and/or recumbency) can occur. As many *Klebsiella* spp.-caused infections become chronic with production loss and high SCC, culling occurs frequently (Ruegg 2017; Kleinhenz et al. 2019). Although it is likely that the effects of antimicrobial treatment will vary according to pathogen-level characteristics, significant heterogeneity and inconsistency of effects of antimicrobial treatment were noted for clinical mastitis caused by *Klebsiella* spp. (Nobrega et al. 2020).

The immune response to lipopolysaccharide (LPS) from the outer cell membrane of the coliform bacteria is activated leading to high levels of the pro- and anti-inflammatory cytokines in milk. Milk cytokine concentrations are greater after infection with *Klebsiella* spp. than *E. coli*, with longer persistence of systemic cytokines in

Klebsiella-caused cases. Bacteria, with more acyl groups in the LPS molecule, are more aggressive, causing lethality more frequently. Some indicia was given for *K. pneumoniae* studied on murine model (Clements et al. 2007), where the authors concluded that lipopolysaccharide acylation plays an important role in providing Gram-negative bacteria some resistance to structural and innate defenses.

As already noted above, in serious mastitis, caused by *E. coli* but mainly *Klebsiella* spp., treatment protocols mostly rely on systemic administration of third- and fourth-generation cephalosporins or fluoroquinolones (mostly marbofloxacin). Interestingly, thiamphenicol as a highly bioavailable antibiotic, independently of the administration route, being highly available in the mammary gland, can be considered as an important possibility. It has a bacteriostatic action against a broad range of microorganisms, including Gram-negatives, although it may be bactericidal in concentrations 3–5 times higher than bacteriostatic ones (Mestorino and Errecalde 2012). An example of the possible treatment protocol is listed in Table 11

for moderate and severe mastitis. Correction of dehydration, prevention of bacteraemia and fever reduction are the key considerations; therefore cows with severe mastitis scores need, except antimicrobial treatment, also strong supportive therapy. Therefore, NSAIDs, hypertonic saline (intravenous), calcium (intravenous and oral) and water drenching (oral) are recommended to be administered. Water via an orogastric tube to lower dehydration and hypovolemia due to endotoxic shock is beneficial as well as calcium to correct subclinical hypocalcaemia. Despite all this care, the rate of cure and especially the ability to be back to production are very limited. Prevention therefore plays a key role.

Ten antimicrobials were assessed in the retrospective study (isolates 2008–2019, N = 483 *Klebsiella* spp., USA, according to CLSI standard): ampicillin, penicillin, erythromycin, oxacillin + 2% NaCl, pirlimycin, penicillin/novobiocin, tetracycline, ceftiofur, cephalothin and sulfadimethoxine. The authors summarised that the proportion of isolates resistant to ceftiofur, cephalotin or tetracycline did not increase over time, and decrease was noted for sulfadimethoxine. These results do not demonstrate a trend toward increasing antimicrobial resistance among *Klebsiella* spp. isolates (Fuenzalida et al. 2021). On the other hand, there is the need for further studies to score precisely the correlation of results of in vitro susceptibility and in vivo efficacy, especially for *Enterobacterales*-caused mastitis. Although treatment options, including antimicrobials, exist, the rate of complete resolution (clinical and bacteriological cure and being back to production, even lower than prior diseases) is quite low. Prevention is therefore of utmost importance and includes milking hygiene and bedding hygiene (sawdust considered most risky for *Klebsiella* mastitis, while sandy bedding can be recommended), and fortification of bedding by additives lowering contamination or modifying positively microbiota could be one part of the preventive measures, as well as udder hygiene including removal of hair, clean and heathy top of teats and reduction of udder oedema in periparturient cows by better nutrition (sodium, potassium). Although vaccinations exist, it can decrease the severity of infection but not protect against it. Recently, a vaccine that targets *Klebsiella* spp. by immunisation against siderophore (iron-trapping) proteins that are an important virulence factor has shown promise for mildening of infections caused by this pathogen (Erskine 2022). It is hard to find any reference, but new technologies, using as well autogenous vaccines via specific routes of administration, started to be developed.

7.3.3.10 *Mycoplasma bovis*

As a source, mostly infected (new arriving) animals (cows but heifers/calves as well), positive for *Mycoplasma bovis,* are an issue. Disease spread via cow-to-cow contact, also with colostrum or to suckling calves but also via aerosol from animals with respiratory infection. Bacteria colonising/infecting respiratory tract can be present also in the udder. Prevention and control start in purchasing of the replacement animals to the herd (preferably declared as *Mycoplasma*-free). It is recommended to use the test to check first

Table 11 – Example of the treatment protocols for moderate (*Klebsiella* spp. caused) and severe (*Klebsiella* spp. and *E. coli* caused) mastitis

Mild	No systemic therapy indicated.	Re-evaluate in 24hours.
Moderate ***Klebsiella***	**ANTIMICROBIAL:** **Amoxicillin / clavulanic acid:** - SC or IM, 7.0 mg amox/1.75mg clavulanic; q24 hours for 3-5 days Below antimicrobials belongs to AMEG B (last choice category antimicrobials, classified as CIAs by WHO and WOAH)	According to the need: NSAID: Flunixin meglumine: - IV 2.2 mg/kg bw; q24 ORAL FLUIDS: 35-55 L (considering cow status and approx. weight of the animal)
Severe ***Klebsiella*** ***E. coli***	**Cefquinome** (CEF 4th gen) - IM, 1mg/kg bw; q24 hours for 3-5 days **Marbofloxacin (FQ):** - SC, IM 2 mg/kg bw (first dose can be IV) q24hours for 3-5 days **Enrofloxacin (FQ):** - Single dose formulation: 5mg/kg bw IV - Twice dose formulation: 5mg/kg bw SC (first dose can be administered slowly IV); q24hours 2 days Not authorised for mastitis indication in EU: **Ceftiofur (CEF III):** SC or IM 2.2 mg/kg bw; q24 for 3days	NSAID: Flunixin meglumine: - IV 2.2 mg/kg bw; - q24 for 2-5 days Note: ketoprofen and carprofen containing VMPs are used as well Hypertonic saline 7.2% - 5mL/kg bw - Rapid infusion ORAL FLUIDS: - 35-55 L - (considering cow status and approx. weight of the animal) - Immediately following hypertonic saline admin. - Specific additives mixed to the oral fluids Calcium gluconate 23% - 250 – 500 mL - IV- slow infusion

milking/bulk tank (PCR). Separate towels for wash and dry hygiene, teat dipping, checking cows at drying off and re-checking at the early lactation phase are other tools to minimise infections. There is no commercial *M. bovis* mastitis vaccine available. As *Mycoplasma bovis* has been shown to modulate both humoral and cellular immune response during bovine mastitis, future research can be focused on new immunogenic and protective vaccine targets. Isolation of suspected cows and culling of confirmed cows remain vital control measures for *M. bovis* in dairy farms (Gelgie et al. 2022).

7.3.4 Subclinical Mastitis

Among subclinical mastitis characteristics are that milk yield and somatic cell count in the milk are changed. Cows with subclinical mastitis are present in all dairy herds, and prevalence of affected cows varies highly from 5% to 75% (at quarter level 2–40%) (Erskine 2022). Compared to clinical mastitis, the incidence of subclinical mastitis is estimated to be 15–40 times higher (Martin et al. 2018), which increases its economic importance (higher frequency and capacity to reduce milk yields while many times going unnoticed). In many cows, subclinical mastitis caused by various pathogens can develop to chronic infection only occasionally manifested clinically. Considering an example of streptococcus mastitis, the average duration is 12 days but can be prolonged in chronic cases (Martin et al. 2018). There are mostly no visible signs of the disease and no udder and cow's health status alteration. For contagious pathogens, adult lactating cattle are mostly at risk of infection. Such cases are mostly noted as elevation of SCC indicating an issue, and most culturing bacteria are *Streptococci* and *Staphylococci*, which are and were historically the pathogens mostly focused by control measures. During the last decade, herds have been identified that have considerable subclinical mastitis caused by Gram-negatives: *Klebsiella* spp., *Serratia marcescens* and *Pseudomonas aeruginosa* but *Candida* spp. and Prototheca spp. as well. In some herds being rather rare, but in herds importing cattle, *Mycoplasma bovis* has been recognised in cows exhibiting both subclinical and clinical signs. The mammary gland with infection serves as a reservoir for pathogen spreading, via milkers' hands or milking equipment. Teat-end dermatitis caused by *Haematobium irritans* being a vector for *S. aureus* has been associated with an increased risk of infection in heifers, especially in warmer climates (Erskine 2022).

In animals with subclinical mastitis, treatment should be targeted, with knowledge of the history of the cow, and in automatic systems providing cell counting, with knowledge of the dynamics of SCC; culturing and knowledge of resistance patterns present within the herd are an advantage. However, many subclinical intramammary infections are chronic. For certain pathogens, prediction of therapeutic outcome by in vitro testing is unreliable (e.g. *S. aureus*). In subclinical or chronic *S. aureus* infections, where there can occur an extensive fibrosis and micro-abscess formation in the gland affecting pharmacokinetics and availability of therapeutic concentrations capable of targetting the pathogen of concern, an antimicrobial may not be able to resolve an issue. Based on invasiveness and location, as well as the antimicrobial used, only 20–40% of subclinical infections, but less percentage of chronic infections, have the chance to be cured. A combination of increased duration and frequency (5 × 48 h) leads to the best result assessing bacteriological cure rate of antimicrobial treatment of subclinical mastitis in dairy cows with somatic cell count exceeding 500,000 cells/mL. This result was obtained in a study with 200 mg cloxacillin sodium intramammary administration, compared to a no treatment control group ($n = 80$ glands) with groups following regimen 3 × 48 h ($n = 273$ glands), 5 × 24 h ($n = 279$ glands) and 5 × 48 h ($n = 72$ glands), with resampling scheme after 21 and 28 days after treatment initiation. The gland-level cure rate for any pathogen was control (6.2%) < regimen 3 × 48 h (49.8%) < regimen 5 × 24 h (61.6%), and regimen 5 × 48 (80.6%) was the most successful. *Staphylococcus aureus* or *Streptococcus* spp. cure rate was control (7.7%) < regimen 3 × 48 h (42.6%) < regimen 5 × 24 h (52.5%), and regimen 5 × 48 h (75%) was the most successful (McDougall et al. 2022). Depending on susceptibility testing, lipophilic antimicrobial drugs that distribute well into the mammary tissue should be chosen. Culling should be considered for cows with a long duration of infection (>3 months), those with more than one quarter infected, those infected with pathogens resistant to portfolio of intramammary antimicrobials or those not responding to treatment (as detected in subsequent monitoring of SCC and culturing). Intramammary administration of VMPs with amoxicillin or penicillin as the first line or cephalosporins (first generation preferred) can be used in *S. agalactiae* subclinical mastitis. Generally, subclinical infections caused by *S. uberis* and *S. dysgalactiae* are more difficult to treat and should be preferentially treated at the end of lactation with dry cow intramammary VMPs, and dry cow therapy is also more economical and least likely to lead in residues in milk compared with lactating cow therapy. Drying-off quarters, culling or therapy as a means to reduce SCC contributions from infected cows and spreading of pathogen is a palliative approach to mastitis control, which would be better addressed by prevention of infections (Erskine 2022).

7.3.5 Chronic Mastitis

As was noted above, some subclinical and clinical cases may become chronic. *Streptococcus uberis* and *S. agalactiae* are well-known pathogens able to induce chronic mastitis as well as *S. aureus*. Recent studies also indicate that 20–30% of all mastitis cases are recurrent mastitis (Schmenger and Krömker 2020; Wente et al. 2020). Consideration of the clinical mastitis history in the current lactation and persistent elevation in cow SCC allow a determination of cows with a low probability of bacteriological cure following antibiotic treatment (Krömker et al. 2021). Specific are chronic mastitis caused by *S. aureus*, where bacteriological cure rates after antibiotic therapy seldom exceed self-cure rates (Linder et al. 2013). A large proportion of all clinical cases may be attributed to animals with longer-lasting udder infections with clinical flare-ups (Krömker et al. 2021). Infections caused by *S. uberis* (especially well-susceptible strains) resolve well (i.e. have low pathogen and infection persistence frequency), and subsequent infections are largely caused by other strains of *S. uberis* (i.e. have high recurrence of mastitis) as detected by Wente et al. (2020). It means that chronically diseased animals may cure bacteriologically between individual infections, but due to compromised udder parenchyma, they will remain chronically ill and prone to develop clinical signs in turn as well as show persistently elevated quarter SCC (Krömker et al. 2021). It brings new view to the value of use of antibiotics in treatments of such chronic cases and shows that if feasible, those cows should be removed from the herd (Degen et al. 2015) or treated symptomatically in the case of clinical mastitis to avoid useless administration of antibiotics (Krömker and Friedrich 2011). From a farmer's point of view, such cows, especially the high-yielding animals, are still profitable as long as they show no clinical symptoms and the milk fulfil quality limits for selling. Therefore, treatment should focus on decreasing the symptoms of inflammation. Survey and study on implementation of targeted therapy concept indicate practices in chronic mastitis with recurrent clinical flare-ups, which show that farmers tend to prolong antibiotic treatment (Schmenger et al. 2020; Falkenberg et al. 2019). The decision to extend initial antimicrobial treatment was based on persistent flakes in milk (78.6%), persistent udder swelling (51.0%), detection of *S. uberis* (3.5%) and detection of other pathogens (7.9%); on the other hand, reduction of flakes in milk led to treatment termination in 57.7% of cases (study from Germany, 499 completed survey forms, Falkenberg et al. 2019). A situation within the herd and success of management and treatment protocols should be assessed, and the farmer in close cooperation with the veterinarian should accommodate it dynamically. Improving automatic milking systems (AMS) can collect mastitis indicator data as SCC, blood in the milk and conductivity parameters, which together with further on-farm data (parity, milk diversion, time interval between milkings, milk yield and DIM) create input to sensor-based prediction model. It has been demonstrated (Bonestroo et al. 2023) that it is possible with a high accuracy to predict the future chronicity based on subclinical mastitis status using past sensor data and machine learning models. Models for clinical mastitis and possibility of chronicity have also been established for such AMS. Data from AMS, together with data from culturing and AST (identification of pathogen and antibiogram), teat dips and biocides and actives of both groups, drying off methodology and mastitis treatment protocols, can create important variables, which if used properly and their effect evaluated for exact farm conditions and animals can help to reduce the incidence of mastitis and its recurrence and finally chronicity as well.

7.3.6 Specific Considerations on Use of Antimicrobials at Drying Off

Although several decades ago in many countries blanket drying off treatment with antimicrobials was considered as a standard treatment, Nordic guidelines have started decades ago to promote selective use of antimicrobials at drying off. Recently, mastitis pathogens classified as "mainly environmental" are the most common causes for mastitis, and instead of dry Cow IMM antibiotics administration, the emphasis is on other mastitis prevention measures (Klaas and Zadoks 2018). The growing issue of antimicrobial resistance needs action. Minimising the use of antibiotics and drying off has been recognised as one of the possibilities for optimisation of use of antimicrobials in dairy cows. The most recent studies combined selective DCT with concurrent internal teat sealants use, and the results consistently indicate no negative effect on udder health if selective DCT is performed based on proper algorithm tailored in details on the exact farm conditions and herd health and production status (McParland et al. 2019; Rowe et al. 2020). The most challenging questions coming on each farm on proper procedure for drying off these days. In current high-yield dairy cows, timing and stepwise decrease of production are of utmost importance. Tools or algorithms appropriate for certain herd and farm are expected to help in finding a cow with the need of treatment with antimicrobials.

Several approaches are currently used in Europe, and it should be noted that some countries in Europe ban blanket dry cow therapy (Nordic European countries and the Netherlands (Santman-Berends et al. 2021; McCubbin et al. 2022). Also new European regulations are extremely restrictive, prohibiting mass veterinary prophylactic antimicrobial use (AMU) and allowing it only for individual and in exceptional cases (Regulation (EU) 2019/6), and the dairy industry is shifting toward SDCT and limited prophylactic AMU in general. It can be expected that more strategies and algorithms will be developed, and some already working examples or studies comparing the approaches are given below. All approaches could be simplified and visualised as indicated in Fig. 6.

Examples of SDCT (Selective Dry Cow Therapy) (simplified):

1. SDCT—cases when to use antibiotic DCT (Dry Cow Therapy) (Finnish approach summarised by Niemi et al. 2021):
   (a) Delay the treatment of subclinical mastitis until dry-off.
   (b) Administer antibiotic DCT to cows with SCC (>200,000; *note: some systems recommend to differentiate, e.g. SCC 150,000 (for primiparous) and SCC 50,000 (for multiparous) cows; see below approach "C"*), especially during the last 3 months of lactation, based on either DHI measurements or data from automatic milking system.
   (c) Administer antibiotic DCT to cows treated for mastitis during preceding lactation.
   (d) Perform culturing of milk samples prior to drying off (sampling strategies may differ per herd).
2. Two approaches of SDCT—when to use antibiotic + teat sealant (Rowe et al. 2020):
   (a) SDCT based on culturing: treating only quarters from which aseptically collected milk samples tested positive for mastitis-causing pathogens (Minnesota Easy 4Cast plate) after 30–40 h of incubation.
   (b) SDCT-based algorithm—antibiotic treatments were selectively allocated at the cow level, with all quarters receiving antibiotic treatment if the cow had either a Dairy Herd Improvement Association test somatic cell count >200,000 cells/mL during the current lactation or two or more clinical mastitis cases during the current lactation. All quarters of the respective cows are treated.

Conclusion on above 2. a) and b) approaches: Non-inferiority analysis indicated that culture- and algorithm-guided SDCT approaches

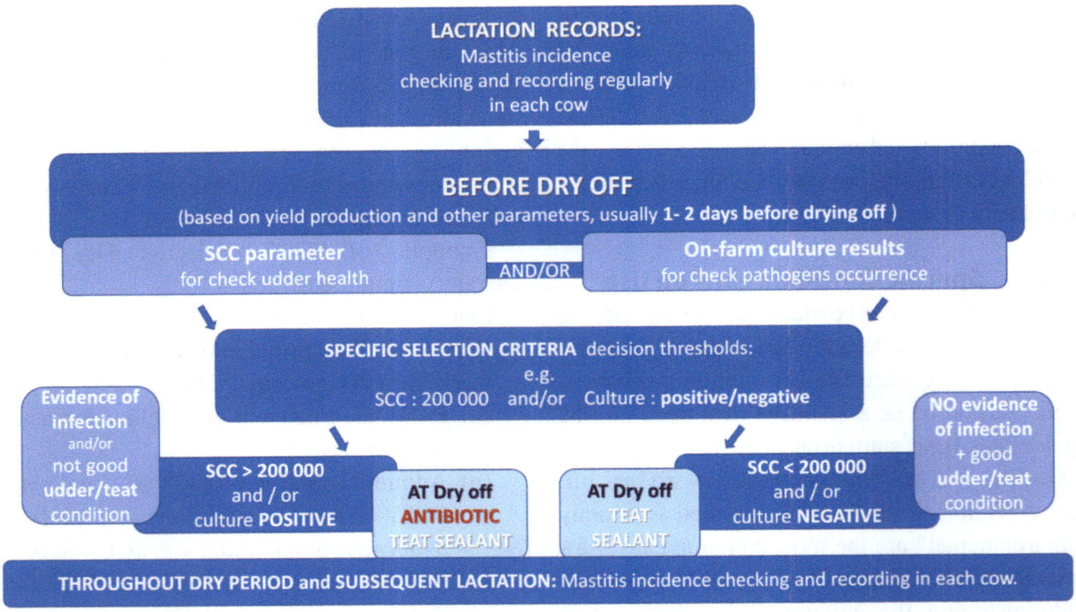

Fig. 6 Simplified decision tree for selective dry cow therapy

performed at least as well as BDCT for dry period IMI cure risk. Each of two above-mentioned approaches reduced antibiotic use by 55% without having a marked effect on the risk of IMI cure and new IMI during the dry period and IMI prevalence at 1–13 DIM (Rowe et al. 2020).

3. SDCT in the Netherlands (The Royal Dutch Veterinary Association cow-level selection criteria based on SCC (brief summary) on using antibiotic in DCT, where an SCC-based approach is used—the cow SCC on the last milk recording before drying off was >50,000 cells/mL for multiparous cows or >150,000 cells/mL for primiparous cows. The last milk recording should occur ≤6 weeks before dry-off.

Microbiological culturing of milk samples: design of sampling scheme tailored for the exact conditions of the herd/group of cows at drying-off characterisation is recommended. Regular and comprehensive microbiological culturing provides essential information on the known causal pathogens in the herd (Gussmann et al. 2018; Ruegg 2018). Better setting of preventive measures (and their effectivity evaluation as well) can be performed in the herds, in which combining information about a known pathogen with the onset of infection in relation to the lactation stage and the season, the infection pathways can be identified. Lipkens et al. (2019) reported that the lowest test-day SCC throughout the next lactation was observed in cows that had a test-day SCC <200,000 cell/mL both before and after dry-off. Several studies reported that higher milk production at period 0–30 days before dry-off was associated with an increased SCC after calving (Madouasse et al. 2012; Gott et al. 2017; Niemi et al. 2021).

Regardless of the choice of dry cow therapy, it is important to be as sterile as possible when administering dry cow therapy to avoid introducing infection into the teat. Clean gloves should always be worn, the teat properly cleaned and hygiene kept to the highest standard. This is even more important when using teat sealants without concurrent antibiotic treatment. The nozzle containing either the antibiotic or teat sealant should only be partially inserted into the teat. Poor infusion technique can cause damage to the teat canal and increases the risk of mastitis, which can result in cows dying.

As for authorised antimicrobial VMPs, which were sold in 2021, eight antimicrobials were used in monocomponent VMPs for dry cow treatment, with highest volumes of sales for cloxacillin, followed by group of first-generation cephalosporins (cefapirin (second mostly sold) cefalexin, cefalonium, cefazolin) and fourth-generation cephalosporin cefquinome (AMEG B category!), which was the third mostly sold substance. Rifaximin and erythromycin were sold in DC intramammary VMPs as well. Among VMPs with fixed combination containing two antibiotics benzylpenicillin, aminopenicillins (ampicillin and amoxicillin), cloxacillin and cefalexin were sold. Aminoglycosides neomycin, streptomycin and dihydrostreptomycin but also erythromycin, novobiocin and, surprisingly also last resort, AMEG B categorised, colistin were sold in VMPs with two actives in fixed combination in 2021 (unpublished results, ESVAC 2021). Three active substances are still present in some fixed DC combinations in VMPs, among them benzylpenicillin and nafcillin, which are mostly in combination with two aminoglycosides (dihydrostreptomycin, neomycin, gentamicin, framycetin) and were sold in 2021 within EU member states (unpublished results, ESVAC 2021). Susceptibility of pathogens occurring within the herd should be considered together with clinical experience and AMR trend analysis, and whenever possible narrow-spectrum, first-line antimicrobial monocomponent VMP should be used. Second-line, synergic combination of two substances should be preferred to the last resort (AMEG-B categorised) antimicrobials either alone (cefquinome) or in combination (aminopenicillin with colistin). Blanket, non-evidence-based, three-active substances containing fixed combinations of DC intramammary VMPs should be avoided.

8 Alternatives to Well-Established Antimicrobials Used in Legacy Veterinary Medicinal Products

Despite a plethora of research activities in the last decades, there are still not many alternatives that were finally approved as treatment-supportive products or authorised as veterinary medicinal products actively preventing or being powerful tools for treatment of mastitis and therefore equally replace authorised antimicrobials. Except those already authorised VMPs, still a lot of new alternative tools are in the pipeline of experiments. Some examples that can be perceived as promising are listed below. But it should be highlighted that also alternatives/alternative tools should be thoroughly assessed as for quality, safety and efficacy. Among alternative tools are not only those listed in below subsections as possible alternative “compounds and components” of treatment or prevention or support to healthy udder but also a complex of other measures. Examples of them can involve proper care on animals in all stages of their production (from mothers, calves, heifers to highly production dairy cows), with proper cleaning regimens, including disinfection in appropriate manner, biosecurity, hygiene of milking at the level of animals, udder and teats as well as hygiene of milking parlours and devices. The complexity is excellently described by “udder health pyramid” that was designed by experts from Sweden as an advisory support for farmers and veterinarians to improve udder health (Landin et al. 2020), as described in Fig. 7. This pyramid together with a signal system, which accurately and early detects also subclinical signs of udder health disorders in the cow, is of importance to minimise health and welfare issues.

8.1 Specific, Non-antimicrobial Veterinary Medicinal Products Authorised

8.1.1 Dry Cow Therapy Non-antimicrobial Teat Sealants

Naturally, sealing of the teat canal by the natural keratin plug formed during the dry period creates the basis of the protection against new intramammary infection in the late dry period. Teat sealants are among the non-antibiotic treatments commonly used for selective dry cow therapy. Internal bismuth subnitrate teat sealants have no antimicrobial activity and are therefore recommended for use alone only in the healthy/uninfected udder. Some herbal additives were tested with bismuth subni-

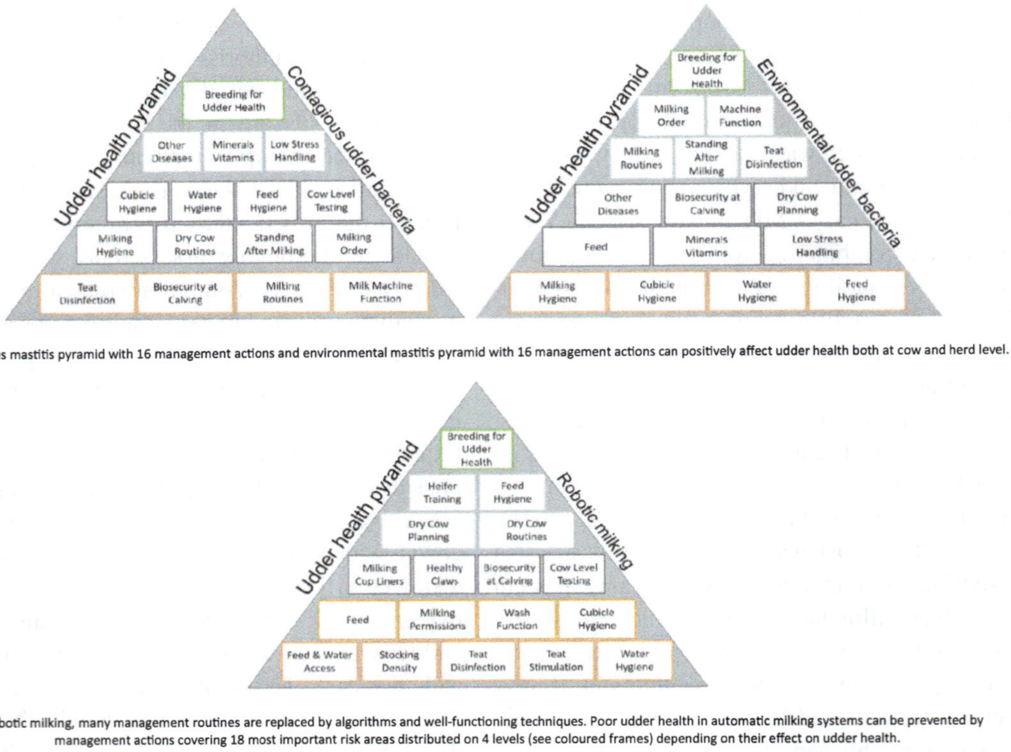

Fig. 7 Udder health pyramid measures start from the base of the pyramid and goes upwards in four levels towards the goal of ≤150,000 herd average BTSCC

trate in internal teat sealants (Pearce et al. 2023). Cows screened as negative for mastitis during lactation, having the scores of SCC (please refer to Sect. 7.3.6) and have good total health status can be managed at dry-off with teat sealant alone without antimicrobial and without detrimental effects on milk microbiome and limiting as well bacterial load at first week post-partum. When used alone in uninfected quarters, internal teat sealants have been shown to prevent significantly more new infections than using no treatment at all. It is paramount that the very best hygienic practices are adopted when infusing the teat sealant and great care should be taken to proper administration of the teat sealants (more detail in Sect. 7.3.6) to protect teat from contamination with bacteria. Improper administration can introduce pathogenic bacteria, which can finally cause fatal mastitis.

Another method to help cow's defences is coat the teat with external sealant by dipping. These products are administered after antimicrobial infusion. Mostly in external teat sealants, the duration of persistency on teat ends is not as long as it will be necessary, as only fully covered teat ends provide protection from bacteria entering the teat canal and finally the mammary gland. Visual inspection is therefore highly recommended, and if needed re-application every 5–7 days throughout the dry period or at least in the late (transition) period. Aseptically and properly administered internal teat sealants are favoured to external teat sealants.

8.1.2 Non-steroidals

Non-steroidal anti-inflammatory drugs (NSAIDs) can be used to treat painful conditions. They are authorised in many countries for the adjunctive treatment of clinical mastitis and frequently used in mild, moderate and severe mastitis, being used as well to relieve other diseases and occasions (as, e.g. post-operative pain) in both cows and calves (Brown et al. 2022; Krömker et al. 2021). Non-steroidal anti-inflammatory drugs are the only family of analgesics available for food-producing animals in the European Union that provide long acting, i.e. 24–72 h, pain relief per dose (Hudson et al. 2008). NSAIDs inhibit the enzymes cyclooxygenases (COX), which, as a key enzyme, are involved in inflammation by producing inflammatory mediators such as prostaglandins. They also act as potent anti-pyretics, reducing body temperature and increasing appetite as well as providing pain relief. Differences in the selectivity of COX inhibition between the different NSAIDs available for use in cows mean that certain products theoretically have fewer side effects (such as gastrointestinal irritation, renal toxicity and inhibition of blood clotting). For example, meloxicam is a selective inhibitor and exhibits preferential binding to cyclooxygenase-2 (COX-2) receptor and consequently generates fewer negative gastrointestinal side effects than nonspecific COX inhibitors such as flunixin meglumine and ketoprofen, which are non-selective COX inhibitor. In addition to improving the welfare of the dairy cow through reduced pain, NSAID pain relief has an effect also on preventing milk starvation due to insufficient feed intake. Moreover, NSAIDs have positive effects on bacteriological cure rate, and cows regain physiological milk secretion earlier as was proven for carprofen (Krömker et al. 2011) and meloxicam, while in the latter it was also shown to reduce the SCC increase in milk and the risk of culling during and after mastitis (McDougall et al. 2009). NSAIDs in optimised dose would likely support the recovery and improvement of milk composition. However, an overdose of NSAID would likely cause tissue irritation and, in turn, a delayed recovery of the barrier permeability (Sintes et al. 2020).

Recently results of the trial (mastitis with mostly detected *S. uberis*, *S. aureus* and *E. coli*) investigating the efficacy of NSAIDs (ketoprofen INJ 3 × 24 h) and IMM treatment with antimicrobials of mild to moderate clinical mastitis (CM) associated with long-lasting udder disease have not proved inferiority; however they indicate that in mild to moderate CM cases exclusive treatment with ketoprofen may constitute an alternative to antimicrobial IMM therapy. Probability of achieving no CM recurrence 60 days after the end of treatment was also investigated, and interestingly, almost numerically identical results were gained for the group treated by IMM antimicrobials (63.3%) and group with NSAID only (64.9%). As very low bacteriological cure rates were gained within the above-mentioned trial for antibiotic, in longer-lasting mastitis history of cows, antibiotic treatment can hardly be justified (Krömker et al. 2021). Nevertheless, farmers many times consider the milk of these cows (without clinical signs) as still saleable and are rather targeted on clinical cure score, a low recurrence rate and, lastly mentioned but of high economic importance, short time of discarding milk due to withdrawal period for milk.

Despite proven roles of various NSAIDs in pain relief and inflammation reduction, decrease of SCC, diminishing risk of culling, positive effect on fertility, positive effect on bacteriological cure rate and recovery into milk production, unfortunately, it still seems that the role of NSAIDs is underestimated, even if in cases of chronically diseased cows it could help to decrease the volume of antibiotics used.

8.1.3 Enzymatic

In some European countries are authorised veterinary medicinal products (VMPs) indicated for non-infectious intramammary inflammation or as supportive treatment concomitantly with antimicrobials, which contain enzymes and vitamins, either in the pharmaceutical form of ointment or injection. Intramammary VMP contain proteolytic enzymes (chymotrypsin, trypsin and papain) as well as vitamins (retinol as palmitate and tocopherol as acetate) and except the above-mentioned indications can also be used for damaged skin and wound healing in animals (Summary of Product Characteristics Masti Veyxym®). Injection VMP contains proteolytic enzymes chymotrypsin, trypsin and papain and after intramuscular administration can also be supportive (in more serious infectious inflammation together with antimicrobial) to successful treatment (Summary of Product Characteristics Necro Veyxym®).

High fibrin production accompanies mammary gland inflammation with fibrin slow and delayed degradation. Excessive amount of fibrin is not only present at the site of inflammatory process but also in surrounding tissue, which is therefore insufficiently supplied with oxygen, and the inflammation process can spread to it. The cow's organism in such a way creates a non-specific barrier, which should be broken down and replaced by regenerated tissue after the inflammation has been managed. Also, some bacteria can be bound to fibrin and therefore hardly reachable to the effect of antimicrobials, which can lead to recurrence or the transition of the infection process to chronic form. It has been documented that in many etiological agents complete bacteriological cure after treatment with antimicrobials without support of enzymatic fibrin degradation can be achieved in 30% of cases only.

In Table 12 the mechanism of action of enzymes in inflammation process, tissue oedema as well as fibrinolytic effect and analgesic effect is commented.

It should be also mentioned further supportive action of enzymes, which can have either bacteriostatic or bactericidal effect (based on the nature of the pathogens). Enzymes can also help to release pathogens from udder parts and protect against epithelium adhesion as well as biofilm formation due to damage of bacterial adhesins. Fibrinolytic effect helps for more effective action of antimicrobials in cases of combined treatment.

A study made by Ziesch et al. (2017) showed small differences in cure rates ranging from 62.7% for the reference treatment with antimicrobial only to 63.5% for antimicrobial with combined enzyme/vitamin VMP and 61.5% for the non-antibiotic combined enzyme/vitamin VMP treatment. A total of 174 mastitis cases (different antimicrobials cefquinome (57), ampicillin/cloxacillin (25), cefoperazone (20), cephalexin/kanamycin (5), procaine benzylpenicillin (4), lincomycin/neomycin (3), cephalexin (1) and fixed combination VMP (7)) were used in 122 cases: sole antimicrobial (59) or antimicrobial with enzyme/vitamin VMP (63) vs. 52 treated by combined enzyme/vitamin VMP. Further parameter assessed within this study was the probability that treated mastitic cows will not suffer from recurrent clinical mastitis within 60 days after treatment. The investigation leads to similar results for both antimicrobial VMP (53.6%) and non-antibiotic combined enzyme/vitamin VMP (52.2%); in combined treatment antimicrobial with combined enzyme/vitamin VMP, 58.1% of non-recurrence was achieved by this study compared to Pinzón-Sánchez and Ruegg (2011), where 80% of non-recurrence in 60 days was achieved. This can be caused by the fact commented by Ziesch et al. (2017) that in the non-inferiority study, cows with higher SCC and history of clinical mastitis were enrolled to the trials. Despite that some results were not indisputably conclusive, it can be summarised that sole treatment with combined enzyme/vitamin VMP in non-severe CM cases may constitute an alternative therapy to reduce the use of antimicrobials.

Table 12 Mechanism of action of enzymes in inflammation process

Pharmacological action	Mechanism of action
Anti-inflammatory effect	• Degrading toxic products of inflammation
	• Eliminating inflammatory-promoting immunocomplexes
Anti-oedematous effect	• Increasing blood flow speed, e.g. by reducing the viscosity of blood
	• Improving microcirculation and restoring of homeostasis by removing fibrin and protein-containing exudate
Analgesic effect	• Relieving pain by compensating for the lack of oxygen
	• Reducing tissue tension due to more speedy blood circulation
	• Relieving pain due to degradation and flow out of the inflammatory process elements causing pain
Fibrinolysis	• Degrading fibrin filaments
	• Restoring the physiological drainage channels due to clearing the vascular pathway by removing of the microthrombi

8.1.4 Vaccines

Vaccination should be listed among the important tools for prevention of infectious diseases. Also, in the dairy cattle both scientists and pharmaceutical companies try to find vaccines as a tool that could be used to prevent specifically mastitis

Table 13 Overview of the target causative mastitis pathogens and principle of vaccines available in the USA/Europe

Vaccine: causative agent targeted	Characterisation	Protective effect
Escherichia coli; trade name: ENVIRACOR J-5	J5 bacterins, formulated with an Rc mutant strain of *Escherichia coli* O111:B4 that is lacking the "O" antigen capsule of the cell wall but with the core lipopolysaccharide (endotoxin) intact form	*N* = 530 cows vaccinated. Rate of coliform clinical mastitis 0.444 per 100 cow-days (20 *E.coli* cases, 3 *Klebsiella pneumoniae*); rate of total clinical mastitis 0.1138 per 100 cow-days; percentage of animals culled or died (4.4%), diagnosed pregnant after first service (33.9%) and experienced pregnancy loss (7.1%) (note the study by Bovilis J-5, Merck AH, Intervet 2020)
Escherichia coli; trade name: Bovilis J-5	J5 bacterin formulated with *Escherichia coli*	*N* = 506 cows vaccinated. Rate of coliform clinical mastitis 0.0183 per 100 cow-days (9 *E.coli* cases); rate of total clinical mastitis 0.0690 per 100 cow-days; percentage of animals culled or died (3.8%), diagnosed pregnant after first service (35.8%) and experienced pregnancy loss (5.8%) (note the study by Bovilis J-5, Merck AH, Intervet 2020)
Escherichia coli; trade name: J-VAC (2023)	Bacterin toxoid	*N* = total 733 dairy cows (203 DC,530 LC) vaccinated vs. placebo; various ages, parities, milk yield. Incidence total coliform clinical mastitis 3.27% (24 cases), from it J-VAC (4 cases, 1.07%) vs. placebo (20 cases, 5.57%). Within DC protocol (*n* = 203) vaccinated/placebo: 109/9, with 0/7.4% coliform clinical mastitis. Within LC (*n* = 530), vaccinated/placebo: 265/265, with 1.1%/4.9% coliform clinical mastitis (note the study by J-Vac, Boehringer Ingelheim)
Escherichia coli and other endotoxins creating as E. coli, Salmonella, Pasteurella multocida and Mannheimia haemolytica; trade name: Endovac Dairy/IMMVAC	The genetically engineered Gram-negative bacterin toxoid from mutant Re-17 *Salmonella typhimurium*, comprised of the universal core antigen common to all Gram-negative bacteria, containing also immunostimulant component, antimastitis and anti-septicaemia	Declared spectrum *E.coli* and *Salmonella* (mastitis) + *Pasteurella* and *Mannheimia* (respiratory), studies demonstrating efficacy against *E. coli* mastitis and endotoxemia associated with diseases caused by above-mentioned pathogens submitted in authorisation
Escherichia coli; other VMPs authorised due time against coliform mastitis: Pfizer Animal Health's Upjohn J-5 Bacterin®, Bayer's Mastiguard™, Biogenesis-Bago's Rotatec® J-5, Merial's Imocolibov® and J-Vac®, IMMVAC's Endovac-Dairy® and Novartis' Animal Health J-5 Shield™ (Note: exact names and marketing authorisation holders may change during the time course)		
Escherichia coli and *Staphylococcus aureus;* trade names: StartVac/ToPVac/Mastivac	Bacterins. Inactivated *Escherichia coli* (strain J5) in combination with inactivated *Staphylococcus aureus* (CP8, strain SP 140), with expression of the slime-associated antigenic complex (SAAC)	Moderate reduction of new IMI, decreased duration of IMI, transmissibility of *S. aureus*, coliforms and CNS (Schukken et al. 2014). No reduction in incidence and prevalence of clinical and subclinical mastitis but significant reduction of severity of clinical cases and milk losses (Bradley et al. 2015)
Staphylococcus aureus; trade names: Lysigin®/*Somato Staph*®	Bacterin. Lysed culture of highly antigenic polyvalent somatic antigen containing phage types I, II, III and IV and capsular serotypes 5, 8 and 336 of *S. aureus*	Reduced SCC, clinical mastitis (Nickerson et al. 1999), reduced chronic IMI. Field-based studies concluded no such effect (Middleton et al. 2009)

Streptococcus uberis; trade name: UBAC®	Subunit vaccine containing lipoteichoic acid from adhesive component of biofilm creating *Streptococcus uberis* strain 5616	Reduction in milk production losses and incidence of clinical mastitis cases; study by Collado et al. (2018); 25 gestating Holst-Fries heifers randomly assigned to G 1 ($n = 13$, IM vaccinated + G 2; $n = 12$, control IM phosphate-buffered saline), both immunised 60 and 21 days before the expected parturition (75 and 36 days before challenge on day 14 after calving). All challenged quarters developed clinical mastitis; significant reduction of the clinical signs, bacterial count, rectal temperature and daily milk yield losses after intramammary infection; significant increase of the *n* of quarters with no bacterial isolation and SCC <200,000 cells/mL at the study end (days 19, 20 and 21 after challenge). Field study on six farms (Puig et al. 2018; HIPRA): Incidence of animals with *S. uberis* CM significantly lower in UBAC® ($n = 401$) than in the placebo ($n = 380$), vaccinated group incidence reduced by 52.5% + reduction of the number of antibiotic treatments (56% less treatments in the UBAC® than in the placebo group)
Streptococcus uberis; trade name: DUNPREBAC	*Streptococcus uberis* bacterin	USDA APHIS safety data only: https://www.aphis.usda.gov/aphis/ourfocus/animalhealth/veterinary-biologics/product-summaries/vet-label-data/9c844005-ccfc-467c-95f8-856d00d03b63
Klebsiella pneumoniae; trade name: KLEB Vax SRP	Vaccine using siderophore receptor and porin technology made from *Klebsiella pneumoniae* cell wall extract containing antigens	569 dairy cows and heifers (Holstein, Jersey), randomised into 2 groups, vaccinated/placebo, monitored in 90 days: vaccinated (total 288/after vaccination: 221 negative/4 positive) vs. placebo (total 281/placebo: 211 negative/14 positive)

erinary medicinal product (EC) 2019/6 (in force since January 2022); autogenous vaccines are introduced in the European regulation, it means for the first time in the EU legal provisons. In this regulation recital #70 and Articles 2(3), 4(44), 94, 105, 106(5), 108, 117, 120, 123, 134 and 159 shall apply to autogenous vaccines. Autogenous vaccines are generally accepted as a useful tool in the control of infectious diseases in animals by filling prophylactic gaps left by licensed vaccines.

Such vaccines include, but are not limited to, vaccines against *Mycoplasma* spp., *Klebsiella* spp. and *Streptococcus uberis*. Autogenous vaccines allow for farm/epidemiologically related unit tailored approach, where safety is the primary factor for use and efficacy is expected based on the fact that bacterins/antigens are made from pathogens coming from the exact farm.

8.1.5 Colony-Stimulating Factors

Pegylated bovine granulocyte colony-stimulating factor (PEG bG-CSF), called also pegbovigrastim, is a modified form of the naturally occurring immunoregulatory cytokine. Bovine granulocyte colony-stimulating factor occurred naturally as a protein produced by mononuclear leukocytes, endothelial cells and fibroblasts. The production and functional activities of immune cells are regulated by PEG bG-CSF. The immunoregulatory activities concern notably cells of the neutrophilic granulocyte lineage which bear cell surface receptors for the protein and lead to increase of the number of circulating neutrophils. Microbicidal capabilities of neutrophils, mediated by myeloperoxidase hydrogen peroxide halide, are enhanced by the action of PEG bG-CSF, Direct or indirect functions on other cells/receptors and cytokine pathways are expected as further functions (SPC Imrestore). Treatment with PEG reduced the occurrence of a first case of clinical mastitis during the first 30 days in milk by 24.6% with particular effect in cows at risk of elevated lipid mobilisation. Hazard of a first case and the rate of total cases of clinical mastitis during the full lactation were also reduced. Prepartum body condition score together with administration PEG act in synergy in reduction of the occurrence of a first case of clinical mastitis during the first 30 days in milk by 49.5%. All PEG treatment effects were independent of parity (Barca et al. 2021).

8.2 New Alternatives Studied

8.2.1 Antimicrobial Peptides

Antimicrobial peptides (AMPs) are molecules playing a significant role in microbial ecology via contributing to the innate immunity of organisms by destroying invading microorganisms (Moravej et al. 2018; Chung and Khanum 2017). Among their properties are production of chemokines, neutralisation of toxins, synergy with conventional antibiotics, stimulation of angiogenesis, acceleration of wound healing processes and influence of apoptosis in multicellular organisms (Chung and Khanum 2017; Moravej et al. 2018). Due to their promising activity against many strains of Gram-positive and Gram-negative bacteria, including drug-resistant strains, fungi and microbes formatting biofilms, they are subject of research also as mastitis treatment options. Unfortunately, yet the therapeutic use of AMPs is very much limited due to their high production cost, enzymatic degradation, short half-life and the fact that bacteria can also develop resistance to them as well as cytotoxic effects on the eukaryotic cells (Moravej et al. 2018). Although some of AMPs seem promising, of importance are mechanisms of resistance existing in target pathogens as activation of efflux pumps, trapping proteins, modification of cell surface charge and altering the membrane fluidity as well as protease production and even more sophisticated systems of selective gene expression in biofilms (Moravej et al. 2018). One potential strategy to surmount this resistance problem could involve derivatisation of bacterin, which has been tested with nisin to counteract such systems in recognising it as a substrate (Medeiros-Silva et al. 2018). Despite the fact that these substances are ranked as "alternatives", there exists a question if these substances can be considered as alternatives to antimicrobials being in fact also naturally produced antimicrobials (but not "conventional" one with well-established use in VMPs, some of them of natural origin and some semi-synthesised or chemically synthesised).

8.2.2 Bacteriocins

Bacteriocins as nisin, bovicin, lysostaphin and other related compounds have been already proven as promising in mastitis treatment. Among bacteriocins belong classes of lantibiotics—bacterially derived ribosomally synthesised antimicrobial peptides containing post-translationally modified amino acids including lanthionine and/or methyllanthionine (Field et al. 2021). Nisin is a bacteriocin, a prototypical lantibiotic produced by *Lactococcus lactis* ssp. *lactis*, with bactericidal activity against strains of *Staphylococcus* spp. and *Streptococcus* spp. isolated from bovine mastitis resistant to conventional antibiotics (Castelani et al. 2019). Nisin A M17Q, nisin A T2L and nisin A HTK were identified as three promising derivatives that can be used in mastitis treatment (Field et al. 2021). A nisin-based treatment was effective in curing mastitis caused by several different pathogens in a field trial consisting of 139 cows with subclinical mastitis (Cao et al. 2007). For dosing schedule with 2 500,000 IU for 3 days as intramammary infusion for subclinical mastitis cases, bacteriological cure rate was demonstrated in *S. agalactiae* (90.1%) and less for staphylococci: coagulase negative (58.8%) and *S. aureus* (50%) (Wu et al. 2007). Nisin A has been used as an anti-mastitis product in the form of nisin wipes to clean and disinfect the teat (pre- and post-milking) and also as an intramammary infusion product (Immucell Corp., Portland, ME, USA). The veterinary medicinal product indicated for the treatment for subclinical mastitis with zero withdrawal period has been indicated as involved in the FDA approval process (Roy et al. 2016). Nisins A M17Q, A HTK and in particular nisin A T2L warrant further consideration as novel anti-mastitis antimicrobials (Field et al. 2021). Among the other substances from bacteriocins group, also bovicin HC5 was reported to have well-established therapeutic effect against microorganisms responsible for mastitis (Castelani et al. 2019; Godoy-Santos et al. 2019). Transgenic technologies have been proposed for the engineering of cows resistant to mastitis caused by *S. aureus*. Such cows express lysostaphin in their mammary epithelium and secrete this antimicrobial peptide into milk (Donovan et al. 2005).

8.2.3 Cathelicidins

Cathelicidins belong among antimicrobial peptides, and they are produced by neutrophils. The infiltrating neutrophils present in mastitis milk release cathelicidin by degranulation in high concentration. This is typical for epithelial cells of the mammary gland only once intramammary infection is present; therefore cathelicidin can be a mastitis marker (Cubeddu et al. 2017). Direct infection control and regulation of host defences make them potent broad-spectrum antimicrobials acting against bacteria, fungi and enveloped viruses (Langer et al. 2017; Young-Speirs et al. 2018). There have been already tested two bovine cathelicidins on strains isolated from diseased dairy cows suffering from clinical and subclinical mastitis—one of them, BMAP-27, showed lower MICs for *E. coli*, and the other one, BMAP-28, showed lower MICs for *S. aureus* in in vitro testing system (Langer et al. 2017). Further evaluation and characterisation are needed to gain more results useful for further utilising these substances in clinical practice.

8.2.4 Bacteriophages and Bacteriophage Endolysins

A study made by Song et al. (2021) has proven that bacteriophage vB_SauM_SDQ (SDQ) has the ability to remove biofilm from polystyrene plastic plates, glass coverslips, milk and mammary tissues. Authors used the in vitro experiment mimicking the conditions of mammary ducts and mammary alveoli utilising the methodology by Milho et al. (2019) for evaluation of the effect of SDQ bacteriophage on the *S. aureus* biofilm. The number of bacteria in the biofilm decreased significantly, and the biofilm structure was destroyed after SDQ treatment, having also an effect on the elimination of *S. aureus* biofilm from milk. The findings of Song et al. (2021) show that SDQ can be used in prevention and treatment of bovine mastitis. Moreover this phage expresses polysaccharide depolymerase, an enzyme that may enhance the antibiofilm activity of phages by lytic effect on bacteria.

Endolysins derived from the bacteriophages were experimentally proven to be effective against Gram-positive pathogens. They are the proteins (peptidoglycan hydrolases) degrading the peptidoglycan layer of bacterial cell wall (Breyne et al. 2017). Novel bacteriophage-derived peptidase disrupting staphylococcal biofilm was described by Fenton et al. (2013). Reduction of the pathogen and return to standard milk appearance were reported by Fan et al. (2016) when lysin trx-SA1 was introduced by intramammary infusion for treatment of mastitis caused by S. *aureus*. The future seems to need engineered lysins, which have the ability to be transported across the eukaryotic cell membrane and internalised into mammalian cells, where some pathogens (e.g. *S. aureus*) are located.

8.2.5 Microbial Extracts

Some works occur also with respect to possible antimicrobial activity of microbial extracts, where actinomycetes-derived crude extracts (Caat1-54 (active compound lysophilin I) and Caat P5–8 (active compound kalafungin)) were proven in in vitro tests to inhibit bacterial growth obtained from clinical and subclinical mastitis. It was demonstrated that these extracts presented an in vitro antimicrobial activity against isolates of *S. aureus*, *S. dysgalactiae*, *S. chromogenes* and *S. uberis* (Leite et al. 2018).

8.2.6 Immunostimulants

8.2.6.1 Specific Antibodies

Streptococcus uberis belongs among the pathogens seriously affecting mammary gland. The work of Almeida et al. (2015) carried out an experiment where mammary quarters of healthy cows were infused with *S. uberis* UT888 opsonised with affinity-purified anti-rSUAM antibodies. After the administration with these antibodies, clinical symptoms of mastitis were mild to undetectable in treated dairy cows, which moreover showed lower milk bacterial counts and less infected mammary quarters compared to control animals. The study indicates protective effect against intramammary infections caused by *Streptococcus uberis*, with tentative explanation of the mechanism either by blocking adherence and internalisation of pathogen into host cells and/or in combination with removal of *Streptococcus uberis* via phagocytosis by neutrophils and macrophages.

Microbeads carrying specific antibodies to mastitis-causing bacteria and an enhancer of phagocytosis are among other new immunostimulating technologies, called Y-complex. Leitner et al. (2013) treated 32 mastitic cows naturally infected with *E. coli*, *S. dysgalactiae* or coagulase-negative staphylococci with Y-complex and compared the results to treatments of 8 cows with fixed combinations of sulfadiazine/trimethoprim or procaine penicillin/streptomycin and with a non-steroidal anti-inflammatory drug administration in 19 cows. The Y-complex was as effective as antibiotics and superior to NSAID in eliminating the bacteria. Interestingly use of NSAIDs like meloxicam in addition to the antimicrobials improved subsequent reproduction of the cows (McDougall et al. 2016). Less discarded milk was evidenced after the treatment with Y-complex in relation to the treatment group with administration of antimicrobials. Within bovine mastitis challenge with *E. coli*, a vigorous immune response was induced resulting into a systematic immunomodulation in in vivo models, whereas *S. aureus* induces a weak response (Petzl et al. 2018).

8.2.6.2 Interleukin

Interleukin-17A (IL-17A) was proven to mediate crucial interlink among the immune system and various epithelial tissues, initiating a series of defensive mechanisms against bacterial and fungal infections of mammary gland. An elevated IL-17 level is key during mastitis, as the anti-IL-17A antibody blockade protects against LPS-induced mammary gland inflammation (Hu et al. 2019).

A single injection of interleukin-2 (IL-2) into the skin region drained by the supramammary lymph node (3–5 days after calving) induced significant increase of several milk markers related to white blood cell and epithelial cell functions including SCC, serum amyloid A (SAA), lactoferrin and NAGase (Zecconi et al. 2009). As milk markers concentrations increased, the activity of IL-2 on epithelial cells can be expected, which results in a higher resistance to invading pathogens. Concomitant administration of IL-2 with regular dry cow antimicrobial targeting *S. aureus* enhanced the effectiveness of the treatment. Prophylactic administration of IL-2 was shown to protect the mammary gland from subsequent intramammary challenge with *S. aureus*, while therapeutic administration into *S. aureus* infected quarters was less effective (Daley et al. 1992). Despite beneficial effect of IL-2 especially on mastitis prevention, abortion was observed in 7.5% of the cows that received IL-2 (Resende et al. 2009).

8.2.6.3 Immunogenic Yeast Extracts

Fermentation products of the strains of *Saccharomyces cerevisiae* (SCFP) were reported to increase leukocyte function both in vitro and in vivo (Magalhães et al. 2008; Jensen et al. 2008) as well as contribute to further beneficial health and performance effects in dairy cows due to being involved in the fight against immunological and physiological challenges (Alugongo et al. 2017, Vailati-Riboni et al. 2021). Different *Saccharomyces cerevisiae* fermentation products and routes of administration were used in several studies (Ferguson et al. 2018; Vailati-Riboni et al. 2021; Sharun et al. 2021).

Infusion of extract of *Saccharomyces cerevisiae* yeast into the mammary gland during the dry period caused an increase in the activity of immune cells in the gland, which could respond immediately to a new infection. It was suggested that this treatment could minimise the risk of a new IMI during the periparturient period (Resende et al. 2009).

Within the studies published (Ferguson et al. 2018; Vailati-Riboni et al. 2021), SCFP is characterised as a dried product with active components as vitamins, antioxidants and β-glucans but also polyphenols and further not specified bioactive compounds including fermentation end products and other components of the yeast cell. Exact components, their ratio and composition of the final products are considered proprietary and not publicly available (Vailati-Riboni et al. 2021). Within a very recent study performed by Vailati-Riboni et al. (2021), the efficacy of SCFP from the commercially available products as NutriTek® (Diamond V, Cedar Rapids, IA, USA) was investigated in case of subclinical mastitis, modelled via an intramammarily infection challenge involving *Streptococcus uberis*. One part of the study was targeted on addressing the molecular mechanism of action (at the tissue level) via transcriptomic and bioinformatic analysis. The results of the study discussed that the SCFP complex can modulate the immune response of animals by priming the innate and adaptive immune response through activation of immune cells. A greater pathogen killing capacity as well as cytoprotective activity in the udder is suggested as one possible mechanism of maintaining the integrity of the mammary gland epithelial tissue, preventing damage through attack of pathogen. Moreover, the higher calcium concentration in peripheral blood suggested a lower presence of pyrogen in cows supplemented by SCFP. There was also recorded reduced rectal temperature, somatic cell score and temperature of the infected quarter during the challenge. It should not be overlooked also the heat shock protein response mechanism in the liver, which might explain the broad ameliorating effect of these yeast fermentation products' specific supplementation against other environmental, immunological and physiological stresses reported (Vailati-Riboni et al. 2021).

8.2.6.4 Egg Yolk Immunoglobulins

Specific egg yolk immunoglobulin (IgY) enhancement of phagocytosis and inhibition of *S. aureus* as a causal agent of mastitis were determined by several in vitro studies (Zhen et al. 2008; Wang et al. 2011). Specific IgY acting against *E. coli* (*Ec*) and *S. aureus* (*Sa*) was produced by immunising hens with formaldehyde inactivated *Ec* and *Sa* bacteria in a long-standing immunisation response. Promising results from in vitro studies were followed by in vivo trials (Zhen et al. 2009), where small groups of 18 cows were treated with IgY treatment (83.3% and 50% in experimental and clinical *S. aureus* mastitis, respectively) indicating egg yolk IgY use as a therapeutic agent in veterinary medicine, including mastitis treatment (Grzywa et al. 2023). If also economically viable, this approach could come into more broad practice.

8.2.7 Probiotics

Probiotics are defined as live microorganisms, which, when administered in adequate amounts, confer health benefits on the host. Mostly used probiotics contain lactic acid bacteria. Products declaring mastitis protection and potential immunomodulatory activity can be in the form of feed supplements, teat dips, teat sprays as well as intramammary pharmaceutical forms (Pellegrino et al. 2017; Yu et al. 2017; Rainard and Foucras 2018; Sharun et al. 2021). Speaking about action at the site of the future infection of the mammary gland, one of the mechanisms of preventive action by lactic bacteria can be via forming a protective biofilm, which inhibits the growth of mastitis causing pathogens (Rainard and Foucras 2018; Wallis et al. 2018). The extent of biofilm formation strongly depended upon the strain of bacteria as well as on conditions in the mammary gland. *Lactobacillus brevis* 1595, *L. brevis* 1597 and *L. plantarum* 1610 were proven to have high colonisation capacities of bovine mammary epithelial cells and therefore can compete with mastitis pathogens as for the binding sites in the mammary gland and prevent pathogen adhesion and invasion (Bouchard et al. 2015). Not only competition but also inhibitory activity against bovine mastitis pathogens by means of coaggregation and at the same time adhering to the bovine teat canal epithelial cells was confirmed for two strains of lactic acid bacteria isolated from bovine milk: *L. lactis* subsp. *lactis* CRL 1655 and *L. perolens* CRL 1724 (Pellegrino et al. 2017). The last two mentioned strains also increased the amount of immunoglobulin (IgG isotypes) in the blood and milk when administered at drying off (Pellegrino et al. 2017). Another strain of *Lactobacillus casei* (BL3) inhibits the internalisation of *S. aureus* to mammary gland epithelial cells and modulates the innate immune response of bovine mammary epithelial cells. *Lactobacillus casei* BL3 caused also a decrease of the expression of several pro-inflammatory cytokines (IL-6, IL-8, IL-1a, IL-1b) and tumour necrosis factor-alpha (TNF-a) in *S. aureus*. Interestingly this strain does not interfere with defensins induction and does not alter the expression of the pattern recognition receptor nucleotide-binding oligomerisation domain proteins (Souza et al. 2018). Further information come from the study by Rainard and Foucras (2018) summarising that intramammary infusion of lactic acid bacteria exhibited pro-inflammatory activity that induced an influx of neutrophils into the milk both during lactation and at drying off; this is further supported by Nagahata et al. (2020) who performed a study with *Bifidobacterium breve* intramammary infusion with a similar conclusion.

Another form of administration—teat dip—was found to be superior to the commercial disinfectant when containing novel lactobacilli mixture in reducing somatic cell count (Yu et al. 2017). Another from teat spray with lactobacilli was tested in a pilot study that in conclusion suggests that lactobacillus-based product treatment could improve teat-end sphincter functions and udder health (Alawneh et al. 2020). The ability of certain probiotics to alter the teat apex microbiota helps to prevent the colonisation of teat canal by mastitis-causing pathogens (Rainard and Foucras 2018). Probiotics along with or without repeated milk-out are not recommended for efficient management of clinical mastitis in dairy lactating cows (Francoz et al. 2017).

Due to the fact that cows with damaged/imbalanced gut microbiome, not having appropriate metabolic activity, are prone to suffer from several diseases, including mastitis, probiotics can play a beneficial role also in gut microbiome improvement and finally in non-specific health-promoting effects (Rainard and Foucras 2018). Short-chain fatty acids and rumen-derived lipopolysaccharides (LPS) are among two key parameters. Increased production of rumen-derived LPS leads to the blood circulation entry, and once reaching the mammary gland, it increases the blood-barrier permeability causing finally inflammation of the mammary gland. On the other hand, the fermentation product of rumen microbiota rich for short-chain fatty acids is proven to have a protective effect on the mammary gland (Hu et al. 2019). The incorporation of lactic acid bacteria in the animal feed can be considered as an effective tool for the prevention of bovine mastitis as well (Pellegrino et al. 2017).

8.2.8 Herbs and Substances Extracted from Plants

Herbs as well as their components and finally mostly extracts, essential oils or some other form containing active substances presented in herbs can be used as an antibacterial, anti-inflammatory, anti-pyretic, antioxidant and immunomodulatory agent for the treatment of mastitis (Mushtaq et al. 2018; Muluye et al. 2014). Searching for "conventional antimicrobials" alternatives, bibliographical references can be found on the anti-inflammatory and antibacterial effects of some herbs, e.g. Chinese herbs being used effectively in the treatment of bovine mastitis (Muluye et al. 2014; Yang et al. 2019). A variety of phytochemicals was identified from in vitro-tested plants such as saponins, glycosides, alkaloids, carbohydrates flavonoids and terpenoids (Tariq et al. 2017). Some work pay attention also to essential oils (Nehme et al. 2021). An article by Alves-Barroco and colleagues in 2019 showed that certain strains of *S. dysgalactiae* subsp. *dysgalactiae* have the ability to form biofilms on hydrophilic surfaces and that the protein fisetin (flavonol compound) induced a decrease in biofilm formation. Neurotrophic, antioxidant, anti-inflammatory and anti-angiogenic effects were proven for fisetin (Shukla et al. 2019). Although in vitro data exists, in vivo experiments, especially targeted on (multidrug) resistant bacteria causing mastitis, are not available. Therefore there is a need to identify active phytocompounds and carry out in vivo studies to confirm the efficacy of traditionally used medicinal plants. In addition, the toxicology and mechanism of action of these plants extracts against mastitis pathogens need to be studied in detail (Amber et al. 2018).

Herbal compounds include also essential oils (EO), for which factors influencing their success in mastitis therapy are still discussed. The controversial results of EO treatments described in various studies have been linked to the differences between in vitro experiments to in vivo trials. Tentative explanations could be that milk can interact with the EO, altering antibacterial properties of the EO and influencing the final concentration of the EO so that the concentration achieved in in vivo use conditions is not sufficient to inhibit bacteria. Also, the excipients used in the EO product could modify the physicochemical characteristics and thus influence the final efficacy in clinical practice (Corre 2021).

The current bibliography enumerates a plethora of plant species—e.g. *Achillea millefolium*, *Allium sativum*, *Alternanthera brasiliana*, *Atractylodis macrocephalae*, *Baccharis trimera*, *Chenopodium ambrosioides*, *Cuphea carthagenensis*, *Foeniculum vulgare*, *Lavandula angustifolia*, *Laurus nobilis*, *Ocimum basilicum*, *Origanum vulgare*, *Oxytropis glabra*, *Parapiptadenia rigida*, *Phytolacca dioica*, *Rosmarinus verbenone*, *Sambucus nigra*, *Sida rhombifolia*, *Solanum mauritianum*, *Solidago chilensis* and *Thymus vulgaris.* Some more examples of their use as well as effects demonstrated in different studies are given below (Table 14). The topical route (Hase et al. 2013), oral administration (Dash et al. 2016) and intramammary routes of administration (Yang et al. 2019) are the commonly used methods.

Except antimicrobial activity, some herbal extracts can pose also analgesic, anti-pyretic, antioxidative as well as anti-inflammatory effects. The analgesic activity of some herbal extract was found to be comparable to that of ibuprofen and indomethacin as mentioned by Ranjith et al. (2018), who also reported that the methanolic extracts of herbal preparation containing leaves from *Diploclisia glaucescens* and rhizomes of *Curcuma longa* (equal proportions) produced analgesic activity along with anti-inflammatory activity. Moringa extract ameliorating inflammatory mediators and increasing antioxidant systems in bovine udder epithelial cells has been investigated for the exact mechanism of action, which was described as an inhibition of pro-inflammatory cytokine (TNF-a, IL-1b and IL-6) expression, inhibition of cyclooxygenase-2 expression and downregulation of NF-jb, upregulated heme-oxygenase-1 NAD(P)H and quinone oxidoreductase-1. Moreover, increased expres-

Table 14 Plant species: examples of their use and effects demonstrated

Herbs	Comment	Reference
Oxytropis glabra	Western China herbal formulation form the family Fabaceae	Ren et al. (2020)
	Tests in vitro, inhibition of biofilm formation of *S. epidermidis*	
Thymus vulgaris *Lavandula angustifolia*	Intramammary infusion and external administration of 10% mixture, massaging the udder after administration, substantial decrease in colony count	Abboud et al. (2015)
Origanum vulgare	Intramammary infusion of essential oils, administered twice a day for 3 days *S. aureus* and *E. coli* were not detected in milk after the treatment	Cho et al. (2015)
Rosmarinus verbenone *Laurus nobilis* *Thymus satureoides*	Mixture of essential oils in sunflower oil at a concentration of 1.5% of volume of each herb or mixture of *Thymus* and *Rosmarinus* essential oils in oil at a concentration of 6% of each Recovery rate of mastitis approx. 40%	Kammerer et al. (2009)
Terminalia chebula	Ethyl acetate extract effect on isolates of *S. aureus*, *E. coli*, *Pseudomonas aeruginosa* and *Bacillus megaterium* was tested within in vitro experiment, and it was determined that 500 mg/mL concentration of the extract had the same antibacterial efficacy as that of standard amoxicillin	Kher et al. (2019)
Allium sativum *Bunium persicum* *Oryza sativa* *Triticum aestivum*	In tests of in vitro antibacterial activity at a concentration of 50 mg/mL against *S. aureus*, *E. coli* and *K. pneumoniae*, all extracts significantly inhibit the activity of bacterial strains examined	Amber et al. (2018)

sion of casein proteins has been detected (Cheng et al. 2019).

8.2.9 Homeopathics

Controversy exists as for the effectiveness of homeopathy in mastitis treatment, as noted also by Doehring and Sundrum (2014) within a review revealing that various clinical studies testing the efficacy or effectiveness of homeopathy returned heterogeneous results. As in other studies, methodology, design and assessment should be thoroughly followed once considering the results.

The European Regulations (EC 2008) on organic agriculture, except many other tools for sustainable, organic husbandries, promote also the use of homeopathy: homeopathic products shall be used in preference to chemically synthesised veterinary products. In order to ensure that only effective remedies are administered to diseased animals, medicinal products need proven therapeutic efficacy. Randomised controlled trials are widely accepted as the gold standard for clinical research on the efficacy or effectiveness of medicinal products, and two recent studies used it for homeopathy effectiveness comparison with placebo (Ebert et al. 2017) and a second one additionally also with conventional (antibiotic) treatment of mastitis (Keller and Sundrum 2018). A study (Keller and Sundrum 2018) was performed with 180 lactating dairy cows, where culture-positive cases (*S. uberis* (n = 45), *E. coli* (n = 16), *S. dysgalactiae* (n = 13), *Klebsiella* spp. (n = 9), other aesculin-positive streptococci (n = 8), non-aureus staphylococci (n = 7), *S. aureus* (n = 7), coliforms (n = 5), *Enterococcus* spp. (n = 5), *Corynebacterium bovis* (n = 2), yeasts (n = 2) and *Serratia* spp. (n = 1)) were detected and rest of samples were culture negative. Three groups: the antibiotic treatment vs. homeopathy (individualised, digital repertory of homeopathic products, experienced homeopathy veterinarian) vs. placebo. Suboptimal treatment, which was evaluated ($P \leq 0.05$) in antibiotic (60–81%), was better than for individualised homeopathy (33–43%) and placebos (45–47%). In cytology evaluation, ineffectiveness was for all three approaches ($P \leq 0.05$; $P = 0.13$): antibiotic (2–21%), individualised homeopathy (0–8%)

and placebo (3–13%). The study results implied that the effectiveness of individualised homeopathy does not go beyond a placebo effect and successful treatment is highly dependent on the specific mastitis pathogen. Also, the frequency of non-responsive animals was generally higher in the homeopathy and placebo groups than in the antibiotic group.

In another randomised clinical trial (Ebert et al. 2017), a triple-blind, randomised controlled trial was performed, and dairy cows with acute mastitis were randomly allocated to homeopathy (n = 70) or placebo (n = 92), for a total of 162 animals. The homeopathic treatment was selected based on clinical symptoms but most commonly consisted of a combination of nosodes with *Streptococcinum*, *Staphylococcinum*, *Pyrogenium* and *Escherichia coli* at a potency of 200c/once per day/5 days. The same regimen was for placebo group with lactose globules without active ingredients. If necessary, allopathic drugs were used (e.g. antibiotics, udder creams and anti-inflammatory drugs) in both groups due to welfare reasons. Follow-up period was for up to 200 days after clinical recovery. Base-level data did not differ between the homeopathy and placebo groups. Mastitis lasted for an average of 6 days in both groups, and no significant differences were detected in time to recovery, somatic cell count, risk of clinical cure within 14 days after disease occurrence, mastitis recurrence risk or culling risk.

Despite scarce strong evidence, homeopathy is still considered as the tool possible at least for management (Hektoen 2005). The advantages or disadvantages of homeopathy should be carefully assessed for individual animals and in individual farm conditions. If use of antimicrobials decreases and animal welfare and health are not compromised, this tool can be still considered as useful.

8.2.10 Stem Cells

Bovine mammary stem cell therapy can be used for regenerating mammary tissues by means of either repairing or replacing the damaged tissue. The stem cells have the ability to differentiate into epithelial, myoepithelial and/or cuboidal/columnar cells of the udder tissue. Allogenic mesenchymal stem cells were recently reported to have, except regeneration function, also an antibacterial effect (reducing count of cells) on experimentally induced *Staphylococcus aureus* mastitis (Peralta et al. 2020). Since the mammary stem cells are responsible for the growth, renewal and turnover of mammary epithelial cells, they can be utilised for tissue repair and improving milk yield (Capuco et al. 2012). Despite promising results, the driving factor of the use in practice, except medicinal benefit and returning to high milk yields via udder reparation, will be also the price of such medicinal products.

8.2.11 Nanoparticles

Nanoparticle-based products in different (pharmaceutical) forms as liposomes, nanogels, polymeric nanoparticles, inorganic nanoparticles and solid lipid nanoparticles seems will gain more and more popularity as tools for managing *S. aureus* (including possible targeting of MRSA) mastitis and fighting against biofilm formation (Algharib et al. 2020). Different types of nanoparticles have already been evaluated for the treatment of mastitis with positive results as, e.g. silver and copper nanoparticles tested with promising results on human and bovine mammary epithelial cells model (Kalińska et al. 2019), chitosan nanoparticles in vitro testing demonstratinge bacterial cells damage (*S. aureus* as well as coagulase-negative *Staphylococci*) and inhibition of biofilm formation, without affecting the viability of bovine mammary epithelial cell culture (Orellano et al. 2019), propolis nanoparticles in vitro studies against *S. aureus*, where cytotoxicity on mammary epithelial cell culture was desribed (Pinheiro Machado et al. 2019) and cationic nisin-lipid nanoparticles (Castelani et al. 2019). Also, conventional antibiotics' (e.g. tilmicosin and amoxicillin) effect can be enhanced by nanoformulation, and many of antimicrobials were tested with liposomal nanoparticle complexes (Yang et al. 2009; Zhu et al. 2018, Algharib et al. 2020). The solid lipid nanoparticles of tilmicosin were reported to exhibit prolonged and enhanced antibacterial effect against *S. aureus* and *S. agalactiae* in injection-dosed experimental product in a murine model (Zhu et al. 2018). Amoxicillin nanoparticles enhanced the postantibiotic effect when used against pathogenic bacteria causing bovine mastitis (Yang et al. 2009). The liposomes decrease the drug toxicity, increase the pharmacological action of the drug by changing its pharmacokinetics and biodistribution and are safe for parenteral administration but have limited stability and demanding manufacture and high costs (Algharib et al. 2020). Also, metals (silver, gold, copper) and metal-containing compounds (sonochemically synthesised capped zinc oxide nanoparticles) showed antibacterial activity in the intramammary treatment of clinical bovine mastitis caused by different pathogens. Most studies were performed with *S. aureus*, but zinc oxide capped nanoparticles were effective also against *S. aureus*, *E. coli and K. pneumoniae* (Hozyen et al. 2019). As for dairy cows that produce milk for human consumption, even though all of the study results seem promising, it should be as well investigated and proved safety for consumer, and in studies targeted on the fate of nanoparticles in milk and consequences linked with consumption of milk with nanoparticles, new designs of studies and models will therefore be needed (Ahmad 2022).

9 Conclusion

Diseases of cattle from those at the earliest stages of age in both beef and dairy cattle to those in veal calves and finally dairy cows with the most frequent disease (mastitis) all not only have harmful effect from the perspective of health and welfare but also have a significant impact on beef meat production and milk yield, which have an impact on the profitability of the whole cattle sector. Farmers and veterinarians are therefore targeted to find the most effective, easy and cheap prevention tools, demanding minimum specific knowledge and equipment as well as being not demanding for human labour to keep their herds healthy. Many years ago, antimicrobials were used as a preventive tool for a row of diseases, but nowadays this approach is no more considered sustainable due to antimicrobial resistance issues. Moreover, new pieces of veterinary legislation approved in 2019 for the whole European Union in fact ban the prophylactic use of antibiotics at herd and groups level and call for individualisation of any use of VMP containing such antibiotics. Producers have therefore started to more intensively seek effective methods to minimise the occurrence of diseases in their herds. This chapter attempts to summarise considerations on advances in detection of diseases, with basic summary of signs and brief notes with regard to the management and disease prevention aiming to minimise the need for use of antimicrobials. If, despite all effort, treatment with antimicrobials is finally necessary, it should be based on evidence, rationalised as much as possible, and with using of narrow-spectrum first-line antimicrobials and minimising unnecessary usage, whenever possible. Due the fact, that portfolio of antimicrobials available for veterinary use will not be, with high probability, broadened, last part of this chapter is dedicated to "alternatives" to antimicrobials. From data published so far, it seems that more intensive research is needed for understanding of infections and the role of microbiota in diseases and for discovering of possibilities of alternative tools or "alternatives" to antimicrobials via developing innovative medicinal products using either novel technologies or new active substances, some of them natural (herbal, bacterial, animal) originating or repurposing of already known substances or components. However, even with these alternatives, as with conventional drugs and antimicrobials, we need to think not only about efficacy but alsko about safety. Safety for the animals but for consumer as well. Although new technologies used in animal husbandry and monitoring of wellbeing or disease signals could help to keep animals healthy, in a good condition and at high production status, at the same time attention should be paid to considerations reflecting experiences gained by generations of farmers, natural animal habits, biological limits of production and perception that animals help us not only to gain food but also to keep the balance with nature.

References

AABP (2022) American Association of Bovine Practitioners: Key elements for implementing antimicrobial stewardship plans in bovine veterinary practices working with beef and dairy operations. Revision March 2022. Available online at: https://aabp.org/resources/AABP_Guidelines/

AntimicrobialStewardship0322Final.pdf. Accessed 4 July 2023

Abbeloos E, Pyörälä S, Rajala-Schultz P, Myllys V (2018) Determination of the intramammary dose of benzylpenicillin required to maintain an adequate concentration in the milk to inhibit Gram-positive bacteria in the clinically normal udder for 24 hr. J Vet Pharmacol Ther 41(5):691–698

Abboud M, El Rammouz R, Jammal B, Sleiman M (2015) In vitro and in vivo antimicrobial activity of two essential oils *Thymus vulgaris* and *Lavandula angustifolia* against bovine Staphylococcus and Streptococcus mastitis pathogen. Middle East J Agric Res 4:975–983

Abell KM, Theurer ME, Larson RL, White BJ, Apley MA (2017) Mixed treatment comparison meta-analysis of metaphylaxis treatments for bovine respiratory disease in beef cattle. J Anim Sci 95(2):626–635

Aghamohammadi M, Haine D, Kelton DF, Barkema HW, Hogeveen H, Keefe GP, Dufour S (2018) Herd-level mastitis-associated costs on Canadian dairy farms. Front Vet Sci 5:100

AHDB (2019) Agriculture and Horticulture Development Board: 3 Qs of feeding colostrum. https://ahdb.org.uk/knowledge-library/3-qs-of-feeding-colostrum. Accessed 4 July 2023

Ahmad A (2022) Safety and toxicity implications of multifunctional drug delivery nanocarriers on reproductive systems *in vitro* and *in vivo*. Front Toxicol 4:895667

Alawneh JI, James AS, Phillips N, Fraser B, Jury K, Soust M, Olchowy TWJ (2020) Efficacy of a lactobacillus-based teat spray on udder health in lactating dairy cows. Front Vet Sci 7:584436

Alfonseca-Silva E, Cruz-Villa JC, Gutierrez L, Sumano H (2021) Successful treatment of recurrent subclinical mastitis in cows caused by enrofloxacin resistant bacteria by means of the sequential intramammary infusion of enrofloxacin HCl-2H2O and ceftiofur HCl: a clinical trial. J Vet Sci 22:e78

Algharib SA, Dawood A, Xie S (2020) Nanoparticles for treatment of bovine *Staphylococcus aureus* mastitis. Drug Deliv 27(1):292–308

Alim MA, Sun D, Zhang Y, Zhang Y, Zhang Q, Liu L (2015) DNA polymorphisms in the lactoglobulin and K-casein genes associated with milk production traits on dairy cattle. Biores Commun 1:82–86

Almeida RA, Dego OK, Headrick SI, Lewis MJ, Oliver SP (2015) Role of Streptococcus uberis adhesion molecule in the pathogenesis of *Streptococcus uberis* mastitis. Vet Microbiol 179(3–4):332–335

Alugongo GM, Xiao JX, Chung YH, Dong SZ, Li SL, Yoon I, Wu ZH, Cao ZJ (2017) Effects of *Saccharomyces cerevisiae* fermentation products on dairy calves: Performance and health. J Dairy Sci 100(2):1189–1199

Alves-Barroco C, Roma-Rodrigues C, Balasubramanian N, Guimarães MA, Ferreira-Carvalho BT, Muthukumaran J, Nunes D, Fortunato E, Martins R, Santos-Silva T, Figueiredo AMS, Fernandes AR, Santos-Sanches I (2019) Biofilm development and computational screening for new putative inhibitors of a homolog of the regulatory protein BrpA in *Streptococcus dysgalactiae* subsp. *dysgalactiae*. Int J Med Microbiol 309:169–181

Amber R, Adnan M, Tariq A, Khan SN, Mussarat S, Hashem A, Al-Huqail AA, Al-Arjani AF, Abd Allah EF (2018) Antibacterial activity of selected medicinal plants of northwest Pakistan traditionally used against mastitis in livestock. Saudi J Biol Sci 25(1):154–161

AMCRA (2023) AMCRA Formularium or responsible use of antibacterial agents in animals: Bovine. Available online at: https://formularium.amcra.be/a/4. Accessed 4 July 2023

Andrews T, Neher DA, Weicht TR, Barlow JW (2019) Mammary microbiome of lactating organic dairy cows varies by time, tissue site, and infection status. PLoS One 14(11):e0225001

Angelos JA (2010) *Moraxella bovoculi* and infectious bovine keratoconjunctivitis: cause or coincidence? Vet Clin North Am Food Anim Pract 26(1):73–78

Angelos JA (2015) Infectious bovine keratoconjunctivitis (pinkeye). Vet Clin North Am Food Anim Pract 31:61–79

Astrup LB, Pedersen K, Farre M (2022) Microbiological diagnoses on clinical mastitis-comparison between diagnoses made in veterinary clinics versus in laboratory applying MALDI-TOF MS. Antibiotics (Basel) 11(2):271

AVA-AMA (2022) Antimicrobial prescribing guidelines for dairy cattle 4-01-22. Available online at: https://www.ava.com.au/siteassets/policy-and-advocacy/policies/animal-welfare-principles-and-philosophy/ava-ama-antimicrobial-prescribing-guidelines-for-dairy-cattle-4-01-22-final.pdf. Accessed 4 July 2023

Azevedo D (2019) Latin America: Exploring antibiotic usage across the region, Dairy Global Net, Special Article, Available online at: https://www.dairyglobal.net/specials/latin-america-exploring-antibiotic-usage--across-the-region/. Accessed 4 July 2023

Baggot JD (2001) The bioavailability and disposition of antimicrobial agents in Neonatal animals. In: The physiological basis of veterinary clinical pharmacology. Blackwell Science, Oxford, pp 252–264

Bansal BK, Hamann J, Grabowskit N, Singh K (2005) Variation in the composition of selected milk fraction samples from healthy and mastitic quarters, and its significance for mastitis diagnosis. J Dairy Res 72:144–152

Baptiste KE, Kyvsgaard NC (2017) Do antimicrobial mass medications work? A systematic review and meta-analysis of randomised clinical trials investigating antimicrobial prophylaxis or metaphylaxis against naturally occurring bovine respiratory disease. Pathog Dis 75(7):ftx083

Barca J, Meikle A, Bouman M, Gnemmi G, Ruiz R, Schukken YH (2021) Effect of pegbovigrastim on clinical mastitis and uterine disease during a full lactation in grazing dairy cows. PLoS One 16(5):e0252418

Barkema HW, Schukken YH, Lam TJ, Beiboer ML, Wilmink H, Benedictus G, Brand A (1998) Incidence of clinical mastitis in dairy herds grouped in three cat-

egories by bulk milk somatic cell counts. J Dairy Sci 81(2):411–419
Barkema HW, Schukken YH, Zadoks RN (2006) Invited Review: The role of cow, pathogen, and treatment regimen in the therapeutic success of bovine *Staphylococcus aureus* mastitis. J Dairy Sci 89(6):1877–1895
Bar-Or D, Bar-Or R, Rael LT, Brody EN (2015) Oxidative stress in severe acute illness. Redox Biol 4:340–345
Barry C (2019) Cattle injection sites. EVG Endell Vets Beef Team, Farm Vet News, 27th Dec 2019. Available online at: https://www.endellfarmvets.co.uk/images/EVG-blog2-december-2019.pdf Accessed 25 Jul 2023. Accessed 26 Aug 2023
Bertone I, Bellino C, Alborali GL, Cagnasso A, Cagnotti G, Dappiano E, Lizzi M, Miciletta M, Ramacciotti A, Gianella P, D'Angelo A (2015) Clinical-pathological findings of otitis media and media-interna in calves and (clinical) evaluation of a standardized therapeutic protocol. BMC Vet Res 11:297
Bexiga R, Koskinen MT, Holopainen J, Carneiro C, Pereira H, Ellis KA, Vilela CL (2011) Diagnosis of intramammary infection in samples yielding negative results or minor pathogens in conventional bacterial culturing. J Dairy Res 78:49–55
Biggs A (2018) *Streptococcus uberis*: Environmental, contagious or both? How best to manage it? Cattle Pract 26:61–68
Blagitz MG, Souza FN, Batista CF, Azevedo LF, Benites NR, Melville PA, Diniz SA, Silva MX, Haddad JP, Heinnemann MB, Cerqueira MM, Della Libera AM (2015) The neutrophil function and lymphocyte profile of milk from bovine mammary glands infected with Streptococcus dysgalactiae. J Dairy Res 82(4):460–469
Blum SE, Heller ED, Leitner G (2014) Long term effects of *Escherichia coli* mastitis. Vet J 201:72–77
Blum SE, Heller ED, Sela S, Elad D, Edery N, Leitner G (2015) Genomic and phenomic study of mammary pathogenic *Escherichia coli*. PLoS One 10:e0136387
Blum SE, Goldstone RJ, Connolly JPR, Répérant-Ferter M, Germon P, Inglis NF, Krifucks O, Mathur S, Manson E, Mclean K, Rainard P, Roe AJ, Leitner G, Smith DGE (2018) Postgenomics characterization of an essential genetic determinant of mammary pathogenic *Escherichia coli*. MBio 9:e423–e418
Böker AR, Bartel A, Do Duc P, Hentzsch A, Reichmann F, Merle R, Arndt H, Dachrodt L, Woudstra S, Hoedemaker M (2023) Status of udder health performance indicators and implementation of on farm monitoring on German dairy cow farms: results from a large scale cross-sectional study. Front Vet Sci 10:1193301
Bokma J, Boone R, Deprez P, Pardon B (2019) Risk factors for antimicrobial use in veal calves and the association with mortality. J Dairy Sci 102(1):607–618
Bokma J, Gille L, De Bleecker K, Callens J, Haesebrouck F, Pardon B, Boyen F (2020) Antimicrobial susceptibility of mycoplasma bovis isolates from veal, dairy and beef herds. Antibiotics (Basel) 9(12):882
Bolte J, Zhang Y, Wente N, Krömker V (2020) In vitro susceptibility of mastitis pathogens isolated from clinical mastitis cases on northern German dairy farms. Vet Sci 7(1):10
Bonestroo J, Fall N, Hogeveen H, Emanuelson U, Klaas IC, van der Voort M (2023) The costs of chronic mastitis: A simulation study of an automatic milking system farm. Prev Vet Med 210:105799
Bottinelli M, Merenda M, Gastaldelli M, Picchi M, Stefani E, Nicholas RAJ, Catania S (2020) The pathogen Mycoplasma dispar shows high minimum inhibitory concentrations for antimicrobials commonly used for bovine respiratory disease. Antibiotics (Basel) 9(8):460
Bouchard DS, Seridan B, Saraoui T, Rault L, Germon P, Gonzalez-Moreno C, Nader-Macias FM, Baud D, François P, Chuat V, Chain F, Langella P, Nicoli J, Le Loir Y, Even S (2015) Lactic acid bacteria isolated from bovine mammary microbiota: potential allies against bovine mastitis. PLoS One 10(12):e0144831
Boujenane I, El Aimani J, By K (2015) Incidence and occurrence time of clinical mastitis in Holstein cows. Turk J Vet Anim Sci 39:42–49
BQA (2020) Beef quality assurance: Antibiotic stewardship for beef producers. Available online at: https://www.bqa.org/Media/BQA/Docs/bqa--antibiotics-2020.pdf. Accessed 4 July 2023
Bradley AJ, Breen J, Payne B, White V, Green MJ (2015) An investigation of the efficacy of a polyvalent mastitis vaccine using different vaccination regimens under field conditions in the United Kingdom. J Dairy Sci 98:1706–1720
Braiek M, Rokbani KB, Chrouda A, Mrabet B, Bakhrouf A, Maaref A, Jaffrezic-Renault N (2012) An electrochemical immunosensor for detection of *Staphylococcus aureus* bacteria based on immobilization of antibodies on self-assembled monolayers-functionalized gold electrode. Biosensors 2:417–426
Breen J (2017) The importance of non-steroidal anti-inflammatory drugs (NSAIDs) in mastitis therapeutics. Livestock (Lond) 22:182–185
Breyne K, Honaker RW, Hobbs Z, Richter M, Żaczek M, Spangler T, Steenbrugge J, Lu R, Kinkhabwala A, Marchon B, Meyer E, Mokres L (2017) Efficacy and safety of a bovine-associated *Staphylococcus aureus* phage cocktail in a murine model of mastitis. Front Microbiol 8:2348
Briggs RE, Billing SR, Boatwright WD Jr, Chriswell BO, Casas E, Dassanayake RP, Palmer MV, Register KB, Tatum FM (2021) Protection against *Mycoplasma bovis* infection in calves following intranasal vaccination with modified-live Mannheimia haemolytica expressing Mycoplasma antigens. Microb Pathog 161(Pt A):105159
Brook I (2002) Cutaneous and subcutaneous infections in newborns due to anaerobic bacteria. J Perinat Med 30(3):197–208
Brown SA (1992) Pharmacology in neonatal calves. In: The bovine proceedings—no. 24, pp 96–100

Brown N, Conneely M, Hudson C (2022) Use of non-steroidal anti-inflammatory drugs and attitudes to pain in pasture-based dairy cows: A comparative study of farmers and veterinarians. Front Vet Sci 9

Bruijnis M (2012) Foot disorders in dairy cattle—a socio-economic approach to improve dairy cow welfare. PhD thesis, Wageningen University, The Netherlands, p 176. https://edepot.wur.nl/237390. Accessed 25 July 2023

Buck JM, Brown SA, Weise DW, Gauze JJ, Smith GG, Greco DS (1990) Oral absorption of apramycin in newborn calves. In: Proceedings of the 71st conference of research workers in animal disease A187, p 33

Budd KE, McCoy F, Monecke S, Cormican P, Mitchell J, Keane OM (2015) Extensive genomic diversity among bovine adapted *Staphylococcus aureus*: Evidence for a genomic rearrangement within CC97. PLoS One 10:e0134592

Burt A (2008) Calf health depends on the first 12 hours, beef–livestock–cow-calf. Available online at: https://www.beefmagazine.com/cowcalfweekly/calf-health-depends-first-12-hours. Accessed 25 July 2023

Calvinho LF, Almeida RA, Oliver SP (1998) Potential virulence factors of *Streptococcus dysgalactiae* associated with bovine mastitis. Vet Microbiol 61:93–110

Campbell J (2023) Necrotic laryngitis in calves MSD Manual, Veterinary Manual. Available online at: https://www.msdvetmanual.com/respiratory-system/infectious-respiratory-system-diseases-in-cattle/necrotic-laryngitis-in-cattle. Accessed 4 July 2023

Cao LT, Wu JQ, Xie F, Hu SH, Mo Y (2007) Efficacy of nisin in treatment of clinical mastitis in lactating dairy cows. J Dairy Sci 90(8):3980–3985

Capuco AV, Choudhary RK, Daniels KM, Li RW, Evock-Clover CM (2012) Bovine mammary stem cells: cell biology meets production agriculture. Animal 6(3):382–393

Cardoso M (2017) Antimicrobial use, resistance and economic benefits and costs to livestock producers in Brazil: Breslau OECD Food, Agriculture and Fisheries Papers, No. 135. OECD, Paris, 2019. Available online at: https://doi.org/10.1787/27137b1e-en. Accessed 4 July 2023

Castelani L, Arcaro JRP, Braga JEP, Bosso AS, Moura Q, Esposito F, Sauter IP, Cortez M, Lincopan N (2019) Short communication: Activity of nisin, lipid bilayer fragments and cationic nisin-lipid nanoparticles against multidrug-resistant Staphylococcus spp. isolated from bovine mastitis. J Dairy Sci 102(1):678–683

Chen S, Hao H, Yan X, Liu Y, Chu Y (2019) Genome-wide analysis of mycoplasma dispar provides insights into putative virulence factors and phylogenetic relationships. G3 Genes Genomes Genet 9:317–325

Cheng WN, Jeong CH, Seo HG, Han SG (2019) Moringa extract attenuates inflammatory responses and increases gene expression of casein in bovine mammary epithelial cells. Animals (Basel) 9(7):391

Cheng WN, Han SG (2020) Bovine mastitis: risk factors, therapeutic strategies, and alternative treatments—a review. Asian Australas J Anim Sci 33(11):1699–1713

Chengolova Z, Atanasova M, Godjevargova T (2021) Neutrophil and CD4+milk cell count related to natural incidence of mastitis in Jersey cattle. J Dairy Sci 88:334–336

Chirico J, Jonsson P, Kjellberg S, Thomas G (1997) Summer mastitis experimentally induced by *Hydrotaea irritans* exposed to bacteria. Med Vet Entomol 11(2):187–192

Cho YI, Yoon KJ (2014) An overview of calf diarrhea—infectious ethiology, diagnosis, and intervention. J Vet Sci 15:1–17

Cho BW, Cha CN, Lee SM, Kim MJ, Park JY, Yoo CY, Son SE, Kim S, Lee HJ (2015) Therapeutic effect of oregano essential oil on subclinical bovine mastitis caused by *Staphylococcus aureus* and *Escherichia coli*. Korean J Vet Res 55:253–257

Chung PY, Khanum R (2017) Antimicrobial peptides as potential anti-biofilm agents against multidrug-resistant bacteria. J Microbiol Immunol Infect 50(4):405–410

Clements A, Tull D, Jenney AW, Farn JL, Kim S-H, Bishop RE, McPhee JB, Hancock REW, Hartland EL, Pearse MJ, Wijburg OLC, Jackson DC, Mcconville MJ, Strugnell RA (2007) Secondary acylation of *Klebsiella pneumoniae* lipopolysaccharide contributes to sensitivity to antibacterial peptides. Biol Chem 282:15569–15577

Collado R, Montbrau C, Sitjà M, Prenafeta A (2018) Study of the efficacy of a *Streptococcus uberis* mastitis vaccine against an experimental intramammary infection with a heterologous strain in dairy cows. J Dairy Sci 101(11):10290–10302

Constable PD (2003) Use of antibiotics to prevent calf diarrhea and septicaemia. Bovine Pract 37:137–142

Constable PD (2004) Antimicrobial use in the treatment of calf diarrhea. J Vet Intern Med 18:8–17

Constable PD (2009) Treatment of calf diarrhea: antimicrobial and ancillary treatments. Vet Clin Food Anim 25:101–120

Constable PD, Pyörälä S, Smith GW (2008) Guidelines for antimicrobial use in cattle. In: Guardabassi L, Jennsen LB, Kruse H (eds) Guide to antimicrobial use in animals. Blackwell, Vancouver, pp 143–160

Constable PD, Hinchcliff KW, Done S, Grünberg W (2017a) Disease of the alimentary tract -ruminant: oral and laryngeal necrobacillosis. In: Veterinary medicine. Elsevier, St Louis, MO, pp 534–536

Constable PD, Hinchcliff KW, Done S, Grünberg W (2017b) General systemic states: toxemia in recently calved cows. In: Veterinary medicine. Elsevier, St Louis, MO, pp 67–70

Corre C (2021) L'utilisation de L'aromathérapie Dans les Élevages Français: État des Lieux, Efficacité et Limites (in French). https://hal.inrae.fr/hal-02788515/document. Accessed 4 July 2023

Cresswell E, Remnant J, Butterworth A, Wapenaar W (2017) Injection-site lesion prevalence and potential risk factors in UK beef cattle. Vet Rec 180(3):70

Cubeddu T, Cacciotto C, Pisanu S, Tedde V, Alberti A, Pittau M, Dore S, Cannas A, Uzzau S, Rocca S, Addis

MF (2017) Cathelicidin production and release by mammary epithelial cells during infectious mastitis. Vet Immunol Immunopathol 189:66–70

CZ NMTP (2020) National programme of monitoring antimicrobial resistance of pathogens with veterinary importance, data 2020 (in Czech). Available at: https://www.svscr.cz/wp-content/files/dokumenty-a-publikace/Narodni-program-sledovani-rezistenci-k-antimikrobikum-u-veterinarne-vyznamnych-patogenu-2020.pdf. Accessed 22 Aug 2023

CZ NMTP (2021) National programme of monitoring antimicrobial resistance of pathogens with veterinary importance, data 2021 (in Czech). Available at: https://www.svscr.cz/wp-content/files/dokumenty-a-publikace/Narodni-program-sledovani-rezistenci-k-antimikrobikum-u-veterinarne-vyznamnych-patogenu-2021.pdf. Accessed 22 Aug 2023

CZ NMTP (2022) National programme of monitoring antimicrobial resistance of pathogens with veterinary importance, data 2022 (in Czech). Available at: https://www.svscr.cz/wp-content/files/dokumenty-a-publikace/Narodni-program-sledovani-rezistenci-k-antimikrobikum-u-veterinarne-vyznamnych-patogenu-2022.pdf. Accessed 22 Aug 2023

Daley MJ, Furda G, Dougherty R, Coyle PA, Williams TJ, Johnston P (1992) Potentiation of antibiotic therapy for bovine mastitis by recombinant bovine interleukin-2. J Dairy Sci 75(12):3330–3338

Dash JR, Sar TK, Samanta I, Mandal TK (2016) Effects of herbal extract of Ocimum sanctum as supportive therapy with intravenous ceftriaxone in experimentally induced staphylococcal chronic mastitis in goat. Small Rum Res 137:1–8

De Briyne N, Atkinson J, Borriello SP, Pokludová L (2014) Antibiotics used most commonly to treat animals in Europe. Vet Rec 175:325

De Buck J, Ha V, Naushad S, Nobrega DB, Luby C, Middleton JR, De Vliegher S, Barkema HW (2021) Non-aureus Staphylococci and bovine udder health: current understanding and knowledge gaps. Front Vet Sci 8:658031

De Donder KD, Apley MD, Li M, Gehring R, Harhay DM, Lubbers BV, White BJ, Capik SF, KuKanich B, Riviere JE, Tessman RK (2016) Pharmacokinetics and pharmacodynamics of gamithromycin in pulmonary epithelial lining fluid in naturally occurring bovine respiratory disease in multisource commingled feedlot cattle. J Vet Pharmacol Ther 39(2):157–166

Degen S, Paduch JH, Hoedemaker M, Krömker V (2015) Factors affecting the probability of bacteriological cure of bovine mastitis (in German). Tierarztl Prax Ausg G 43:222–227. https://doi.org/10.15653/TPG-141082

Dego OK (2020) Current status of antimicrobial resistance and prospect for new vaccines against major bacterial bovine mastitis pathogens. In: Aral F (ed) Animal reproduction in veterinary medicine. IntechOpen, London, UK, p 78921

de Jong E, McCubbin KD, Speksnijder D, Dufour S, Middleton JR, Ruegg PL, Lam TJGM, Kelton DF, McDougall S, Godden SM, Lago A, Rajala-Schultz PJ, Orsel K, De Vliegher S, Krömker V, Nobrega DB, Kastelic JP, Barkema HW (2023) Invited review: Selective treatment of clinical mastitis in dairy cattle. J Dairy Sci 106(6):3761–3778

Demuth DC, Müntener CR (2006) Compendium of Veterinary Medicines of Switzerland 2006/2007. Tierarzneimittelkompendium der Schweiz (in German) 2006/2007 7:1–686

Derakhshani H, Fehr KB, Sepehri S, Francoz D, De Buck J, Barkema HW, Plaizier JC, Khafipour E (2018) Invited review: Microbiota of the bovine udder: Contributing factors and potential implications for udder health and mastitis susceptibility. J Dairy Sci 101(12):10605–10625

DG Health and Food Safety (2017) DG Health and Food Safety overview report: welfare of cattle on dairy farms, 21 p

Dobrut A, Wójcik-Grzybek D, Młodzińska A, Pietras-Ożga D, Michalak K, Tabacki A, Mroczkowska U, Brzychczy-Włoch M (2023a) Detection of immunoreactive proteins of *Escherichia coli, Streptococcus uberis*, and *Streptococcus agalactiae* isolated from cows with diagnosed mastitis. Front Cell Infect Microbiol 13:987842

Dobrut A, Młodzińska A, Drożdż K, Wójcik-Grzybek D, Michalak K, Pietras-Ożga D, Karakulska J, Biegun K, Brzychczy-Włoch M (2023b) The two-track investigation of fibronectin binding protein A of Staphylococcus aureus from bovine mastitis as a potential candidate for immunodiagnosis: a pilot study. Int J Mol Sci 24(7):6569

Doehring C, Sundrum A (2014) Efficacy of homeopathy in livestock according to peer-reviewed publications from 1981 to 2014. Vet Rec 179:628. https://doi.org/10.1136/vr.103779

Donahue M, Godden SM, Bey R, Wells S, Oakes JM, Sreevatsan S, Stabel J, Fetrow J (2012) Heat treatment of colostrum on commercial dairy farms decreases colostrum microbial counts while maintaining colostrum immunoglobulin G concentrations. J Dairy Sci 95(5):2697–2702

Donovan DM, Kerr DE, Wall RJ (2005) Engineering disease resistant cattle. Transgenic Res 14(5):563–567

Döpfer D, Barkema HW, Lam TJ, Schukken YH, Gaastra W (1999) Recurrent clinical mastitis caused by Escherichia coli in dairy cows. J Dairy Sci 82(1):80–85

Duse A, Persson-Waller K, Pedersen K (2021) Microbial aetiology, antibiotic susceptibility and pathogen-specific risk factors for udder pathogens from clinical mastitis in dairy cows. Animals 11:2113

Ebert F, Staufenbiel R, Simons J, Pieper L (2017) Randomized, blinded, controlled clinical trial shows no benefit of homeopathic mastitis treatment in dairy cows. J Dairy Sci 100(6):4857–4867

EC (2008) Commission Regulation (EC). No 889/2008 of 5 September 2008 laying down detailed rules for the implementation of Council Regulation (EC) No

834/2007 on organic production and labelling of organic products with regard to organic production, labelling and control

EDF Congress (2017) Dairy farming in the Czech Republic: A key to efficient farm management? Focus on details! Available online at: https://www.thuenen.de/media/institute/bw/Startseite_Aktuelles/Aktuelles_17-07_EDF_Kongress_Prag.pdf. Accessed 4 July 2023

EFSA (2022) EFSA Panel on Animal Health and Welfare: Scientific Opinion on the assessment of listing and categorisation of animal diseases within the framework of the Animal Health Law (Regulation (EU) No 2016/429); antimicrobial-resistant Staphylococcus aureus in cattle and horses. EFSA Journal 20(5):7312, 103 pp

Eibl C, Bexiga R, Viora L, Guyot H, Félix J, Wilms J, Tichy A, Hund A (2021) The antibiotic treatment of calf diarrhea in four European countries: a survey. Antibiotics 10(8):910

Elbers AR, Miltenburg JD, De Lange D, Crauwels AP, Barkema HW, Schukken YH (1998) Risk factors for clinical mastitis in a random sample of dairy herds from the southern part of The Netherlands. J Dairy Sci 81(2):420–426

El Garch F, Youala M, Simjee S, Moyaert H, Klee R, Truszkowska B, Rose M, Hocquet D, Valot B, Morrissey I, de Jong A (2020) VetPath Study Group. Antimicrobial susceptibility of nine udder pathogens recovered from bovine clinical mastitis milk in Europe 2015-2016: VetPath results. Vet Microbiol 245:108644

EMA (2014) Divergent position on a CVMP opinion on an Article 13(2) of Commission Regulation (EC) No 1234/2008, Resflor solution injectable (EMEA/V/A/101) Available online at: https://www.ema.europa.eu/en/documents/referral/divergent-position-cvmp-opinion-article-132-commission-regulation-ec-no-1234/2008-resflor-solution-injectable_en.pdf. Accessed 25 July 2023

EMA (2015) Principles on assignment of defined daily dose for animals (DDDvet) and defined course dose for animals (DCDvet) (EMA/710019/2014) Available online at: https://www.ema.europa.eu/en/documents/scientific-guideline/principles-assignment-defined--daily-dose-animals-dddvet-defined-course-dose-animals-dcdvet_en.pdf. Accessed 4 July 2023

EMA (2020) Committee for Medicinal Products for Veterinary Use Advice on implementing measures under Article 106 (6) of Regulation (EU) 2019/6 on veterinary medicinal products—scientific problem analysis and recommendations to ensure a safe and efficient administration of oral veterinary medicinal products via routes other than medicated feed (EMA/CVMP/508559/2019) Available online at: https://www.ema.europa.eu/en/documents/regulatory-procedural-guideline/advice-implementing-measures-under-article-106-6-regulation-eu-2019/6-veterinary-medicinal-products-scientific-problem-analysis-recommendations-ensure-safe-efficient_en.pdf. Accessed 4 July 2023

Erskine RJ (2022) Mastitis in cattle MSD manual, veterinary manual. Available online at: https://www.msdvetmanual.com/reproductive-system/mastitis--in-large-animals/mastitis-in-cattle. Accessed 26 Aug 2023

Erskine RJ, Wagner S, DeGraves FJ (2003) Mastitis therapy and pharmacology. Vet Clin North Am Food Anim Pract 19(1):109–318

EUROSTAT (2023) Milk market observatory. Available online at: https://agriculture.ec.europa.eu/system/files/2023-03/eu-milk-yield-herds_en.pdf. Accessed 4 July 2023

Falkenberg U, Krömker V, Heuwieser W, Fischer-Tenhagen C (2019) Survey on routines in udder health management and therapy of mastitis on German dairy farms. Milk Sci Int 72:11–15

Fan J, Zeng Z, Mai K, Yang Y, Feng J, Bai Y, Sun B, Xie Q, Tong Y, Ma J (2016) Preliminary treatment of bovine mastitis caused by *Staphylococcus aureus*, with trx-SA1, recombinant endolysin of *S. aureus* bacteriophage IME-SA1. Vet Microbiol 191:65–71

Fecteau G (2023) Neonatal septicemia in bovine neonates, MSD manual, veterinary manual. Available online at: https://www.msdvetmanual.com/management-and-nutrition/management-of-the-neonate/neonatal--septicemia-in-bovine-neonates. Accessed 4 July 2023

Fecteau G, Smith BP, George LW (2009) Septicemia and meningitis in the newborn calf. Vet Clin North Am Food Anim Pract 25(1):195–208

Fenton M, Keary R, McAuliffe O, Ross RP, O'Mahony J, Coffey A (2013) Bacteriophage-derived peptidase CHAP(K) eliminates and prevents staphylococcal biofilms. Int J Microbiol 2013:625341

Ferguson JD, Sattler MA, Hanson DL, Davis CP, Edrington TS, Yoon I (2018) Feeding NutriTek reduces linear scores and clinical mastitis cases. J Dairy Sci 101(Suppl. 2):135

Feuerstein A, Scuda N, Klose C, Hoffmann A, Melchner A, Boll K, Rettinger A, Fell S, Straubinger RK, Riehm JM (2022) Antimicrobial resistance, serologic and molecular characterization of *E. coli* isolated from calves with severe or fatal enteritis in Bavaria, Germany. Antibiotics 11:23

Field D, Considine K, O'Connor PM, Ross RP, Hill C, Cotter PD (2021) Bio-engineered nisin with increased anti-Staphylococcus and selectively reduced anti-lactococcus activity for treatment of bovine mastitis. Int J Mol Sci 22(7):3480

Firth CL, Laubichler C, Schleicher C, Fuchs K, Käsbohrer A, Egger-Danner C, Köfer J, Obritzhauser W (2019) Relationship between the probability of veterinary-diagnosed bovine mastitis occurring and farm management risk factors on small dairy farms in Austria. J Dairy Sci 102:4452–4463

Fleitas Martínez O, Cardoso MH, Ribeiro SM, Franco OL (2019) Recent advances in anti-virulence therapeutic strategies with a focus on dismantling bacterial membrane microdomains, toxin neutralization, quorum-sensing interference and biofilm inhibition. Front Cell Infect Microbiol 9:74

Fox LK, Besser TE, Jackson SM (1996) Evaluation of a coagulase-negative variant of *Staphylococcus aureus* as a cause of intramammary infections in a herd of dairy cattle. J Am Vet Med Assoc 209:1143–1146

Francoz D, Wellemans V, Dupré JP, Roy JP, Labelle F, Lacasse P, Dufour S (2017) Invited review: A systematic review and qualitative analysis of treatments other than conventional antimicrobials for clinical mastitis in dairy cows. J Dairy Sci 100(10):7751–7770

Freick M, Frank Y, Steinert K, Hamedy A, Passarge O, Sobiraj A (2016) Mastitis vaccination using a commercial polyvalent vaccine or a herd-specific Staphylococcus aureus vaccine. Results of a controlled field trial on a dairy farm. Tierarztl Prax Ausg G Grosstiere Nutztiere 44(4):219–229

Friton G, Van Hattum JJC, Hörstermann D (2003) Pharmacokinetics in plasma and milk of benzylpenicillin following repeated intramuscular administration of Mamyzin (penethamate hydriodide) in lactating cows. J Vet Pharmacol Ther 26(Suppl. 1):100–101

Fuenzalida MJ, Furmaga E, Aulik N (2021) Antimicrobial resistance in Klebsiella species from milk specimens submitted for bovine mastitis testing at the Wisconsin Veterinary Diagnostic Laboratory, 2008-2019. JDS Commun 2(3):148–152

Fukushima Y, Kino E, Furutani A, Minamino T, Mikurino Y, Horii Y, Honkawa K, Sasaki Y (2020) Epidemiological study to investigate the incidence and prevalence of clinical mastitis, peracute mastitis, metabolic disorders and peripartum disorders, on a dairy farm in a temperate zone in Japan. BMC Vet Res 16(1):389

FVE (2016) Federation of Veterinarians Europe: Antimicrobial use in food producing animals, FVE input to RONAFA report. EFSA J 15(1)

Gelgie AE, Korsa MG, Kerro Dego O (2022) *Mycoplasma bovis* mastitis. Curr Res Microb Sci 3:100123

Giguere S, Huang R, Malinski TJ, Dorr PM, Tessman RK, Somerville BA (2011) Disposition of gamithromycin in plasma, pulmonary epithelial lining fluid, bronchoalveolar cells, and lung tissue in cattle. Am J Vet Res 72:326–330

Gilleard J (2021) Update on coccidiosis in dairy calves. WCDS Adv Dairy Technol 33:41–49. Available online at: https://wcds.ualberta.ca/wp-content/uploads/sites/57/2021/08/2021WCDS_Gilleard1.3.pdf. Accessed 25 July 2023

Godden SM, Lombard JE, Woolums AR (2019) Colostrum management for dairy calves. Vet Clin North Am Food Anim Pract 35(3):535–556

Godoy-Santos F, Pinto MS, Barbosa AAT, Brito MAVP, Mantovani HC (2019) Efficacy of a ruminal bacteriocin against pure and mixed cultures of bovine mastitis pathogens. Indian J Microbiol 59(3):304–312

Gomes F, Maria JS, Mariana H (2016) Bovine mastitis disease/pathogenicity: evidence of the potential role of microbial biofilms. Pathog Dis 74(3)

Goncalves JL, de Campos JL, Steinberger A, Safdar N, Sethi A, Shutske J, Rodriguez Z, Suen G, Goldberg T, Ruegg PL (2022) Outcomes of first cases of selected bovine disease on 37 large WI dairy farms. National Mastitis Council (NMC), San Diego, CA

Gorden PJ, Ydstie JA, Kleinhenz MD, Wulf LW, Gehring R, Lee CJ, Wang C, Coetzee JF (2016) A study to examine the relationship between metritis severity and depletion of oxytetracycline in plasma and milk after intrauterine infusion. J Dairy Sci 99:8314–8322

Gorden PJ, Kleinhenz MD, Ydstie JA, Brick TA, Slinden LM, Peterson MP, Straub DE, Burkhardt DT (2018) Efficacy of vaccination with a Klebsiella pneumoniae siderophore receptor protein vaccine for reduction of Klebsiella mastitis in lactating cattle. J Dairy Sci 101(11):10398–10408

Gott PN, Rajala-Schultz PJ, Schuenemann GM, Proudfoot KL, Hogan JS (2017) Effect of gradual or abrupt cessation of milking at dry off on milk yield and somatic cell score in the subsequent lactation. J Dairy Sci 100(3):2080–2089

Goulart DB, Mellata M (2022) *Escherichia coli* mastitis in dairy cattle: etiology, diagnosis, and treatment challenges. Front Microbiol 13:928346

Gould S, Dewell R, Tofflemire K, Whitley RD, Millman ST, Opriessnig T, Rosenbusch R, Trujillo J, O'Connor AM (2013) Randomized blinded challenge study to assess association between *Moraxella bovoculi* and Infectious Bovine Keratoconjunctivitis in dairy calves. Vet Microbiol 164:108–115

Gruet P, Maincent P, Berthelot X, Kaltsatos V (2001) Bovine mastitis and intramammary drug delivery: review and perspectives. Adv Drug Deliv Rev 50(3):245–259

Grunert T, Stessl B, Wolf F, Sordelli DO, Buzzola FR, Ehling-Schulz M (2018) Distinct phenotypic traits of Staphylococcus aureus are associated with persistent, contagious bovine intramammary infections. Sci Rep 8:15968

Grzywa R, Łupicka-Słowik A, Sieńczyk M (2023) IgYs: on her majesty's secret service. Front Immunol 14:1199427

Günther J, Petzl W, Bauer I, Ponsuksili S, Zerbe H, Schuberth HJ, Brunner RM, Seyfert HM (2017) Differentiating *Staphylococcus aureus* from *Escherichia coli* mastitis: *S. aureus* triggers unbalanced immune-dampening and host cell invasion immediately after udder infection. Sci Rep 7:1–14

Gurjar A, Gioia G, Schukken Y, Welcome F, Zadoks R, Moroni P (2012) Molecular diagnostics applied to mastitis problems on dairy farms. Vet Clin North Am Food Anim Pract 28(3):565–576

Gussmann M, Steeneveld W, Kirkeby C, Hogeveen H, Nielen M, Farre M, Halasa T (2018) Economic and epidemiological impact of different intervention strategies for clinical contagious mastitis. J Dairy Sci 102(2):1483–1493

GVA (2009) Guidelines for aseptic milk sampling and guidelines to isolate and identify mastitis pathogens, 2nd edn. German Veterinary Association, Gießen

Hadfeld SJ, Robinson G, Elwin K, Chalmers RM (2011) Detection and differentiation of Cryptosporidium spp.

in human clinical samples by use of real-time PCR. J Clin Microbiol 49:918–924
Halasa T, Huijps K, Osteras O, Hogeveen H (2007) Economic effects of bovine mastitis and mastitis management: a review. Vet Q 29:18–31
Hammer JF, Morton JM, Kerrisk KL (2012) Quarter-milking-, quarter-, udder- and lactation-level risk factors and indicators for clinical mastitis during lactation in pasture-fed dairy cows managed in an automatic milking system. Aust Vet J 90(22510075):167–174
Harp JA, Goff JP (1998) Strategies for the control of *Cryptosporidium parvum* infection in calves. J Dairy Sci 81:289–294
Hase P, Digraskar S, Ravikanth K, Dandale M, Maini S (2013) Management of subclinical mastitis with Mastilep Gel and herbal spray (AV/AMS/15). Int J Pharm Pharmacol 2:64–67
Haxhiaj K, Wishart DS, Ametaj BN (2022) Mastitis: what it is, current diagnostics, and the potential of metabolomics to identify new predictive biomarkers. Dairy 3(4):722–746
Hebert A, Sayasith K, Senechal S, Dubreuil P, Lagace J (2000) Demonstration of intracellular *Staphylococcus aureus* in bovine mastitis alveolar cells and macrophages isolated from naturally infected cow milk. FEMS Microbiol. Lett 193:57–62
Heikkilä AM, Liski E, Pyörälä S, Taponen S (2018) Pathogen-specific production losses in bovine mastitis. J. Dairy Sci 101:9493–9504
Hektoen L (2005) Review of the current involvement of homeopathy in veterinary practice and research. Vet Rec 157(2005):224–229. 16113167
Hewson CJ, Dohoo IR, Lemke KA, Barkema HW (2007) Canadian veterinarians' use of analgesics in cattle, pigs, and horses in 2004 and 2005. Can Vet J 48(2):155–164
Hiitiö H, Vakkamäki J, Simojoki H, Autio T, Junnila J, Pelkonen S, Pyörälä S (2017) Prevalence of subclinical mastitis in Finnish dairy cows: changes during recent decades and impact of cow and herd factors. Acta Vet Scand 59:22
Hogan J, Smith K (2003) Coliform mastitis. Vet Res 34:507–519
Hogeveen H, Huijps K, Lam TJGM (2011) Economic aspects of mastitis: New developments. N Z Vet J 59:16–23
Hoque MN, Istiaq A, Rahman MS, Islam MR, Anwar A, Zoaned Siddiki AMAM, Sultana M, Crandall KA, Hossain MA (2020) Microbiome dynamics and genomic determinants of bovine mastitis. Genomics 112(6):5188–5203
Hozyen HF, Ibrahim ES, Khairy EA, El-Dek SI (2019) Enhanced antibacterial activity of capped zinc oxide nanoparticles: A step towards the control of clinical bovine mastitis. Vet World 12(8):1225–1232
Hu X, He Z, Jiang P, Wang K, Guo J, Zhao C, Cao Y, Zhang N, Fu Y (2019) Neutralization of interleukin-17A attenuates lipopolysaccharide-induced mastitis by inhibiting neutrophil infiltration and the inflammatory response. J Interferon Cytokine Res 39(9):577–584
Huang RA, Letendre LT, Banav N, Fischer J, Somerville B (2010) Pharmacokinetics of gamithromycin in cattle with comparison of plasma and lung tissue concentrations and plasma antibacterial activity. J Vet Pharmacol Ther 33(3):227–237
Hubackova M, Rysánek D (2007) Effects of freezing milk samples on the recovery of alimentary pathogens and indicator microorganisms. Acta Vet Brno 76:301–307
Hudson C, Whay H, Huxley J (2008) Recognition and management of pain in cattle. In Pract 30:126–134
Ishiyama D, Mizomoto T, Ueda C, Takagi N, Shimizu N, Matsuura Y, Makuuchi Y, Watanabe A, Shinozuka Y, Kawai K (2017) Factors affecting the incidence and outcome of *Trueperella pyogenes* mastitis in cows. J Vet Med Sci 79:626–631
Izquierdo AC, Guerra Liera JE, Cervantes RE, Inzunza Castro JF, Abel Villa Mancera E, Crispın RH, Juarez Mosqueda ML, Vazquez AG, Perez JO, Aparicio PS, Rodriguez Denis BE (2017) Production of milk and bovine mastitis. J Adv Dairy Res 5(2):2
Jagielski T, Bakuła Z, Gawor J, Maciszewski K, Kusber WH, Dyląg M, Nowakowska J, Gromadka R, Karnkowska A (2019a) The genus *Prototheca* (Trebouxiophyceae, Chlorophyta) revisited: Implications from molecular taxonomic studies. Algal Res 43:101639
Jagielski T, Krukowski H, Bochniarz M, Piech T, Roeske K, Bakuła Z, Wlazło Ł, Woch P (2019b) Prevalence of Prototheca spp. on dairy farms in Poland—a cross-country study. Microb Biotechnol 12(3):556–566
Jamali H, Barkema HW, Jacques M, Lavallée-Bourget EM, Malouin F, Saini V, Stryhn H, Dufour S (2018) Invited review: Incidence, risk factors, and effects of clinical mastitis recurrence in dairy cows. J Dairy Sci 101(6):4729–4746
Jenkins SN, Okello E, Rossitto PV, Lehenbauer TW, Champagne J, Penedo MCT, Arruda AG, Godden S, Rapnicki P, Gorden PJ, Timms LL, Aly SS (2019) Molecular epidemiology of coagulase-negative Staphylococcus species isolated at different lactation stages from dairy cattle in the United States. PeerJ 7:e6749
Jensen GS, Patterson KM, Yoon I (2008) Yeast culture has anti-inflammatory effects and specifically activates NK cells. Comp Immunol Microbiol Infect Dis 31(6):487–500
Jerab J, Jansen W, Blackwell J, van Hout J, Palzer A, Lister S, Chantziaras I, Dewulf J, De Briyne N (2022) Real-world data on antibiotic group treatment in european livestock: drivers, conditions, and alternatives. Antibiotics (Basel) 11(8):1046
J-VAC (2023) Boehringer Ingelheim Animal Health USA Inc., Duluth, GA. All Rights Reserved. US-BOV-0323-2020-V2. https://docs.boehringer-ingelheim.com/AH/JVAC_Field_Efficacy_Tech_Bulletin.pdf. Assessed 26 Aug 2023
Kaartinen L, Lohonen K, Wiese B, Franklin A, Pyörälä S (1999) Pharmacokinetics of sulphadiazine-trimethoprim in lactating dairy cows. Acta Vet Scand 40(3):271–278

(Dego 2020). As the mammary gland with its compartments susceptible to attack of pathogens is largely different from other body sites, development of effective vaccine is biologically challenging. The difficulties also contribute to the fact that environmental mastitis is a disease complex with a variety of causative agents, which would need to be covered by multiple vaccines or multi-strain polyvalent vaccines. For majority of recently used mastitis vaccines, repeated administration is needed to maintain a protective immune response, which influences the practicality and commercial viability of vaccines. Therefore, vaccines of a new generation, with single administration dosing regimen, are desirable (Lofano et al. 2020).

Moreover, despite some studies bringing the information on a certain level of benefit of vaccination either using commercially produced vaccines or autogenous vaccines, the summary of the current knowledge rather indicates that vaccination is not necessarily effective or economical, especially in dairy herds with high mastitis rates. Unfortunately, no such a powerful vaccine being the sole tool for eradication of mastitis pathogens in the herd has been developed yet. Therefore, the combination of vaccination and carrying out infection control programmes including excellent milking hygiene procedures, rational and justified treatment of clinical cases, segregation and culling of known infected cows are essential for a significant reduction in the incidence and duration of intramammary infections (Rainard et al. 2021).

8.1.4.1 Commercially Produced Vaccines

There exist several concepts of the commercially produced anti-mastitis vaccines, for which evidence of efficacy and economic viability is mixed. Generally, it can be commented that for *E. coli* vaccines stronger efficacy evidence is reported than for *S. aureus* vaccines. Independent scientific evidence of vaccine effectiveness is available at a certain scale for *E. coli* vaccines (reduced milk yield losses, reduce risk of mortality, cost-effectiveness), albeit it with variable efficacy and cost-effectiveness at the level of individual farms. On the other hand, similar independent studies in *S. aureus* vaccines are limited and not available yet for *S. uberis* vaccines, where rather studies supported were held. Demand of standardised methods and independent scientific evidence to evaluate the effectiveness of the vaccines under field conditions is therefore warranted. Recently, complex meta-analysis (Mata et al. 2023) working with data available due to decades worldwide, related to the efficacy of vaccines in dairy cattle shown dubious results but some interesting outcomes, has also demonstrated e.g. better (but still "reduced") efficacy when vaccine is administered prior calving comparing to post calving vaccination (with inefficacy). Based on meta-analysis, commercial vaccines analysis does not show their efficacy [log RR 1.07 (0.94, 1.22)], and benefit of vaccination demonstrated was rather a reduction of the severity in clinical cases, lowering the rate of culling and increasing the production of milk and milk solids (Mata et al. 2023).

Both EFSA (2022) and Mata et al.'s (2023) conclusion directs to the fact that vaccination can improve the health and welfare of dairy cattle but does not prevent the disease effectively. It therefore serves rather as an additional supportive tool to the traditional preventive measures, and it can be expected that farmers will put it under scrutiny as for cost-effectiveness ratio with other measures intended as prevention of mastitis.

Vaccines to protect from mastitis caused by *Streptococcus agalactiae*, *Streptococcus dysgalactiae* or *Mycoplasma* spp. have not been yet developed, authorised nor commercialised. Also, development and starting of needle-free and/or automated administration techniques in mastitis vaccination will be beneficial. Table 13 provides an overview of target causative mastitis pathogens and principle of vaccines available in the USA/Europe. Specific technology using siderophores receptor protein is present in the authorised vaccine claiming reduction of *Klebsiella* spp. mastitis in lactating cattle (Gorden et al. 2018).

8.1.4.2 Autogenous Vaccines

In several countries, use of autogenous mastitis vaccines is permitted and with Regulation on vet-

Kabelitz T, Aubry E, van Vorst K, Amon T, Fulde M (2021) The role of *Streptococcus* spp. in bovine mastitis. Microorganisms 9(7):1497

Kalińska A, Jaworski S, Wierzbicki M, Gołębiewski M (2019) Silver and copper nanoparticles—an alternative in future mastitis treatment and prevention? Int J Mol Sci 20(7):1672

Kammerer M, Le Guenic M, Roussel P, Linclau O, Cartaud G, Tainturier D, Larrat M, Bareille N (2009) Le traitement des mammites cliniques de la vache laitière par des huiles essentielles. Innov Agron 4:79–83. (in French)

Kano R (2020) Emergence of fungal-like organisms: prototheca. Mycopathologia 185(5):747–754

Keane OM (2019) Symposium review: Intramammary infections—Major pathogens and strain-associated complexity. J Dairy Sci 102:1–14

Keller D, Sundrum A (2018) Comparative effectiveness of individualised homeopathy and antibiotics in the treatment of bovine clinical mastitis: randomised controlled trial. Vet Rec 182(14):407

Kelly RA, Leedale J, Calleja D, Enoch SJ, Harrell A, Chadwick AE, Webb S (2019) Modelling changes in glutathione homeostasis as a function of quinone redox metabolism. Sci Rep 9(1):6333

Khatun M, Bruckmaier RM, Thomson PC, House J, García SC (2019) Suitability of somatic cell count, electrical conductivity, and lactate dehydrogenase activity in foremilk before versus after alveolar milk ejection for mastitis detection. J Dairy Sci 102:9200–9212

Kher MN, Sheth NR, Bhatt VD (2019) In vitro antibacterial evaluation of *Terminalia chebula* as an alternative of antibiotics against bovine subclinical mastitis. Anim Biotechnol 30(2):151–158

Kibebew K (2017) Bovine mastitis: A review of causes and epidemiological point of view. J Biol Agric Healthc 7:1–14

Kirkwood RM, Remnant JG, Payne RM, Murphy AM, Wapenaar W (2018) Risk of iatrogenic damage to the sciatic nerve in dairy cattle. Vet Rec 182(5):140

Klaas IC, Zadoks RN (2018) An update on environmental mastitis: Challenging perceptions. Transbound Emerg Dis 65(Suppl 1):166–185

Kleinhenz MD, Ydstie JA, Gorden P (2019) Case report: Management of *Klebsiella* spp. mastitis on a dairy farm. American Association of bovine Practitioners. Bovine Pract 53(1)

Koivula M, Pitkälä A, Pyörälä S, Mäntysaari EA (2007) Distribution of bacteria and seasonal and regional effects in a new database for mastitis pathogens in Finland. Acta Agric Scand Sect Anim Sci 57:89–96

Krishnamoorthy P, Suresh KP, Jayamma KS, Shome BR, Patil SS, Amachawadi RG (2021) An understanding of the global status of major bacterial pathogens of milk concerning bovine mastitis: a systematic review and meta-analysis (scientometrics). Pathogens 10:545

Kroker R (2003) Pharmaceuticals for the treatment and prevention of bacterial infections. In: Löscher W, Ungemach FR, Kroker R (eds) Pharmakotherapie bei Haus- und Nutztiere, 6th edn. Parey Buchverlag, im Blackwell, Berlin, pp 208–247. (in German)

Krömker V, Friedrich J (2011) Recommendations for diagnostic measures regarding mastitis control on herd level. Prakt Tierarzt 92:516–524. (in German)

Krömker V, Leimbach S (2017) Mastitis treatment-Reduction in antibiotic usage in dairy cows. Reprod Domest Anim 52(Suppl 3):21–29

Krömker V, Paduch JH, Abograra I, Zinke C, Friedrich J (2011) Effects of an additional nonsteroidal anti-inflammatory therapy with carprofen (Rimadyl Rind®) in cases of severe mastitis of high yielding cows. Berl Munch Tierarztl Wochenschr 124:161–167 (in German)

Krömker V, Reinecke F, Paduch JH, Grabowski N (2014) Bovine *Streptococcus uberis* intramammary infections and mastitis. Clin Microbiol Open Access 3:4

Krömker V, Schmenger A, Klocke D, Mansion-de Vries EM, Wente N, Zhang Y, Leimbach S (2021) Non-inferiority trial investigating the efficacy of non-steroidal anti-inflammatory drugs and antimicrobial treatment of mild to moderate clinical mastitis in dairy cows with long-lasting udder diseases. Front Vet Sci 8:660804

Krupová Z, Zavadilová L, Wolfová M, Krupa E, Kašná E, Fleischer P (2019) Udder and claw-related health traits in selection of Czech Holstein cows. Ann Animal Sci 19:647–661

Kuang Y, Jia H, Miyanaga K, Tanji Y (2009) Effect of milk on antibacterial activity of tetracycline against *Escherichia coli* and *Staphylococcus aureus* isolated from bovine mastitis. Appl Microbiol Biotechnol 84(1):135–142

Kurz JP, Yang Z, Weiss RB, Wilson DJ, Rood KA, Liu GE, Wang Z (2019) A genome-wide association study for mastitis resistance in phenotypically well-characterized Holstein dairy cattle using a selective genotyping approach. Immunogenetics 71:35–47

Lam TJ, Lipman LJ, Schukken YH, Gaastra W (1996) Brand A (1996) Epidemiological characteristics of bovine clinical mastitis caused by *Staphylococcus aureus* and *Escherichia coli* studied by DNA fingerprinting. Am J Vet Res 57(1):39–42

Landin H, Mörk MJ, Larsson M, Waller KP (2015) Vaccination against *Staphylococcus aureus* mastitis in two Swedish dairy herds. Acta Vet Scand 25(57):81

Landin H, Lundberg A, Gyllenswärd M, Carlén E, Lindberg M, Persson N (2020) The Udder health pyramid. Växa Sverige, Stockholm, Sweden, pp 1–55

Langer MN, Blodkamp S, Bayerbach M, Feßler AT, Buhr N, de Gutsmann T, Kreienbrock L, Schwarz S, Köckritz-Blickwede MV (2017) Testing cathelicidin susceptibility of bacterial mastitis isolates: Technical challenges and data output for clinical isolates. Vet Microbiol 210:107–115

Latosinski GS, Amzalak MJ, Pantoja JCF (2020) Efficacy of ketoprofen for treatment of spontaneous, culture-negative, mild cases of clinical mastitis: A randomized, controlled superiority trial. J Dairy Sci 103(3):2624–2635

Lees P, Pelligand L, Illambas J, Potter T, Lacroix M, Rycroft A, Toutain PL (2015) Pharmacokinetic/pharmacodynamic integration and modelling of amoxicillin for the calf pathogens *Mannheimia haemolytica* and *Pasteurella multocida*. J Vet Pharmacol Ther 38(5):457–470

Leite RF, Gonçalves JL, Peti APF, Figueiró FS, Moraes LAB, Santos MV (2018) Antimicrobial activity of crude extracts from actinomycetes against mastitis pathogens. J Dairy Sci 101(11):10116–10125

Leitner G, Pinchasov Y, Morag E, Spanier Y, Jacoby S, Eliau D, Pitcovski J (2013) Immunotherapy of mastitis. Vet Immunol Immunopathol 153(3/4):209–216

Leslie KE, Petersson-Wolfe CS (2012) Assessment and management of pain in dairy cows with clinical mastitis. Vet Clin North Am Food Anim Pract 28(2):289–305

Leuenberger A, Sartori C, Boss R, Resch G, Oechslin F, Steiner A, Moreillon P, Graber HU (2019) Genotypes of Staphylococcus aureus: On-farm epidemiology and the consequences for prevention of intramammary infections. J Dairy Sci 102(4):3295–3309

Li L, Wang R, Huang Y, Huang T, Luo F, Huang W, Yang X, Lei A, Chen M, Gan X (2018) High incidence of pathogenic Streptococcus agalactiae ST485 strain in pregnant/puerperal women and isolation of hypervirulent human CC67 strain. Front Microbiol 9:50

Libisch B, Picot C, Ceballos-Garzon A, Moravkova M, Klimesová M, Telkes G, Chuang ST, Le Pape P (2022) Prototheca infections and ecology from a one health perspective. Microorganisms 10(5):938

Linder M, Paduch JH, Grieger AS, Mansion-de Vries E, Knorr N, Zinke C, Teich K, Krömker V (2013) Cure rates of chronic subclinical *Staphylococcus aureus* mastitis in lactating dairy cows after antibiotic therapy. Berl Munch Tierarztl Wochenschr 126:291–296. (In German)

Lindquist DA, Baynes RE, Smith GW (2015) Short communication: Pharmacokinetics of intramammary hetacillin in dairy cattle milked 3 times per day. J Dairy Sci 98(3):1856–1861

Lipkens Z, Piepers S, Verbeke J, De Vliegher S (2019) Infection dynamics across the dry period using Dairy Herd Improvement somatic cell count data and its effect on cow performance in the subsequent lactation. J. Dairy Sci 102:640–651

Liu J, Ju M, Wu Y, Leng N, Algharib SA, Luo W (2022) Antibacterial activity of florfenicol composite nanogels against *Staphylococcus aureus* small colony variants. J Vet Sci 23(5):e78

Lobell RD, Varma KJ, Johnson JC, Sams RA, Gerken DF, Ashcraft SM (1994) Pharmacokinetics of florfenicol following intravenous and intramuscular doses to cattle. J Vet Pharmacol Ther 17(4):253–258

Lofano G, Mallett CP, Bertholet S, O'Hagan DT (2020) Technological approaches to streamline vaccination schedules, progressing towards single-dose vaccines. NPJ Vaccines 5:88

Lofstedt J, Dohoo IR, Duizer G (1999) Model to predict septicemia in diarrheic calves. J Vet Intern Med 13(2):81–88

Lucas MF (2009) Alternativas terapéuticas para el manejo racional de la mastitis subclínica por *Staphylococcus aureus*. In: Mestorino N and Errecalde J (eds) Pharmacokinetic–pharmacodynamic considerations for bovine mastitis treatment. (in Spanish)

Lucas MF, Moncada Cárdenas A, Marchetti L, Lambertini A, Mestorino N, Errecalde J (2009) Pharmacokinetics and T>MIC of amoxicillin-clavulanic acid in milk of lactating dairy cows with S. aureus subclinical mastitis. Proceeding BII-50. Biocell 33(1):A1–A52. ISSN 0327-9545

Luo W, Liu J, Algharib SA, Chen W (2022) Antibacterial activity of enrofloxacin loaded gelatin-sodium alginate composite nanogels against intracellular *Staphylococcus aureus* small colony variants. J Vet Sci 23(3):e48

Luo Y, Kong Z, Yang B, He F, Huan C, Li J, Yi K (2023) Relationship between microflora changes and mammary lipid metabolism in dairy cows with mastitis. Animals (Basel) 13(17):2773

Lyhs U, Kulkas L, Katholm J, Waller KP, Saha K, Tomusk RJ, Zadoks RN (2016) *Streptococcus agalactiae* serotype IV in humans and cattle, Northern Europe. Emerg Infect Dis 22(12):2097–2103

Maciel-Guerra A, Esener N, Giebel K, Lea D, Green MJ, Bradley AJ, Dottorini T (2021) Prediction of *Streptococcus uberis* clinical mastitis treatment success in dairy herds by means of mass spectrometry and machine-learning. Sci Rep 11(1):7736

Mader R, EU-JAMRAI, Bourély C, Amat JP, Broens EM, Busani L, Callens B, Crespo-Robledo P, Damborg P, Filippitzi ME, Fitzgerald W, Grönthal T, Haenni M, Heuvelink A, van Hout J, Kaspar H, Muñoz Madero C, Norström M, Pedersen K, Pokludova L, Dal Pozzo F, Slowey R, Urdahl AM, Vatopoulos A, Zafeiridis C, Madec JY (2022) Defining the scope of the European Antimicrobial Resistance Surveillance network in Veterinary medicine (EARS-Vet): a bottom-up and One Health approach. J Antimicrob Chemother 77(3):816–826

Madouasse A, Browne WJ, Huxley JN, Toni F, Bradley AJ, Green MJ (2012) Risk factors for a high somatic cell count at the first milk recording in a large sample of UK dairy herds. J Dairy Sci 95(4):1873–1884

Magalhães VJ, Susca F, Lima FS, Branco AF, Yoon I, Santos JE (2008) Effect of feeding yeast culture on performance, health, and immunocompetence of dairy calves. J Dairy Sci 91(4):1497–1509

Makovec JA, Ruegg PL (2003) Results of milk samples submitted for microbiological examination in Wisconsin from 1994 to 2001. J Dairy Sci 86(11):3466–3472

Malcata F, Pepler P, O'Reilly E, Brady N, Eckersall P, Zadoks R, Viora L (2020) Point-of-care tests for bovine clinical mastitis: What do we have and what do we need? J Dairy Res 87(S1):60–66, including supplementary table available online at: https://

static.cambridge.org/content/id/urn:cambridge.org:id:article:S002202992000062X/resource/name/S002202992000062Xsup001.pdf. Accessed 5 Aug 2023

Martin P, Barkema HW, Brito LF, Narayana SG, Miglior F (2018) Symposium review: Novel strategies to genetically improve mastitis resistance in dairy cattle. J Dairy Sci 101(3):2724–2736

Mata F, Jesus MS, Pinto RP, Mata A (2023) A meta-analysis of the mastitis vaccination efficacy in dairy cattle. Open Vet J 13(2):179–187

Mattie H, Zhang L, van Strijen E, Sekh BR, Douwes-Idema AEA (1997) Pharmacokinetic and Pharmacodynamic models of the antistaphylococcal effects of meropenem and cloxacillin in vitro and in experimental infection. Antimicrob Agents Chemother 41(10):2083–2088

McCubbin KD, de Jong E, Lam TJGM, Kelton DF, Middleton JR, McDougall S, de Vliegher S, Godden S, Rajala-Schultz PJ, Rowe S, Speksnijder DC, Kastelic JP, Barkema HW (2022) Invited review: Selective use of antimicrobials in dairy cattle at drying off. J. Dairy Sci 105:7161–7189

McDougall S (1998) Efficacy of two antibiotic treatments in curing clinical and subclinical mastitis in lactating dairy cows. N Z Vet J 46:226–232

McDougall S, Bryan MA, Tiddy RM (2009) Effect of treatment with the nonsteroidal antiinflammatory meloxicam on milk production, somatic cell count, probability of re-treatment, and culling of dairy cows with mild clinical mastitis. J Dairy Sci 92(9):4421–4431

McDougall S, Abbeloos E, Piepers S, Rao AS, Astiz S, van Werven T, Statham J, Pérez-Villalobos N (2016) Addition of meloxicam to the treatment of clinical mastitis improves subsequent reproductive performance. J Dairy Sci 99(3):2026–2042. d

McDougall S, Clausen L, Hintukainen J, Hunnam J (2019) Randomized, controlled, superiority study of extended duration of therapy with an intramammary antibiotic for treatment of clinical mastitis. J Dairy Sci 102(5):4376–4386

McDougall S, Clausen LM, Hussein HM, Compton CWR (2022) Therapy of subclinical mastitis during lactation. Antibiotics (Basel) 11(2):209

McParland S, Dillon PG, Flynn J, Ryan N, Arkins S, Kennedy A (2019) Effect of using internal teat sealant with or without antibiotic therapy at dry-off on subsequent somatic cell count and milk production. J Dairy Sci 102(5):4464–4475

Medeiros-Silva J, Jekhmane S, Paioni AL, Gawarecka K, Baldus M, Swiezewska E, Breukink E, Weingarth M (2018) High-resolution NMR studies of antibiotics in cellular membranes. Nat Commun 9(1):3963

Meganck V, Hoflack G, Opsomer G (2014) Advances in prevention and therapy of neonatal dairy calf diarrhoea: a systematical review with emphasis on colostrum management and fluid therapy. Acta Veterinaria Scandinavica 56(1):75

Mestorino N, Errecalde JO (2012) Pharmacokinetic–pharmacodynamic considerations for bovine mastitis treatment. In: Perez-Marin CC (ed) A bird's-eye view of veterinary medicine. IntechOpen, pp 423–472. Available online at: https://www.intechopen.com/chapters/28685. Accessed 24 July 2023

Mestorino N, Alt M, Errecalde JO (1993) The pharmacokinetics of thiamphenicol in lactating cows. Vet Res Commun 17:295–303

Mestorino N, Speroni J, Errecalde JO (1995) Thiamphenicol concentrations in serum andm ilk after its intramammary administration to Holando Argentino cows. In: Proceedings II of the 3rd international mastitis seminar, ISBN 965-222-636

Mestorino N, Martinez M, Pérsico JR, Garcia S, Buldain D, Buchamer A, Aliverti F, Marchetti L (2018) Pharmacokinetic and milk residues of cloxacillin in dairy cows after intramammary administration for dry cow therapy. In: National Mastitis Council 57th annual meeting, conference paper

Middleton JR, Luby CD, Adams DS (2009) Efficacy of vaccination against staphylococcal mastitis: a review and new data. Vet Microbiol 134:192–198

Mileva R, Karadaev M, Fasulkov I, Rusenova N, Vasilev N, Milanova A (2022) Oxytetracycline persistence in uterine secretion after intrauterine administration in cows with metritis. Animals 12(15):1922

Milho C, Andrade M, Vilas Boas D, Alves D, Sillankorva S (2019) Antimicrobial assessment of phage therapy using a porcine model of biofilm infection. Int J Pharm 557:112–123

Miller LM, Gal A (2017) Cardiovascular system and lymphatic vessels. In: Pathologic basis of veterinary disease (6th edn). Elsevier, pp 576, 606

Moncada Cárdenas A, Marchetti L, Daniele M, Errecalde J, Mestorino N (2009) Effect of pH on the antibacterial activity of azithromycin and penicillin against *Staphylococcus aureus*. Biocell 33(1)

Morandi S, Cremonesi P, Capra E, Silvetti T, Decimo M, Bianchini V, Alves AC, Vargas AC, Costa GM, Ribeiro MG, Brasca M (2016) Molecular typing and differences in biofilm formation and antibiotic susceptibilities among Prototheca strains isolated in Italy and Brazil. J Dairy Sci 99(8):6436–6445

Moravej H, Moravej Z, Yazdanparast M, Heiat M, Mirhosseini A, Moosazadeh Moghaddam M, Mirnejad R (2018) Antimicrobial peptides: features, action, and their resistance mechanisms in bacteria. Microb Drug Resist 24(6):747–767

Morin DE (2004) Brainstem and cranial nerve abnormalities: listeriosis, otitis media/interna and pituitary abscess syndrome. Vet Clin North Am Food Anim Pract 20:243–273. https://doi.org/10.1016/j.cvfa.2004.02.007

Morin DE, Shanks RD, McCoy GC (1998) Comparison of antibiotic administration in conjunction with supportive measures versus supportive measures alone for treatment of dairy cows with clinical mastitis. J Am Vet Med Assoc 213(5):676–684

Mujawar LH, Moers A, Norde W, van Amerongen A (2013) Rapid mastitis detection assay on porous nitrocellulose membrane slides. Anal Bioanal Chem 405:7469–7476

Muluye RA, Bian Y, Alemu PN (2014) Anti-inflammatory and antimicrobial effects of heat-clearing Chinese herbs: a current review. J Tradit Complement Med 4(2):93–98

Mushtaq S, Shah AM, Shah A, Lone SA, Hussain A, Hassan QP, Ali MN (2018) Bovine mastitis: An appraisal of its alternative herbal cure. Microb Pathog 114:357–361

Musser JM, Anderson KL (2001) Bioavailability and disposition of sodium and procaine penicillin G (benzylpenicillin) administered orally with milk to calves. J Vet Pharmacol Ther 24(3):161–169

Nagahata H, Moriyama A, Sawada C, Asai Y, Kokubu C, Gondaira S, Higuchi H (2020) Innate immune response of mammary gland induced by intramammary infusion of Bifidobacterium breve in lactating dairy cows. J Vet Med Sci 82(12):1742–1749

Nash DL, Rogers GW, Cooper JB, Hargrove GL, Keown JF, Hansen LB (2000) Heritability of clinical mastitis incidence and relationships with sire transmitting abilities for somatic cell score, udder type traits, productive life, and protein yield. J. Dairy Sci 83:2350–2360

Naushad S, Naqvi SA, Nobrega D, Luby C, Kastelic JP, Barkema HW, De Buck J (2019) Comprehensive virulence gene profiling of bovine non-aureus staphylococci based on whole-genome sequencing data. mSystems 4:1–23. https://doi.org/10.1128/mSystems.00098-18

Nehme R, Andrés S, Pereira RB, Ben Jemaa M, Bouhallab S, Ceciliani F, López S, Rahali FZ, Ksouri R, Pereira DM, Abdennebi-Najar L (2021) Essential oils in livestock: from health to food quality. Antioxidants (Basel) 10(2):330

Nickell J (2022) Whsper on arrival, Merck Animal Health Announces WHISPER® ON ARRIVAL | Merck Animal Health USA (merck-animal-health-usa.com), Accessed online June 2024

Nickerson SC, Owens WE, Tomita GM, Widel P (1999) Vaccinating dairy heifers with a *Staphylococcus aureus* bacterin reduces mastitis at calving. Large Anim Pract 20:16–28

Niemi RE, Hovinen M, Vilar MJ, Simojoki H, Rajala-Schultz PJ (2021) Dry cow therapy and early lactation udder health problems-Associations and risk factors. Prev Vet Med 188:105268

Nobrega DB, Naqvi SA, Dufour S, Deardon R, Kastelic JP, De Buck J, Barkema HW (2020) Critically important antimicrobials are generally not needed to treat nonsevere clinical mastitis in lactating dairy cows: Results from a network meta-analysis. J Dairy Sci 103(11):10585–10603

Nonnemann B, Svennesen L, Lyhs U, Kristensen KA, Klaas IC, Pedersen K (2019) Bovine mastitis bacteria resolved by matrix-assisted laser desorption/ionization time of flight mass spectrometry. J Dairy Sci 102:2515–2524

Odden A, Enemark HL, Ruiz A, Robertson LJ, Ersdal C, Nes SK, Tømmerberg V, Stuen S (2018) Controlled efficacy trial confirming toltrazuril resistance in a field isolate of ovine Eimeria spp. Parasit Vector 11(1):394

Ødegaard S, Sviland S (2001) Comparison of intramammary antibiotic preparations for the treatment of clinical bovine mastitis caused by bacteria sensitive to penicillin. In: Proc. 2nd int. symp. mast milk quality, Vancouver, Canada, pp 502–503

Ogorevc J, Kunej T, Razpet A, Dovc P (2009) Database of cattle candidate genes and genetic markers for milk production and mastitis. Animal Gen 40:832–851

Oliveira L, Ruegg PL (2014) Treatments of clinical mastitis occurring in cows on 51 large dairy herds in Wisconsin. J Dairy Sci 97(9):5426–5436

Orellano MS, Isaac P, Breser ML, Bohl LP, Conesa A, Falcone RD, Porporatto C (2019) Chitosan nanoparticles enhance the antibacterial activity of the native polymer against bovine mastitis pathogens. Carbohydr Polym 213:1–9

Orr J, Geraghty T, Ellis K (2014) Anti-inflammatories in cattle medicine. Livestock (Lond) 19:322–328

Østerås O, Sølverød L, Reksen O (2006) Milk culture results in a large Norwegian survey—effects of season, parity, days in milk, resistance, and clustering. J Dairy Sci 89:1010–1023

Painter K, Anand S, Philip K (2023) Omphalitis. StatPearls. https://www.ncbi.nlm.nih.gov/books/NBK513338/. Assessed 28 Aug 2023

Pantoja JCF, Correia LBN, Rossi RS, Latosinski GS (2020) Association between teat-end hyperkeratosis and mastitis in dairy cows: A systematic review. J Dairy Sci 103(2):1843–1855

Pas ML, Bokma J, Lowie T, Boyen F, Pardon B (2023) Sepsis and survival in critically ill calves: Risk factors and antimicrobial use. J Vet Intern Med 37(1):374–389

Pearce SD, Parmley EJ, Winder CB, Sargeant JM, Prashad M, Ringelberg M, Felker M, Kelton DF (2023) Evaluating the efficacy of internal teat sealants at dry-off for the prevention of new intra-mammary infections during the dry-period or clinical mastitis during early lactation in dairy cows: A systematic review update and sequential meta-analysis. Prev Vet Med 212:105841

Pellegrino M, Berardo N, Giraudo J, Nader-Macias MEF, Bogni C (2017) Bovine mastitis prevention: humoral and cellular response of dairy cows inoculated with lactic acid bacteria at the dry-off period. Benef Microbes 8(4):589–596

Peralta OA, Carrasco C, Vieytes C, Tamayo MJ, Muñoz I, Sepulveda S, Tadich T, Duchens M, Melendez P, Mella A, Torres CG (2020) Safety and efficacy of a mesenchymal stem cell intramammary therapy in dairy cows with experimentally induced *Staphylococcus aureus* clinical mastitis. Sci Rep 10:2843

Perner J, Winter P, Baumgartner W (2002) Retrospektive Studie zum Einsatz von Synulox® in der Mastitistherapie (Retrospective study using Synulox® in mastitis therapy). Tierärztliche Praxis 30:286–294. ISSN: 0303-6286

Persson Waller K, Hårdemark V, Nyman AK, Duse A (2016) Veterinary treatment strategies for clinical mastitis in dairy cows in Sweden. Vet Rec 178(10):240

Petzl W, Zerbe H, Günther J, Seyfert HM, Hussen J, Schuberth HJ (2018) Pathogen-specific responses in the bovine udder. Models and immunoprophylactic concepts. Res Vet Sci 116:55–61

Pilla R, Schwarz D, König S, Piccinini R (2012) Microscopic differential cell counting to identify inflammatory reactions in dairy cow quarter milk samples. J Dairy Sci 95:4410–4420

Pinheiro Machado GG, Veleirinho MB, Mazzarino LL, Machado Filho LCP, Maraschin M, Cerri R, Kuhnen S (2019) Development of propolis nanoparticles for the treatment of bovine mastitis: in vitro studies on antimicrobial and cytotoxic activities. Can J Animl Sci 99:713–723

Pinzón-Sánchez C, Ruegg PL (2011) Risk factors associated with short-term post-treatment outcomes of clinical mastitis. J Dairy Sci 94(7):3397–3410

Piras C, Greco V, Gugliandolo E, Soggiu A, Tilocca B, Bonizzi L, Zecconi A, Cramer R, Britti D, Urbani A et al (2020) Raw cow milk bacterial consortium as bioindicator of circulating anti-microbial resistance (AMR). Animals 10:13

Pitkäla A, Koort J, Björkroth J (2008) Identification and antimicrobial resistance of Streptococcus uberis and Streptococcus parauberis isolated from bovine milk samples. J Dairy Sci 91:4075–4081

Plessers E, Wyns H, Watteyn A, Pardon B, De Baere S, Sys SU, De Backer P, Croubels S (2016) Immunomodulatory properties of gamithromycin and ketoprofen in lipopolysaccharide-challenged calves with emphasis on the acute-phase response. Vet ImmunolImmunopathol 171:28–37

Poonsuk K, Kordik C, Hille M, Cheng TY, Crosby WB, Woolums AR, Clawson ML, Chitko-McKown C, Brodersen B, Loy JD (2023) Detection of *Mannheimia haemolytica*-specific IgG, IgM and IgA in sera and their relationship to respiratory disease in cattle. Animals (Basel) 13(9):1531

Poutrel B, Stegemann MR, Roy O, Pothier F, Tilt N, Payne-Johnson M (2008) Evaluation of the efficacy of systemic danofloxacin in the treatment of induced acute *Escherichia coli* bovine mastitis. J Dairy Res 75(3):310–318

Prieto A, López-Novo C, Díaz P, Díaz-Cao JM, López-Lorenzo G, Antón C, Remesar S, García-Dios D, López C, Panadero R, Díez-Baños P, Morrondo P, Fernández G (2022) Antimicrobial susceptibility of enterotoxigenic Escherichia coli from diarrhoeic neonatal calves in Spain. Animals (Basel) 12(3):264

Puig A, Perozo E, Roura P, Franquesa O, Herrera D, Urtado L, Calm I, Armengol R, Mallo J, Ponté D, Echevarria JM, Arrieta C, Collado R, Prenafeta T, March R, Sabaté D (2018) Efficacy and safety under field conditions of the new vaccine UBAC® against *Streptococcus uberis* intramammary infections in dairy cows. In: International bovine mastitis conference, NMC 2018, Milano

Pyörälä S (2009) Treatment of mastitis during lactation. Ir Vet J 62(Suppl 4):S40–S44

Pyörälä S, Manner L, Kesti E, Sandholm M (1994) Local tissue damage in cows after intramuscular injections of eight antimicrobial agents. Acta Vet Scand 35(1):107–110

Quinn PJ, Markey BK, Leonard FC, Fitzpatrick ES, Fanning S, Hartigan PJ (2011) Veterinary microbiology and microbial disease. Wiley-Blackwell, Chichester, UK

Rainard P (2017) Mammary microbiota of dairy ruminants: fact or fiction? Vet Res 48(1):25

Rainard P, Foucras G (2018) A critical appraisal of probiotics for mastitis control. Front Vet Sci 5:251

Rainard P, Gilbert FB, Germon P, Foucras G (2021) Invited review: a critical appraisal of mastitis vaccines for dairy cows. J Dairy Sci 104:10427–10448

Ranjith D, Nisha AR, Nair SN, Litty M, Rahman M, Juliet S (2018) Evaluation of analgesic and anti-inflammatory activity of herbal formulation used for mastitis in animals. Int J Appl Sci Eng IJASE 6(1):37–48

Rasmussen F (1959) Mammary excretion of benzylpenicillin, erythromycin, and penethamate hydroiodide. Acta Pharmacol Toxicol (Copenh) 16:194–200

Reboud J, Xu G, Garrett A, Adriko M, Yang Z, Tukahebwa EM, Rowell C, Cooper JM (2019) Paper-based microfluidics for DNA diagnostics of malaria in low resource underserved rural communities. Proc Natl Acad Sci U S A 116:4834–4842

Regulation (EU) 2019/6 of the European Parliament and of the Council of 11 December 2018 on veterinary medicinal products and repealing Directive 2001/82/EC. Off J Eur Union 4:43–167

Reinoso EB (2017) Bovine mastitis caused by *Streptococcus uberis*. Virulence factors and biofilm. J Microb Biochem Technol 9:5

Ren X, Wang L, Chen W (2020) Oxytropis glabra DC. Inhibits biofilm formation of staphylococcus epidermidis by down-regulating ica operon expression. Curr Microbiol 77(7):1167–1173

Resende D, Bilby T, Bruno R, Lager K, Jordan E (2009) Dry cow mastitis and immunotherapy: The use of natural body defenses, Texas A&M AgriLife Extension Service The Texas A&M University System, Available online at: http://agrilife.org/texasdairymatters/files/2012/09/6-09-MastitisImmunotherapy-Updated-01-17-1.pdf. Accessed 23 Sept 2023

Rhodes A, Evans LE, Alhazzani W, Levy MM, Antonelli M, Ferrer R, Kumar A, Sevransky JE, Sprung CL, Nunnally ME, Rochwerg B, Rubenfeld GD, Angus DC, Annane D, Beale RJ, Bellinghan GJ, Bernard GR, Chiche JD, Coopersmith C, De Backer DP, French CJ, Fujishima S, Gerlach H, Hidalgo JL, Hollenberg SM, Jones AE, Karnad DR, Kleinpell RM, Koh Y, Lisboa TC, Machado FR, Marini JJ, Marshall JC, Mazuski JE, McIntyre LA, McLean AS, Mehta S, Moreno RP, Myburgh J, Navalesi P, Nishida O, Osborn TM, Perner A, Plunkett CM, Ranieri M, Schorr CA, Seckel MA, Seymour CW, Shieh L, Shukri KA, Simpson

SQ, Singer M, Thompson BT, Townsend SR, Van der Poll T, Vincent JL, Wiersinga WJ, Zimmerman JL, Dellinger RP (2017) Surviving sepsis campaign: international guidelines for management of sepsis and septic shock: 2016. Intensive Care Med 43(3):304–377

Rivera JD, Woolums AR, Giguère S, Johnson JT, Lutz AG, Tipton PN, Crosby WB, Hice I, Thoresen M (2019) Pharmacokinetics of tulathromycin following administration to stocker cattle with remote delivery devices. J Anim Sci 97(11):4482–4487

Robertson LJ, Campbell AT, Smith HV (1992) Survival of *Cryptosporidium parvum* oocysts under various environmental pressures. Appl Environ Microbiol 58:3494–3500

Rogers DG, Cheville NF, Pugh GW (1987) Pathogenesis of corneal lesions caused by *Moraxella bovis* in gnotobiotic calves. Vet Pathol 24:287–295

Rollin E, Dhuyvetterb KC, Overton MW (2015) The cost of clinical mastitis in the first 30 days of lactation: An economic modeling tool. Prev Vet Med 122:257–264

Rowe SM, Godden SM, Nydam DV, Gorden PJ, Lago A, Vasquez AK, Royster E, Timmerman J, Thomas MJ (2020) Randomized controlled non-inferiority trial investigating the effect of 2 selective dry-cow therapy protocols on antibiotic use at dry-off and dry period intramammary infection dynamics. J Dairy Sci 103:6473–6492

Roy SM, Riley MA, Crabb JH (2016) The bacteriocins: current knowledge and future prospects. Caister Academic Press, Poole, UK. Treating bovine mastitis with nisin: a model for the use of protein antimicrobials in veterinary medicine, pp 127–140

Ruegg PL (2017) A 100-year review: mastitis detection, management, and prevention. J Dairy Sci 100:10381–10397

Ruegg PL (2018) Making antibiotic treatment decisions for clinical mastitis. Vet Clin North Am Food Anim Pract 34(2018):413–425

Ruegg PL (2022) Realities, challenges and benefits of antimicrobial stewardship in dairy practice in the United States. Microorganisms 10(8):1626

Ruegg PL, Pantoja JCF (2013) Understanding and using somatic cell counts to improve milk quality. Ir J Agric Food Res 52:101–117

Ruiz-Romero RA, Vargas-Bello-Pérez E (2023) Non-aureus staphylococci and mammaliicocci as a cause of mastitis in domestic ruminants: current knowledge, advances, biomedical applications, and future perspectives—a systematic review. Vet Res Commun 47(3):1067–1084

Saleh A, Macia Ortega A, Muotri AR (2019) Transposable elements, inflammation and neurological disease. Front Neurol 10:894

Santman-Berends IMGA, van den Heuvel KWH, Lam TJGM, Scherpenzeel CGM, van Schaik G (2021) Monitoring udder health on routinely collected census data: Evaluating the short- to mid-term consequences of implementing selective dry cow treatment. J Dairy Sci 104:2280–2289

Savoia S, Brugiapaglia A, Pauciullo A, Di Stasio L, Schiavon S, Bittante G, Albera A (2019) Characterisation of beef production systems and their effects on carcass and meat quality traits of Piemontese young bulls. Meat Sci 153:75–85

Schmenger A, Krömker V (2020) Characterization, cure rates and associated risks of clinical mastitis in Northern Germany. Vet Sci 7:170. https://doi.org/10.3390/vetsci7040170

Schmenger A, Leimbach S, Wente N, Zhang Y, Biggs AM, Krömker V (2020) Implementation of a targeted mastitis therapy concept using an on-farm rapid test: antimicrobial consumption, cure rates and compliances. Vet Rec 187:401. https://doi.org/10.1136/vr.105674

Schnitt A, Tenhagen BA (2020) Risk factors for the occurrence of methicillin-resistant staphylococcus aureus in dairy herds: an update. Foodborne Pathog Dis 17(10):585–596

Schukken YH, Wilson DJ, Welcome F, Garrison-Tikofsky L, Gonzalez RN (2003) Monitoring udder health and milk quality using somatic cell counts. Vet Res 34:579–596

Schukken YH, Bronzo V, Locatelli C, Pollera C, Rota N, Casula A, Testa F, Scaccabarozzi L, March R, Zalduendo D, Guix R, Moroni P (2014) Efficacy of vaccination on *Staphylococcus aureus* and coagulase-negative staphylococci intramammary infection dynamics in 2 dairy herds. J Dairy Sci 97:5250–5264

Schwarz D, Diesterbeck US, Failing K, Konig S, Brugemann K, Zschock M, Wolter W, Czerny CP (2010) Somatic cell counts and bacteriological status in quarter foremilk samples of cows in Hesse, Germany—a longitudinal study. J Dairy Sci 93:5716–5728

Scott DW (2007) Color atlas of farm animal dermatology, 1st edn. Blackwell Publishing Professional, Ames, IA

Seegers H, Fourichon C, Beaudeau F (2003) Production effects related to mastitis and mastitis economics in dairy cattle herds. Vet Res 34:475–491

Serieys F, Raguet Y, Goby L, Schmidt H, Friton G (2005) Comparative efficacy of local and systemic antibiotic treatment in lactating cows with clinical mastitis. J Dairy Sci 88:93–99

Shahbandeh M (2023) Cattle population worldwide 2012-2023. Statista. Available online at: https://www.statista.com/statistics/263979/global-cattle-population-since-1990/. Accessed 4 July 2023

Sharma C, Rokana N, Chandra M, Singh BP, Gulhane RD, Gill JPS, Ray P, Puniya AK, Panwar H (2017) Antimicrobial resistance: its surveillance, impact, and alternative management strategies in dairy animals. Front Vet Sci 4:237

Sharun K, Dhama K, Tiwari R, Gugjoo MB, Iqbal Yatoo M, Patel SK, Pathak M, Karthik K, Khurana SK, Singh R, Puvvala B, Amarpal R, Singh KP, Chaicumpa W (2021) Advances in therapeutic and managemental approaches of bovine: a comprehensive review. Vet Q 41(1):107–136

Shave CD, Millyard L, May RC (2021) Now for something completely different: Prototheca, pathogenic algae. PLoS Pathog 17(4):e1009362

Sheldon IM, Cronin J, Goetze L, Donofrio G, Schuberth HJ (2009) Defining postpartum uterine disease and the mechanisms of infection and immunity in the female reproductive tract in cattle. Biol Reprod 81(6):1025–1032

Shimazaki KI, Kawai K (2017) Advances in lactoferrin research concerning bovine mastitis. Biochem Cell Biol 95:69–75

Shinozuka Y, Morita T, Watanabe A, Kawai K (2018) Live bacteria in clots from bovine clinical mastitis milk with no growth in conventional culturing. Asian J Anim Vet Adv 13:197–200

Shpigel NY, Levin D, Winkler M, Saran A, Ziv G, Böttner A (1997) Efficacy of cefquinome for treatment of cows with mastitis experimentally induced using Escherichia coli. J Dairy Sci 80(2):318–323

Shpigel NY, Winkler M, Ziv G, Saran A (1998) Relationship between in vitro sensitivity of coliform pathogens in the udder and the outcome of treatment for clinical mastitis. Vet Rec 142:135–137

Shukla R, Pandey V, Vadnere GP, Lodhi S (2019) Role of flavonoids in management of inflammatory disorders. In Watson R.R., Preedy V.R (ed) Bioactive food as dietary interventions for arthritis and related inflammatory disease. Elsevier, Amsterdam. pp. 293–322.

Sintes GF, Bruckmaier RM, Wellnitz O (2020) Nonsteroidal anti-inflammatory drugs affect the mammary epithelial barrier during inflammation. J Dairy Sci 103(11):10742–10753

Smith G (2015) Antimicrobial decision making for enteric diseases of cattle. Vet Clin North Am Food Anim Pract 31(1):47–60

Smola J, Prášek J, Novák P, Malá G (2022) Volby antmkrobiálních látek pro raconální léčbu infekčních onemocnění hospodářských zvířat (skotu, prasat a drůbeže), Veterinární univerzita Brno, Výzkumný ústav živočišné výroby, v. v. i., Praha Uhříněves,. ISBN 978-80-7403-284-4 (Czech language) 2022_CM_VUZV_Volby antimikrobiálních látek pro racionální léčbu infekčních onemocnění hospodářských zvířat. pdf (agronavigator.cz) Accessed June 2024

Smulski S, Gehrke M, Libera K, Cieaslak A, Huang H, Patra AK, Szumacher-Stragel M (2020) Effects of various mastitis treatments on the reproductive performance of cows. BMC Vet Res 16:99

Song J, Ruan H, Chen L et al (2021) Potential of bacteriophages as disinfectants to control of Staphylococcus aureus biofilms. BMC Microbiol 21(1):57

Sora VM, Panseri S, Nobile M, Di Cesare F, Meroni G, Chiesa LM, Zecconi A (2022) Milk quality and safety in a one health perspective: results of a prevalence study on dairy herds in Lombardy (Italy). Life 12(6):786

Sørensen UBS, Poulsen K, Ghezzo C, Margarit I, Killan M (2010) Emergence and global dissemination of host-specific *Streptococcus agalactiae* clones. MBio 1:e00178

Souza RFS, Rault L, Seyffert N, Azevedo V, Le Loir Y, Even S (2018) *Lactobacillus casei* BL23 modulates the innate immune response in *Staphylococcus aureus*-stimulated bovine mammary epithelial cells. Benef Microbes 9(6):985–995

Staněk S (2023) Effective rearing of calves (Efektivní odchov telat), vol XII, pp 1–8 (in Czech). Available online at: https://www.mikrop.cz/magazin/odchov--telat-xii~m1516. Accessed 25 June 2023

Steele N, McDougall S (2014) Effect of prolonged duration therapy of subclinical mastitis in lactating dairy cows using penethamate hydriodide. N Z Vet J 62(1):38–46

Stewart N, Nordstrom S, Champaign J, Hurwitch A, Hogan J (2022) Field trial to compare efficacy of two commercially available *Escherichia coli* J-5 vaccines against clinical coliform mastitis. In: Oral—Udder health and mastitis, published in IVIS with the permission of the WBC, World Buiatric Congress, Madrid

Stockler RM, Morin DE, Lantz RK, Constable PD (2009) Effect of milking frequency and dosing interval on the pharmacokinetics of cephapirin after intramammary infusion in lactating cows. J Dairy Sci 92:4262–4275

Suojala L, Kaartinen L, Pyörälä S (2013) Treatment for bovine Escherichia coli mastitis—an evidence-based approach. J Vet Pharmacol Ther 36(6):521–531

Svennesen L, Skarbye AP, Farre M, Astrup LB, Halasa T, Krömker V, Denwood M, Kirkeby C (2023) Treatment of mild to moderate clinical bovine mastitis caused by gram-positive bacteria: A noninferiority randomized trial of local penicillin treatment alone or combined with systemic treatment. J Dairy Sci 106(8):5696–5714

Swinkels JM, Lam TJ, Green MJ, Bradley AJ (2013) Effect of extended cefquinome treatment on clinical persistence or recurrence of environmental clinical mastitis. Vet J 197(3):682–687.

Taponen S, Liski E, Heikkilä AM, Pyörälä S (2017) Factors associated with intramammary infection in dairy cows caused by coagulase-negative staphylococci, *Staphylococcus aureus, Streptococcus uberis, Streptococcus dysgalactiae, Corynebacterium bovis,* or *Escherichia coli.* J Dairy Sci 100(1):493–503

Tariq A, Adnani M, Ahmad M, Shinwari S, Amber R, Sadia S, Mussarat S (2017) Pakistani anti-mastitis medicinal plants and their scientific validation against multidrug resistance mastinogen *Staphylococcus aureus*. Pak J Bot 49(6):2395–2404

Teijeira Á, Garasa S, Gato M, Alfaro C, Migueliz I, Cirella A, Melero I (2020) CXCR1 and CXCR2 chemokine receptor agonists produced by tumors induce neutrophil extracellular traps that interfere with immune cytotoxicity. Immunity 52:856–871

Thomson S, Hamilton CA, Hope JC, Katzer F, Mabbott NA, Morrison LJ, Innes EA (2017) Bovine cryptosporidiosis: impact, host-parasite interaction and control strategies. Vet Res 48:42

Toutain PL, Bousquet-Mélou A, Damborg P, Ferran AA, Mevius D, Pelligand L, Veldman KT, Lees P (2017) En route towards european clinical breakpoints for veterinary antimicrobial susceptibility testing: a position paper explaining the VetCAST approach. Front Microbiol 8:2344

UBC (2020) Giving injections: subcutaneous (SC), intramuscular (IM) and intravenous (IV); SOP University of British Columbia, Centre for Dairy Education and Research, pp 1–9. Available online at: https://dc.sites.olt.ubc.ca/files/2021/01/SOP-Cow-012-Giving-IV--SC-IM-injections.pdf. Accessed 25 July, 4 Aug 2023

Vailati-Riboni M, Coleman DN, Lopreiato V, Alharthi A, Bucktrout RE, Abdel-Hamied E, Martinez-Cortes I, Liang Y, Trevisi E, Yoon I, Loor JJ (2021) Feeding a *Saccharomyces cerevisiae* fermentation product improves udder health and immune response to a Streptococcus uberis mastitis challenge in mid-lactation dairy cows. J Anim Sci Biotechnol 12(1):62

Van Amstel SR, Shearer JK (2006) Manual for treatment and control of lameness in cattle, 1st edn. Blackwell, Iowa

van Biervliet J, Perkins GA, Woodie B, Pelligrini-Massini A, Divers TJ, de Lahunta A (2004) Clinical signs, computed tomographic imaging, and management of chronic otitis media/interna in dairy calves. J Vet Intern Med 18(6):907–910

Van den Borne BHP, van Schaik G, Lam TJGM, Nielen M (2010) Variation in herd level mastitis indicators between primi-and multiparae in Dutch dairy herds. Prev Vet Med 96(20541822):49–55

van Soest FJS, Santman-Berends IMGA, Lam TJGM, Hogeveen H (2016) Failure and preventive costs of mastitis on Dutch dairy farms. J Dairy Sci 99(10):8365–8374

Vezina B, Al-harbi H, Ramay HR, Soust M, Moore RJ, Olchowy TWJ, Alawneh JI (2021) Sequence characterisation and novel insights into bovine mastitis-associated *Streptococcus uberis* in dairy herds. Sci Rep 11:3046

Vissio C, Dieser SA, Agnelli HL, Odierno LM, Larriestra AI (2014) Accuracy of the composite somatic cell count to detect intra-mammary infection in dairy cows using latent class analysis. Prev Vet Med 113:547–555

Wallis JK, Krömker V Paduch JH (2018) Biofilm formation and adhesion to bovine udder epithelium of potentially probiotic lactic acid bacteria. AIMS Microbiol 4(2):209–224

Wang LH, Li XY, Jin LJ, You JS, Zhou Y, Li SY, Xu YP (2011) Characterization of chicken egg yolk immunoglobulins (IgYs) specific for the most prevalent capsular serotypes of mastitis-causing *Staphylococcus aureus*. Vet Microbiol 149(3/4):415–421

Wente N, Krömker V (2020) *Streptococcus dysgalactiae*: contagious or environmental? Animals (Basel) 10(11)

Wente N, Grieger AS, Klocke D, Paduch JH, Zhang Y, Leimbach S, Tho Seeth M, Mansion-De Vries EM, Mohr E, Krömker V (2020) Recurrent mastitis-persistent or new infections? Vet Microbiol 244:108682

WOAH (2022) Annual report on antimicrobial agents intended for use in animals, 6th edn. Available online at: https://www.woah.org/app/uploads/2022/06/a--sixth-annual-report-amu-final-1.pdf. Accessed 4 July 2023

Wolf J, Wolfova M, Stipkova M (2010) A model for the genetic evaluation of number of clinical mastitis cases per lactation in Czech Holstein cows. J Dairy Sci 93:1193–1204

Wu J, Hu S, Cao L (2007) Therapeutic effect of nisin Z on subclinical mastitis in lactating cows. Antimicrob Agents Chemother 51(9):3131–3135

Wu Z, Zhang Z, Lei Z, Lei P (2019) CD14. Biology and role in the pathogenesis of disease. Cytokine Growth Factor Rev 48:24–31. https://doi.org/10.1016/j.cytogfr.2019.06.003

Yang X, Ouyang W, Sun J, Li X (2009) Post-antibiotic effect of amoxicillin nanoparticles against main pathogenic bacteria of bovine mastitis in vitro. J Northwest A F Univ Nat Sci Ed 37(6):1–6

Yang WT, Ke CY, Wu WT, Lee RP, Tseng YH (2019) Effective treatment of bovine mastitis with intramammary infusion of angelica dahurica and rheum officinale extracts. Evid Based Complement Alternat Med 2019:7242705

Yeruham I, Elad D (2004) Acute bilateral suppurative otitis media in a dairy cow. Can Vet J 45:79

Young-Speirs M, Drouin D, Cavalcante PA, Barkema HW, Cobo ER (2018) Host defense cathelicidins in cattle: types, production, bioactive functions and potential therapeutic and diagnostic applications. Int J Antimicrob Agents 51(6):813–821

Yu J, Ren Y, Xi X, Huang W, Zhang H (2017) A Novel lactobacilli-based teat disinfectant for improving bacterial communities in the milks of cow teats with subclinical mastitis. Front Microbiol 8:1782

Zadoks RN, Allore HG, Barkema HW, Sampimon OC, Wellenberg GJ, Gröhn YT, Schukkent YH (2001) Cow- and quarter-level risk factors for Streptococcus uberis and Staphylococcus aureus mastitis. J Dairy Sci 84(12):2649–2663

Zadoks RN, Gillespie BE, Barkema HW, Sampimon OC, Oliver SP, Schukken YH (2003) Clinical, epidemiological and molecular characteristics of Streptococcus uberis infections in dairy herds. Epidemiol Infect 130(2):335–349

Zambriski JA, Nydam DV, Bowman DD, Bellosa ML, Burton AJ, Linden TC, Liotta JL, Ollivett TL, Tondello-Martins L, Mohammed HO (2013) Description of fecal shedding of *Cryptosporidium parvum* oocysts in experimentally challenged dairy calves. Parasitol Res 112:1247–1254

Zecconi A, Piccinini R, Fiorina S, Cabrini L, Daprà V, Amadori M (2009) Evaluation of interleukin-2 treatment for prevention of intramammary infections in cows after calving. Comp Immunol Microbiol Infect Dis 32(5):439–451

Zecconi A, Dell'Orco F, Rizzi N, Vairani D, Cipolla M, Pozzi P, Zanini L (2020) Cross-sectional study on the prevalence of contagious pathogens in bulk tank milk and their effects on somatic cell counts and milk yield. Ital J Anim Sci 19(1):66–74

Zemanova M, Langova L, Novotná I, Dvorakova P, Vrtkova I, Havlicek Z (2022) Immune mechanisms, resistance genes, and their roles in the prevention of mastitis in dairy cows. Arch Anim Breed 65(4):371–384

Zhen YH, Jin LJ, Guo J, Li XY, Lu YN, Chen J, Xu YP (2008) Characterization of specific egg yolk immunoglobulin (IgY) against mastitis-causing *Escherichia coli*. Vet Microbiol 130(1/2):126–133

Zhen YH, Jin LJ, Li XY, Guo J, Li Z, Zhang BJ (2009) Efficacy of specific egg yolk immunoglobulin (IgY) to bovine mastitis caused by staphylococcus aureus. Vet Microbiol 133:317–322

Zhou K, Wang X, Chen D, Yuan Y, Wang S, Li C, Yan Y, Liu Q, Shao L, Huang L, Yuan Z, Xie S (2019) Enhanced treatment effects of tilmicosin against *Staphylococcus aureus* cow mastitis by self-assembly sodium alginate-chitosan nanogel. Pharmaceutics 11(10):524

Zhu L, Cao X, Xu Q, Su J, Li X, Zhou W (2018) Evaluation of the antibacterial activity of tilmicosin-SLN against *Streptococcus agalactiae*: in vitro and in vivo studies. Int J Nanomed 13:4747–4755

Ziesch M, Wente N, Zhang Y, Zaremba W, Engl S, Krömker V (2017) Noninferiority trial investigating the efficacy of a nonantibiotic intramammary therapy in the treatment of mild-to-moderate clinical mastitis in dairy cows with longer lasting udder diseases. J Vet Pharmacol Ther 41(28449183):11–21. https://doi.org/10.1111/jvp.12415

Zigo F, Vasil' M, Ondrašovičová S, Výrostková J, Bujok J, Pecka-Kielb E (2021) Maintaining optimal mammary gland health and prevention of mastitis. Front Vet Sci 8:607311

Ziv G (1980a) Practical pharmacokinetic aspects of mastitis therapy - 3: intramammary treatment. Vet Med Small Anim Clin 75(4):657–670

Ziv G (1980b) Drug selection and use in mastitis: systemic vs local therapy. J Am Vet Med Assoc 176(10):1109–1115

Ziv G, Storper M (1985) Intramuscular treatment of subclinical staphylococcal mastitis in lactating cows with penicillin G, methicillin and their esters. J Vet Pharmacol Ther 8(3):276–283

Zouharova M, Nedbalcova K, Slama P, Bzdil J, Masarikova M, Matiasovic J (2022) Occurrence of virulence associated genes in *Streptococcus uberis* and *Streptococcus parauberis* isolated from bovine mastitis. Vet Med Czech 67:123–130

Antimicrobial Considerations for Horses

Keith Edward Baptiste and Lucie Pokludová

Abstract

The treatment and management of bacterial diseases in horses require a portfolio of antimicrobials with good efficacy and safety profiles. This portfolio has been narrowed in the last decades, due to the overuse of antimicrobials both in human and veterinary medicine leading to a constant selection pressure of antimicrobial-resistant clinical bacterial isolates. Following the tools of antimicrobial stewardship can help to preserve the efficacy of the available portfolio, via minimising the need for antimicrobials. Efforts should be focused on minimising the of use of critically important antimicrobials. The biggest barrier to prudent use of antimicrobials in equine medicine is 'tradition'. With the shortcomings of robust evidence-based selection of relevant antimicrobials or combination antimicrobials, including knowledge of factors such as pathogen/infection/target site of infection/animal (age) category, the authors provide an update and advice commentary, which can serve as a basis for responsible choices of antimicrobials. This is key for treatment, but conditions and individual characteristics of the animal treated should be considered as well. Some traditional uses of antimicrobials in equine medicine are challenged and refuted. Indications covered represent the most common diseases of respiratory, gastrointestinal, urinary and reproductive tract, as well as covering musculoskeletal, ophthalmologic, cardiovascular, neurological and skin infections. Recommendations are provided regarding dose, interval and route of administration together with specific remarks, where considered appropriate. The theme of the chapter is responsible and judicious use of antimicrobials, including perioperative prophylaxis. Being aware of equine medicine specifics then off-label and cascade use is covered and commented. Final remarks target the difficulties but also opportunities for the future use of antimicrobials.

Keywords

Antimicrobials in horses · Responsible use · Treatment options · Perioperative · Prophylaxis

K. E. Baptiste (✉)
Department of Veterinary Medicine, Danish Medicines Agency (Lægemiddelstryelsen), Copenhagen South, Denmark
e-mail: keb@dkma.dk

L. Pokludová
Institute for State Control of Veterinary Biologicals and Medicines, Brno, Czech Republic

L. Pokludová (ed.), *Antimicrobials in Livestock 2: Choices of Prevention and Treatment*,
https://doi.org/10.1007/978-3-031-60807-0_5

1 Introduction

Antimicrobial stewardship involves practices that promote the prudent use of antimicrobials, improve patient outcomes, reduce antimicrobial resistance and decrease the spread of infections caused by multidrug-resistant organisms. Unfortunately, equine medicine is far less progressive on antimicrobial stewardship issues compared to other branches of veterinary medicine, relying more on tradition than evidence-based medicine. Over-reliance on broad-spectrum antimicrobial combinations (e.g. penicillin/gentamicin, trimethoprim/sulphonamide, rifampicin/macrolide), prophylaxis and empirical therapy has contributed to increased and spread of multidrug-resistant organisms, especially in equine hospitals. Even when evidence emerges that traditional combination therapy is contrary to antimicrobial stewardship principles (e.g. rifampicin/macrolide combinations), recommendations remain unchanged.

Although the basic principles of equine antimicrobial therapy are not different than in other animal species, there are some special considerations. Some horses are food-producing animals, and inherent concerns about antimicrobial residues and antimicrobial resistance in food products need to be considered if equine meat is destined for human consumption. As hindgut fermenters, horses are particularly susceptible to adverse gastrointestinal consequences from antimicrobial administration. As a result, relatively few antimicrobials are safe for adult horses, compared to other animal species. The fragility as well as the economic and emotional value of neonatal foals encourages broad-spectrum antimicrobial treatment, despite neonatal foal pharmacodynamics being poorly understood. The size and temperament of a horse may influence the selection of treatment options based on the ability of veterinarians or owners to safely administer drugs via different routes. In some countries, lay people have ready access to over-the-counter antimicrobials, and veterinarians may encounter cases that have been treated with one or more drugs, often with inappropriate dosing regimens. Furthermore, both the limited portfolio of authorised equine veterinary medicinal products and the unmet clinical need of some unauthorised administration routes (e.g. intra-articular administration) are further complicated by a lack of availability on the market, especially in small countries (Knych and Magdesian 2021). This can lead to the regular use of unproven dosing regimens with limited predictability of the adverse reactions as well as overuse of certain antimicrobials. Moreover, in 2022 the European Commission Implementing Regulation (EU) 2022/1255 designated antimicrobials or groups of antimicrobials reserved for treatment of certain infections in human, which will impact certain antimicrobials used occasionally in horses (e.g. ticarcillin use in mares or foals). All these factors need to be taken into consideration when prescribing or administering antimicrobials to a horse.

2 Antimicrobial Stewardship in Equine Practice

Antimicrobial stewardship, originating from human medicine, is further promoted in veterinary medicine, especially in companion animals and specialised clinics. According to the EU legislation, all animals belonging to *Equidae* are considered food-producing animals, but in fact, a major proportion are kept and declared in the "horse passport" as non-food producing (never to be slaughtered for human consumption).

As pointed out by Prescott (2020), antimicrobial stewardship is a complex, dynamic and multifaceted approach that involves continuous improvements based on reducing, improving, monitoring and evaluating the use of antimicrobials to ensure future efficacy, both in human and veterinary areas. Adapting antimicrobial stewardship for equine practice is quite challenging. Until now, there was a popular programme, PROTECT me (British Equine Veterinary Association—BEVA; www.beva.org.uk/Protect-Me) containing ToolKit, modelled on the BSAVA PROTECT guidelines. The name of the programme is the abbreviation compiled reflecting the complexity of essential factors that can be used in equine stewardship: *p*ractice policy, *r*educe prophylaxis, *o*ther options, *t*ypes of drug and bacteria, employ narrow spectrum, *c*ulture and sensitivity, *t*reat effectively, *m*onitor use, compliance and resistance and *e*ducate.

The authors' perception is that several clinically relevant principles can be defined for equine practice, with an emphasis of some aspects:

Practice policy: Decision/s on antimicrobial use should rely on the veterinarian only. Prescribers should be veterinarians with knowledge of the animal's history and previous conditions relevant to the care of that individual horse.

Reduce prophylaxis: Antimicrobial use only when absolutely necessary, following valid legal provisions as well as prudent use principles.

Other options: Consider non-antimicrobial options, whenever appropriate (e.g. reduce bacterial load by debridement or lavage where this is possible, optimise patient immunity, maintain infection control: isolate from contagious conditions, ensure aseptic protocols, perform barrier nursing and hand/working tool hygiene).

Types of drug and bacteria: Carefully consider proper diagnostic procedures and expected causal agent/s. Collect proper sample/s and explain to the owner the necessity of these steps and payment for them.

Employ a narrow-spectrum antimicrobial for "first-line" treatment for the diagnosed cause, whenever possible. Broad-spectrum antimicrobials should be reserved for those cases with evidence for their need (e.g. thorough diagnostics/culture/susceptibility).

Culture and susceptibility: Perform culture and susceptibility testing, whenever possible and reasonable (not only for the sake of the individual patient but also for establishing a pool of results for the same/similar aetiology cases). Do not forget to collect samples prior to the start the administration of antimicrobials, ensure proper sampling technique and transport of samples and communicate results with laboratory, especially in case of doubts.

Treat effectively: Use an evidence-based approach to antimicrobial medicinal product administration, disease prevention, hygiene and biosecurity measures to minimise spread of antimicrobial resistance.

Monitor use, compliance and resistance: Whenever possible, identify common clinical scenarios in your practice and establish protocols formulated for antimicrobial therapy (first, second and last resort treatment). Completely avoid any use of antimicrobials from the list of antimicrobials reserved for the treatment of human only (EU Commission Implementing Act 1255/2022). Ensure a dosing regimen (dose, frequency, treatment duration) for the route of administration most relevant for the condition being treated. Record all antimicrobial administrations, document the use and effect and report any cases of failed efficacy of authorised medicines to the relevant authorities.

Educate: Ensure education, an evidence-based approach and regular updating of knowledge not only for antimicrobial use but also for prevention (hygiene, biosecurity, care on animal, etc.) and accommodate communication to all people involved (veterinarians, zootechnicians, farmers/owners).

3 Adverse Effects of Antimicrobials

A variety of adverse effects can occur as a result of antimicrobial therapy, including colitis or generally less serious dysbiosis, hypersensitivity reactions (e.g. urticaria), immune-mediated disease (e.g. immune-mediated haemolytic anaemia/thrombocytopenia), arthropathy, nephrotoxicity, ototoxicity, dysrhythmia, anorexia as well as neurologic abnormalities (Khusro et al. 2020). Characteristics of the antimicrobial substance/s alone are not the only reason for adverse effects, but severity and possible resolution can also be linked with the route of administration (intra-articular, intramuscular can be linked to local reactions at the injection site, as well systemic reactions) and pharmaceutical forms (better tolerated solution for injection than suspensions, with systemic reactions related to the properties of the antimicrobial). Toutain (2014) commented that local tolerance is poor with aminoglycosides, sulphonamide/trimethoprim, macrolides and tetracyclines. Better local tolerance is known for penicillin, where procaine penicillin is better than penicillin G. Clinically, the most serious adverse effect is antimicrobial-associated colitis (AAC). This syndrome develops in temporal association with antimicrobial therapy and may be related to proportional changes in the intestinal microflora or the proliferation and toxin production of *Clostridioides difficile*. Although the relative risk of colitis from different antimicrobials is not known, there are important individual factors. Thus, prevention of AAC is difficult to predict, but there is some degree of risk with any antimicrobial administered via any route. Drugs that have low oral bioavailability or excreted in bile or enterocytes pose a higher risk because of the drug levels achieved within the gastrointestinal tract. Drugs with activity against anaerobes are also considered more likely to cause colitis. To date, individual horse risks for AAC are difficult to define. Some new insights come from new methods for understanding the microbiome and metabolome (Liepman et al. 2022). Horses with AAC have presumably greater disturbances of the bacterial communities within the hindgut and subsequent metabolic changes that result in more severe clinical symptoms. (Arnold et al. 2021). If *Clostridioides difficile* is present at a low level within the gastrointestinal tract of a horse, then the risk is high, but it is not possible to know this beforehand.

Thus, if oral antimicrobials are chosen, then proven safety and efficacy in horses are important. Drugs such as lincomycin, clindamycin and oral penicillins/aminopenicillins are considered a very high risk and should never be used in adult horses. Macrolides are used typically in foals for treatment of *Rhodococcus equi* but can cause fatal colitis for accompany adult mares, even following minimal exposure (Baverud et al. 1998; Gustafsson et al. 1997). However, this is not a simple association between macrolide use in foals and AAC in accompanying mares, since foals are also treated typically with concomitant rifampicin. It is well known that rifampicin significantly blocks the oral bioavailability of macrolides, via the activation of p-glycoproteins on the enterocyte cell membranes, and thus rifampicin is likely the indirect cause of AAC in mares. Fatal colitis has also been reported following oral doxycycline administration, and one out of six horses died of acute colitis in a study on doxycycline pharmacokinetics (Davis et al. 2006). Also, other selected antimicrobials including fluoroquinolones, tetracyclines (Barr et al. 2013), trimethoprim-sulphonamides and beta-lactams (penicillins, aminopenicillins, cephalosporins) (Baverud et al. 1997) as well as metronidazole (McGorum et al. 1998) have been reported with AAC in horses. The likelihood and consequences of AAC should be carefully considered when deciding whether antimicrobials are necessary.

Enrofloxacin has been further associated with arthropathy in young foals. However, complete scientific documentation on the adverse effects of enrofloxacin in foals has never been published.

4 Drug Interactions

Negative in vivo drug interactions in horses do occur and should be considered when selecting antimicrobial therapy. In horses, oral bioavailability of drugs shows great individual variations. Absorption of most antimicrobial agents is significantly hindered by feeding, and thus 2 – 4 h fasting is essential before administration of these drugs. Even in these cases, systemic availability can show wide variations between individuals, as the case of metronidazole between 60% and 90% (Baggot et al. 1988). Despite some attempts to introduce enteric coating pharmaceutical forms in equine medicine, most orally administered antimicrobial veterinary medicinal products are oral pastes or powder, where interactions occur with food.

Some negative drug interactions that should be avoided in equine medicine include:

- β-lactams (penicillins/cephalosporins) with tetracyclines. The inhibition of cell wall synthesis exerted by β-lactams requires bacterial replication and is affected by the bacteriostatic effect of tetracyclines.
- Procaine penicillin and trimethoprim/sulphonamides. Trimethoprim/sulphonamides

inhibit bacterial folic acid synthesis, but many bacteria can break down the procaine portion of penicillin to para-aminobenzoic acid, a precursor of folic acid, thus counteracting the effect of trimethoprim/sulphonamides.

- Trimethoprim/sulphonamides and alpha-2 agonist drugs. This combination enhances fatal cardiac arrhythmias.
- Rifampin/Rifampicin:
  - Profound in vivo negative drug interactions between rifampicin and macrolides, either administered orally or injectable macrolides, preventing either the macrolide or rifampicin to reach minimum inhibitory concentrations (MIC) at the target site of infection.
  - Fluoroquinolones and rifampin. Rifampicin results in inhibition of bacterial autolysin synthesis, which is necessary for the antibacterial effect of fluoroquinolones (Michelet et al. 1997).
  - Trimethoprim/sulphonamides and rifampin. The latter drug increases trimethoprim/sulphonamide elimination.
  - Rifampicin further antagonises metronidazole and aminoglycosides.
- Phenylbutazone increases plasma concentrations of penicillin and sulphonamides but lowers tissue concentrations by displacing drugs from their protein-binding sites. Phenylbutazone decreases the gentamicin elimination half-life by 23% and decreases volume of distribution (V_d) by 26%.

5 Antimicrobial Resistance in Bacteria Isolated from Horses

The emergence of multidrug-resistant bacteria is a global problem in horses, as with other animals. Moreover, as horses are kept in close contact with their owners, then a "One Health perspective" should be considered (Lönker et al. 2020), from the perspective of zoonosis as well as multiresistance pathogens. *Staphylococcus aureus* resistant to methicillin (Baptiste et al. 2005) but also carrying resistance to other antimicrobials or *Enterobacterales* with extended spectrum beta-lactamases (ESBL), *Enterobacter cloacae*, *Escherichia coli*, *Klebsiella oxytoca* and *Citrobacter freundii* with both ESBL and *mcr-9* (genetically determining resistance to colistin) have been reported in horses (Börjesson et al. 2020). Also, these pathogens and types of resistance increase the risk of treatment failure leading to higher costs to horse owners as a consequence of prolonged hospitalisation and use of other antimicrobials, including drugs of critical importance in human medicine.

Extended-spectrum beta-lactamase (ESBL) producing *Enterobacterales*, multiresistant *Salmonella* spp. as well as methicillin-resistant *Staphylococcus aureus* (MRSA) also belong among "nosocomial" pathogens of significant public health concern reported from many horse clinics (Weese et al. 2005). These pathogens imply a poor prognosis for foals and adult horses as well as pose risks for veterinarians, nurses and other employees working at clinics (Weese et al. 2006; Dolejska et al. 2008; Maddox et al. 2012). Of concern, especially within referral hospitals (comparing to first opinion submissions), are isolates with multidrug resistance (MDR), as commented in the study by Isgren et al. (2021), where the prevalence of MDR was 31.7% for *E. coli* and 25.3% for *Staphylococcus* spp. isolates (UK clinic; 6018 isolates tested). MDR isolates (n = 347) were sampled from surgical site infections, catheter-related infections or associated with orthopaedic indications. For *Enterococcus* spp. there was no readily available treatment for adult horses in the UK in 30.2% of positive submissions. In contrast, *Actinobacillus* spp., *Pasteurella* spp. and β-haemolytic *Streptococcus* spp. do not belong among isolates with MDR profiles (Isgren et al. 2021).

Some programmes exist in referral horse clinics (e.g. systems for standardised registration, surveillance and evaluation), including cause-of-effect investigations of (nosocomial) infections (Rohwedder et al. 2021). New introduced systems manage clustering of bacterial infections associated with certain cases, with profiling and evaluating of isolates (especially MDR), monitoring hospital facilities, including critical points identified, and monitoring of the animals (e.g. foal admitted with mare, monitor not only diseased foal but in-contact mare as well). This is important especially for some bacterial species from both Gram-negatives (*Pseudomonas* spp., *Acinetobacter* spp., *E. coli*) and Gram-positives (*Enterococcus* spp., *Staphylococcus* spp.) which, due to their high intrinsic resistance, have limited treatment options for horses (Isgren 2021).

Recently, the European Commission requested an expert panel from EFSA on Animal Health and Welfare (AHAW) to assess current scientific evidence for resistant animal pathogens that cause transmissible animal diseases and identify the most relevant antimicrobial-resistant bacteria in the EU (Nielsen et al. 2021). A literature review as well as experiences gained from three European countries monitoring systems (FINRES, SWEDRES and RESAPATH) resulted in considering the following equine pathogens (no. of studies in bracket; total "*n*" of isolates from which data were considered/weighted arithmetic average calculated) and main findings. For *Escherichia coli*, selected hospitals and referral clinics studies show that isolates from hospitalised patients are more commonly resistant than isolates from the community (especially for third-/fourth-generation cephalosporins, where data from 7 studies (n = 1677) were involved), due to either an over-representation of complicated cases, some already treated before submission, or nosocomial hospital strain infections. For *Staphylococcus aureus* (MRSA, 8; n = 3365) results varied, but one common feature from MRSA strain typing studies was that horses carry a specific host-adapted CC398 lineage. Studies testing *Klebsiella* spp. (studies differ as per antimicrobial number of isolates tested; the EU: 3–4 studies; n = 251–420 isolates; the USA: 1–2 studies; n = 24–65 isolates) demonstrated a higher proportion of resistance among US than EU isolates. For *Pseudomonas aeruginosa*, a high proportion of resistance was reported for fluoroquinolones in both EU (2; n = 1384) and US (1; n = 36) studies. From studies performed mostly in the USA, for erythromycin (4; n = 1 716) and (3; n = 1 686) rifampicin isolates, both antimicrobials' weighted arithmetic average proportion of resistance was 15.6% and 15.7%, respectively, for *Rhodococcus equi*. Despite other species of bacteria tested, very limited datasets were obtained for some pathogens (*Streptococcus equi* subsp. *equi* (3), *Pasteurella* spp. (2), *Streptococcus dysgalactiae* subsp. *dysgalactiae/equisimilis* (2), *Enterococcus* spp. (1)), while for *Actinobacillus equuli* (0) and *Dermatophilus congolensis*, no relevant data were obtained fulfilling the criteria (limited also by the fact that only English bibliography were involved).

6 Empirical Antimicrobial Therapy and Diagnostic Submissions

Empirical antimicrobial therapy is still routinely employed in horses, although it is generally recognised that submission of culture and antimicrobial susceptibility testing (AST) are critical for successful treatment of individual patients, identification of population (herd, region) problems and detection of changes in pathogen distribution or resistance patterns. Unless the nature of the disease is such that there is no contraindication to delaying therapy, empirical therapy at the first visit is a common practice in equine medicine. In some situations, it may not be required or even appropriate to collect a diagnostic specimen. For example, nasal swabs (except in suspected cases of strangles with the presence of purulent exudate) and swabs from contaminated wounds have limited diagnostic value due to the presence of high numbers of bacterial contaminants. In other cases, representative diagnostic specimens are challenging due to difficulties to access the infection site (i.e. abdominal abscesses) or economic constraints. Also, false-negative results can occur due to intermittent shedding of the pathogen, presence of fastidious or unculturable organisms, failure to use specialised microbiological media, improper sample collection, improper sample storage or shipping or prior antimicrobial therapy. However, in most circumstances, the importance of AST cannot be disregarded. Furthermore, it has become essential that WHO high/highest-priority critically important antimicrobials represented by ceftiofur, enrofloxacin, rifampicin and polymyxin B (not always used for infections) are used prudently in horses and ideally based on AST. In some European countries (e.g. Germany or Czechia), based on their national law, there is an obligation for mandatory AST for certain antimicrobials (e.g. third-/fourth-generation cephalosporins and fluoroquinolones). This national policy could, through laboratory AST, further highlight some emerging profiles of multidrug resistance, especially in hospital settings, where nosocomial strains can be an issue.

Disease conditions that are not life-threatening but prone to relapse or treatment failure (e.g. cystitis) and caused by pathogens with unpredictable antimicrobial susceptibility (e.g. Gram-negative bacteria) should be treated empirically with narrow-spectrum antimicrobials, and culture specimens should be collected prior to initiation of therapy. A variety of potentially debilitating/life-threatening infections can be encountered, such as severe pneumonia, pleuropneumonia, peritonitis, septic arthritis and neonatal septicaemia. Only in these situations is the empirical use of broad-spectrum antimicrobials or antimicrobial combinations (i.e. penicillin and aminoglycoside) justified while awaiting culture results. There are also disease conditions in which empirical antimicrobial therapy is not recommended. Mild upper respiratory tract infection (i.e. likely viral in origin), diarrhoea in adult horses and superficial wounds (not involving a joint or tendon sheath) do not typically require antimicrobials. Supportive care, close monitoring and wound management should be adequate.

Recently, especially due to high proportions of resistance in some pathogens and also due to the high value of individual horses or outbreaks on breeding horse farms (e.g. *Rhodococcus equi*), more advanced, rapid methods have become available. Single-plex and multiplex qPCR assays useful for *R. equi* and rapid diagnostic accompanies with detection of MLS_B resistance (via specific *erm* gene detection) were designed and validated (Narváez et al. 2022) and are promising tools to improve diagnosis and targeted use of antimicrobials. Another example is loop-mediated isothermal amplification (LAMP) assay and PCR, followed by high-resolution melt (HRM) curve analysis (Garner et al. 2023), for mares suspected of endometritis from *Streptococcus zooepidemicus*. Comparing standard culturing methods taking at least 3 days, these methods allow short turnaround time and high level of sensitivity/specificity for detection of *S. zooepidemicus* (6 h for PCR-HRM and 2 h for LAMP, with slightly lower specificity equal to 89.5%). Advantages of these assays are the avoidance of unnecessary treatment of mares, minimising antibiotic resistance selection pressure.

7 Disease and Pathogen-Specific Guidelines

Equine guidelines are not always based on objective and fully valid scientific studies. Clinical reports, antimicrobial susceptibility data (limited by unavailability of interpretive criteria) and clinical experiences are often used to formulate published recommendations because of the relative paucity of scientific data. As such, equine guidelines are often dominated by recommendations that promote broad-spectrum antimicrobials use as well as antimicrobial combinations. There is an unmet need for more, well-designed and evaluated, research to rationalise the use of antimicrobials in equine medicine (Weese et al. 2008).

The accompanying disease pathogen Table 1 provides recommendations for specific disease/pathogen treatment possibilities. Being aware of the above shortcomings and searching available recent bibliography and amending information, the tables have been drafted considering a combination of factors, including expected pathogens, expected susceptibility patterns and typical patient factors. The scientific quality of the literature on which these tables are based is highly variable, due to general lack of well-controlled studies on antimicrobial efficacy in horses over time. As already mentioned, combinations are frequently used, typically penicillin and gentamicin, or even more broad, triple combination of penicillin, gentamicin and metronidazole. Such triple combinations are sometimes used for the

Table 1 Antimicrobial advice for various equine medical/surgical conditions

Respiratory					
Disease/procedure	Pathogen(s)	First choice	Second choice	Last choice	Comments
Upper airways					
Guttural pouch empyema	Various, especially *Streptococcus* spp.	None	Penicillin		Guttural pouch lavages more important. Systemic antibiotics rarely indicated. Important to ensure no *S. equi* carriers
Strangles	*Streptococcus equi* subsp. *equi*	None	Penicillin	Ampicillin	Typical first-time cases do *not* require antibiotic treatment
Laryngoplasty Tie forward	Skin bacteria contaminates	Penicillin	SulfaTMP		
Laser surgery (GP fenestration, epiglottic entrapment, subepiglottic cyst)		None	Penicillin	SulfaTMP	Controlling inflammation is more important
Sinusitis	*Streptococcus* spp. *Staphylococcus* spp.	None	Penicillin	SulfaTMP	Antimicrobials are not necessary with adequate removal and flushing of purulent debris
Lower airways—All suspected lower airway infections should have diagnostic samples (e.g. tracheal aspiration and/or bronchoalveolar lavage) submitted for culture and sensitivity before starting antibiotic treatment					
Bacterial pneumonia	β-haemolytic streptococci (incl. *Streptococcus equi* ssp. *zooepidemicus*), *Staphylococcus* spp., *Actinobacillus* spp., anaerobes	Penicillin[c,d] or SulfaTMP	Cephalosporin	Ampicillin[c,d] or marbofloxacin[d]	
	E. coli, *Klebsiella* spp.	SulfaTMP	Ampicillin	Cephalosporin or marbofloxacin[d]	
	Stenotrophomonas maltophilia	SulfaTMP	Tetracycline	Gentamicin[d] or marbofloxacin[d]	Prolonged treatment duration may be required
Lung abscess	β-haemolytic streptococci, *Actinobacillus* spp., *E. coli*	Penicillin[c,d]	Ampicillin	Macrolides[a]	Consider surgical drainage in refractory cases. Penicillin does not penetrate into the abscess but prevents spread
Mycoplasma pneumonia	*Mycoplasma felis* *Mycoplasma equirhinis*	Tetracycline	Marbofloxacin[d]	Macrolides[a]	Long-term treatment may be needed. Can also present as pleuropneumonia
Pleuropneumonia	β-haemolytic streptococci, *Staphylococcus* spp., *Actinobacillus* spp., *E. coli*, *Mycoplasma*, anaerobes	Penicillin and gentamicin[d,e]	Cephalosporin or macrolides	Penicillin and marbofloxacin[d,e]	Supportive care and pleural drainage are important. Gentamicin is less effective in chronic cases due low O_2 tension and inflammatory debris. Also, consider anaerobic infection

Rhodococcus equi[b]	Subclinical Abs Score ≤10 cm	None	Doxycycline		High self-cure rate. Monitor clinically and thoracic ultrasound.
	Mild to moderate signs Abs Score 10-15 cm	Macrolides[a]	Macrolides[a] and doxycycline		Treatment for 3–4 weeks. Monotherapy just as effective. Monitor clinically and thoracic ultrasound. Doxycycline monotherapy performed poorly in randomised clinical trials
	Severe signs	Macrolides[a] and doxycycline	Enrofloxacin/ marbofloxacin		Treatment for 3–4 weeks. Poor prognosis. Macrolides and tetracyclines are considered synergistic
Pneumonia in foals	*Bordetella bronchiseptica*	Gentamicin[d]	Amikacin	Macrolides[a]	Prolonged treatment may be required
	Pneumocystis carinii	SulfaTMP			Associated with immunocompromised foals and *R. equi*

Gastrointestinal

Disease/procedure	Pathogen(s)	First choice	Second choice	Last choice	Comments
Exploratory laparotomy	Celiotomy—clean	Penicillin	Ampicillin		Important to check skin suture line regularly
	Celiotomy—contaminated or enterotomy	Penicillin and gentamicin[e,d]	Ampicillin and gentamicin[e,d]	Penicillin and enrofloxacin[e,d]	Supportive care is more important. Important to check skin suture line regularly
Abdominal abscess	*Streptococcus* spp., *Staphylococcus* spp., *Rhodococcus* spp.	Penicillin or tetracycline	Macrolides[a]		Long-term treatment required. Poor prognosis. Penicillin does not penetrate into the abscess but prevents spread
Antibiotic-associated diarrhoea	*Clostridioides difficile*	Metronidazole[f]	Bacitracin		Poor prognosis
Clostridium-associated diarrhoea	*Clostridium perfringens*	Metronidazole	Penicillin	Bacitracin	
Septic peritonitis	*Enterobacteriaceae*, streptococci, staphylococci, *Rhodococcus* spp., *Actinobacillus* spp., *Bacillus* spp., *Bacteroides* spp.	Penicillin and gentamicin[e,d]	Cephalosporin[e]	Penicillin and enrofloxacin[e,d]	Difficult to find cause. Culture peritoneal fluid. Consider other causes of peritonitis (e.g. nematode larval migration). Gentamicin less effective in chronic cases due to low O_2 tension and inflammatory debris
Rectal tear (partial thickness)	Mixed GI flora	SulfaTMP or penicillin	Ampicillin	Ampicillin and gentamicin[d]	NSAIDs more important. Full-thickness rectal tear is the same Tx as septic peritonitis

(continued)

Table 1 (continued)

Gastrointestinal					
Disease/procedure	Pathogen(s)	First choice	Second choice	Last choice	Comments
Proliferative enteropathy	*Lawsonia intracellularis*	Oxytetracycline or doxycycline	Macrolides[a,d]		Difficult to prove as cause
Proximal enteritis	*Clostridium perfringens*, perhaps others	Metronidazole (IV)[f]	Penicillin		Efficacy unknown
Salmonellosis	*Salmonella* typhimurium, *Salmonella* anatum, *Salmonella* newport	None	SulfaTMP or ampicillin with gentamicin[d]	Cephalosporin or enrofloxacin	Antibiotics for immune-compromised patients or neonates, based on antibiotic susceptibility

Hepatic					
Disease/procedure	Pathogen(s)	First choice	Second choice	Last choice	Comments
Cholangiohepatitis Cholangitis	Enteric bacteria, especially *E. coli*	SulfaTMP or oxytetracycline	Ampicillin and gentamicin[d]	Cephalosporin or enrofloxacin[d]	Consider causes other than bacterial infection (e.g. nematode larval migration). Long-term therapy required. Poor prognosis if several obstructing calculi are present
Liver abscess	β-haemolytic streptococci, *R. equi*, *E. coli*	Penicillin or tetracycline	Macrolides[a,d]		May occur with other abdominal abscesses. Poor prognosis
Listeriosis	*Listeria monocytogenes*	Penicillin or ampicillin	Oxytetracycline	Cephalosporin[d]	Especially Icelandic horses fed silage

Musculoskeletal					
Disease/procedure	Pathogen(s)	First choice	Second choice	Last choice	Comments
Superficial wounds	*Streptococcus* spp., *Staphylococcus* spp.	Topical disinfectant	Penicillin	SulfaTMP	Lavage, debridement and local therapy are more important than systemic antibiotics
Deep wounds[g]	Various, including anaerobes	Topical wound dressings	Cephalosporin[e] or SulfaTMP[e]	Penicillin and gentamicin[d,e]	Poultice and wound dressing products can draw out infection and purulent debris. Heavy metal-coated bandage dressings have potent antibacterial properties. Systemic antibiotics indicated if wound involves important structures (e.g. joint, tendon). Check tetanus vaccine status
Wounds: deep penetrating foot wounds	Various, including anaerobes	Penicillin and gentamicin[d,e]	Cephalosporin and gentamicin[d,e]	Tetracycline[e]	Consider regional perfusion. Check tetanus vaccine status
Septic arthritis	*Staphylococcus* spp., *S. zooepidemicus*, *E.coli*, *Actinobacillus* spp.	Intra-articular gentamicin[d] or cephalosporin[d]	Penicillin and gentamicin[d]	Enrofloxacin[d]	Intra-articular therapy and lavage are important. Can also consider regional perfusion

Synovial structures	Synovial injuries	Penicillin and gentamicin[d]			Intra-articular therapy and lavage are important. Can also consider regional perfusion
	Arthroscopy	None	Penicillin	Penicillin and gentamicin[d]	Critical to have sterile technique
Septic tenosynovitis		Penicillin and gentamicin[d,e], intra-synovial gentamicin[d], regional limb perfusion	Cephalosporin	Penicillin and enrofloxacin[d]	Lavage is crucial. Check tetanus vaccine status. Poor prognosis. Gentamicin maybe less effective in chronic cases due to low O_2 tension and inflammatory debris
Clostridial myositis	*C. perfringens*, *Clostridium septicum*, *Clostridium chauvoei*	Penicillin (IV)	Metronidazole[f]	Tetracycline	Surgical debridement/aeration is critical. Check tetanus vaccine status. Poor prognosis
Lymphangitis	*P. aeruginosa*, *Streptococcus* spp., *Staphylococcus* spp., *Pasteurella* spp.	Penicillin and gentamicin[d,e]	Cephalosporin	Enrofloxacin[d]	Check tetanus vaccine status
Osteomyelitis	Staphylococci, streptococci, *Salmonella* spp., *E. coli*, *Klebsiella* spp., *Acinetobacter* spp., *Enterobacter* spp.	Penicillin (IV) or ampicillin ± gentamicin[d,e]	Cephalosporin	Penicillin and enrofloxacin[d,e]	Regional limb perfusion, intra-osseous perfusion or antibiotic-impregnated materials may be useful
Fractures	Fractures with implants	Penicillin and gentamicin[d]			Supportive care is more important
	Closed fractures	Penicillin			Supportive care is more important
	Open fractures	Penicillin and gentamicin[d]			Supportive care is more important
	Long bone fractures (open or closed)	Penicillin and gentamicin[d]	Ampicillin and gentamicin[d]	Penicillin and enrofloxacin[d]	Can also consider regional perfusion
	Head fractures	SulfaTMP	Penicillin and gentamicin[d]	Penicillin and enrofloxacin[d]	
	Arthrodesis—implants	None	Penicillin and gentamicin[d]		

(continued)

Table 1 (continued)

Dermatology

Disease/procedure	Pathogen(s)	First choice	Second choice	Third choice	Comments
Mass removal (sarcoid)		None	Penicillin	SulfaTMP	Electrocautery or superficial wound management are alternatives. Check tetanus vaccine status
Removal/lancing abscess		Drainage, flushing with antiseptic			Check tetanus vaccine status
Cellulitis	*S. aureus*, *E. coli*	Penicillin and gentamicin[i]	Cephalosporin or SulfaTMP	Enrofloxacin[i]	Surgical drainage may be required. Check tetanus vaccine status
Dermatophilosis	*Dermatophilus congolensis*	Chlorhexidine shampoo	Penicillin	SulfaTMP	Often self-limiting. Important to keep affected skin dry
Folliculitis and furunculosis	*Staphylococcus* spp., mainly *S. aureus*	Povidone-iodine *or* chlorhexidine shampoo	SulfaTMP *or* cephalosporin	Topical mupirocin[i]	Often self-limiting. Remove underlying risk factors
Pastern dermatitis (scratches, mud fever, grease heal)	May involve *S. aureus*, *D. congolensis* and β-haemolytic streptococci	Topical treatment with povidone-iodine *or* chlorhexidine shampoo	SulfaTMP or penicillin		Multifactorial disease, not always involving bacterial infection. Need to address other factors. Important to keep affected skin dry
Staphylococcal pyoderma	Mainly *S. aureus*	Topical antimicrobials; antibacterial shampoos	SulfaTMP	Cephalosporin	Important to have culture and susceptibility

Urinary tract disease

Disease/ procedure	Pathogen(s)	First choice	Second choice	Last choice	Comments
Cystitis	*E.coli, Proteus* spp., *Pseudomonas* spp., *Klebsiella* spp., *Enterobacter* spp., *Streptococcus* spp., *Staphylococcus* spp.	Penicillin or SulfaTMP	Ampicillin or cephalosporin	Penicillin and gentamicin[i] or enrofloxacin[i]	Sulfamethoxazole should not be used since excreted in an inactive form. Beware that *biofilms* may develop on bladder wall that antibiotics cannot penetrate. Culture/susceptibility with CFU counts is advisable
Pyelonephritis	Various, particularly Gram-negative	Penicillin or SulfaTMP	Ampicillin or cephalosporin	Penicillin and gentamicin[i] or enrofloxacin/ marbofloxacin[i]	Minimum 2 weeks treatment
Perineal urethrostomy	Skin bacteria contaminants	SulfaTMP	Penicillin	Ampicillin	

Ruptured bladder		Penicillin and gentamicin[i]	Ampicillin and gentamicin[i]		Emergency surgical repair is vital. Aftercare very important
Leptospirosis	*Leptospira hardjo*, *Leptospira pomona*, *Leptospira bratislava*, *Leptospira icterohaemorrhagiae*	Penicillin	Oxytetracycline		Possible immune-mediated complications
Ophthalmology					
Disease/procedure	Pathogen(s)	First choice	Second choice	Last choice	Comments
Conjunctivitis	Various	None	Topical antibiotics[j]		Conjunctivitis is not usually caused by bacterial infection
Corneal laceration/ perforation	Various	Penicillin (IV) and gentamicin[d]	Cephalosporin	SulfaTMP	Concurrent aggressive medical and surgical treatment required
Corneal ulceration	Various	Topical antibiotics[j]	Topical antibiotics[j] and penicillin (IV) ± gentamicin[d]	Doxycycline if keratomalacia is present	Conjunctival flaps are treatment of choice for deep ulcers. Also consider fungal infection as cause
Stromal abscess	Various	Topical antibiotics[j] with penicillin (IV)	Topical antibiotics[j] with penicillin (IV) ± gentamicin[d]	Topical ciprofloxacin[d] with penicillin (IV)	Not always caused by bacterial infection. Both medical and surgical procedures are needed
Leptospirosis	*L. hardjo*, *L. pomona*, *L. icterohaemorrhagiae*, *L. bratislava*	Penicillin	Tetracycline		Also NSAIDS. Ocular signs may be immune mediated
Foreign body	Various	Topical antibiotics[j]	Penicillin (IV) and gentamicin[d]	Ampicillin (IV) and gentamicin[d]	Systemic antimicrobials indicated if anterior chamber penetrated or infected peri-orbital tissues. Check tetanus vaccine status
Enucleation with implant		Penicillin (IV) and gentamicin[d]	Ampicillin (IV) and gentamicin[d]		
Corneal graft		Topical antibiotics[j]	Penicillin or SulfaTMP and topical antibiotics[j]		

(continued)

Table 1 (continued)

Cardiovascular

Disease/procedure	Pathogen(s)	First choice	Second choice	Last choice	Comments
Endocarditis	*Streptococcus* spp., *Actinobacillus* spp., *Pasteurella* spp.	Penicillin (IV) and gentamicin[i]	Ampicillin and gentamicin[i]	Penicillin and enrofloxacin[i]	Blood culture, using enrichment media, should be performed. Poor prognosis
Pericarditis	*Streptococcus* spp., *Actinobacillus* spp.	Penicillin (IV) and gentamicin[i]	Ampicillin and gentamicin[i]	Penicillin and Enrofloxacin[i]	Pericardial drainage and lavage are important adjunctive treatments. Poor prognosis
Thrombophlebitis	Various opportunists	None	Penicillin	Penicillin ± gentamicin[i]	Most cases are inflammatory reactions (e.g. catheter material, endotoxemia) and *not* infection. NSAIDs are useful. Surgical drainage if abscessation. Very unlikely antibiotics will penetrate target site but might protect against endocarditis
Neonatal septicaemia	*E. coli*, *Klebsiella* spp., *Actinobacillus equuli*, *Streptococcus* spp., *Staphylococcus* spp.	Penicillin and gentamicin[i] or ampicillin and gentamicin[i]	Cephalosporin	Cefoxitin and gentamicin[i]	Blood culture with enrichment media should be performed. Neonates should receive higher end of antimicrobial doses

Reproduction

Disease/procedure	Pathogen(s)	First choice	Second choice	Last choice	Comments
Castration/ cryptorchid		None	Penicillin	SulfaTMP	Check tetanus vaccine status
Castration-contaminated scirrhous cord		Penicillin	SulfaTMP		Check tetanus vaccine status
Seminal vesiculitis	*Staphylococcus* spp., *Streptococcus* spp., *Pseudomonas* spp.	Infuse buffered gentamicin[i] into seminal vesicular openings via a flexible endoscope	Penicillin (IV) and gentamicin[i,k]	Enrofloxacin (IV)[i]	
Contagious equine metritis	*Taylorella equigenitalis*	Mares: topical gentamicin[i,k] and 4% chlorhexidine for 5 days Positive uterine culture: intrauterine penicillin	Gentamicin (IV)[i] or penicillin		Best if mares are treated during oestrus. Concurrent uterine lavage may be required
		Stallions: topical gentamicin[i,k] and 4% chlorhexidine for 5 days			External washing of genitalia important for treatment

Endometritis	Various	Intrauterine penicillin and gentamicin[i,k]	Cephalosporin or SulfaTMP (intrauterine antimicrobials optional)	Enrofloxacin (IV)[i]	Intrauterine therapy for 3–5 days during oestrus or early dioestrus. Caslick may be indicated in most cases. Intrauterine antiseptics may induce chemical irritation. Uterine lavage (saline, Ringer's) and others (e.g. oxytocin, PGF2α) are adjunct treatments. Systemic antibiotics indicated if endometrial biopsy suggests deep endometrial infection *or* systemic clinical signs present
Intramammary infection	*Streptococcus* spp., *Staphylococcus* spp., *E. coli*	Penicillin	Intramammary cephalosporin	Cephalosporin	Concurrent stripping of udder is important. Cytology important to rule out neoplasia
Nocardioform placentitis	Gram-positive filamentous branching bacteria	None	SulfaTMP	Cephalosporin	Effectiveness of any treatment uncertain. Most mares do not require treatment and do not have further problems
Placentitis	*S. zooepidemicus*, *E. coli*, *Klebsiella* spp., *P. aeruginosa*	SulfaTMP	Penicillin and gentamicin[i]	Cephalosporin	Gentamicin may not cross the placenta. Caslick may be indicated as a preventative measure
Retained placenta	*E. coli*, *Klebsiella* spp., *S. zooepidemicus*	Penicillin (IV) with gentamicin[i]	Cephalosporin	SulfaTMP	Bacterial infections are more common with prolonged retention (>6 h). Uterine lavage, oxytocin and other medical therapies are required, after the placenta passes. Systemic antibiotics indicated if early treatment with oxytocin fails. Monitor for signs of laminitis
Vaginitis	*S. zooepidemicus*, *E. coli*, others	SulfaTMP	Penicillin and gentamicin[i]	Cephalosporin	Eliminate underlying causes. Caslick may be indicated. Pain management important
Ovariectomy		Penicillin	SulfaTMP		
Rectovaginal fistula	Mixed GI flora	SulfaTMP	Penicillin and gentamicin[i]	Ampicillin and gentamicin[i]	Important to perform a full reproduction examination

(continued)

Table 1 (continued)

Neurology

Disease/procedure	Pathogen(s)	First choice	Second choice	Last choice	Comments
Bacterial meningitis	*E. coli*, *Actinobacillus* spp., *Streptococcus* spp., others	Ampicillin (IV) and gentamicin	Cephalosporin	Penicillin with enrofloxacin[d]	Ceftiofur does not cross the intact blood-brain barrier
Brain abscess	*Streptococcus* spp., others	Penicillin[d,h]	Oxytetracycline *or* macrolides[a,d]	Penicillin with enrofloxacin[d]	Poor prognosis
Spinal abscess	*Streptococcus* spp., others	Penicillin[d,h]	Oxytetracycline *or* macrolides[a,d]	Penicillin with enrofloxacin[d]	
Temporohyoid osteoarthritis/otitis interna-media	*Streptococcus* spp., *Actinobacillus* spp., others	SulfaTMP	Cephalosporin	Penicillin and enrofloxacin[d]	Long-term treatment required. Also consider trauma ± fracture
Tetanus	*Clostridium tetani*	Penicillin (IV)	Metronidazole		Supportive care most important. Antibiotics can eliminate infection. Tetanus antitoxin to neutralise unbound toxin. Poor prognosis

Miscellaneous

Disease/procedure	Pathogen(s)	First choice	Second choice	Comments
Ehrlichiosis	*Anaplasma phagocytophilum*	Oxytetracycline	Doxycycline	Risk of colitis
Lyme disease	*Borrelia burgdorferi*	Oxytetracycline	Doxycycline	Difficult to prove as cause since most horses have *Borrelia* antibody titres
Omphalophlebitis	Various, especially those listed under neonatal septicaemia	Penicillin and gentamicin[i]	Ampicillin and gentamicin[i]	Antimicrobials alone may not be effective. Surgical resection or drainage should be considered

Herniorrhaphy		Penicillin	SulfaTMP	
Apical tooth infection	Typically mixed infection	SulfaTMP	Penicillin	Also consider surgical curettage or tooth repulsion
Tooth extraction		SulfaTMP	Penicillin	Supportive care is more important

SulfaTMP, trimethoprim/sulphonamides; Abs Score, combined thoracic ultrasound abscess measurements

[a]Azithromycin or clarithromycin. Erythromycin has been associated with fatal colitis in adult horses

[b]Rifampicin treatment not recommended, especially in monotherapy for reason of single-step resistance development and risk of resistance transfer. Also, a strong negative drug interaction was identified between rifampicin and macrolides, where no added clinical benefit has been shown in randomised controlled clinical trials between monotherapy with macrolides and rifampicin/macrolide combinations

[c]Combination with a gentamicin is optional if Gram-negative pathogens are suspected

[d]It is important to respect national rules/laws regarding culture/sensitivity before using critically important antimicrobials

[e]Combination with metronidazole is optional if anaerobes are suspected (typically with chronic infections)

[f]Metronidazole is a prohibited substance for horses intended for human consumption

[g]Definition: "deep" or "major" wound infection is "a purulent discharge complicated by spreading cellulitis, systemic signs or wound breakdown". Peel AL, Taylor EW. Proposed definitions for the audit of postoperative infection: a discussion paper. Surgical Infection Study Group. *Ann R Coll Surg Engl* 1991; 73: 385–8

[h]Combination with gentamicin is optional

[i]It is important to respect national rules/laws regarding culture/susceptibility before using critically important antimicrobials

[j]A wide range of topical antimicrobials may be used, including fusidic acid, oxytetracycline/polymyxin-B, triple antimicrobial (bacitracin/neomycin/polymyxin B), aminoglycosides and ciprofloxacin. The most appropriate formulation should be based on bacteriological culture/susceptibility. Certain ophthalmic antimicrobials target specific bacterial groups. For example, fusidic acid and polymyxin are mainly active against *S. aureus* and Gram-negative bacteria, respectively. Some commercial topical antimicrobial preparations contain hydrocortisone (e.g. hydrocortisone, oxytetracycline/polymyxin-B), that cannot be used if a corneal ulcer is present. Topical autologous serum is a valuable adjunct to treatment. Subpalpebral lavage systems are a very useful method for delivering several or long-term topical treatments

[k]Gentamicin infusion is optional, buffered with an equal volume 7.5% sodium bicarbonate

treatment of life-threatening conditions such as pleuropneumonia and peritonitis being considered the most broad-spectrum coverage possible for equine pathogens, with the exception of resistant microorganisms and *Mycoplasma* spp. This broad-spectrum coverage that minimises fears of missing a possible pathogen involved (lack of diagnosis) together with the horse's economic value, or lack of knowledge about the disease, promotes the overuse of antimicrobial combinations. The combination of a β-lactam with an aminoglycoside is a very broad-spectrum combination for susceptible organisms. However, some anaerobes, notably some strains of *Clostridium* and *Bacteroides*, are not affected by β-lactams. Metronidazole treatment improves anaerobic coverage with better pharmacodynamic/pharmacokinetic characteristics for long-acting penetration into difficult-to-reach body sites. Since aerobic Gram-positive and Gram-negative bacteria are causal pathogens for most infections in horses, this triple antimicrobial combination does not represent improved coverage and belongs therefore among non-justified overuse of antimicrobials in many cases. Thus, it would be more prudent to put thought and effort into finding the cause(s) and choose antimicrobials with better pharmacodynamic/pharmacokinetic characteristics against Gram-positive and Gram-negative infections, especially considering new rapid diagnostic methods (e.g. different types of PCR techniques). It should be noted that most cases of equine anaerobic infections are more likely associated with mixed chronic infections (e.g. >5 days) in body sites that support low oxygen tension (e.g. pleura, peritoneum, deep wounds) (Weese et al. 2008).

Recommendations should be considered general guidance that do not supersede information obtained through culture and susceptibility testing of the target pathogen associated with disease of the individual horse. Most are disease-specific recommendations that can be used in situations where the specific agent or its susceptibility pattern is unknown, when samples are not submitted to laboratory diagnosis or while culture results are pending and a timely start to treatment is needed. Only few pathogen-specific recommendations are provided to guide antimicrobial selection when the causative agent has been identified and in vitro susceptibility data are available, as this situation is rather infrequent in clinical practice. In some situations, multiple options are presented in each category (first, second). This is for the reason that recommended drugs within the same category are presumed to be similarly appropriate and other factors are considered such as route of administration and patient factors (e.g. age, concurrent disease); also, immediate availability of medicinal products and cost should be considered for selecting the best antimicrobial option. Furthermore, not all of the suggested antimicrobials can be used in accordance with all jurisdictions, and the use of certain antimicrobials (i.e. chloramphenicol and metronidazole) is banned according to the legal provisions (residues related to food producing species, here horses not declared in passport as non-food producing) in the European Union as well as in some non-EU countries.

Treatment success is not only based on the selection of the appropriate antimicrobial but also the dosing schedule. Dosing schedule consists of the dose, frequency and duration but also linked to the volume administered (IM, SC) of injectable medicinal products (target animal safety, local tolerance, which can differ for products even with the same active substance, based on excipients and formulation composition). Individual aspects to be considered include age (care taken for neonates, with specific care to foals after premature birth), concomitant disease, severity of infection and acute or chronic course of infection, which preclude universal recommendations. Since doses are expressed in mg/kg of bw, then attention should be given to establishment of the correct weight of the animal treated. Despite these limitations, recommendations in Table 2 can be the starting point for considering the individual patient's dosing schedule.

Table 2 Dose recommendations for selected antibiotics/antimicrobials in horses

Drug	Dose(s)	Route	Interval	Comment
Amoxicillin	40 mg/kg bw	IV	q8 - 12 h	Efficacy unknown
Ampicillin sodium	10-20 mg/kg bw	IV	q6 - 8 h	
	25 - 40 mg/kg bw:for intermediate resistant bacteria or natural tissue barriers (e.g. CNS)	IV	q6 - 8 h	Risk of gastrointestinal side effects with higher doses
	Intrauterine: 1 -3 g *pro toto*	IUT	SA	Typically, SA to avoid introduction of infection via repeated administration
Ampicillin trihydrate	20 mg/kg bw	IM	q8 h	
Azithromycin	10 mg/kg bw	PO	q24 h for 5 days and then q 48 h	Reduced GI absorption, if combined with rifampicin (azithromycin/rifampicin is no longer considered a justified combination)
Cefazolin	20 mg/kg bw	IV	q6 h	
	20 - 25 mg/kg bw	IM	q8 h	Efficacy unknown
	500 mg/d/joint	Intra-articular	SA	Typically, SA to avoid introduction of infection via repeated administration
Cefepime	Foals: 11mg/kg bw	IV	q8 h	
	Adults: 6mg/kg bw	IV	q8 h	
Cefotaxime	25–40 mg/kg bw	IV	q6 - 8 h	
Cefoxitin	20 mg/kg bw	IM/IV	q8 h	
Cefquinome	Foals: 1mg/kg bw	IM/IV	q12 h	Doses based on effcacy aganst *S. zooepdemicus* (i.e. MIC<0.25 µg/mL). More frequent dosing required for less sensitive bacteria (e.g. *Staphylococcus*)
	Adults: 1mg/kg bw	IM/IV	q24 h	
Ceftofur	Foals: 2.2 - 6.6 mg/kg bw	IV/IM	q12 - 24 h	≥4.4 mg/kg for Gram-negative infections. Less effective for *Staphylococcus* or *Proteus* . No value for *Pseudomonas*
	Adults: 2.2 - 4.4 mg/kg bw	IV/IM	q12 h	
	Regional perfusion 20 ml of 50 mg/mL	Regional perfus	q24 h	Synovial infecton. Effiicacy unknown
	Intrauterine: 1 g *pro toto*	IUT	SA	Typically, SA to avoid introduction of infection via repeated administration
Ceftriaxone	25 mg/kg bw	IV	q12 h	
Cephalexin	30 mg/kg bw	PO	q8 h	Safety unclear
Clarithromycin	7.5 mg/kg bw	PO	q12 h	Greatly reduced GI absorption, if combined with rifampicin
Doxycycline	10 mg/kg bw	PO	q12 h	Efficacy unknown. Oral bioavailability = 17% after intragastric administration and 6% as top dressing in adult horses. Does not penetrate CSF.
	20 mg/kg bw	PO	q24 h	Risk of colitis with 20 mg/kg in some areas
Enrofloxacin	5.0 - 7.5 mg/kg bw	IV	q24 h	Not for use in growing foals. Not recommended with rifampicin, No value for *Streptococccus, Enterococcus* and Dermatophilosis.
	7.5 - 10 mg/kg bw	PO	q24 h	
	7.5 - 10 mg/kg bw for *Pseudomonas* infections	IV	q24 h	
Erythromycin	Estolate: 25 mg/kg bw	PO	q6 h	Hyperthermia may develop in hot weather.
	Phosphate: 37.5 mg/kg bw	PO	q12 h	High risk of colitis in adult horses.
Gentamicin	6.6 mg/kg bw	IV/IM	q24 h	Nephrotoxic. Not recommended with phenylbutazone.
	Aerosol 20 ml of 50 mg/ml	INHAL	q24 h	For treatment of bacterial pneumonia. Efficacy unknown.
	Intraosseous perfusion: 2.2 mg/kg in 0.1 mL/kg saline	Intraosseous perfusion		Infections of synovial structure and bones in distal limb.
	Intra-articular: 150 mg/d/joint	Intra-articular	SA	Typically, SA to avoid introduction of infection via repeated administration.
	Intra-uterine: 1-2 g *pro toto*	IUT	SA	Buffer with equal volume 7.5% sodium bicarbonate. Typically, SA to avoid introduction of infection via repeated administration
Marbofloxacin	2 mg/kg bw	PO/IM/IV	q24 h	Not for use in growing foals. Might not be optimal for treating S. aureus infections
Metronidazole	Colitis: 15 mg/kg bw	PO	q8 h	Teratogenic. Stop if anorexia develops.
	20 mg/kg bw	per rectum	q12 - 24 h	Clostridial infections should be treated q12h.
	Other: 20 mg/kg bw	PO/IV	q12 - 24 h	Forbidden for use in food-producing horses.
Oxytetracycline	6.6 - 15 mg/kg bw	slow IV	q12 - 24 h	Risk of colitis. No value for *Pseudomonas* .
	Foals (contracted tendons): 40-50 mg/kg bw	IV	Typically SA	Controversial, as non-infectious indication. Most efficient in first 3 days of life. In severe cases: 2-3 administrations with risk of nephrotoxic and hepatotoxic adverse effects. Induces tendon laxity for 12-18hrs. Must be combined with supportive therapy (e.g. splints).
Penicillin G: procaine	20,000 - 25,000 IU/kg bw	IM	q12 - 24 h	Once daily dosing acceptable for streptococcal infections. Maximum 10ml per injection site. Neck injections are absorbed more rapidly than gluteal injections. **NEVER GIVEN IV !!!**
Benzylpenicillin: sodium/potassium	20,000 - 25, 000 IU/kg bw	IV/IM	q 6 h	No value for *Pseudomonas* . Q8h acceptable for *Streptococcus* .
	Intrauterine: 5 - 10 million IU *pro toto*	IUT	SA	Typically, SA to avoid introduction of infection via repeated administration.
Trimethoprim-sulphonamide	Horses: 24 - 30 mg/kg bw	IV/PO	q12 h	Slow IV infusion, due to risk of hypotension. Not recommended with detomidine or other α_2-agonists.
	Ponies: 30 mg/kg bw	IV/PO	q12 h	Not recommended to treat purulent infections. Not recommended in combination with procaine penicillin.
	Donkeys: 30 mg/kg bw	IV/PO	q12 h	Decreased PO absorption if given with food. No value for *Pseudomonas* . Conflicting reports of efficacy against *S. zooepidemicus* .

Routes of administration: IV intravenous; IM intramuscular, PO oral; IUT intrauterine; SA single administration;
Further terms: GI GastroIntestinal; CSF CerebroSpinal Fluid; MIC Minimum Inhibitory Concentration
pro toto = per (entire) animal

7.1 Respiratory Infections

Respiratory tract diseases are common in horses and one of the most frequent reasons for antimicrobial administration. Clinical signs are often indistinguishable from infectious versus non-infectious causes. Recommended treatments are variable, and some references still endorse broad-spectrum antimicrobials for equine respiratory tract diseases, even to cover the possibility of secondary bacterial pneumonia from a primary viral infection, despite the absence of data supporting these approaches. Nevertheless, many upper respiratory diseases in horses are non-infectious or viral and do not require antimicrobial treatment. Furthermore, secondary bacterial pneumonia is rare. Broad-spectrum antimicrobial use for respiratory tract diseases, based on clinical signs alone (e.g. fever, cough, nasal discharge), can no longer be justified. There are a number of diagnostic techniques (e.g. trans-tracheal wash, tracheal aspirate, bronchoalveolar lavage, lung biopsy) that are simple and safe to perform, whereby representative samples can be obtained for culture and sensitivity (Hoffman and Viel 1997). Regardless of the causal agent of a lower respiratory tract infection, adequate stall rest in a well-ventilated stable and supportive care, including good quality hay and water, are the most important components of recovery. Exercise during clinical disease and recovery may worsen the clinical disease. Stall rest for 3 weeks after clinical resolution is a common recommendation for bacterial lower airway diseases.

The direct delivery of antimicrobials to the lower airways through nebulisation is an enticing method to deliver maximal drug concentrations to the site of infection, a rapid onset of action, while minimising systemic exposure. Aerosol particle sizes between 1 and 5 μm are thought to be ideal for therapy, using ultrasonic or jet nebulisers. Aminoglycosides are the most commonly reported aerosolised antimicrobial agents because they remain bioactive when aerosolised and poorly absorbed across epithelial surfaces, thus remaining within the pulmonary tree where they exert concentration-dependent effects (McKenzie and Murray 2004). However, inhalation antimicrobial therapy has remained controversial in human medicine because of the potential risk of pulmonary contamination with environmental bacteria and poor drug delivery to consolidated regions of the lung. Irritation from the drug may induce bronchoconstriction, and aerosol administration on surfaces containing large numbers of diverse bacteria (especially producing biofilm) may select for antimicrobial-resistant bacteria. For most nebuliser systems, it has been estimated that approximately 10% of the drug reaches the lungs during disease, due to the excessive mucus secretions, bronchospasm and higher, more turbulent air flow rates with tachypnoea.

7.1.1 Strangles

Streptococcus equi subspecies *equi (S. equi)* causes 'strangles', which is a highly contagious endemic disease of equids. Spread is mainly through direct contact, and thus strict control measures are essential to limit the impact on horse populations. This includes testing of in-contact horses, routine testing of apparently healthy horses at the time of entry to a new facility, isolation of infected horses, quarantine of exposed horses or facilities and testing for release from quarantine (Boyle et al. 2018). Some non-clinical horses may shed *S. equi* or have positive PCR tests for extended periods, which further complicates control measures.

Risk of transmission is dependent on shedding of adequate numbers of viable bacterium to susceptible hosts. Testing may not provide accurate answers, since the bacterium is unlikely homogenously distributed within the upper respiratory tract, and standard specimens from either the nasopharynx or guttural pouch are not from sites directly exposed to other horses or surfaces. Shedding and spread of disease are a dynamic process, where knowledge gaps with diagnostic testing and poor understanding of the infectious dose of *S. equi* make it difficult to conduct a full risk assessment.

Culture and PCR are the main tests used for clinical diagnosis or surveillance. Culture detects viable *S. equi*, whereas polymerase chain reaction (PCR) may detect DNA from viable or non-viable bacteria. PCR testing is three times more sensitive than culture, and quantitative PCR (qPCR) can provide an estimate of the bacterial load. Increased diagnostic sensitivity from PCR is enticing, but if it detects horses that are not infectious, then PCR testing could result in unnecessary restrictions or misuse of antimicrobials. Unfortunately, qPCR does not correlate well with viable/culturable *S. equi* (Weese et al. 2023). Furthermore, if reusable medical devices (e.g. endoscopes) are used for sample collection, then false-positive PCR results may stem themselves from the medical device. For example, the manufacturers of reusable medical devices (e.g. endoscopes) provide disinfection protocols but cannot promise PCR-negative medical devices afterwards, including the biopsy channel.

The majority of strangles cases require no antibiotic treatment and just proper rest and a dry, warm stall as well as good-quality feed and water (Boyle et al. 2018). Veterinary opinion is strongly divided on the issue of whether or not to use antibiotics for certain strangles cases or stages of infection. It remains unclear as to any clinical benefit of antibiotic use in strangles cases. Pressure from horse owners could be the major reason for antimicrobial prescriptions. The use of non-steroidal anti-inflammatory medications provides more clinical benefits.

7.1.2 Rattles

Rhodococcus equi pneumonia is a common reason for critically important antimicrobial use in foals. Culture/susceptibility is performed less commonly, in favour of thoracic ultrasound. The popularity of thoracic ultrasound in foals has resulted in substantial increases in the misuse of critically important antimicrobials, for prevention and treatment. Based on thoracic ultrasound, foals with rhodococcal pneumonia can be characterised into three groups:

- Subclinical rhodococcal pneumonia: Most common type
  - Peripheral pulmonary consolidation or abscessation without manifesting clinical signs
- Mild-to-moderate clinical signs of rhodococcal pneumonia: <50% lung involvement
  - No evidence of other body sites infected
- Severe rhodococcal pneumonia: >50% lung involvement
  - Respiratory distress and other body sites could be infected

Although many antimicrobials are active in vitro against *Rhodococcus equi*, this in vitro activity does not correlate with in vivo antimicrobial efficacy, since antimicrobials must penetrate three sites ((1) lungs, (2) encapsulated pyogranulomatous abscess/es and (3) intracellular location of *R. equi* within alveolar macrophages) to above MICs for *R. equi*.

A tradition was born in the 1980s with the introduction of macrolide-rifampicin concomitant therapies for foal rhodococcal pneumonia. The most commonly quoted reference is from Hillidge (1987) where survival rates increased from 20% to nearly 90%, with this approach. Later on, in vitro studies found macrolides/rifampicin to be synergistic against *Rhodococcus equi*.

However, the treatment of *R. equi* infections is no longer simple given the wealth of in vivo evidence in foals of profound negative drug interactions between macrolide-rifampicin concomitant therapies. This has led to the emergence of highly prevalent macrolide and rifampicin resistance strains (Huber et al. 2019; Giguère et al. 2017; Burton et al. 2013), as well as the difficulty of resolving severe infections. Furthermore, certain geographic regions (e.g. Scandinavia) also contend with the real risk of fatal colitis in mares from the standard macrolide-rifampicin concomitant therapies in foals.

Resistance to rifampicin develops very quickly, by single-step point substitutions in *R. equi*, and therefore, rifampicin is always used concomitantly with other antimicrobials (standard practice). Pharmacodynamic in vivo studies in foals have revealed that rifampicin almost completely abolishes the oral absorption of macrolides through the activation of p-glycoproteins on the cell membrane of enterocytes (Peters et al. 2011, 2012; Berlin et al. 2016). This prevents macrolides from reaching MIC levels at the target site of *R. equi* infection, selecting for MDR *R. equi*, both for macrolides and rifampicin (due to unintentional rifampicin monotherapy). Injectable macrolides (e.g. tulathromycin, gamithromycin) also exhibit profound negative drug interactions with rifampicin, due to mechanisms that are not fully understood; however, this also prevents either macrolides or rifampicin from reaching MIC levels at the target site of *R. equi* infection (Venner et al. 2010; Berlin et al. 2018). Clearly, macrolide-rifampicin concomitant therapies are not synergistic in vivo and should be avoided.

Based on the results of six randomised clinical trials (RCTs), with a negative control group (placebo), the following recommendations can be given for *R. equi* in foals:

- RCTs have revealed a strong 'placebo' affect (self-cure of up to 88%) for foals with subclinical, mild-to-moderate bronchopneumonia, with a median ultrasound abscess score around 6 cm. This threshold has been recently (2023) elevated to 10–15 cm; antimicrobial therapy does not significantly accelerate lesion resolution relative to the placebo group (Venner et al. 2012, 2013a, b). Thus, antimicrobials are not justified, and clinical/ultrasound monitoring should be done.
- Mild-to-moderate *R. equi* bronchopneumonia cases (ultrasound abscess score 10–15 cm) warrant antimicrobial therapy; RCTs confirm that monotherapy with macrolides perform equally as well (non-inferior) as rifampicin concomitant therapies (Wetzig et al. 2019; Hildebrand et al. 2015; Rutenberg et al. 2017). The use of rifampicin concomitant therapies provides no added clinical advantage for the treatment of these cases and thus not justified.
- Foals with severe *R. equi* bronchopneumonia require antimicrobial therapy (poor prognosis) but have not been evaluated in clinical studies. The combination of azithromycin/doxycycline or clarithromycin/doxycycline has been evaluated and considered synergistic.

7.2 Gastrointestinal Infections

Serious problems may originate in horses from the disruption of the normal intestinal microflora. Although the overgrowth of pathogens is an important component of the pathophysiology, the exact mechanisms as well as the means to 'restore' the normal balance are poorly understood. The use of broad-spectrum antimicrobial therapy for colitis in adult horses is controversial and in general not indicated. Some clinicians use antimicrobials routinely, while others restrict treatment to severely compromised and/or neutropenic animals, and some rarely, if ever, use parenteral antimicrobials. It is important to remember that broad-spectrum therapy is unlikely to have an effect on the pathogen within the gastrointestinal tract. Rather, the goal of therapy is to prevent/treat bacterial translocation, something that is identified uncommonly in adult, immunocompetent horses. Using antimicrobials may cause further microflora disruption, adversely affecting the re-establishment of a 'normal' microflora and lead to antimicrobial resistance. In the case of antimicrobial-associated colitis, if bacterial infection is not likely, or not of such severity that withdrawal of treatment for a period of time would adversely affect the outcome, then cessation of all antimicrobials is recommended. Antimicrobial therapy is not indicated to eliminate colonisation with pathogens such as *Salmonella* because there is no evidence of efficacy.

Oral antimicrobial therapy, mainly using metronidazole, is a common treatment of *idiopathic* and *clostridial colitis* (McGorum et al. 1998). Some clinicians have used vancomycin for the treatment of *C. difficile*-associated diarrhoea. Considering the importance of vancomycin in the treatment of serious infections in humans, the recent emergence of vancomycin-resistant enterococci in horses and the lack of safety or efficacy data in horses, the authors consider the use of vancomycin in horses inappropriate. Moreover, in the EU, vancomycin is on the list of antimicrobials reserved for human use only and since February 2023 is therefore banned for any use in animals.

Abdominal abscessation and *peritonitis* provide different challenges. Abdominal abscesses may be caused by a variety of pathogens, and obtaining samples for culture and susceptibility testing is difficult. Furthermore, incision and drainage, the most effective form of abscess treatment, are rarely an option. Therefore, long-term empirical antimicrobial therapy is required, and drugs chosen must have the ability to penetrate abscesses and have some activity in the presence of organic debris. Peritonitis may be caused by the same range of pathogens as abdominal abscesses; however, diagnostic sample collection is easy, adjunctive therapy (abdominal lavage) is feasible, and therapeutic antimicrobial levels are more easily achieved at the site of infection.

7.3 Musculoskeletal Infections

The musculoskeletal system can present a variety of challenges for antimicrobial therapy. It is difficult to reach therapeutic antimicrobial levels in many tissues, such as bone, tendon, tendon sheath and joints, following oral or parenteral antimicrobial administration. Alternative approaches such as intra-articular therapy, intra-osseous infusion and regional perfusion may be useful in many situations to provide very high local antimicrobial levels. The tendency for abscess formation in soft tissues presents further challenges that inhibit antimicrobial penetration and activity.

Septic arthritis presents several challenges. Direct culture is often unrewarding. The use of enrichment culture is important, and synovial fluid samples should be inoculated into blood culture broth shortly after collection. Samples in blood culture broth are incubated for up to 1 week, thereby greatly increasing sensitivity. Sterile technique is critical because this enrichment process can detect very low numbers of bacteria, including contaminants. Empirical therapy is important because of the low sensitivity of direct culture, the time delay associated with enrichment culture and possible consequences of disease. Gram staining of synovial fluid should be performed to provide basic information about the likely pathogen. Prompt treatment is required, including antimicrobials and ancillary procedures such as joint lavage, to reduce the risk of permanent damage within the infected site. A distinct advantage for the treatment of septic arthritis is the ease of local (intra-articular) therapy in most situations. Intra-articular injection of antimicrobials is a very common practice when treating septic arthritis because of the ability to provide very high drug levels at the infected site. This procedure is often combined with joint lavage. Some antimicrobials are irritating and may produce chemical synovitis, so only drugs known to be safe and effective (e.g. amikacin, gentamicin, ceftiofur, cefazolin, sodium/potassium penicillin) should be injected into joints. Since antimicrobials injected into joints will be absorbed into the circulation, then the total amount injected should be considered, particularly when multiple joints are being treated and when the same drug is being used parenterally. In situations where neonatal foals are being treated for multiple septic joints, aminoglycosides should not be used both parenterally and intra-articularly because excessive drug levels may result in acute tubular nephrosis.

Osteomyelitis is more difficult to treat because of the difficulty in producing high-enough antimicrobial levels and presence of organic debris at the infection site. The deep site of many infections may complicate collection of culture specimens. Bone biopsy is the best method to obtain a positive culture. Combination therapy may be required including parenteral therapy plus surgical interventional and local therapy. Antimicrobials can be impregnated into a variety of materials, including polymethylmethacrylate (PMMA) beads or plaster of Paris (Cruz et al. 2006). These materials can then be surgically implanted in an infected area with the antimicrobial being eluted over time. This can result in prolonged high drug levels at the site of infection with minimal systemic exposure and lower drug cost compared to systemic therapy. However, elution rates are variable and depend on the antimicrobial, dose and characteristics of the implant site. These are most often used in cases of osteomyelitis, deep wound infections and fracture site infections.

Regional perfusion involves the administration of antimicrobials into the temporarily occluded vasculature (e.g. via tourniquet) of the infected limb, resulting in high local tissue antimicrobial levels including the synovial fluid, soft tissues and bone. Antimicrobials are injected either into a superficial vein or into the medullary bone cavity. Therapeutic antimicrobial levels may be achieved in poorly vascularised tissues where therapeutic levels cannot be reached with systemic treatment. β-lactams and aminoglycosides are used most commonly. Irritating drugs should not be used. While uncommonly used, subcutaneous placement of an infusion pump delivers high antimicrobial levels at the site of infection for prolonged periods. Pumps can be filled weekly and produce therapeutic levels for weeks or months. This approach is probably most useful for osteomyelitis and fracture-site infections.

Biofilms can complicate certain infections, particularly those involving orthopaedic implants and other invasive devices (Trampuz and Widmer 2006). Biofilms are communities of bacteria that adhere to inert surfaces (i.e. implants) and dead tissue (i.e. bony sequestra). Sessile bacteria that are resident within biofilms are largely resistant to antimicrobials, phagocytes and antibodies. The potential role of biofilms in infections of these types should be considered, particularly if there is poor response to initial treatment.

7.4 Ophthalmologic Infections

Ophthalmologic conditions present some unique challenges, based on the types of infection and ability of drugs to penetrate certain areas. Topical therapy facilitates the best treatment in many cases. Blood-ocular barriers (i.e. blood-retinal equivalent to the blood-brain barrier and blood-aqueous barrier) affect the ability of most systemic drugs to penetrate into the eye (posterior and anterior segments, respectively), thereby limiting the usefulness of this route. Certain antimicrobials (i.e. chloramphenicol, sulphonamides) penetrate this barrier better than others. Most drugs will penetrate better in the presence of inflammation, but many will also bind to proteins present in inflammatory exudates. Systemic, subconjunctival, intra-ocular and topical administrations are potential routes of administration. Topical therapy alone may involve direct application to the eye, administration via a subpalpebral lavage system (Giuliano et al. 2000) or through the use of an antimicrobial-impregnated collagen shield or contact lenses. Topical therapy alone is often adequate for corneal ulcers; however, a combination of different routes may be indicated for more serious conditions. Not all antimicrobials are safe for topical or intra-ocular use. Intravenous administration is preferred over intramuscular or oral administration because higher plasma levels attained by this route of administration may result in higher ocular levels.

A variety of bacteria and fungi can be part of the normal ocular microflora. These include *Staphylococcus aureus*, coagulase-negative staphylococci, *Moraxella equi*, *Streptococcus zooepidemicus*, *Corynebacterium* spp., *Bacillus* spp., *Aspergillus* spp., *Penicillium* spp., *Alternaria* spp. and *Cladosporium* spp. (Andrew et al. 2003).

7.5 Urinary Tract Infections

Urinary tract infections are less common in horses than other animals. Cystitis is the most common bacterial infection of the urinary tract and often associated with bladder dysfunction or other predisposing factors. As such, recurrent infections can be encountered. Pyelonephritis is uncommon, but appropriate therapy is essential because of the potential consequences of disease. The advantage of treating urinary tract disease is the ability of many antimicrobials to reach high concentrations in urine, including penicillin, cephalosporins and trimethoprim-sulphonamides. As a result, pathogens that are reported as resistant in vitro may be susceptible in vivo. However, the underlying bladder wall (i.e. biophase for infection) is protected against xenobiotics, including drugs, by the uppermost cells of the urothelium at the inner surface of the bladder, known as umbrella cells. One important point to note is that while many sulphonamides concentrate in urine, this is not true for all. Sulfamethoxazole is largely metabolised before urinary excretion and therefore less likely to be effective in urinary tract disease. In addition, urine pH may influence local antibacterial activity.

Another important advantage with lower urinary tract disease is the relative ease of collecting culture specimens. Aseptically collected catheterised samples should be used for culture and submitted in a sterile container. Urine swabs should be avoided because of the much smaller volume of material for culture and they do not allow for quantitative culture. Quantitative culture is useful to determine the clinical relevance of results, as contamination can occur even with catheterised samples. Growth from a catheterised sample of >1000 colony forming units (CFU/mL) is considered abnormal, while a suspicious growth is 500–1000 CFU/mL. For free-flow samples, >40,000 CFU/mL is considered abnormal with 20–40,000 CFU/mL as suspicious (MacLeay and Kohn 1998).

Urinary tract disease often requires longer treatment than other body sites, possibly due to the location of bacteria within the bladder wall or biofilm. Cystitis is often treated for 7–10 days initially, and re-culture is indicated a few days following cessation of treatment. In refractory or severe cases, culturing a few days after the onset of antimicrobial therapy can be useful to detect early treatment failure, with the understanding that negative results do not necessarily indicate successful treatment. Pyelonephritis should be treated for a minimum of 2 weeks.

While less common in horses compared to household pets, recurrent urinary tract infections may be problematic. It is essential to determine the underlying cause for recurrent disease and differentiate relapse from re-infection. If there is an underlying problem such as an anatomical defect, urolith or neurological dysfunction, antimicrobial therapy alone is unlikely to be successful.

7.6 Cardiovascular Infections

Bacterial infections of the cardiovascular system are uncommon. Injection or intravenous catheter-associated thrombophlebitis is likely the most common problem; however, most cases of thrombophlebitis are inflammatory versus infection. Therefore, antimicrobials are not indicated in all cases and should be reserved for situations where there is a high likelihood of infectious thrombophlebitis or associated abscessation. Bacterial endocarditis and pericarditis are rare but potentially life-threatening conditions requiring appropriate therapy. Blood culture should be performed in cases of endocarditis to identify the cause, while culture of blood and pericardial fluid should be performed in cases of suspected bacterial pericarditis. Pericardial fluid should be inoculated into blood culture broth immediately after collection. Pericardial drainage and lavage are important adjunctive treatments for pericarditis. Surgical incision and drainage are required if thrombophlebitis progresses to abscessation.

7.7 Neurological Infections

Bacterial infections of the central nervous system are uncommon but can have a devastating impact. Accessibility to the site of infection for collection of diagnostic specimens and administration of antimicrobials is highly variable. The blood-brain barrier (BBB) and blood-cerebrospinal fluid barriers have a major impact on the penetration of most drugs. In general, drugs that are lipid soluble, non-ionised, not highly protein bound and with small molecular size penetrate better. Nevertheless, even drugs with these properties often poorly penetrate the CNS because the main determinant of the BBB passage is the presence of efflux pumps (i.e. P-glycoprotein, MRP, etc). Inflammation can result in increased drug penetration; however, this cannot necessarily be relied on, so treatment choices should be based on knowledge of the etiologic agent and drug penetration. Aminoglycosides penetrate poorly even in the presence of inflammation; however, they are often used in combination therapy. Unlike other third generation cephalosporins, CNS penetration of ceftiofur is poor (Cervantes et al. 1993).

Bactericidal drugs are preferable because of the poor immune response within the CNS. As a result, intravenous administration is required because of the ability to provide high peak blood levels. Intrathecal administration of antimicrobials has been described, but there is little evidence of efficacy. Bacteriostatic drugs may be useful in some cases of brain abscess and spinal abscess, particularly drugs such as chloramphenicol that have other desirable properties.

7.8 Reproductive Infections

Bacterial infections of the reproductive tract are relatively common, particularly in brood mares. Many infections are associated with breeding, parturition, uterine motility defects and conformational defects. Underlying risk factors for infection must be considered and addressed.

Systemic and local (intrauterine) approaches may be practical in some cases. Infections confined to the uterine lumen and superficial endometrium are best treated by intrauterine therapy. A variety of antimicrobials may be infused into the uterus, including sodium/potassium penicillin, gentamicin, amikacin, ceftiofur and ampicillin (Pedersoli et al. 1985; Love et al. 1990). It should be noted that within the EU, ticarcillin belongs to antimicrobials which are reserved for use in human medicine and therefore banned for use in animals. Typically, a volume of 50–250 mL is infused. Irritating drugs should be avoided to reduce the risk of causing chemical endometritis. Buffering of acidic drugs such as aminoglycosides with sodium bicarbonate has been recommended. Systemic therapy, with or without local therapy, is indicated if deeper tissues are involved.

Uterine lavage is often an important component of treatment because excessive uterine fluid can result in a marked dilutional effect, as well as containing organic debris that decreases the activity of most antimicrobials. Another benefit of lavage is to remove bacteria and bacterial by-products. Other adjunctive medical therapies, such as oxytocin administration, can be considered.

The external reproductive tract is not a sterile site, and care must be taken to prevent contamination during sampling of the uterus, which should be performed during oestrus. In order to minimise contamination, the perineum should be carefully washed, and sampling should be performed with a gloved hand in the vagina and double-guarded, occluded swabs. The culture swabs are transported in transport media to the lab.

7.9 Skin Infections

Normal skin has a complex endogenous microflora, whereby disease is prevented by a combination of factors, not the least of which is the physical barrier of the skin. Disruption of this normal barrier by a variety of means creates the potential for a secondary bacterial infection. Primary bacterial skin infections are much less common but may also occur.

Horses and humans share the utilisation of sweating as their primary cooling mechanism. Sweating is also part of the disease response to pain (e.g. colic) or fever. However, this comes with the cost that any systemic antimicrobial therapy leads to sub-therapeutic antimicrobial concentrations secreted in sweat (Høiby et al. 2000). This can select for AMR skin microflora for any future skin infections. Also, this may be the biological explanation for the close temporal relationship between the marketing of antibiotics and increase in AMR in horses (especially *Staphylococcus*).

Cytological examination is often used for a presumptive diagnosis of bacterial skin disease. For example, identification of intracellular cocci is interpreted typically as a clinically relevant tentative diagnosis of coagulase-positive staphylococci or streptococci. Culture is often used for severe or refractory cases. The skin is a readily accessible site for collection of samples; however, interpretation of results can be difficult because of the complex normal microflora, including many potential opportunistic pathogens. The conundrum is to assess whether culture results reveal the cause or contaminants. Cultures of moist lesions and crusts are not typically recommended because of the likelihood of growing contaminants. Cultures of superficial lesions can lead to false-positive results. More reliable results can be obtained from intact pustules or furuncles, if samples are collected by sterile aspiration. As opposed to direct sampling of superficial lesions, samples should be collected from the surface of plaques, nodules and fistulous tracts by skin biopsy, after aseptic preparation of the site.

An advantage of dermatological disease is the ability to treat the affected area topically with antiseptics or antimicrobials (often antimicrobials are unnecessary). This route allows for delivery of high antimicrobial levels at the affected site while reducing systemic exposure. Drugs such as mupirocin, fusidic acid, bacitracin/neomycin/polymyxin B and silver sulfadiazine can be effective in refractory cases. Recently, questions have been raised over the use of mupirocin in animals because of the importance of this drug in MRSA decolonisation therapy in humans and emergence of resistance in MRSA. It is unclear whether the use of topical antimicrobials such as mupirocin for short-term therapy of local infections contributes to resistance in human isolates. However, due to the increasing importance of MRSA in both humans and horses, this aspect should be considered when choosing a topical antimicrobial.

Topical therapy has some drawbacks, as it can be difficult, time-consuming and not properly or effectively applied by some owners. For example, if the skin infection involves deeper layers of the epidermis and subdermal layers, then topical antimicrobials may be ineffective. Local irritation is also a potential problem in some cases. Factors including the type of disease, pathogen involved, severity and ability of the owner to treat must be considered when deciding whether to use systemic, topical or combination therapy. Removal of debris by appropriate bathing is an important aspect of topical therapy as it facilitates contact of the antimicrobial or antiseptic with the infected skin surface. Clipping of the haircoat may also be indicated to facilitate drug contact.

Many skin infections are self-limiting or respond to topical antiseptic (povidone iodine, chlorhexidine) therapy. The severity and chronicity of disease are used often to determine whether antimicrobials are indicated. If systemic antimicrobials are chosen, then an adequate duration of therapy is important. Skin infections tend to require longer treatment than many other types of infection, and 3–8 weeks of therapy is often necessary. It is stated often that treatment should extend 7–10 days beyond apparent resolution of superficial infections and 14–21 days beyond resolution of deep infections.

7.10 Neonatal Septicaemia

Antimicrobial therapy may be indicated for treatment of a variety of neonatal conditions. Among the most important of these is neonatal septicaemia, due to the incidence of disease, mortality rate and the potential for performance-limiting complications such as septic arthritis.

For these reasons, broad-spectrum antimicrobial therapy is indicated, once septicaemia is diagnosed. The use of inaccurate surrogate biomarkers (e.g. serum amyloid A) is not justified to assist in the diagnosis of foal neonatal septicaemia. Previously, it was found that blood culture of clinical cases also had issues with accuracy. However, both equine and human medicine has revisited techniques to improve the accuracy of blood cultures (Furr and McKenzie 2020; Theelen et al. 2019; Lamy et al. 2016). Some of these improved recommendations include:

- Repeat cultures to increase detection of intermittent bacteraemia.
- Duplicate cultures in order to better identify contaminants.
- Larger blood volumes for cultures (up to 20 mL) can increase culture sensitivity.

Antimicrobial therapy for a minimum of 10–14 days in foals with documented bacteraemia is often necessary provided that no localised areas of spread of infection develop requiring more prolonged treatment. Specific sites of infection (e.g. pneumonia, meningitis, arthritis and osteomyelitis) require prolonged antimicrobial therapy for 30 days or longer.

Optimal drugs or drug combinations may vary greatly with geographic region. Knowledge of local susceptibility patterns and clinical experience can guide optimal use, both in terms of clinical effect and prudent use.

8 Unmet Medical Need

For perioperative prophylaxis in equine medicine, the evidence base is weak due to a lack of large-scale, randomised, clinical trials, despite recommendations restricting antimicrobial use for surgical procedures where the incidence of infection exceeds 5% without prophylactic antimicrobial use (Esposito 1999). Years after this survey, antimicrobial prophylaxis continues to be routinely given in 39–98% of equine surgeries, including elective procedures (Olds et al. 2006; Weese and Cruz 2009). Even for clean procedures performed on horses, this practice remains controversial, especially considering prudent use of antimicrobials (Borg and Carmalt 2013). Documented heavy use of perioperative prophylactic antimicrobials is in spite of the fact that the incidence of post-operative infections is very low (0–0.9%) for common elective surgeries (e.g. carpal arthroscopy) (McIlwraith et al. 1987; Olds et al. 2006; Ridge 2011; Weese and Cruz 2009). The most frequently used is the combination of intravenously administered sodium (or potassium) penicillin and gentamicin (Southwood 2014; Stöckle et al. 2021). This combination was reported mostly for equine colic surgeries (Traub-Dargatz et al. 2002), where surprisingly also the last resort fourth-generation cephalosporin, cefquinome, was used (Widmer et al. 2009; Damborg et al. 2012). This practice of broad-spectrum antimicrobial prophylaxis has been linked high rates of faecal shedding of CTX-M producing *E. coli* in horses as well as nosocomial post-operative infections (Damborg et al. 2012). As reported above for cefquinome, in some cases, highest-priority critically important antimicrobials (WHO, 2024) including enrofloxacin and ceftiofur/cefquinome are used for surgical prophylaxis, despite being reserved for the cases where the use of gentamicin was contraindicated due to its nephrotoxic side effects (Ceriotti et al. 2021; Pezzanite et al. 2021).

Redpath et al. (2021) also reported results of an electronical survey on the common use of gentamicin empirically without antimicrobial susceptibility testing (across the continents, as respondents were veterinary surgeons from Europe, North America and Australia/Asia). Gentamicin use only after susceptibility testing was more frequently reported for generalists than specialists, for a range of indications (respiratory diseases, septic peritonitis, acute febrile diarrhoea, cellulitis and contaminated limb wounds). Gentamicin was more often used in high-risk procedures or contaminated surgeries (86% and 74%, respectively) compared with clean surgery (32%; $p < 0.0001$). Gentamicin was often used perioperatively in horses undergoing exploratory celiotomy and more commonly used in horses undergoing an enterotomy (90%) than without and enterotomy (79%; $p = 0.04$). The authors of this study emphasised the need for further improvement, including the existence of guidelines for perioperative prophylaxis as well as introduction of antimicrobial stewardship to equine practices.

For the future, more real-time clinical data could be obtained by auditing clinics, not only to check the data to be provided to authorities in relation to obligations to monitor the use of antimicrobials (phase starting for horses in 2026, according to the regulation EC/6/2019) but to gain information on adherence to current guidelines (usually rather based on experience extrapolations than evidence) and most importantly to build the evidence pool necessary to formulate species-specific evidence-based guidelines.

8.1 Antimicrobial (Perioperative) Prophylaxis

General principles of perioperative prophylaxis also apply to horses. There is no need to administer antimicrobials for uncomplicated procedures such as castration and some orthopaedic procedures, including those performed in field situations. It is important to emphasise that antimicrobials should not be used in place of good surgical technique, a proper surgical environment, good management and optimal infection control practices.

Stöckle (2019) assessed retrospectively datasets including a total 652 patients, of which 259 (39.7%) received antimicrobial perioperative prophylaxis (APP), compared to 393 (60.3%) horses without prophylaxis (CG). Penicillins were administered to 171 horses, and trimethoprim/sulfadiazine was given to 36 horses. Combinations of gentamicin and penicillins were administered in 43 cases, while cephalosporins were dosed to seven horses. Other combinations (cefquinome and gentamicin; trimethoprim-sulfadiazine and gentamicin) were used in two patients undergoing arthroscopy. Antibiotic administration started in the evening prior to surgery and continued for 3–5 days. The majority of surgeries were arthroscopy, and others included fasciotomy, splint bone extraction, tenotomy, neurectomy, tendovaginoscopy/bursoscopy or combined surgical interventions. There were no significant differences for post-operative complications between the 97/259 (37.5%) antibiotic-receiving patients and 158/393 (40.2%) controls. Therefore, perioperative antibiotic prophylaxis in uncomplicated, elective orthopaedic surgical procedures in equines should be thoroughly revisited, as it is not justified in most cases.

It is likely that the various surgical procedures in horses can be subdivided similar to human medicine into categories representing level of possible contamination (Table 3):

Clean surgeries do not require antimicrobials, unless mitigating factors are encountered. Surgical mitigating factors that could influence the need for antimicrobials include a surgical duration >90 min, surgery involving an implant/s or surgical site infection that would be a major threat to the patient.

Table 3 Categories representing level of possible contamination to be considered in surgical procedures

Clean	Non-traumatic, uninfected. No break in aseptic technique, no inflammation encountered. Elective, closed primarily and no drain used
Clean contaminated	Controlled entering of a hollow muscular viscus, minor break in aseptic technique
Contaminated	Open, fresh traumatic wound. Incision into a site with acute, non-purulent inflammation. Major break in aseptic technique
Dirty	Purulent debris encountered during surgery. Perforated viscus found. Traumatic wound with devitalised tissue, foreign material or faecal contamination or of more than 4-h duration

Clean-contaminated, contaminated and dirty surgical procedures require antimicrobials. Tissue concentrations of antimicrobials are required at the time of first incision to confer protection from a surgical site infection. Examples include intravenous antimicrobials administered 30–60 min prior to surgery and intramuscular procaine penicillin administered 3.5 h prior to surgery. For the particular antimicrobial, T_{max} can be used to assess the optimal timing to achieve peak serum levels at the time of the first incision. The dosing interval should be measured from the time of the preoperative dose. The dosing interval can be calculated as twice the elimination half-life of the antimicrobial. For example, the half-life for sodium penicillin is under 1 and 3 h for gentamicin. Thus, in surgeries of long duration, a second dose of penicillin should be given during surgery. The recommendation is to repeat the dosage after two half-lives (2 h for penicillin; 6 h for gentamicin).

Often, a single perioperative dose is all that is required, and prolonged treatment after surgery is often not necessary. Commencing antimicrobial prophylaxis after surgery is generally regarded as inefficacious. As a rule, one to three injections during 24 h after the end of surgery are sufficient.

9 Off-Label and Cascade Use

Prescriptions of antimicrobials outside of the terms of marketing authorisation ('off label') are common in equine practice. Frequent use of antimicrobial combinations, including combinations containing WHO highest-priority critically important antimicrobials with noticeable amounts of non-authorised products for horses, is evidenced with the below mentioned studies.

A large postal questionnaire was conducted including 740 veterinarians that treat horses in the UK (Hughes et al. 2013), with a return rate of 38%. Approximately 11% percent of prescriptions were for antimicrobial drugs not licensed for use in horses in the UK. Approximately 5% of prescriptions for licensed antimicrobials were under the recommended dose rate and 56% over the recommended dose rate. Fluoroquinolones and third-

and fourth-generation cephalosporins accounted for 1% and 3% of prescriptions, respectively. Veterinary surgeons working at referral practices were more likely to prescribe third- and fourth-generation cephalosporins and fluoroquinolones and off-label antimicrobials, whereas those working in first-opinion practices were more likely to prescribe potentiated sulphonamides.

In another larger American survey of 761 hospitalised horses, a total of 511 (67.2%) received an inappropriate amount of antimicrobial/s pre-operatively (Dallap Schaer et al. 2012). The majority of these horses underwent colic surgery. Under-dosing was the most common inaccuracy observed. In addition to this, timing of antimicrobial administration was inadequate (e.g. >one hour before surgery), with 88 (11.6%) of horses receiving the antimicrobial at the appropriate time (Dallap Schaer et al. 2012). In the majority of cases, antimicrobial therapy was continued for an average of 3.8 days. Out of the 761 horses followed, 680 received the combination of penicillin and gentamicin (89.3%), 16 received ceftiofur and gentamicin (2.1%), and only 22 horses (2.9%) received a single antimicrobial.

In a recent German study (Schnepf et al. 2020), data from 2017 were presented for 1773 horses with 34,432 medicines administrations, including 6489 (18.85%) administrations of antimicrobials (including combinations) in 837 (47.21%) horses. Of the total 9402 entries of antimicrobial active ingredients, the most frequently used were sulphonamides [n = 2798 (29.76%)] and trimethoprim [n = 2757 (29.76%)], mostly in combination, followed by aminoglycosides [n = 1381 (14.69%)] and penicillin [1362 entries (14.49%)]. There were documented 324 cases of polymyxin B administrations (3.45%), 85 cases of fluoroquinolone administrations (enrofloxacin, marbofloxacin, moxifloxacin and ofloxacin) altogether 0.90%, 61 cases of administrations of the fourth-generation cephalosporin cefquinome (0.65%) and 52 cases of macrolide (azithromycin—0.55%)—all these antimicrobials belong to the highest-priority critically important antimicrobials according the WHO classification and together account for 5.55% of administrations. Treatments (292) using metronidazole accounting for 3.11% were also recorded, despite use of this antimicrobial being prohibited in EU food-producing animals (including horses, primarily being considered as food-producing animals). Monotherapy antimicrobials were administered in 51.64% (n = 472) of the cases, combination (with 2 antimicrobials) administered in 22.65% (n = 207) and triple in 17.61% (n = 161), and alarming is that between 4 and 8 different antimicrobials were administered in 8.1% (n = 74). One third of medicines (32.34%) were not licensed for horses (human, other animals or magistral formula); among human medicinal products used in horses, intravenous benzylpenicillin (79.35%) and gentamicin (13.55%) accounted for the majority.

9.1 News Influencing Cascade Use of Human Medicinal Products in Equine Medicine

From a European perspective, current legal provisions, which came into action in January 2022 (Regulation on Veterinary medicinal products (EU) 2019/6), will have an impact on critically important antimicrobials available for equine practice. For example, Commission Implementing Regulation (EU) 2022/1255 designating antimicrobials or groups of antimicrobials reserved for treatment of certain infections in humans, which came into action in February 2023, will stop certain antimicrobials from use in individual equine cases. This is the case of some antimicrobials, previously reported as used within the cascade conditions as, e.g. ticarcillin-clavulanate, imipenem and vancomycin or combination of cephalosporines with beta-lactamase inhibitors. Table 4 specifies antibiotics and antivirals that are subject of this above noted restrictions. In 2024, Commission Implementing Regulation (EU) 2024/1973 has been published, which establish a list of antimicrobials, which shall not be used in accordance with Articles 112 and 113 of Regulation (EU) 2019/6 of the European Parliament and of the Council or which shall only be used in accordance with those Articles subject to certain conditions, This legal provision shall apply from August 2026.

Table 4 List of antimicrobials or groups of antimicrobials reserved for treatment of certain infections in humans as stipulated by Commission Implementing Regulation (EU) 2022/1255

Antibiotics	Antivirals
Carboxypenicillins	Amantadine
Ureidopenicillins	Baloxavir marboxil
Ceftobiprole	Celgosivir
Ceftaroline	Favipiravir
Combinations of cephalosporins with beta-lactamase inhibitors	Galidesivir
Siderophore cephalosporins	Lactimidomycin
Carbapenems	Laninamivir
Penems	Methisazone/ metisazone
Monobactams	Molnupiravir
Phosphonic acid derivates	Nitazoxanide (+ as antiprotozoal)
Glycopeptides	Oseltamivir
Lipopeptides	Peramivir
Oxazolidinones	Ribavirin
Fidaxomicin	Rimantadine
Plazomicin	Tizoxanide
Glycylcyclines	Triazavirin
Eravacycline	Umifenovir
Omadacycline	Zanamivir

9.2 Alternative Routes of Administration

Alternative routes of administration are common in equine medicine. These include intra-synovial antimicrobial injections, regional limb perfusion as well as others (e.g. inhalation, intrauterine antimicrobials, guttural pouch antimicrobial lavages). Also, recommendations are available for antimicrobial-impregnated beads for local administration into surgical sites (Cruz et al. 2006). In a survey of 761 hospitalised horses, mostly undergoing colic surgery, additional antimicrobial therapy was used during surgery, including intra-operative abdominal lavage antimicrobials or antimicrobials placed along the incision during closure (Dallap Schaer et al. 2012).

Instillation of penicillin into the equine guttural pouches, following infections with *Streptococcus equi*, has become common practice. This is thought to help eliminate the bacteria, as well as prevent future carriers of the bacteria (Verheyen et al. 2000). However, the justification and efficacy of this practice have neither been critically evaluated nor followed up if there has been a positive impact to prevent future outbreaks. This practice also represents a misuse of antimicrobials.

Intra-articular administration of antimicrobials, especially aminoglycosides, is common practice with horses. This can be either prophylactic (e.g. risk based), with concurrent administration of other drugs (e.g. hyaluronic acid), or based on clinical signs of septic arthritis. Furthermore, joint culture has poor sensitivity/specificity and less commonly performed in practice.

Local therapy with concentration-dependent antimicrobials has been shown to achieve significantly (5–100 times) higher concentrations compared to systemic administration. Local administration methods include intra-synovial injections, regional intravenous perfusion, regional osseous perfusion or implantation of various antibiotic-impregnated material (e.g. gentamicin-impregnated PMMA pearls).

9.3 Individual Patient Characteristics Influencing Use

Due to the practicalities and challenges of handling and medicating horses, there is a bias towards oral antimicrobials in horses. As hindgut fermenters, there are very few safe options for oral medications in horses. For example, trimethoprim-sulphonamide medications are used commonly in equine practice. There is also regular, off-label use of doxycycline in equine practice because it can be given orally, in spite of very poor oral bioavailability in adult horses (Winther et al. 2011).

The age of the horse can be reason for off-label antimicrobial usage. In one survey, older horses undergoing a celiotomy were predisposed to developing infection compared with yearlings (43% versus 15%) (Wilson et al. 1991).

Neonates and foals represent a common population for off-label antimicrobial usage, especially WHO critically important antimicrobials.

Foals are not (yet) hindgut fermenters, and antimicrobials that can cause severe colitis in mature horses do not carry the same risk in foals. Next, given that a foal weighs approximately 1/10th of a mature horse, antimicrobials that are cost prohibitive in mature horses can be used in foals. In neonatal foals, for many drugs, the dosage given is considerably higher than adult horses. This is explained partly by higher percentage of body water content in neonatal foals, as compared to adults. In general, the dosage of hydrophilic drugs (such as the aminoglycosides) is higher in foals, whereas the dosage is the same as adults for lipophilic drugs. The higher incidence of bacterial infections in neonates has led to the practice of prophylactic administration of antimicrobials in the first days of life, by some veterinarians. However, a study found no difference in the incidence of infectious disease between neonatal foals treated with prophylactic antimicrobials and those that were not treated (Wohlfender et al. 2009).

9.4 Unavailability of Medicines

Equine medicine is frequently faced with unavailability of veterinary medicinal products, among them also those containing antimicrobials. Unavailability is not only an issue for small markets, like horses, or markets with low economic power but also an issue of the limited portfolio of veterinary medicinal products authorised for horses with proven, evidence-based efficacy established for exact dosing schedules linked to indications. The lack of interpretive criteria for antimicrobial susceptibility testing of equine pathogens further complicates responsible use of antimicrobials.

There is a lack of effective veterinary antimicrobials approved for *Rhodococcus equi* infection in young foals, a common condition in some geographic locations. Traditionally, drugs of first choice for the treatment of *Rhodococcus equi* infection in foals are the combination of macrolides (e.g. erythromycin, azithromycin, clarithromycin) and rifampicin (Giguère et al. 2004, 2008), for a minimum of 4 weeks of treatment (see Sect. 7.1.2). Azithromycin and rifampicin are endorsed currently for *Rhodococcus equi* infections by the CVMP in the "Essential Substances for Horses" updated list (Official Journal of the European Union, 2013). These antimicrobials are currently only approved for human medicine. Other antimicrobials have been investigated including tulathromycin (Venner et al. 2013b) and doxycycline (Venner et al. 2013a) but considered less effective. These drugs are also used prophylactically, and in one randomised study, azithromycin prophylaxis for the first 2 weeks of life reduced *Rhodococcus equi* incidence from approximately 20–5%, while the authors recognised more investigations were needed (Chaffin et al. 2008). The EU new veterinary regulation describes the conditions for antimicrobial prophylaxis under Article 107(3) ("Antimicrobial medicinal products shall not be used for prophylaxis other than in exceptional cases, for the administration to an individual animal or a restricted number of animals when the risk of an infection or of an infectious disease is very high and the consequences are likely to be severe"). *Rhodococcus equi* infection in foals does not fulfil Article 107(3) for prophylaxis, since the risk of infection is highly variable, where a complex of risk factors includes the type of management on farms (types of stalls, stocking density), seasonal conditions (dry, hot spring/summer) and foal characteristics (immunity and other factors). Consequences of *R. equi* infection are also highly variable. Foals progress through different stages of infection (subclinical, mild, moderate, severe), where self-cure is possible or other antibiotics can be highly effective at stages before severe pneumonia. Alternatives for prophylaxis include administering hyperimmune plasma, cranial lung lobe ultrasound screening/monitoring and farm management changes. Previous experiences of mass antimicrobial administration on *R. equi* endemic farms for both prevention and subclinical *R. equi* lead to irrefutable evidence of emerging antimicrobial resistance in *R. equi* with alarming prevalence (Huber et al. 2019). Due to environmental persistence of *R. equi*, then resistant clinical strains can persist year after year on endemic farms.

Another example is that *Clostridioides* and *Clostridium* spp. disease conditions (e.g. *C. difficile*, *C. perfringens*) associated with colitis (e.g. colitis X, duodenitis-jejunitis syndrome, antimicrobial-associated diarrhoea) in horses are being increasingly recognised. As in human medicine, *Clostridioides difficile* diarrhoea carries a grave prognosis without treatment (Cohen and Woods 1999; Magdesian et al. 2002). There are no approved medicines for this condition, and thus many horses are treated with metronidazole as the drug of choice. However, this is likely a temporary situation, since up to 43% of metronidazole-resistant *C. difficile* isolates from horses have been reported in certain geographic locations (Jang et al. 1997; Magdesian et al. 2002).

Other examples of the unavailability of medicines include the indications of anaplasmosis (*Anaplasma phagocytophila*), mycoplasma (*M. felis, M. equirhinis*), contagious equine metritis (*Taylorella equigenitalis*), Lyme's disease (*Borrelia burgdorferi*), proliferative enteropathy in foals (*Lawsonia intracellularis*), dermatophilosis (*Dermatophilus congolensis*) and *Pneumocystis carinii* in foals and leptospirosis in horses (*L. hardjo*, *L. pomona*, *L. bratislava*, *L. icterohaemorrhagiae*). These pathogens present further challenges due to the fact that most of them are hardly culturable and in vitro antimicrobial susceptibility testing using routine microbiological methods is not available; therefore, empirical therapy is more frequent for the pathologies caused by them.

9.5 Equine Antimicrobial Use for Non-antimicrobial Indications

There are examples in equine medicine of antimicrobials endorsed for non-antimicrobial purposes. For example, it has become common practice to inject neonatal foals born with contracted tendons with 1–2 high doses of oxytetracycline (40–60 mg/kg) (Kasper et al. 1995). The use of oxytetracycline for this purpose is due to a unique side effect that causes temporary tendon relaxation in foals, possibly related to calcium chelation.

Polymyxins are critically important in human medicine as a last resort antimicrobial for carbapenem-resistant Gram-negative bacteria (Li et al. 2006). However, polymyxin B is used in equine medicine for non-antibacterial purposes and at an anti-endotoxic dose for the treatment of systemic inflammatory response syndrome (SIRS), due to endotoxemia in horses (Morresey and Mackay 2006). Polymyxin B has a property of binding to bacterial toxins and therefore reducing activation of the proinflammatory cascade. Horses and humans are a particularly sensitive to the effects of endotoxemia and thus a common problem with sick horses. Endotoxins (free-floating) are produced commonly in the equine gastrointestinal tract and absorbed systemically secondary to a gastrointestinal disease or also due to a bacterial infection. Polymyxin B doses for of SIRS in horses vary between 5000 and 10,000 IU/kg IV every 8–12 h (Parviainen et al. 2001; Barton et al. 2004; Morresey and Mackay 2006; Werners 2017). Doses up to 25,000 IU/kg/day are needed to reach an antibacterial effect in humans, compared to lower dose to reach anti-endotoxic effects in horses. There is evidence that this approach can improve survival associated with SIRS (Barton et al. 2004), if used prophylactically. However, polymyxin B is used when there is already widespread evidence of SIRS. Administering polymyxin B intravenously to horses can lead to serious side effects (Schwarz et al. 2013; van Spijk et al. 2022). A study showed neurotoxicosis in healthy horses receiving seven doses of polymyxin B 6000 IU/kg IV q12h. Mild-to-moderate ataxia was reversible after cessation of therapy. The number of polymyxin B doses and co-administration of gentamicin increased the severity of ataxia. Nephrotoxicity from polymyxin B was only evident when gentamicin was co-administered (van Spijk et al. 2022). Adverse events would be expected to be more severe in clinical cases of endotoxemia. Conversely, polymyxin B is not used in human medicine for endotoxemia due to serious side effects.

The example subtherapeutic doses of polymyxin for non-antibacterial purposes are identical to the scenario of polymyxin E (colistin) used as growth promoters in food animals that resulted in global transmissible colistin resistance in animal and human populations. Plasmid-mediated colistin-resistant genes have been identified in horses (Börjesson et al. 2020; Fernandes et al. 2018). With widespread global colistin resistance and the critical importance of polymyxins for human medicine, equine clinicians must question if the clinical benefit of using polymyxin B in the treatment of SIRS outweighs the potential global contribution to colistin resistance (Isgren 2021). Alternatives to low-dose polymyxin B for SIRS include hydration fluids, NSAIDs or intravenous lidocaine (Peiró et al. 2010).

Doxycycline has been promoted in horses as a treatment for osteoarthritis (Maher et al. 2014). The authors report that low-dose, low-frequency oral administration of doxycycline can attain in vivo synovial fluid concentrations and capable of chondroprotective effects through reduction of matrix metalloproteinase (MMP)-13 activity while remaining below MIC_{90} of most equine pathogens.

An unfortunate trend in equine medicine is the use of surrogate biomarkers, namely, the acute phase protein, serum amyloid A (SAA), to indicate an infection somewhere in the horse's body. Acute-phase proteins, like SAA, are secreted during the acute phase of inflammation. SAA is also an acute phase marker that responds rapidly and non-specifically (e.g. inflammation versus infection), whereby relatively trivial inflammatory stimuli can lead to elevated SAA blood levels. However, SAA is promoted as a specific marker of infection (Jacobsen and Andersen 2007; Jacobsen et al. 2006), whereby horses are treated with antibiotics, based on SAA levels, without culture and susceptibility results, as well as treated both to prevent high SAA levels (Busk et al. 2010) and until SAA returns to normal (Jacobsen and Andersen 2007). There are neither robust studies that have shown that SAA can differentiate between inflammatory and infectious conditions in horses nor any credible scientific basis that SAA can reliably indicate an infectious process (Isgren 2021). Treatment based on SAA levels is both off-label as well as a non-antimicrobial indication and represents a gross misuse of antimicrobials. Following antimicrobial administration, serial SAA blood measurements could be utilised to evaluate declining inflammation and thus indirectly the effectiveness of the response to treatment (e.g., antimicrobial and other treatments) but not to indicate starting an antimicrobial treatment once causal infectious agent has not been recognised. To a lesser extent, fibrinogen has also been used in equine medicine as a marker of *Rhodococcus equi* (Leclere et al. 2011) infections in foals, and similarly as SAA it should not be used as indicator for use of antimicrobials.

10 Conclusions

Equine medicine is typically focused on the individual patient and demands placed on the veterinarian, which are many times driven not only by personal but also the emotional wishes of horse owners. Treatment costs and expectations are intricately linked to the probability of full return to sport performance or owners' hobby racing. Horses encompass the full range of types of domestic animals including companion animals, food animals and sport performance animals. The different roles of horses in society are further challenged by the availability of the veterinary medicinal products authorised for horses, where veterinarians are often faced with finding suitable antimicrobials for certain indications/pathogens that are safe for horses. This is further complicated by the need for responsible use of antimicrobials due to widespread antimicrobial resistance in several equine pathogens. These factors as well as considering unmet medical needs, as commented within this chapter, confirm an urgent need to gain new valid pharmacokinetic data (PK) reflecting doses used in real practice that can be utilised together with relevant, recent results of culture and antimicrobial susceptibility testing (PD data). Both PK and PD datasets can serve as variables to be filled into well-designed and valid PK/PD models, as one of the future tools for equine practitioners. Such a vision could work, if well, robustly designed and managed studies are available, so as to move away from traditional uses of antimicrobials in equine practice towards more evidence-based practices. Hopefully more targeted and responsible use, antimicrobial stewardship and attempts to find alternative tools to minimise the use of antimicrobials, together with recording of use electronically and allowing further better analysis of the data on use, could help to preserve efficacy of molecules available for the treatment of bacterial infections in horses.

References

Andrew SE, Nguyen A, Jones GL, Brooks DE (2003) Seasonal effects on the aerobic bacterial and fungal conjunctival flora of normal thoroughbred brood mares in Florida. Vet Ophthalmol 6:45–50

Arnold C, Pilla R, Chaffin K, Lidbury J, Steiner J, Suchodolski J (2021) Alterations in the fecal microbiome and metabolome of horses with antimicrobial-associated diarrhea compared to antibiotic-treated and non-treated healthy case controls. Animals (Basel) 11(6):1807

Baggot JD, Wilson WD, Hietala S (1988) Clinical pharmacokinetics of metronidazole in horses. J Vet Pharmacol Ther 11(4):417–420

Baptiste KE, Williams K, Williams NJ (2005) Methicillin-resistant staphylococci in companion animals. Emerg Infect Dis 11:1942–1944

Barr BS, Waldridge BM, Morresey PR, Reed SM, Clark C, Belgrave R, Donecker JM, Weigel DJ (2013) Antimicrobial-associated diarrhoea in three equine referral practices. Equine Vet J 45:154–158

Barton MH, Parviainen A, Norton N (2004) Polymyxin B protects horses against induced endotoxaemia in vivo. Equine Vet J 36:397–401

Baverud V, Gustaffsson A, Franklin A, Lindholm A, Gunnarsson A (1997) *Clostridium difficile* associated with acute colitis in mature horses treated with antibiotics. Equine Vet J 29:279–284

Baverud V, Franklin A, Gunnarsson A, Gustafsson A, Hellander-Edman A (1998) *Clostridium difficile* associated with acute colitis in mares when their foals are treated with erythromycin and rifampicin for Rhodococcus equi pneumonia. Equine Vet J 30:482–488

Berlin S, Spieckermann L, Oswald S, Keiser M, Lumpe S, Ullrich A, Grube M, Hasan M, Venner M, Siegmund W (2016) Pharmacokinetics and pulmonary distribution of clarithromycin and rifampicin after concomitant and consecutive administration in foals. Mol Pharm 13(3):1089–1099

Berlin S, Wallstabe S, Scheuch E, Oswald S, Hasan M, Wegner D, Grube M, Venner M, Ullrich A, Siegmund W (2018) Intestinal and hepatic contributions to the pharmacokinetic interaction between gamithromycin and rifampicin after single-dose and multiple-dose administration in healthy foals. Equine Vet J 50(4):525–531

Borg H, Carmalt JL (2013) Postoperative septic arthritis after elective equine arthroscopy without antimicrobial prophylaxis. Vet Surg 42:262–266. https://doi.org/10.1111/j.1532-950X.2013.01106.x

Börjesson S, Greko C, Myrenås M, Landén A, Nilsson O, Pedersen K (2020) A link between the newly described colistin resistance gene mcr-9 and clinical Enterobacteriaceae isolates carrying blaSHV-12 from horses in Sweden. J Glob Antimicrob Resist 20:285–289

Boyle AG, Timoney JF, Newton JR, Hines MT, Waller AS, Buchanan BR (2018) *Streptococcus equi* infections in horses: guidelines for treatment, control, and prevention of strangles—revised consensus statement. J Vet Intern Med 32:633–647

Burton AJ, Giguère S, Sturgill TL, Berghaus LJ, Slovis NM, Whitman JL, Levering C, Kuskie KR, Cohen ND (2013) Macrolide- and rifampin-resistant *Rhodococcus equi* on a horse breeding farm, Kentucky, USA. Emerg Infect Dis 19(2):282–285

Busk P, Jacobsen S, Martinussen T (2010) Administration of perioperative penicillin reduces postoperative serum amyloid A response in horses being castrated standing. Vet Surg 39:638–643

Ceriotti S, Westerfeld R, Bonilla AG, Pang DSJ (2021) Use of clinical audits to evaluate timing of preoperative antimicrobials in equine surgery at a veterinary teaching hospital. Front Vet Sci 8:630111. https://doi.org/10.3389/fvets.2021.630111

Cervantes CC, Brown MP, Gronwall R, Merritt K (1993) and concentrations of ceftiofur sodium in body fluids and endometrium after repeated intramuscular injections in mares. Am J Vet Res 54:573–575

Chaffin MK, Cohen ND, Martens RJ (2008) Chemoprophylactic effects of azithromycin against *Rhodococcus equi*-induced pneumonia among foals at equine breeding farms with endemic infections. J Am Vet Med Assoc 232:1035–1047

Cohen ND, Woods AM (1999) Characteristics and risk factors for failure of horses with acute diarrhea to survive: 122 cases (1990-1996). J Am Vet Med Assoc 214:382–390

Cruz AM, Rubio-Martinez L, Dowling T (2006) New antimicrobials, systemic distribution, and local methods of antimicrobial delivery in horses. Vet Clin North Am Equine Pract 22:297–322, vii–viii

Dallap Schaer BL, Linton JK, Aceto H (2012) Antimicrobial use in horses undergoing colic surgery. J Vet Intern Med 26:1449–1456

Damborg P, Marskar P, Baptiste KE, Guardabassi L (2012) Faecal shedding of CTX-M-producing *Escherichia coli* in horses receiving broad-spectrum antimicrobial prophylaxis after hospital admission. Vet Microbiol 154:298–304

Davis JL, Salmon JH, Papich MG (2006) Pharmacokinetics and tissue distribution of doxycycline after oral administration of single and multiple doses in horses. Am J Vet Res 67:310–316

Dolejska M, Duskova E, Machkova G, Kohoutova L, Cizek A (2008) Detection of ESBL-producing *Escherichia coli* in horses, personnel and the environment within a veterinary teaching hospital. In: Conference proceedings; antimicrobial resistance in zoonotic bacteria and foodborne pathogens. American Society for Microbiology, Copenhagen, pp 57–58

Esposito S (1999) Is single-dose antibiotic prophylaxis sufficient for any surgical procedure? J Chemother 11:556–564

Fernandes MR, Cerdeira L, Silva MM, Sellera FP, Muñoz M, Junior FG, Lincopan N (2018) Novel *mcr*-5.3 variant in a CTX-M-8-producing *Escherichia coli* ST711 isolated from an infected horse. J Antimicrob Chemother 73:3520–3522

Furr M, McKenzie H (2020) Factors associated with the risk of positive blood culture in neonatal foals presented to a referral center (2000-2014). J Vet Intern Med 34:2738–2750

Garner C, Stephen C, Pant SD, Ghorashi SA (2023) Comparison of PCR-HRM, colorimetric LAMP and culture based diagnostic assays in the detection of endometritis caused by *Streptococcus equi* subsp. *zooepidemicus* in mares. Vet Res Commun 47(2):495–509

Giguère S, Jacks S, Roberts GD, Hernandez J, Long MT, Ellis C (2004) Retrospective comparison of azithromycin, clarithromycin, and erythromycin for the treatment of foals with *Rhodococcus equi* pneumonia. J Vet Intern Med 18:568–573

Giguère S, Lee E, Cohen N, Chaffin M, Halbert N, Martens R, Franklin R, Clark C (2008) Prevalence of *Rhodococcus equi* isolates resistant to macrolides or rifampin and outcome of infected foals. J Vet Intern Med 737

Giguère S, Berghaus LJ, Willingham-Lane JM (2017) Antimicrobial resistance in *Rhodococcus equi*. Microbiol Spectr 5(5):ARBA-0004-2016

Giuliano EA, Maggs DJ, Moore CP, Boland LA, Champagne ES, Galle LE (2000) Inferomedial placement of a single-entry subpalpebral lavage tube for treatment of equine eye disease. Vet Ophthalmol 3:153–156

Gustafsson A, Baverud V, Gunnarsson A, Rantzien MH, Lindholm A, Franklin A (1997) The association of erythromycin ethylsuccinate with acute colitis in horses in Sweden. Equine Vet J 29:314–318

Hildebrand F, Venner M, Giguère S (2015) Efficacy of gamithromycin for the treatment of foals with mild to moderate bronchopneumonia. J Vet Intern Med 29(1):333–338

Hillidge CJ (1987) Use of erythromycin-rifampin combination in treatment of *Rhodococcus equi* pneumonia. Vet Microbiol 14(3):337–342

Hoffman AM, Viel L (1997) Techniques for sampling the respiratory tract of horses. Vet Clin North Am Equine Pract 13:463–475

Høiby N, Pers C, Johansen HK, Hansen H (2000) Excretion of β-lactam antibiotics in sweat—a neglected mechanism for development of antibiotic resistance? Antimicrob Agents Chemother 44(10):2855–2857

Huber L, Giguère S, Cohen ND, Slovis NM, Hanafi A, Schuckert A, Berghaus L, Greiter M, Hart KA (2019) Prevalence and risk factors associated with emergence of *Rhodococcus equi* resistance to macrolides and rifampicin in horse-breeding farms in Kentucky, USA. Vet Microbiol 235:243–247

Hughes LA, Pinchbeck G, Callaby R, Dawson S, Clegg P, Williams N (2013) Antimicrobial prescribing practice in UK equine veterinary practice. Equine Vet J 45:141–147

Isgren CM (2021) Improving clinical outcomes via responsible antimicrobial use in horses. Equine Vet Educ 34(1). https://doi.org/10.1111/eve.13502

Isgren CM, Williams NJ, Fletcher OD, Timofte D, Newton JR, Maddox TW, Clegg PD, Pinchbeck GL (2021) Antimicrobial resistance in clinical bacterial isolates from horses in the UK. Equine Vet J 54(12). https://doi.org/10.1111/evj.13437

Jacobsen S, Andersen P (2007) The acute phase protein serum amyloid A (SAA) as a marker of inflammation in horses. Equine Vet Educ 19:38–46

Jacobsen S, Thomsen MH, Nanni S (2006) Concentrations of serum amyloid A in serum and synovial fluid from healthy horses and horses with joint disease. Am J Vet Res 67:1738–1742

Jang SS, Hansen LM, Breher JE, Riley DA, Magdesian KG, Madigan JE, Tang YJ, Silva J Jr, Hirsh DC (1997) Antimicrobial susceptibilities of equine isolates of Clostridium difficile and molecular characterization of metronidazole-resistant strains. Clin Infect Dis 25(Suppl 2):S266–S267

Kasper CA, Clayton HM, Wright AK, Skuba EV, Petrie L (1995) Effects of high doses of oxytetracycline on metacarpophalangeal joint kinematics in neonatal foals. J Am Vet Med Assoc 207:71–73

Khusro A, Aarti C, Buendía-Rodriguez G, Arasu MV, Al-Dhabi NA, Barbabosa-Pliego A (2020) Adverse effect of antibiotics administration on horse health: an overview. J Equine Vet 97:103339. https://doi.org/10.1016/j.jevs.2020.103339

Knych HK, Magdesian KG (2021) Equine antimicrobial therapy: Current and past issues facing practitioners. J Vet Pharmacol Ther 44(2):270–279. https://doi.org/10.1111/jvp.12964

Lamy B, Dargère S, Arendrup MC, Parienti JJ, Tattevin P (2016) How to optimize the use of blood cultures for the diagnosis of bloodstream infections? A state-of-the art. Front Microbiol 12(7):697

Leclere M, Magdesian KG, Kass PH, Pusterla N, Rhodes DM (2011) Comparison of the clinical, microbiological, radiological and haematological features of foals with pneumonia caused by *Rhodococcus equi* and other bacteria. Vet J 187:109–112

Li J, Nation RL, Turnidge JD, Milne RW, Coulthard K, Rayner CR, Paterson DL (2006) Colistin: the re-emerging antibiotic for multidrug-resistant Gram-negative bacterial infections. Lancet Infect Dis 6:589–601

Liepman RS, Swink JM, Habing GG, Boyaka PN, Caddey B, Costa M, Gomez DE, Toribio RE (2022) Effects of intravenous antimicrobial drugs on the equine fecal microbiome. Animals (Basel) 12(8):1013

Lönker NS, Fechner K, Wahed AAE (2020) Horses as a crucial part of one health. Vet Sci 7(1):28. https://doi.org/10.3390/vetsci7010028. PMID: 32121327; PMCID: PMC7157506

Love CC, Strzemienski PJ, Kenney RM (1990) Endometrial concentrations of ampicillin in mares after intrauterine infusion of the drug. Am J Vet Res 51:197–199

MacLeay JM, Kohn CW (1998) Results of quantitative cultures of urine by free catch and catheterization from healthy adult horses. J Vet Intern Med 12:76–78

Maddox TW, Clegg PD, Diggle PJ, Wedley AL, Dawson S, Pinchbeck GL, Williams NJ (2012) Cross-sectional study of antimicrobial-resistant bacteria in horses. Part 1: Prevalence of antimicrobial-resistant Escherichia coli and methicillin-resistant Staphylococcus aureus. Equine Vet J 44(3):289–296

Magdesian KG, Hirsh DC, Jang SS, Hansen LM, Madigan JE (2002) Characterization of *Clostridium difficile* isolates from foals with diarrhea: 28 cases (1993-1997). J Am Vet Med Assoc 220:67–73

Maher MC, Schnabel LV, Cross JA, Papich MG, Divers TJ, Fortier LA (2014) Plasma and synovial fluid concentration of doxycycline following low-dose, low-frequency administration, and resultant inhibition of matrix metalloproteinase-13 from interleukin-stimulated equine synoviocytes. Equine Vet J 46:198–202

McGorum B, Dixon P, Smith D (1998) Use of metronidazole in equine idiopathic toxaemic colitis. Vet Rec 142:635–638

McIlwraith CW, Yovich JV, Martin GS (1987) Arthroscopic surgery for the treatment of osteochondral chip fractures in the equine carpus. J Am Vet Med Assoc 191:531–540

McKenzie HC, Murray MJ (2004) Concentrations of gentamicin in serum and bronchial lavage fluid after once-daily aerosol administration to horses for seven days. Am J Vet Res 65:173–178

Michelet C, Avril JL, Arvieux C, Jacquelinet C, Vu N, Cartier F (1997) Comparative activities of new fluoroquinolones, alone or in combination with amoxicillin, trimethoprim-sulfamethoxazole, or rifampin, against intracellular Listeria monocytogenes. Antimicrob Agents Chemother 41:60–65

Morresey PR, Mackay RJ (2006) Endotoxin-neutralizing activity of polymyxin B in blood after IV administration in horses. Am J Vet Res 67:642–647

Narváez SÁ, Fernández I, Patel NV, Sánchez S (2022) Novel quantitative PCR for Rhodococcus equi and macrolide resistance detection in equine respiratory samples. Animals (Basel) 12(9):1172

Nielsen SS, Bicout DJ, Calistri P, Canali E, Drewe JA, Garin-Bastuji B, Gonzales Rojas JL, Gortazar Schmidt C, Herskin M, Michel V, Miranda Chueca MA, Padalino B, Pasquali P, Roberts HC, Sihvonen LH, Spoolder H, Stahl K, Velarde A, Viltrop A, Winckler C, Dewulf J, Guardabassi L, Hilbert F, Mader R, Baldinelli F, Alvarez J (2021) Scientific opinion on the assessment of animal diseases caused by bacteria resistant to antimicrobials: Horses. EFSA J 19(12):7112, 43. https://doi.org/10.2903/j.efsa.2021.7112

Olds AM, Stewart AA, Freeman DE, Schaeffer DJ (2006) Evaluation of the rate of development of septic arthritis after elective arthroscopy in horses: 7 cases (1994-2003). J Am Vet Med Assoc 229:1949–1954

Parviainen AK, Barton MH, Norton NN (2001) Evaluation of polymyxin B in an ex vivo of endotoxemia in horses. Am J Vet Res 62:72–76

Pedersoli WM, Fazeli MH, Haddad NS, Ravis WR, Carson RL Jr (1985) Endometrial and serum gentamicin concentrations in pony mares given repeated intrauterine infusions. Am J Vet Res 46:1025–1028

Peiró JR, Barnabé PA, Cadioli FA, Cunha FQ, Lima VMF, Mendonça VH, Santana AE, Malheiros EB, Perri SHV, Valadão CAA (2010) Effects of lidocaine infusion during experimental endotoxemia in horses. J Vet Intern Med 24(4):940–948

Peters J, Block W, Oswald S, Freyer J, Grube M, Kroemer HK, Lämmer M, Lütjohann D, Venner M, Siegmund W (2011) Oral absorption of clarithromycin is nearly abolished by chronic comedication of rifampicin in foals. Drug Metab Dispos 39(9):1643–1649

Peters J, Eggers K, Oswald S, Block W, Lütjohann D, Lämmer M, Venner M, Siegmund W (2012) Clarithromycin is absorbed by an intestinal uptake mechanism that is sensitive to major inhibition by rifampicin: results of a short-term drug interaction study in foals. Drug Metab Dispos 40(3):522–528

Pezzanite LM, Griffenhagen GM, Krause DM, Hendrickson DA (2021) Retrospective evaluation of association between perioperative antimicrobial protocol and complications following elective equine synovial endoscopy. Vet Med Sci 7(3):609–620

Prescott JF (2020) Outpacing the resistance tsunami: antimicrobial stewardship in equine medicine, an overview. Equine Vet Educ 33:539–545

Redpath A, Hallowell GD, Bowen IM (2021) Use of aminoglycoside antibiotics in equine clinical practice; a questionnaire-based study of current use. Vet Med Sci 7(2):279–288. https://doi.org/10.1002/vms3.382

Ridge PA (2011) A retrospective study of the rate of postoperative septic arthritis following 353 elective arthroscopies. J Small Anim Pract 52:200–202

Rohwedder L, Lehmann NJ, Furmanczuk A, Junk O, Taxis R, Röder Y et al (2021) VISKIT—Standardization of the surveillance of nosocomial infections in veterinary medicine. In: 2020 IEEE international conference on E-health networking, application & services (HEALTHCOM), Shenzhen, China, pp 1–5

Rutenberg D, Venner M, Giguère S (2017) Efficacy of tulathromycin for the treatment of foals with mild to moderate bronchopneumonia. J Vet Intern Med 31(3):901–906

Schnepf A, Bienert-Zeit A, Ertugrul H, Wagels R, Werner N, Hartmann M, Feige K, Kreienbrock L (2020) Antimicrobial usage in horses: the use of electronic

data, data curation, and first results. Front Vet Sci 7:216
Schwarz B, Anen C, van den Hoven R (2013) Preliminary data of a retrospective study on neurological side effects after administration of polymyxin B to endotoxaemic horses. Equine Vet J 45:18–19
Southwood LL (2014) Perioperative antimicrobials: should we be concerned about antimicrobial drug use in equine surgical patients? Equine Vet J 46:267–269
Stöckle SD (2019) Pre- or perioperative antibiotic prophylaxis in horses undergoing aseptic, elective orthopaedic surgery. Inaugural Dissertation, VVB Laufersweiler Verlag, Giessen. Available on line at: http://geb.uni-giessen.de/geb/volltexte/2019/14448/pdf/StoeckleSabita_2019_03_15.pdf. Accessed 30 June 2023
Stöckle SD, Kannapin DA, Kauter AML, Lübke-Becker A, Walther B, Merle R, Gehlen H (2021) A pilot randomised clinical trial comparing a short-term perioperative prophylaxis regimen to a long-term standard protocol in equine colic surgery. Antibiotics 10:587
Theelen MJP, Wilson WD, Byrne BA, Edman JM, Kass PH, Magdesian KG (2019) Initial antimicrobial treatment of foals with sepsis: Do our choices make a difference? Vet J 243:74–76
Toutain PL (2014) Antimicrobial therapy in horses: A pharmacologist perspective; oral presentation. Department of Veterinary Disease Biology, University of Copenhagen, 30th September 2014
Trampuz A, Widmer AF (2006) Infections associated with orthopaedic implants. Curr Opin Infect Dis 19:349–356
Traub-Dargatz JL, George JL, Dargatz DA, Morley PS, Southwood LL, Tillotson K (2002) Survey of complications and antimicrobial use in equine patients at veterinary teaching hospitals that underwent surgery because of colic. J Am Vet Med Assoc 220:1359–1365
van Spijk JN, Beckmann K, Wehrli Eser M et al (2022) Adverse effects of polymyxin B administration to healthy horses. J Vet Intern Med 1–10. https://doi.org/10.1111/jvim.16470
Venner M, Peters J, Höhensteiger N, Schock B, Bornhorst A, Grube M, Adam U, Scheuch E, Weitschies W, Rosskopf D, Kroemer HK, Siegmund W (2010) Concentration of the macrolide antibiotic tulathromycin in broncho-alveolar cells is influenced by comedication of rifampicin in foals. Naunyn Schmiedebergs Arch Pharmacol 381(2):161–169
Venner M, Rödige A, Laemmer M, Giguère S (2012) Failure of antimicrobial therapy to accelerate spontaneous healing of subclinical pulmonary abscesses on a farm with endemic infections caused by *Rhodococcus equi*. Vet J 192(3):293–298
Venner M, Astheimer K, Lämmer M, Giguère S (2013a) Efficacy of mass antimicrobial treatment of foals with subclinical pulmonary abscesses associated with *Rhodococcus equi*. J Vet Intern Med 27(1):171–176
Venner M, Credner N, Lämmer M, Giguère S (2013b) Comparison of tulathromycin, azithromycin and azithromycin-rifampin for the treatment of mild pneumonia associated with *Rhodococcus equi*. Vet Rec 173(16):397
Verheyen K, Newton JR, Talbot NC, de Brauwere MN, Chanter N (2000) Elimination of guttural pouch infection and inflammation in asymptomatic carriers of Streptococcus equi. Equine Vet J 32:527–532
Weese JS, Cruz A (2009) Retrospective study of perioperative antimicrobial use practices in horses undergoing elective arthroscopic surgery at a veterinary teaching hospital. Can Vet J 50:185–188
Weese JS, Caldwell F, Willey BM, Kreiswirth BN, McGeer A, Rousseau J, Low DE (2005) An outbreak of methicillin-resistant Staphylococcus aureus skin infections resulting from horse to human transmission in a veterinary hospital. Vet Microbiol 114(1–2):160–164
Weese JS, Rousseau J, Willey BM, Archambault M, McGeer A, Low DE (2006) Methicillin-resistant *Staphylococcus aureus* in horses at a veterinary teaching hospital: frequency, characterization, and association with clinical disease. J Vet Intern Med 20:182–186
Weese JS, Baptiste KE, Baverund V, Toutain PL (2008) Guideline for antimicrobial use in horses. In: Guardabassi L, Jenses LB, Kruse H (eds) Antimicrobial use in animals. Blackwell, Oxford, pp 161–182
Weese JS, Saab M, Moore A, Cai H, McClure JT (2023) Relationship between quantitative real-time PCR cycle threshold and culture for detection of *Streptococcus equi* subspecies *equi*. Can Vet J 64:549–552
Werners AH (2017) Treatment of endotoxaemia and septicaemia in the equine patient. J Vet Pharmacol Ther 40:1–15
Wetzig M, Venner M, Giguère S (2019) Efficacy of the combination of doxycycline and azithromycin for the treatment of foals with mild to moderate bronchopneumonia. Equine Vet J. https://doi.org/10.1111/evj.13211
Widmer A, Kummer M, Eser MW, Fürst A (2009) Comparison of the clinical efficacy of cefquinome with the combination of penicillin G and gentamicin in equine patients. Equine Vet Educ 21:430–435
Wilson WD, Spensley MS, Baggot JD, Hietala SK, Pryor P (1991) Pharmacokinetics and bioavailability of ticarcillin and clavulanate in foals after intravenous and intramuscular administration. J Vet Pharmacol Ther 14:78–89
Winther L, Honore Hansen S, Baptiste KE, Friis C (2011) Antimicrobial disposition in pulmonary epithelial lining fluid of horses, part II. Doxycycline. J Vet Pharmacol Ther 34:285–289
Wohlfender FD, Barrelet FE, Doherr MG, Straub R, Meier HP (2009) Diseases in neonatal foals. Part 1: The 30 day incidence of disease and the effect of prophylactic antimicrobial drug treatment during the first three days post partum. Equine Vet J 41:179–185
WHO (2024) WHO's List of Medically Important Antimicrobials: a risk management tool for mitigating antimicrobial resistance due to non-human use. Geneva: World Health Organization; Licence: CC BY-NC-SA 3.0 IGO.

Index

L. Pokludová (ed.), *Antimicrobials in Livestock 2: Choices of Prevention and Treatment*,
https://doi.org/10.1007/978-3-031-60807-0

C

J

K

L

U

V

W

Y

Z

MIX
Papier aus verantwortungsvollen Quellen
Paper from responsible sources
FSC® C105338

If you have any concerns about our products,
you can contact us on
ProductSafety@springernature.com

In case Publisher is established outside the EU,
the EU authorized representative is:
Springer Nature Customer Service Center GmbH
Europaplatz 3, 69115 Heidelberg, Germany

Printed by Libri Plureos GmbH
in Hamburg, Germany